Chemical and Biological Applications of Relaxation Spectrometry

NATO ADVANCED STUDY INSTITUTES SERIES

Proceedings of the Advanced Study Institute Programme, which aims at the dissemination of advanced knowledge and the formation of contacts among scientists from different countries

The series is published by an international board of publishers in conjunction with NATO Scientific Affairs Division

A	Life Sciences	Plenum Publishing Corporation
B	Physics	London and New York
C	Mathematical and Physical Sciences	D. Reidel Publishing Company Dordrecht and Boston
D	Behavioral and Social Sciences	Sijthoff International Publishing Company Leiden
E	Applied Sciences	Noordhoff International Publishing Leiden

Series C – Mathematical and Physical Sciences

Volume 18 – Chemical and Biological Applications of Relaxation Spectrometry

Chemical and Biological Applications of Relaxation Spectrometry

Proceedings of the NATO Advanced Study Institute held at the University of Salford, Salford, England, 29 August–12 September, 1974

edited by

E. WYN-JONES

Department of Chemistry and Applied Chemistry, University of Salford, Salford, England

D. Reidel Publishing Company

Dordrecht-Holland / Boston-U.S.A.

Published in cooperation with NATO Scientific Affairs Division

Library of Congress Cataloging in Publication Data

NATO Advanced Study Institute, University of Salford, 1974.
Chemical and biological applications of relaxation spectrometry.

(NATO advanced study institutes series : C, mathematical and physical sciences ; 18)
Bibliography: p.
1. Relaxation spectroscopy—Congresses. I. Wyn-Jones, Evan. II. Title.
III. NATO advanced study institutes series : Series C, mathematical and physical sciences ; 18.
QD96.R44N37 1974 543'.085 75-17765
ISBN 90-277-0624-7

Published by D. Reidel Publishing Company
P.O. Box 17, Dordrecht, Holland

Sold and distributed in the U.S.A., Canada, and Mexico
by D. Reidel Publishing Company, Inc.
306 Dartmouth Street, Boston, Mass. 02116, U.S.A.

Printed in The Netherlands by D. Reidel, Dordrecht

CONTENTS

PREFACE

Chemical relaxation spectrometry involves the application of several different relaxation techniques to investigate the kinetics and mechanisms of fast chemical reactions and also dynamic molecular processes that occur in liquids in the time range 1 - 10^{-10} secs. These methods have been used widely in several disciplines of the natural sciences including molecular biology, biochemistry, organic stereochemistry, detergent chemistry and inorganic chemistry.

The purpose of the Advanced Study Institute was to provide a forum for scientists to discuss the role, scope and limitations of the various applications of chemical relaxation methods in order to stimulate discussion and interaction between workers in these different fields. The papers described in this volume are a summary of the contributions that were discussed at the meeting. The brief given to the authors was to attempt to prepare an article containing a good supply of references so that the book can be used not only by those starting in the field, but also by the specialist and research worker. These contributions cover a varied range of topics summarizing the achievements, the current state of knowledge and possible application in many disciplines. It is to be hoped that this volume will help to point out some new directions towards which research efforts are required and to attract new researchers with fresh points of view. In a collection of this kind it is inevitable that the style of presentation and length of the articles are varied and also that a certain amount of overlap will occur. In some discussion sessions, particularly the kinetics of micelle formation, this latter point provided an invaluable pointer towards understanding detailed reaction mechanisms especially when the same problem is approached using several different experimental methods.

In the preparation of this volume I am deeply grateful for the invaluable assistance of my colleague John Gettins. I would also like to thank the following people who assisted in the preparation of the manuscript:- Mrs. J. Broadhead, Miss L. Cook, Miss C. Lowe, Mr. Walter Owen and Mrs. G. Wyn-Jones.

Salford.
March 1975.

Evan Wyn-Jones

ULTRASONIC RELAXATION SPECTROMETRY

Jørgen Rassing

Roskilde University Center, Denmark

1. INTRODUCTION

The application of elastic waves of different frequencies (20 KHz - 1 GHz) to the investigation of molecules and their interactions is often referred to as ultrasonic spectrometry. The method has been employed mainly as a tool for studying fast physical and chemical reactions and offers access to the determination of time constants ranging from 10^{-5} - 10^{-10} sec. for the approach to equilibrium by the reaction in question.

For a given compound several physical and chemical reactions with different time constants may exist. These reactions may increase the absorption of the ultrasonic energy. Since the reactions have different time constants, **relaxation times, one type of a** reaction may predominate over a certain frequency range whereas another type may predominate in a different portion of the acoustic spectrum, thus giving rise to the concept of "ultrasonic relaxation spectrometry".

The aim of the experimental technique is to provide measurements of the absorption of ultrasonic energy at different frequencies for a sound wave which travels through the system in which the reaction in question takes place.

In the following presentation of the ultrasonic method emphasis has been placed on showing the different assumptions involved in the theoretical development of the relaxation equations and the problems, connected with a detailed kinetic interpretation of the processes which cause the ultrasonic relaxation spectra.

Wyn-Jones (ed.), Chemical and Biological Applications of Relaxation Spectrometry, 1–16.

2. NATURE OF ULTRASONIC WAVES

(i) Non-reacting systems

When a longitudinal sound wave passes through a liquid it creates local pressure and temperature variations in the liquid. These variations shift at a rate which is equal to the sound frequency. The pressure, P, at time, t, and at distance, x, from the sound transducer is:

$$P(x,t) = P_o \exp(i\omega(t-x/U) - \alpha x) \quad (1)$$

where ω is the cyclic sound frequency, U is the sound velocity, and α is the sound absorption coefficient.

The sound frequencies used in the ultrasonic technique are always larger than 100 KHz. Since heat flow is a slow reaction compared with this frequency there will be no heat flow from the compressed part of the liquid to the surroundings. Consequently, the compression and decompression take place adiabatically. At the adiabatic conditions the temperature variation which follows the pressure change is:

$$T(x,t) = (\gamma-1)\ \beta_S/\ell P(x,t) \quad (2)$$

where γ is the ratio of the heatcapacities at constant pressure and volume respectively, β_S is the isentropic compressibility, and ℓ is the thermal expansion coefficient. For an ultrasonic wave which propagates through water at a frequency of 1MHz the velocity is c.a. 10^5cm s^{-1}, and the wave length c.a. 0.1 cm. The wave length is considerably larger than the size of the molecular aggregates which are going to be studied. Thus **the sound wave essentially** travels through a continuous medium. Usually the effect of the applied sound waves is about 1 mW cm^{-2} leading to local pressure variations of about 0.03 atm. and temperature variations in the region 0.002 deg. Since the temperature and the internal pressure in a liquid are c.a. 10^2K and 10^3 atm. respectively the variations introduced by the ultrasonic technique are negligible.

(ii) Reacting systems

The local temperature and pressure variations may affect a given chemical equilibrium in the system. The pressure and temperature dependence of the equilibrium constant, K, is given by the following equation

$$dK = K\{(\Delta H^o/(RT^2))_P dT - (\Delta V^o/(RT))_T dP\} \quad (3)$$

where ΔH° and ΔV° are the reaction enthalpy and reaction volume respectively. A necessary condition for the equilibrium constant to change with pressure and temperature is that the bracket in eqn (3) is different from zero. However this condition is not sufficient in order for the reaction to affect the absorption of the propagating sound wave. The actual magnitude of the equilibrium constant is also important in order to produce a change of the sound wave amplitude. Consider the following equilibrium between two different conformational states of a molecule in a volume element of the liquid through which the sound wave propagates.

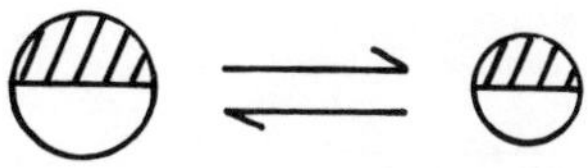

(4)

Before the sound wave propagates through the system matter is equally distributed between the two different conformations as indicated by the shading of half of the two circles which denote the conformational states in eqn. (4). When the sound wave propagates through the system the volume element oscillates between a hot (compressed) and a cold (decompressed) or expanded situation as indicated on Fig. 1. In the compressed position the equilibrium shifts as indicated on the figure because the sound wave invests energy in the chemical equilibrium with the result that the compressed volume element becomes less hot than if the chemical equilibrium did not take place. In the expanded position it is the other way around. At sufficiently low oscillation frequencies the energy invested in the chemical equilibrium is given back to the sound wave in time and consequently the sound wave does not register the shift taking place. By increasing the oscillating frequency a frequency region is reached in which the chemical equilibrium is not able to adjust itself fast enough to give back the invested energy in time. The sound wave registers this as a loss of energy. The frequency range in which this happens obviously is related to the rates involved in the chemical reaction. At higher frequencies the chemical equilibrium is unable to respond to the pressure and temperature variation anymore. At this state the reaction is "frozen out" and does not affect the sound wave.

This simple example shows that a given chemical equilibrium increases the sound absorption coefficient in a limited frequency range only and that this frequency range contains information about the rates of the chemical reaction. Furthermore it becomes evident that the mechanism of sound absorption is completely different from that of light absorption.

The phenomenon described is called a chemical relaxation and the sound frequency at which the absorption coefficient divided by the sound frequency passes through a maximum is called the relaxation frequency.

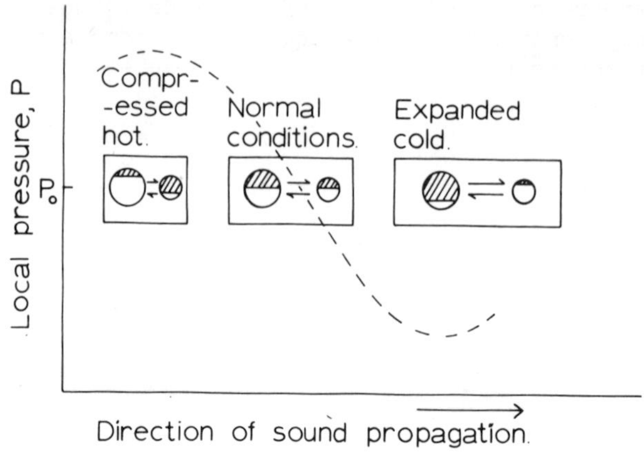

Fig. 1. An illustration of the variation in a two state conformational change during the passage of a sound wave.

3. MEASURING TECHNIQUE

(i) Pulse technique

The pulse technique is the method that has been mostly used for measuring sound absorption coefficients in the frequency range 10 - 800 MHz. The sound wave is produced by means of a transducer made of a piezoelectric material which convertes electrical energy to ultrasonic energy and vice versa. Quartz is very often used because of its mechanical and chemical stability. A quartz transducer is a disc cut from crystalline quartz in such a way that the greatest dimension of the disc is along the Y-axis and the least (the thickness of the disc) along the X-axis of the crystal. This disc produces longitudinal waves when excited at resonance by an alternating electric potential applied across the disc. The fundamental frequency of the disc is given by the following equation.

$$f_o = U/(2\,l) \tag{5}$$

where U is the sound velocity in quartz and l is the thickness of the transducer. Beside the fundamental frequency, the odd harmonic overtones can be produced even up to the fiftieth, but with diminishing efficiency making them less and less suitable for ultrasonic experiments. By means of a disc with a thickness of 0.05 cm sound waves with the frequencies of 5 , 15 , 25 , etc. MHz can be produced. If the alternating potential across the disc consists of sinusoidal pulses ultrasonic wave trains are produced. The measuring cell can be designed in a way that these sound trains are allowed to travel through the solution in question to a reflector and back again to the transducer. The equipment is adjusted in such a way that the time it takes for the wave train to travel all the way through the solution is long compared with the pulse duration time. Consequently, when the wave train returns to the transducer, the transducer is unaffected and now acts as a detector by converting the received sound energy to an electric pulse. The amplitude of this pulse is then compared with the amplitude of the initial pulse which generated the sound train. The ratio of these amplitudes is related to the overall loss of sound energy. Although a great deal of this loss is caused by the reflections these can be eliminated by carrying out the experiment at different distances between reflector and transducer provided that the geometry of the cell is independent of the distance. This means that the absorption coefficient can now be calculated directly. Thus the pulse technique has the very big advantage of being a technique which gives absolute values of the sound absorption coefficients. The basic requirements of the technique are, however, that a) the sound wave is plane, b) the loss of energy due to scattering from the transducer and the reflector is negligible and c) the distance between transducer and reflector can be changed without affecting the levelling of the transducer and reflector surfaces. In order to fulfil the first requirement the surface of the transducer and the reflector must be highly polished. The second requirement is related to the radius, r, of the transducer, the sound frequency, and the distance between the transducer and the reflector. The scattering angle, θ, is given by the following equation.

$$\sin \theta = 0.61\ U/(fr) \tag{6}$$

For example if the 1 MHz sound train at a distance of 5 cm from the transducer is supposed to cover an area less than 10% larger than that of the transducer then the diameter of the transducer must be larger than 5 cm. As the operating frequency gets lower this distance gets larger and as a result a bigger diameter transducer must be used. The third requirement is a technical problem which limits the distance variation. In order to obtain reliable results the distance variation must be large enough to produce an amplitude ratio change at least 20 db. If a decent experiment at 200 KHz has to be carried out by means of the pulse technique it requires a cell with a path length of several meters and a volume of several liters. This immediately sets a practical lower frequency limit for this method. The upper frequency limit is determined mainly by the difficulties of getting sufficient sound energy from the transducer when the high overtones are used.

The pulse technique is sometimes applied in situations where the above conditions are not fulfiled. The method is then reduced to a calibration technique. At frequencies lower than 5 MHz it must be considered as a calibration technique and in this situation a different technique is prefered, for instance, the so called resonance method.

(ii) Resonance technique[2]

The resonance technique makes use of a resonator constructed from two X-cut quartz crystals with the liquid sample introduced between them. The aim of this technique is to measure the quality factor, Q, of the liquid sample column at different resonance frequencies. The quality factor can be calculated from the resonance curvature and is given by the frequency at which the resonance curvature has a maximum divided by the width of the curvature at the 3 db point.

$$Q = f_o/(f_2 - f_1) \tag{7}$$

The resonance curvature is obtained experimentally by exciting the transmitting crystal with a frequency signal over a suitable bandwidth. The resulting sound waves produced in the system are detected by the second crystal and recorded as a voltage against frequency on a display oscillograph.

Theoretical considerations show that the quality factor is related to the sound absorption coefficient at the peak frequency by the following equation:

$$Q^{-1} = \alpha U/f_o \tag{8}$$

provided that Q is the quality factor for the liquid column only. This requires that only the liquid column resonates, that the resonator is ideal with rigid end walls, and that plane sound waves are used. Experiments carried out on systems with well characterized sound absorption coefficients show that the measured value of α is larger than expected due to the fact that the cell is not an ideal resonator. However, the technique can be used as a calibration technique by means of the following equation.

$$Q^{-1}_{solution} - Q^{-1}_{solvent} = \alpha_{excess} U/f_o \tag{9}$$

where α_{excess} is the sound absorption coefficient of the solute. This equation involves implicitly that the damping due to the cell itself is independent of the actual size of α, and that α for the solvent is equal to that of the solution at very high frequencies. These requirements limit the application of the technique to certain systems. However the technique permits low frequency measurements (200 KHz - 10 MHz) to be carried out with small sample volumes of about 10 ml.

Recently an attempt has been made to reduce the volume to

0.02 cc by constructing a different resonance cell[3]. The cell consists of a X-cut quartz crystal with a fundamental frequency of about 300 KHz and a diameter of 3 cm. A small cylindrical hole is bored through this crystal in which 0.02 cc of the sample can be placed. This amount of sample affects the resonance curvature of the quartz crystal in a way which empirically may be related to the acoustic properties of the sample. Although the measurements by this cell are in some respects qualitative the cell design is the only one which, at present, permits such small sample volumes to be measured and this may be important for certain investigations of biological systems.

4. THEORY[4-6]

(i) Relaxation equation

In section 2, Fig. 1, an oscillating volume element was used to illustrate the interaction between the energy of the sound wave and the energy stored in a given chemical equilibrium in the system. In order to carry out a detailed theoretical analysis this picture is modified to the motion of a particle in the medium. The relation between the frequency dependence of the sound absorption coefficient and the kinetic and thermodynamic parameters of a relaxing chemical reaction can then be obtained by considering the particle movement, ξ, which occurs in the medium through which the sound wave propagates.

$$\xi = A\exp(i\omega(t-x/U)) \tag{10}$$

where A is the amplitude of the sound wave. The sound velocity is given by the following equation

$$U^2 = (\partial P/\partial\rho)_S = -(\partial P/\partial V)_S V/\rho = C_p/(C_V\rho\beta_T) \tag{11}$$

where β_T is the isothermal compressibility. Equation (10) describes the particle movement only when no sound absorption occurs. If the system absorbs sound energy the particle movement is given by

$$\xi = A\ \exp(i\omega(t-x/U_c)) \tag{12}$$

where U_c is the complex velocity of sound. The velocity becomes complex when absorption occurs because the absorption means a phase lag, Φ, between the pressure and the density wave. The complex velocity is given by

$$U_c^2 = A'\ (\cos\Phi + i\sin\Phi) \tag{13}$$

where A' is the ratio of the amplitude of the pressure wave to that of the density wave. A different way of describing the particle movement when sound absorption occurs is:

$$\xi = A \exp(i\omega(t-x/U) - \alpha x) \tag{14}$$

By identifying the exponents of eqns (12) and (14) the following relation between the sound absorption coefficient and the complex sound velocity is obtained.

$$U_c^2 = (\omega/(\omega/U - i\alpha))^2 \tag{15}$$

By identifying the ratio of the imaginary part to the real part of eqn. (13) with the corresponding ratio of eqn. (15), equation (16) is obtained.

$$\omega/(U\alpha) = \tan\Phi^{-1} + (1 + \tan\Phi^{-2})^{\frac{1}{2}} \tag{16}$$

If the theory is limited to small values of Φ,

$$\tan\Phi << 1 \tag{17}$$

then eqn. (16) can be approximated to

$$\omega/(U\alpha) = 2\tan\Phi^{-1} \tag{18}$$

or by introducing

$$\omega = 2\pi f \tag{19}$$

$$\alpha/f^2 = \pi/Uf \tan\Phi = \text{Im}(\partial P/\partial V)_S/\text{Re}(\partial P/\partial V)_S \pi/Uf) \tag{20}$$

Equation (20) correlates the frequency dependence of α/f^2 with the kinetic and thermodynamic parameters for a given chemical reaction. Thus $(\partial P/\partial V)_S$ and hence the ratio of the imaginary part to the real part of this quantity can be obtained for a given reaction by means of the procedure commonly used in the field of irreversible thermodynamics. The approximations involved are as follows: (a) The sound absorption is assumed to be caused by the perturbation of the chemical equilibrium in question only. (b) This absorption is usually obtained by subtracting the absorption of the solvent from that of the solution. This means that additivity of the different relaxation phenomena is also assumed. (c) The reaction scheme used to derive the expressions for $(\partial P/\partial V)_S$ is assumed to be the detailed reaction mechanism of the chemical reaction. (d) The concentration fluctuation produced by the sound wave can be described by linear response viz.

$$[N_i]=[N_i]^o + \text{constant}\ \exp(i\omega t) \tag{21}$$

(e) The perturbations of the different species are negligible such that higher order terms can be neglected and (f) the sound absorption is so small that the sound wave propagates isentropically.

If the procedure is carried out for a reaction mechanism which consists of a single elementary reaction step of the type

$$a_1A_1 + a_2A_2 + .. + a_nA_n \underset{k_{-1}}{\overset{k_1}{\rightleftharpoons}} b_1B_1 + b_2B_2 + .. + b_nB_n \tag{22}$$

the following equation, which is often called the relaxation equation, is obtained

$$\alpha/f^2 = \{A/(1+(2\pi\tau f)^2)\} + B \tag{23}$$

where the relaxation time, τ, is given by expression (24)

$$1/\tau = k_1 \chi \prod_{i=1}^{i=n} [A_i]^{o\ a_i} \tag{24}$$

where

$$\chi = \sum_{i=1}^{i=n} (a_i^2/[A_i]^o + b_i^2/[B_i]^o) \tag{25}$$

The relaxation strength, A, is given by

$$A = (2\pi^2 R\bar{V}(C_p-C_V)/UC_pC_V)(\Delta H^o/RT)^2\ \tau/\chi \tag{26}$$

provided that the reaction volume is negligible. $\bar{X}$ denotes the molar quantity of X. B is the value of α/f^2 for the solvent. If n different elementary reaction steps contribute to the relaxation the relaxation equation can be written as

$$\alpha/f^2 = \sum_{i=1}^{i=n} A_i/(1 + (2\pi\tau_i f)^2) + B \tag{27}$$

If one or more of the chemical species are involved in several of the elementary reaction steps with the result that a strong coupling occurs, then eqn. (27) turns out to be a rough approximation only.

(ii) Relaxation spectrum

Equation (23) can be rewritten as

$$F(f) = (\alpha - \alpha_o)/f = f/(1 + (2\pi\tau f)^2) \tag{28}$$

where α_o is the absorption coefficient of the solvent. A plot of F(f) vs. f gives a curvature with the following maximum coordinates

$$(f_r, \mu_m) = (1/(2\pi\tau), A/(4\pi\tau)) \tag{29}$$

and with

$$F(f) \rightarrow 0 \text{ for } f \rightarrow \infty \text{ and } 0 \tag{30}$$

f_r is the relaxation frequency and μ_m the amplitude factor. This plot is the relaxation spectrum. The half value width, Δf, for the spectrum of a single elementary reaction step is given by

$$\Delta f = 3/(\pi\tau) \tag{31}$$

Sometimes the relaxation spectrum is given as F(f) vs. log f which gives a curvature with symmetry about the line

$$\log f = \log (1/(2\pi\tau)) \tag{32}$$

and with the half value width

$$\Delta \log f = 1.1 \tag{33}$$

Equation (33) gives an impression of the size of the frequency range in which a single elementary reaction step contributes to the relaxation.

For kinetic investigations the concentration dependence of the relaxation parameters is often desired. Relaxation measurements carried out over a wide concentration range may not be described by eqn. (28) because a dilution effect may also contribute to the change in absorption. In this situation the B-parameter cannot be considered as the value of α/f^2 of the solvent, but must be considered as a background absorption defined as

$$\alpha/f^2 = B \text{ for } f \rightarrow \infty \tag{34}$$

Then eqn. (23) becomes more convenient and a plot of α/f^2 vs. log f is also called a relaxation spectrum. This curvature is symmetric about the line given by eqn. (32) and has an inflection point for

$$f = f_r = (1/2\pi\tau) \qquad (35)$$

5. TREATMENT OF ULTRASONIC DATA[6]

(i) Analysis of ultrasonic relaxation spectra

The ultrasonic experiment results in values of α/f^2 at certain frequencies for different solute concentrations. The analysis of the data involves finding the minimum value of i required in order for equation (27) to describe the experimental data within experimental error. The analysis thus results in values of i, τ_i, A_i and B. In order to make such an analysis it is necessary to define a criterion which decides whether or not a certain value of i makes eqn. (27) describe the experimental data. This criterion must depend on the experimental error in α/f^2, the actual number of measured points on the relaxation spectrum, the frequency range covered, the location of the relaxation frequency in this range, and the size of the amplitude involved. A criterion can be defined, however, in terms of the goodness of fit parameter, F, which is commonly used to show the degree to which a theoretical expression describes a given set of experimental observations. The problem is to find the actual number of F which must be exceeded in order for us to state, that i=1 does not describe the data within experimental error any more. This value of F can be determined by means of computer simulations.[7]

(ii) Kinetic interpretation of the relaxation parameters

The ultrasonic experiments do not give information about the type of phenomenon which actually causes the observed relaxation. An interpretation of the relaxation parameters involves this knowledge. The first step in a kinetic interpretation is, therefore, to formulate all possible phenomena which may give rise to the observed relaxation. Theoretical considerations and results obtained by means of other experimental techniques for the system in question very often give rise to several plausible phenomena. The next step in the interpretation is to suggest detailed reaction mechanisms for the different phenomena. The theoretical expressions for the relaxation parameters for the different reaction mechanisms are then constructed by means of the procedure mentioned in section 4. The final step is to compare the concentration dependence of the relaxation parameters predicted from the different mechanisms with the concentration dependence actually measured for the different relaxation parameters. Agreement between the predicted concentration dependence of the relaxation parameters and the actually observed concentration dependence makes the mechanism and the phenomenon plausible. The comparison usually gives the values of the rate constants and the energy functions for the mechanism involved.

It must be emphasised, however, that if a given mechanism describes the experimental observations in detail **it does not prove** that the mechanism actually causes the relaxation in question.

6. REACTION MECHANISMS

(i) Formulation of reaction mechanisms

There are in principle two different points of view which can lead to a suggestion of a reaction mechanism for chemical reactions. The first point of view is based on the minimum number of relaxation times (i in eqn. (27)) which is found experimentally. If for instance the experiments are described by a single relaxation time then we attempt to describe the kinetics of the system in terms of a mechanism which consists of a single elementary reaction step according to a transition between **two different energy states in** the system. Thus this point of view leads to the formulation of a two state mechanism. The second point of view is based on describing the kinetics of the system of aggregation number n in terms of a mechanism which consists of n-1 bimolecular elementary reaction steps. Thus this point of view leads to the formulation of a multistep mechanism.

The two state mechanism and the multistep mechanism represent the two extremes of all the possible reaction mechanisms for chemical reactions. For simple systems such as dimerization, the two treatments are identical. For more complicated systems the relation between the two treatments is of the same nature as that between a statistical thermodynamical consideration and a classical thermodynamical consideration of the same system. The two state treatment is prefered as the description of complicated system for which general information on the molecular level is not available. However, if this data exists the multistep mechanism is prefered because it leads to a better understanding of what is going on at the molecular level. On the other hand, the two state mechanism may reflect some physical features of the system. The relaxation equation derived from the two state mechanism can also be obtained from the multistep mechanism provided reasonable approximations are involved. However, it may turn out that different sets of approximations can lead to the same relaxation equation and, in this case, it is very difficult to determine which set of approximations are prefered.[8]

It is important to notice that both above mechanistic treatments require that the associates formed can be considered as polymer molecules with well defined stoichiometry. If the stoichiometry of the associates is badly defined and the energies of formation of the different bonds involved differ considerably, the chemical approach must be given up and replaced by the approach of concentration fluctuation or other physical treatments which explain the relaxation spectra but which do not give kinetic information of the type desired by chemists.

(ii) Application of the two state mechanism, acetic acid/acetone mixtures[9].

The basic knowledge of the rates of hydrogen bond formation has been obtained from ultrasonic investigations of systems in which the mechanism of hydrogen bond formation is simple, for example, the dimerization of benzoic acid dissolved in inert solvents. In order to study the solvent effect on hydrogen bond formation, several compounds dissolved in active solvents such as dimethyl formamide, dimethyl acetamide, methyl propionate, dioxane and acetone have been investigated. Despite the fact that these solvents are strong hydrogen bond acceptors, the reaction mechanisms which are commonly used to interpret the relaxation spectra do not involve solvent molecules. The same reaction mechanisms are used to describe the hydrogen bond formation of the compounds dissolved in both inert and active solvents. As this mechanism explains the experiments it can be argued that there is no need to test a second mechanism which also involves the solvent molecules, especially as this may cause a more complicated expression involving more unknown parameters. A two state treatment, however, involves the same number of unknown parameters as the dimerization mechanism but takes into account interaction of solvent molecules.

Recent investigations of acetic acid/acetone mixtures have shown that the relaxation time for this system increases with increasing concentration of acetic acid.[10] Any type of a reaction mechanism which involves hydrogen bond self association of acetic acid leads to the conclusion that the relaxation time is either constant or decreases with increasing concentration of acetic acid. Since the relaxation spectrum of all concentrations can be described by a single relaxation time it seems reasonable to apply the two state mechanism which describes the kinetics of the system in terms of a transition between two states of different energy, a hydrogen-bonded state and a non-hydrogen-bonded state according to the following elementary reaction step.

$$\mathrm{Ac} + \mathrm{Don} \rightleftharpoons \mathrm{Bond} \qquad (36)$$

where Ac and Don denote an acceptor and donor site for hydrogen bonding respectively and Bond denotes a hydrogen bond. The relaxation time for this mechanism is given by the following equation.

$$1/\tau = k_1((\mathrm{Ac}) + (\mathrm{Don})) + k_{-1} \qquad (37)$$

where k_1 and k_{-1} are the forward and reverse rate constants for hydrogen bond formation respectively. The actual concentrations of acceptors, donors and bonds can be obtained by counting the number of acceptor sites, donor sites and bonds present in the equilibrium situation. It turns out that a complete description of the experimental data which involve the concentration dependence of the relaxation time and relaxation strength requires the presence of three different hydrogen bonded associates, namely **closed dimers of acetic acid, open dimers and monomers of acetic acid linked to acetone by a hydrogen bond. These compounds are defined as the bonded state while monomers of acetic acid and acetone molecules constitute the non-bonded state.**

The advantages of this mechanism are (a) The mechanism takes into account that hydrogen bonds formed in acetic acid/acetone mixtures involve both types of molecules. (b) Despite the fact that several hydrogen bonded complexes are involved, the mechanism agrees with the experimental observation of a single relaxation. (c) The mechanism predicts the observed concentration dependence of the relaxation time and the relaxation strength. It is a disadvantage, however, that the treatment implicitly involves the assumption that the rates by which the hydrogen bonds are formed are independent of the complexes from which the donors and acceptors are made. Although this may be a poor approximation for certain systems it may well reflect the chemistry of many hydrogen-bond forming systems which show a single relaxation even though several hydrogen-bonded aggregates exist in the solution.

(iii) Application of the multistep mechanism, hydrogen-bond polymerization in NMA[11]

The hydrogen-bond formation between NMA (N-methyl acetamide) molecules has been investigated by a variety of different techniques, mainly because this molecule has been considered as a convenient model to use for studies of the hydrogen bond entity of the peptide group.

The ultrasonic relaxation spectra of NMA dissolved in different solvents show a single relaxation time.

The general conclusion is that the amide group in NMA is planar, that NMA exists in trans form predominantly and that this configuration combined with a strong tendency to form hydrogen bonds results in chain associations of NMA molecules. The donor and acceptor sites of the hydrogen bonds are those of the peptide bond.

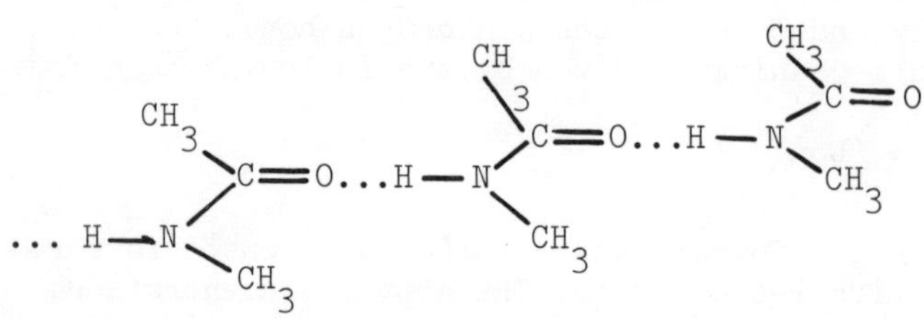

(38)

The scheme which has been used to describe the equilibrium distribution of NMA molecules on the different chains is :

$$
\begin{array}{l}
N + N \rightleftharpoons N_2 \\
N + N_2 \rightleftharpoons N_3 \\
N + N_3 \rightleftharpoons N_4 \\
\cdot \quad \cdot \quad \cdot \\
\cdot \quad \cdot \quad \cdot \\
\cdot \quad \cdot \quad \cdot \\
N + N_n \rightleftharpoons N_{n+1}
\end{array}
\tag{39}
$$

where N_x denotes a polymer with x molecules of NMA. The above given scheme may be taken as a detailed reaction mechanism for the hydrogen bond polymerization. This corresponds to a multistep mechanism with the approximation introduced that collisions between polymers are negligible. In order to carry out the kinetic analysis of the ultrasonic relaxation spectra it is necessary to decrease the number of independent variables in the multistep scheme. The following approximation gives the **simplest description.**

$$K_1 = K_2 = K_3 = .. = K_n \tag{40}$$

where K is the equilibrium constant. In terms of the two state mechanism this approximation of the equilibrium distribution leads to a relaxation time expression equal to that obtained for a simple dimerization. The multistep mechanism leads to a far more complicated expression. However, neither of the mechanisms are consistent with the experimental data because the approximation given by eqn. (40) is not fulfilled. Extended equilibrium investigations have shown that the following approximation is more reasonable:

$$K_1 < K_2 = K_3 .. = K_n \tag{41}$$

This approximation in the multistep treatment leads to a complete description of the ultrasonic data. Although several other approximations exist that given by eqn. (41) seems to be the most realistic **because, even on simple statistical considerations it predicts** two different equilibrium constants, one for the bimolecular equilibrium and one for the remaining equilibria.

The complete analysis which is relatively complicated leads to the conclusion that hydrogen bond polymerization of NMA is a nucleation mechanism, the dimer being the **nucleus.**

The formation rate constant for the dimer is 1.4 $10^{-9}M^{-1}s^{-1}$ and that of the polymers is 1.2 $10^{-10}M^{-1}s^{-1}$. It is important to point out that a reasonable quantitative analysis of the ultrasonic relaxation spectrum in terms of the multistep mechanism requires the actual values of $K_1, K_n, \Delta H_1^o, \Delta H_2^o$, and n. For NMA these quantities are known from numerous sophisticated experimental studies of the equilibrium situation. For other complicated hydrogen bond forming systems these parameters are not known.

7. CONCLUDING REMARKS

Taking the majority of available ultrasonic publications into account it may be concluded that the current understanding of ultrasonic absorption processes in solution is unsatisfactory. This is due in part to the complexity of the systems under consideration, and in part to the limitations imposed by the precision, volumetric requirement and the frequency range for the current measuring technique. In spite of this it is to be expected that the ultrasonic technique in future will be applied to more and more complicated systems. The analysis of the relaxation spectra for these systems requires extensive computational facilities in order to compare the experimental data with the predictions based on mechanisms of progressively increasing complexity.

It may be relevant to state, however, that the analysis of ultrasonic absorption data offers a very critical test of molecular models in solution. Not only is α/f^2 related to the thermodynamic and kinetic parameters of the system, but is also explicitly related to the relaxational increments in these quantities. Thus a mechanism which adequately predicts the behaviour of the ultrasonic absorption coefficient, as well as the behaviour of less demanding parameters, may be regarded with some degree of confidence.

REFERENCES

1. J.R. Pellam and J.K. Galt, J. Chem. Phys. 14, 608 (1946), see also A.J. Matheson, "Molecular Acoustics" Wiley-Interscience, London 1970.
2. F. Eggers, Acustica 19 323 (1968).
3. J. Rassing and B.N. Jensen, Acta Chem. Scand. 27, 1, (1973).
4. K. Herzfeld and T.A. Litovitz "Absorption and Dispersion of Ultrasonic Waves", Academic Press, London (1959).
5. M. Eigen and L. De Maeyer in S.L. Friess, E.S. Lewis and A. Weisberger (Editors) "Technique of Organic Chemistry", Vol.8 Part 2, Interscience, New York (1963).
6. J. Rassing, Advances Mol. Relaxation Processes, 4, 55 (1972).
7. J. Rassing and H. Lassen, Acta Chem. Scand. 23, 1007 (1969).
8. J. Rassing and E. Wyn-Jones, Ber. Bunsenges. physik. Chem. 651 (1974).
9. J. Rassing, J. Chem. Phys. 56, 3225 (1972).
10. D. Corsaro and G. Atkinson, J. Chem. Phys. 54, 4090 (1971)
11. J. Rassing, Ber. Bunsenges. physik. Chem. 75, 334 (1971).

DIELECTRIC RELAXATION

A.M. North

Department of Pure and Applied Chemistry,
University of Strathclyde, Thomas Graham Building,
Cathedral Street, Glasgow, G1 1XL

INTRODUCTION

This contribution takes the form of an introductory review. However, since it is obviously impossible to cover the whole field a general introduction to some of the important concepts in dielectric relaxation is given. These are illustrated with selected topics which are of personal interest to the author.

In general, any relaxation phenomenon can be defined as the time-dependent return to equilibrium of a system which has experienced a change in the constraints acting upon it. In practical terms some observable property is determined as a function of time or frequency, figure 1. In the case of dielectric relaxation the applied constraint is an electric field and the observable property is the macroscopic electric polarization. The relationship between the observed polarization and the applied field can be considered as the response to either a step function or periodic change in the field. As a result of the phase lag between the application of the field and the polarization current, the response of this system (permittivity) is mathematically complex. The real (storage) and imaginary (loss) components vary with frequency as shown in figure 2. Most frequently in dielectric studies we are concerned with polarization processes such as charge movement, and then the molecular change responsible for the polarization occurs at a rate which is temperature dependent. Since the rate of such processes generally increases with increasing temperature (at least in dielectric materials) the dielectric relaxation can be considered as a function of temperature, as is also illustrated in figure 2.

A relaxation process is said to be 'ideal' if the rate of return to equilibrium is proportional to the extent of perturbation

Wyn-Jones (ed.), Chemical and Biological Applications of Relaxation Spectrometry, 17–33.

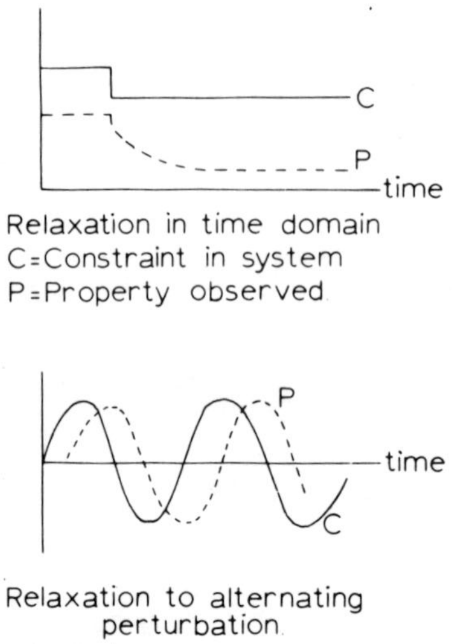

Fig. 1. Relaxation as the change in some property following upon a changing constraint.

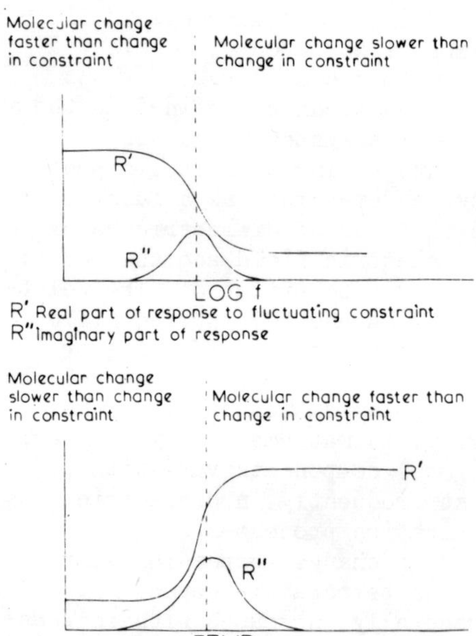

Fig. 2. Real and imaginary components of the response of a system to a periodic constraint.

from equilibrium. Under these circumstances the time-dependent behaviour is described by a linear first order differential equation for which the solution is a simple exponential function of time. The behaviour in the frequency domain is related to that in the time domain through the Laplace transform of the time-decay function, so that the complex permittivity has the familiar form portrayed in equation 1.

Equation (1)

Ideal Relaxation

$$\phi(t) = \exp(-t/\tau)$$

$$\varepsilon^{*} = \varepsilon_{\infty} + \frac{\varepsilon_{0} - \varepsilon_{\infty}}{1 + i\omega\tau}$$

$$\frac{\varepsilon' - \varepsilon_{\infty}}{\varepsilon_{0} - \varepsilon_{\infty}} = \frac{1}{1 + \omega^{2}\tau^{2}}$$

$$\frac{\varepsilon''}{\varepsilon_{0} - \varepsilon_{\infty}} = \frac{\omega\tau}{1 + \omega^{2}\tau^{2}}$$

Here ε^{*} is the complex permittivity, ε_{0} and ε_{∞} represent the permittivity of frequencies much below and much above the relaxation frequency, ε' and ε'' represent the real and imaginary parts of the complex permittivity, ω is the observation frequency and τ is the ideal relaxation time.

NON-IDEAL RELAXATION

Experimental observations of dielectric relaxation very frequently yield values for the real and imaginary permittivity which cannot be fitted to the ideal equations presented above. Generally, in the frequency domain, the decrease in the real permittivity and the maximum in the imaginary component extend over a wider frequency range than predicted by the ideal Debye equations. Quite often, too, the observations are not symmetrical in the log frequency plane, the relevant graphs present a 'skew' dependence. For many years now it has been conventional to fit such observations to empirical functions similar in form to the ideal expressions, but containing a new parameter as a measure of the departure from ideal behaviour. The best known of these are presented below, equation 2.

The departures from ideal behaviour are often ascribed to the existence of a distribution of characteristic relaxation times, and so the adjustable parameter is often referred to in these terms.

Equation (2).

Empirical Distribution Functions

Frequency Domain

Cole-Cole $\varepsilon^* = \varepsilon_\infty + \dfrac{\varepsilon_o - \varepsilon_\infty}{1 + (i\omega\tau)^{1-\alpha}}$

Cole-Davidson $\varepsilon^* = \varepsilon_\infty + \dfrac{\varepsilon_o - \varepsilon_\infty}{(1 + i\omega\tau)^{\alpha}}$

Time Domain

Williams $\phi(t) = \exp\,(-t/\tau_o)^{\beta}$

The symmetry (or lack of it) associated with these empirical functions is seen when an Argand diagram of ε'' against ε' is constructed. In this representation ideal Debye relaxation appears as a semicircle, the symmetrical Cole-Cole distribution as the arc of a semi-circle of centre depressed below the $\varepsilon'' = 0$ axis, and the Cole Davidson distribution as a skewed arc. These are illustrated in figure 3.

The representation of non-ideal relaxation can be carried back from the frequency domain to the underlying time-dependent decay function, by giving the latter a non-exponential form. One such has been proposed by Williams[1], and it has been pointed out that when the empirical 'distribution parameter', $\beta = 0.5$, the decay function generates Cole-Davidson skewed arc behaviour in the frequency domain. It must be stressed that introducing such a parameter in the decay function is every bit as empirical as in the complex permittivity but does aid the search for an explanation in molecular terms.

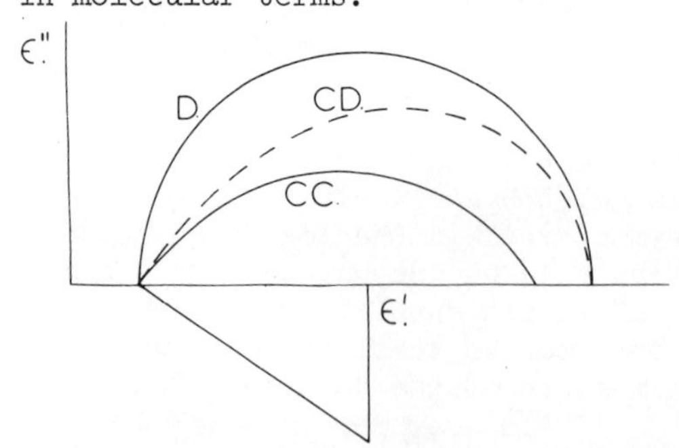

Fig. 3. Argand diagram of real and imaginary permittivity. D, ideal Debye relaxation; CD Cole-Davidson skewed-arc relaxation; CC Cole-Cole symmetric relaxation.

Turning to explanations of non-ideal behaviour other than those based upon a simple distribution of times for a single process, we find that many exhibit the common feature of a basis in two inter-related molecular processes of different time-dependence.

The most widely used of these is due to Glarum[2], and is based upon the premise that the relaxation process of interest can occur only when some 'defect' diffuses to the neighbourhood of the molecule about to relax. In this way the overall process is generated as a combination of diffusion (the time-dependence of which is naturally non-exponential) and relaxation (which may be ideal). The concept is certainly attractive when relaxation requires molecular translation or rotation and when a defect can be visualised as some vacancy or molecular free volume which is required for the appropriate movement. The type of relationship which can be derived from this model is illustrated in equation 3, which show too, that depending on the relative times required for diffusion and relaxation, ideal Debye, Cole-Cole, or Cole-Davidson behaviour can be observed.

Equation (3)

Defect Diffusion Model (Glarum)

$$\phi(t) = \exp(-t/\tau_o) \; \{1 - p(t)\}$$

p(t) Probability that arrival of defect by time t has caused relaxation

If defect diffusion characterised by τ_D

$\tau_o << \tau_D$ Debye relaxation

$\tau_o = \tau_D$ 'Skewed Arc' Relaxation, $\alpha = 0.5$

$\tau_o >> \tau_D$ Circular arc, $\alpha = 0.5$

Another quite basic explanation of non-ideal relaxation is based upon a consideration of the reaction field which surrounds any charged species in a dielectric environment. The basis of this consideration is that the reaction field, due to the charged species, in its turn affects the movement of that species. Jonscher[3] has considered the effect on the translational movement of charge carriers and shown that the loss permittivity can be described as a power-law function of frequency and that the loss tangent is a constant independent of frequency. Measurements on a wide variety of semi-conducting materials suggest that this approach may have considerable validity when dielectric behaviour is associated with carrier migration. The important equations are presented as equation 4.

Equation (4)

Relaxation of Reaction Field

1. Hopping Mechanism (Johnscher)

$$\varepsilon'' = K\,(\omega/\omega_c)^{n-1} \qquad (0 < n < 1)$$

$$\frac{\varepsilon''(\omega)}{\varepsilon'(\omega)} = \tan\left[(1 - n)\pi/2\right]$$

2. Dielectric Friction (Böttcher)

For Dipole, μ, Reaction field r, Torque T_r,

Moment of inertia I

$$\frac{\langle \dot{\mu} \cdot \dot{r} \rangle}{\langle \dot{\mu} \cdot \dot{\mu} \rangle} = \frac{\langle \mu \cdot r \rangle}{\langle \mu \cdot \mu \rangle} - \frac{1}{I}\,\frac{\langle T_r^{\,2} \rangle}{\langle \mu \cdot \mu \rangle}$$

Leads to non-exponential decay function

An alternative approach is to consider polarization due to the rotation of molecular dipoles. Here, the reaction field due to a particular dipole exerts what is known as 'dielectric friction' on its rotational behaviour and this in turn leads to non-exponential decay functions. Considerable progress in this line has been made by Böttcher[4]. In both of these approaches, the same basic concept as before, is differing time-dependent behaviour in an interacting variable.

As a final example of such an approach we consider again dielectric behaviour associated with the migration of charge carriers, but in which liberation of the carriers at the electrodes is inhibited in some way. Under these circumstances charge builds up at the 'blocking electrodes' and the time-dependent behaviour of this charge interacts with the migration behaviour of the carriers. A large number of workers have considered this phenomenon, and useful equations have been derived by Doremus[5] and are presented below, equation 5.

The descriptions of non-ideal behaviour presented below illustrate the way in which the search for an explanation has moved from an empirical consideration of results in the frequency domain, through the responsible decay function in the time-domain to molecular explanations responsible for generating a differential equation other than the ideal linear first order equation of which the decay function represents a solution.

The profitability of commencing any theoretical or interpretive

Equation (5)

Blocking Electrodes (Doremus)

$$\frac{\varepsilon' - \varepsilon_\infty}{\varepsilon_0 - \varepsilon_\infty} = \left\{\frac{(\nu^2 + 1)^{\frac{1}{2}} + 1}{2\,(\nu^2 + 1)}\right\}^{\frac{1}{2}}$$

$$\frac{\varepsilon''}{\varepsilon_0 - \varepsilon_\infty} = \left\{\frac{(\nu^2 + 1)^{\frac{1}{2}} - 1}{2\,(\nu^2 + 1)}\right\}^{\frac{1}{2}}$$

where $\nu = 2\pi f\,\varepsilon_0\varepsilon_\infty/\sigma$

$$f_{max} = \frac{\sqrt{3}\,\sigma}{2\pi\varepsilon_0\varepsilon_\infty}$$

Approximates to 'skewed arc' distribution

study on the appropriate differential equation becomes very obvious for studies of phenomena occuring at very high frequencies. The equation of motion of any body will normally contain inertial (second order differential), viscous (first order differential) and elastic (linear) terms. At low frequencies, when dielectric relaxation has a molecular origin, inertial terms are insignificant and the equation of motion reduces to the familiar linear first order form. Equally, at very high frequencies the inertial terms predominate and the familiar second order differential equations descriptive of resonance phenomena emerge. However, at intermediate frequencies both inertial and viscous terms may be important when the resulting equation of motion will not have a solution in the form of a simple single exponential. Consequently, another type of non-ideal behaviour is observed at these frequencies where the macroscopic observation of relaxation takes on increasingly a resonance character. This particularly affects the real permittivity which passes through a maximum and minimum rather than exhibiting a simple decrease as frequency is increased.

The inter-relating of molecular behaviour and macroscopic relaxation is further aided by the use of time correlation functions. The application of the dipole moment correlation function to dielectric phenomena has been set out in a most useful review by Williams[6]. In such an approach one is examining the time-dependent decay in the correlation of the angle made by a reference dipole with its orientation at zero time (the auto-correlation terms) or the angle made with some other dipole at zero time (cross-correlation terms). The appropriate dipole moment correlation function is presented below, equation 6.

Equation (6)

Correlation Function Approach

General

$$\Gamma(t) = \langle A(t).\ B(t')\rangle$$

Dielectric

$$\Gamma(t) = \frac{\sum_{i=1}^{N} \langle \mu_i(o).\mu_i(t)\rangle + \sum_{i=2}^{N} \sum_{i=1}^{i-1} \langle \mu_i(o).\mu_i'(t)\rangle}{2\sum_{i=1}^{N} \mu_i^2 + 2\sum_{i=2}^{N} \sum_{i=1}^{i-1} \langle \mu_i(o).\mu_i'(o)\rangle}$$

Decay of $\mu^2 \langle \cos \theta_{hk}(t)\rangle$

SOME EXAMPLES OF DIELECTRIC RELAXATION

1. MOLECULAR ORIGINS OF POLARIZATION

Three polarization processes have been selected to illustrate the scope of dielectric relaxation studies. These all have their origin in some form of molecular motion, but it must be appreciated that polarization can arise in a variety of other phenomena including chemical reaction.

The first process to be considered is the molecular polarization which can arise when molecules become distorted during intermolecular collisions. Since these are occuring all the time in condensed phase, any macroscopic sample contains a large number of fluctuating dipoles, the fluctuation frequency of which corresponds to the molecular collision frequency. The fluctuating dipoles can interact with electromagnetic radiation and so dielectric phenomena can be observed. The second phenomenon to be considered is dipole orientation. The original considerations by Debye dealt with systems of isolated molecular dipoles, but some interesting phenomena are observed when a number of dipoles are linked as in a linear polymer chain. Again orientation of the molecular dipoles in the electric field leads to macroscopic polarization and so to the observation of dielectric relaxation phenomena.

The third phenomenon to be considered is the polarization which arises when charge carriers, migrating due to the influence of the applied field, become localised at some interface in the sample. These processes are presented diagrammatically in figure 4.

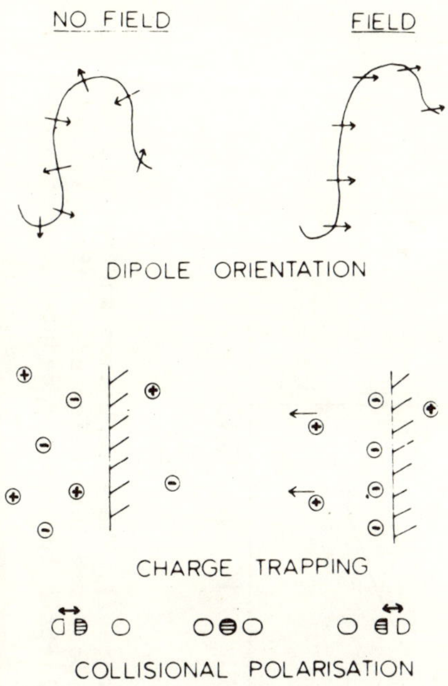

Fig. 4. Three polarization mechanisms.

2. DIELECTRIC PHENOMENA IN LIQUIDS AT VERY HIGH FREQUENCIES

When observations are made on liquids at sub-millimeter microwave (or very far infrared) frequencies a dielectric loss process is observed for which one possible explanation is the collisional polarization referred to earlier. Of course, for polar molecules a number of other explanations are possible, one being restricted libration of the dipole in a non-spherically symmetric intermolecular force field. Whatever the explanation, it is always necessary to consider whether or not inertial 'second order differential' terms contribute to the observed phenomenon. As an illustration of this measurements made on 1,4-dioxan[7] are illustrated on figure 5. In 5(a) the data are represented in the spectroscopic fashion as refractive index and absorption coefficient while in 5(b) is the dielectric loss factor. Up to about 40 $cm.^{-1}$ the real (refractive index) and imaginary (loss or absorption) components exhibit the features expected for relaxation. However, at higher frequencies a minimum can be observed in the refractive index indicating increasing resonance character to the observation.

Similar studies have been carried out on a large number of liquids and also on polymeric solids. As plastic materials become more widely used as dielectric waveguides and as fillers in high

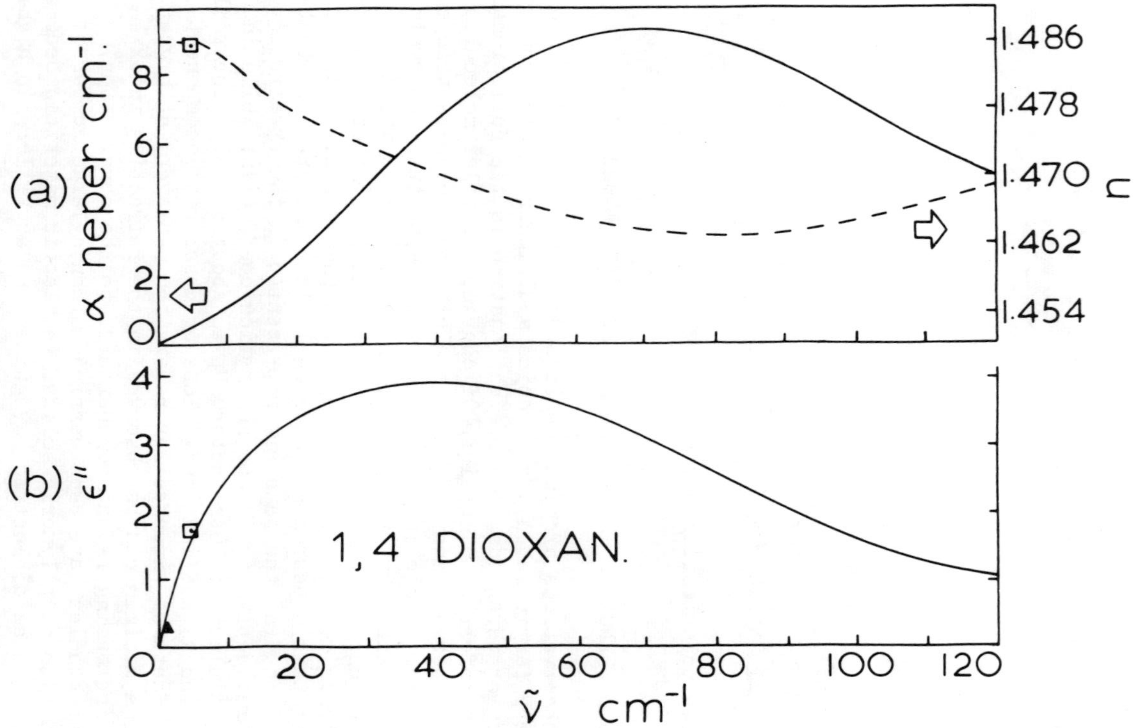

Fig. 5. Dielectric relaxation (reproduced from reference 7). The attenuation and refractive index for liquid 1,4-dioxan at 22 ± 2°C. (a) ———, observed absorption coefficient (α) and - - -, observed refractive index (n); (b) ———, dielectric loss factor deduced from n and α

frequency communication waveguides so an understanding of these dielectric properties assumes a technological significance. In general, it is found that the magnitude of the losses in this frequency region are related to the polarizability of the units in the polymer chain. Thus, the loss magnitude rises in the series polyethylene, polypropylene and poly(4-methylpent-1-ene) all of which show smaller losses than polar polymers. This has been explained in terms of restricted 'collisions' of the side groups attached to the main polymer backbone.

3. DIPOLE ORIENTATION IN DISSOLVED POLYMERS

When the orienting dipoles in a fluid system are attached to a polymer chain, the dipole relaxation can be used as a measurement of the inherent chain flexibility. One of the problems, here, is that macroscopic experiment measures reorientation of the resultant dipole vector for a whole chain. If the chain is very 'stiff' reorientation of the whole molecule occurs more rapidly than the series of internal conformational changes necessary to reorient the resultant. This whole-molecule rotation, however, varies inversely with the chain molecular weight, and so usually it is possible to study polymers sufficiently large that the conformational change can be observed. Some examples of this are illustrated in figure 6 where chains of increasing stiffness (decreasing relaxation frequency) are poly(ethylene oxide), poly(methyl methacrylate), poly(N-vinyl carbazole). Stiff chains for which it has not been possible to observe the internal conformational change are represented on this diagram as poly(γ-benzyl L-glutamate). In this way a quantitative measure of chain 'flexibility' can be made in terms of a time-dependent property rather than in terms of a time-averaged property such as chain dimensions.

4. DIPOLE ORIENTATION IN SOLID POLYMERS

When observations of dielectric relaxation due to the dipole orientation are made in solid polymers, a number of differences from solution behaviour become evident. Probably the most significant of these is that a number of different relaxation processes become obvious. When these are examined as a function of temperature, they can be related to the onset of different modes of dipole motion. Thus, typical polar linear polymers usually exhibit a major transition associated with the onset of gross rotational motion of chain segments (the so-called α- or glass-transition) and a lower temperature transition known as the β-process. The type of behaviour which may be met is illustrated diagramatically in figure 7.

In many uses of plastics, particularly where dielectric properties are concerned it is desirable that energy loss phenomena

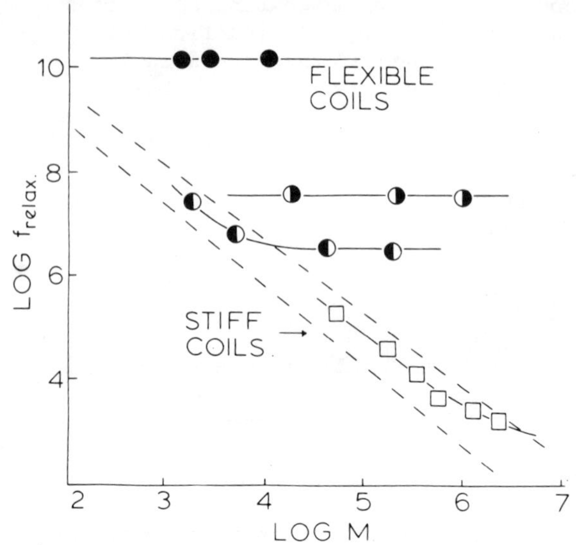

Fig. 6. Dielectric relaxation in dissolved polymers of differing molecular weights. ●, polyethylene oxide; ◑, poly(methylmethacrylat ◐, poly(n-vinyl carbazole); □ poly(γ-benzyl L-glutamate).

Fig. 7. Diagramatic representation of temperature dependence of dielectric loss in two typical polymers.

do not vary widely over the temperature 'working range' of the material. In this way, the useful working range can be defined in terms of the temperature difference between the energy losses associated with the α- and β-processes. Should such criteria be important then polymers of the type portrayed in figure 7(b) would be superior to those whose behaviour is of the type illustrated in figure 7(a). (Examples of the two types are polycarbonates and polystyrene respectively).

An interesting aspect of these two transitions can be seen when we examine the temperature-frequency behaviour. Since both processes have their origin in molecular motion, they will be observed at the higher frequencies at the higher temperatures. However, since the higher temperature process requires more thermal energy to take place, it will exhibit the larger temperature dependence. As a result of this, the two processes will tend to merge when observed at high temperature/frequencies. This is illustrated in figure 8, where it can be seen that the working range of the material, ΔT, decreases for high frequency usage.

The molecular origin of the β-process remains something of a mystery, but is currently thought to arise in restricted libration of segment dipoles, in contrast to the gross rotational behaviour occurring at the α-transition.

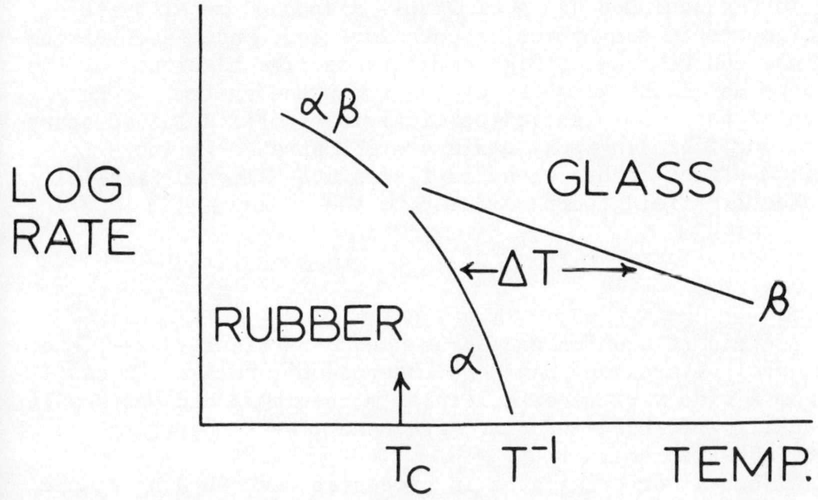

Fig. 8. Diagramatic representation of frequency - temperature behaviour of α and β loss processes.

5. INTERFACIAL POLARIZATION IN POLYMERS

A number of polymers which consist of two phases (for example thermoplastic elastomers or various composites) exhibit phenomena which can be ascribed to the trapping of charge carriers at the interfaces in the material. This interfacial polarization is usually referred to as Maxwell-Wagner-Sillars polarization. It can give rise to very high capacitances and losses and is usually observed at frequencies which relate to the charge-carrier mobilities. Thus, in essentially non-conducting polymers the carrier mobilities are small and the relaxation occurs at very low frequencies. This is illustrated by measurements on a polyurethane elastomer in which the chain is composed of poly(propylene oxide) and aromatic urethane segments[8], figure 9. In such thermoplastic elastomer 'cross-linking' is achieved by phase separation of the urethane blocks into glassy or crystalline domains. Impurities in the polyether elastic phase then cause the observed dielectric behaviour. Very often, such behaviour leads to the failure of plastic materials in technical service. Thus, the phenol formaldehyde-cellulose laminate used as base-board for much electronic circuitry exhibits a deterioration in electrical properties when used in warm humid atmospheres. This has been traced to interfacial polarization of charge carriers at the resin-cellulose boundary brought about by water absorption. The loss characteristics of an electrical grade laminate with different amounts of absorbed water are illustrated in figure 10.

When the phase containing the charge carriers has a reasonable conductance (i.e. the charge carrier mobility is high) high capacitance and low loss are evidenced up to much higher frequencies. Indeed, for an occluded phase of finely dispersed metal these frequencies are in the optical spectral region. Under such circumstances use can be made of high capacitance. The magnitude of the effect depends on the geometry of the two-phase system. As an illustration of this the observed permittivities of a dilute dispersion of carbon fibre in epoxy resin are illustrated in Table 1. The maximum effect occurs when the fibres are alligned parallel to the electric field (perpendicular to the electrodes).

CONCLUSION

Dielectric relaxation can be seen as a particular example of quite general relaxation phenomena. Macroscopic relaxation can arise from a wide variety of molecular mechanisms, and can result in an equally wide range of time, frequency, or temperature dependent observations.

Ideal behaviour originates in processes described by simple first order linear differential rate equations, and explanations of non-ideal behaviour are often best persued by examination of more complex differential equations.

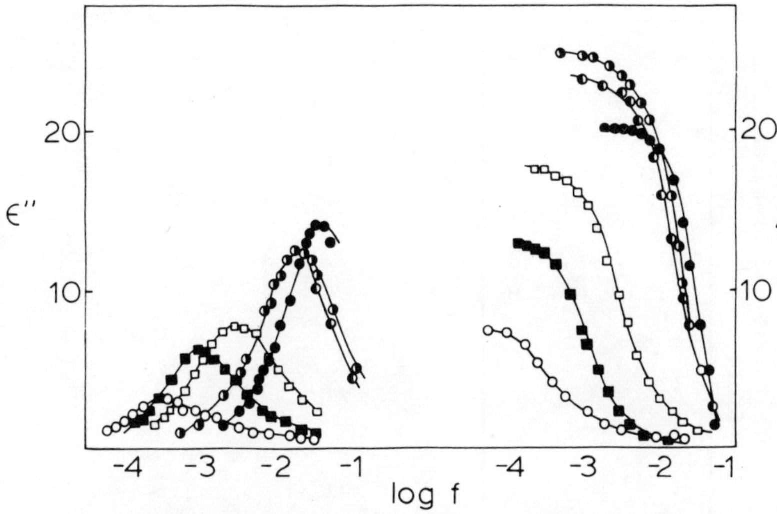

Fig. 9. Maxwell-Wagner-Sillars polarization in poly(propylene oxide) -polyurethane. P400. ● 63°, ◑ 58° (measured by charging transient), ◐ 58° (measured by discharging transient), □ 49°, ■ 45°, ○ 40°.

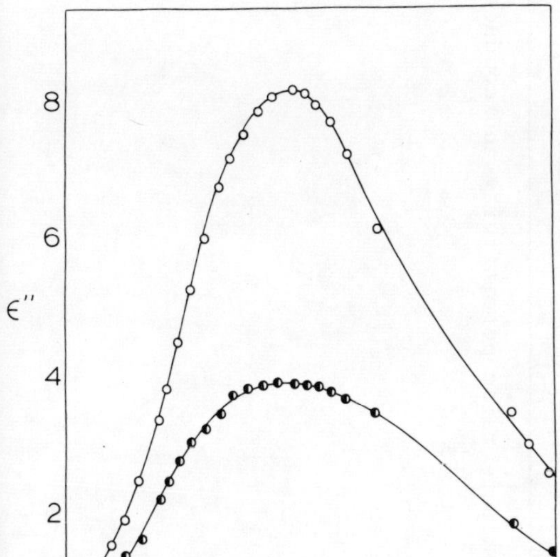

Fig. 10. Interfacial loss in phenol-formaldehyde resin-cellulose laminate with water content at 303 K. O, 0.8% absorbed water; ◐, 0.6% absorbed water; ◑, dried under vacuum for 5 days.

TABLE 1. PERMITTIVITY OF COMPOSITES: 200 μ x 10 μ CARBON FIBRE IN EPOXY RESIN AT 303 K

Fibre Volume Fraction	Permittivity, random orientation	Permittivity, fibres parallel to field
0.00	3.0	3.0
0.01	4.05	4.9
0.02	4.9	6.9
0.05	(12.6)*	12.8

*Inter-fibre contacts raise permittivity and conductivity.

Examples have been presented which illustrate the complexities of molecular behaviour which can be observed, as well as indicating the technological significance of some of the effects.

REFERENCES

1. G. Williams and D.C. Watts, Trans. Fraday Soc., **66**, 80, (1970).
2. S.H. Glarum, J. Chem. Phys., 33, 1371 (1960).
3. A.K. Jonscher, '1972 Ann. Report on Conference on Electrical Insulation and Dielectric Phenomena', Nat. Acad. Sci., Washington, 418 (1973).
4. T.H. Tjia, P. Bordewijk and C.J.F. Bottcher, Adv. Mol. Relax. Processes, **6**, 19 (1974).
5. M. Tomozawa and R.H. Doremus, J. Non-Crystalline Solids, **14**, 54 (1974).
6. G. Williams, Chem. Rev., **72**, 55 (1972).
7. M. Davies, G.F.W. Pardoe, J. Chamberlain and H.A. Gebbie, Trans. Faraday Soc., **66**, 273 (1970).
8. A.M. North and J.C. Reid, Eur. Polymer J., **8**, 1129 (1972).

DIELECTRIC RELAXATION AT MICROWAVE FREQUENCIES

John Crossley

Chemistry Department, Lakehead University, Thunder Bay, Ontario, Canada

Two other articles in these proceedings, by North and Williams, have presented the basic theory of dielectric relaxation and some of its applications. Obviously, there are many areas of interest which have not been covered, although reference has been given to more detailed texts. A good deal of work has been carried out on relatively simple organic liquids and solutions which, at normal temperatures, absorb in the microwave region. This paper will survey the type of information which may be obtained from microwave measurements of permittivity, ε^{1}, and loss, ε^{11}.

1. EXPERIMENTAL METHODS

At low frequencies, $<$ IMHz, it is generally possible to obtain ε^{1} and ε^{11} values at a large number of frequencies using capacitance - resistance bridges. Because the signal generators, usually klystrons, used at microwave frequencies are only tuneable over quite narrow frequency ranges it is often necessary to construct individual pieces of apparatus for each frequency. Consequently, many investigators have been forced to attempt to describe dielectric dispersions with a small number of data points, and it is rare to find laboratories employing more than ten frequencies in the range 1 - 140 GHz. This is generally quite adequate for systems described by a single or mean relaxation time, but analyses for more than one relaxation time are always suspect unless the data is well separated into identifiable dispersion regions. Vaughan[1] has given a review of the experimental methods available. In the author's laboratory three types of apparatus, suitable for low-medium loss measurements, cover the

Wyn-Jones (ed.), Chemical and Biological Applications of Relaxation Spectrometry, 35–39.

range 1 - 140 GHz[2-4]. The relatively new Time Domain Spectroscopy methods provide essentially continuous frequency measurements of ε^1 and ε^{11}, and the experiments can be carried out very quickly[5]. Unfortunately, the technique is, at present, generally restricted to an upper frequency of about 10 GHz and is not suitable for low loss measurements. It is possible, within reason, to bring a dispersion into an available frequency range by raising or lowering the temperature, although in some cases this may defeat the object of the experiment.

Once ε^1, ε^{11} and the low frequency or static permittivity, have been determined an attempt is made to fit the data to various equations to obtain relaxation times, distribution parameters and very high frequency permittivities. The majority of systems do not conform to a single Debye relaxation and are commonly fitted to either a symmetrical Cole-Cole distribution or a Cole-Davidson skewed arc distribution. In certain cases it is possible to analyse the data in terms of contributions from two overlapping Debye relaxations.[6,7]

2. RIGID POLAR MOLECULES

Much of the early microwave work was sensibly aimed at understanding the factors which govern the magnitude of the relaxation time, τ, for a reorientating dipole. In such experiments it is essential to use rigid polar molecules which do not associate, thus avoiding complicating contributions from intramolecular and intermolecular relaxation mechanisms.

For non-associated systems the dielectric relaxation time may be considered as the reciprocal of a rate constant and plots o $\ln\tau T$ against $1/T$ are linear. Thus the free energy, enthalpy and entropy of activation for the relaxation process can be determined by measuring τ at several temperatures.

Many fundamental attempts have been made to understand dielectric relaxation at the molecular level, and to obtain relationships which predict the relaxation time from the molecular dimensions and the viscosity of the medium. In general, such treat ments have met with only limited success. Although these equations fail quantitatively they do provide the variables which influence the relaxation time for a rigid dipolar molecule i.e. molecular size and shape, direction of the molecular dipole, viscosity and temperature.

For rigid polar molecules of similar shape, having their molecular dipole moments in the same direction e.g. the four halobenzenes, measured in the same solvent at the same temperature there is a linear relationship between molecular volume and relaxation time.[8] This is not true for an assortment of molecules or for pure liquids due to the different shapes, dipole moments, viscosities and internal fields.

The direction of the dipole moment can have a considerable influence upon the relaxation time. The dipole moment in 4-iodobiphenyl is directed along its long axis and relaxation occurs by rotation about the short axes. This involves a considerably greater displacement of neighbouring molecules than rotation about the long axis which is the main orientation mechanism in 2-iodobiphenyl. The 2- and 4- iodobiphenyl molecules have very similar shape and size but, due to the difference in the direction of their dipole moments, the relaxation time of the former is much shorter than the latter.[9]

The relationship between macroscopic viscosity and dielectric relaxation time is quite empirical. In general, the relaxation time lengthens with increased viscosity but in most cases the increase in relaxation time lags far behind the increase in viscosity. The effect depends to a large extent upon the shape and size of the polar molecule. For small spherical molecules the relaxation time is almost independent of viscosity, whereas for systems in which the solute molecule is much larger than the solvent molecule a more linear relationship exists.[10]

3. INTRAMOLECULAR RELAXATION

For an aromatic molecule which contains a rotatable polar group there is a possibility of contributions from dipole reorientation by rotation of the molecule as a whole and by intramolecular rotation of the polar group. In more flexible molecules a number of intramolecular processes may be possible. However, the usually inadequate number of data points in the absorption region generally restricts analyses to two relaxation time systems. If the bond angles and group dipole moments are known for a substituted benzene molecule, it is possible to calculate the contributions from molecular and group relaxations and thus have a check on the analysis. A large number of substituted benzenes and naphthalenes have been examined.[11] Aliphatic molecules are far more flexible and it is generally not possible to treat them in terms of a molecular and an intramolecular relaxation. A review of intramolecular relaxation in a large number of straight and branched chain polar aliphatic molecules has been published recently.[12]

4. MOLECULAR INTERACTIONS

The dielectric relaxation time is very sensitive to intermolecular attraction and can provide a useful means of studying molecular interactions in simple systems. Monohydric aliphatic alcohols have been studied more extensively than any other group of compounds, and the literature on the subject is considerable.

The dielectric dispersion of pure liquid primary aliphatic alcohols may be characterized by three relaxation times with the low frequency process dominating. Measurements of straight and branched chain alcohols, a wide range of temperature, and concentration in a variety of non-polar solvents, have been carried out in order to obtain a molecular model capable of explaining the origin of the relaxation.[13-15]

The relaxation time of chloroform in p-dioxane solution is about five times longer than its value in cyclohexane solution. Cyclohexane may be regarded as an inert solvent which shows no significant interaction with the polar solute. In contrast, the attraction between the protonic hydrogen atom in chloroform and the oxygen lone pair electrons in p-dioxane appreciably affects the rate of reorientation. Several interesting studies of polar solute - non-polar solvent interactions have been reported and the strength of such interactions are consistent with values obtained by other techniques.[16-18]

Dielectric relaxation data for mixtures of non-polar electron acceptor molecules such as tetracyanoethylene in non-polar solvents capable of donating electrons i.e. benzene, dioxane have been reported.[19] Since both components are non-polar any significant dielectric loss indicates the formation of a polar complex and the relaxation times may be related to the lifetime of the complex.

REFERENCES

1. N.E. Hill, W.E. Vaughan, A.H. Price and M. Davies, "Dielectri Properties and Molecular Behaviour" Van Nostrand Reinhold Company, London, 1969.
2. W.F. Hassell, M.D. Magee, S.W. Tucker and S. Walker, Tetrahedron, 20, 2137 (1964).
3. S.K. Garg, H. Kilp and C.P. Smyth, J. Chem. Phys., 43, 2341 (1965).
4. S.E. Keefe and E.H. Grant, Rev. Sci. Instrum., 39, 1800 (1968
5. A. Suggett, Chapter 4, Chemical Society Specialist Periodical Report "Dielectric and Related Molecular Processes" Volume 1, (1972).
6. J. Crossley, S.P. Tay and S. Walker, Advan. Mol. Relaxation Processes, 6, 69 (1974.
7. J. Crossley, S.P. Tay and S. Walker, Advan. Mol. Relaxation Processes, 6, 79 (1974).
8. W.F. Hassell and S. Walker, Trans. Faraday Soc., 62, 2695 (19
9. E.N. DiCarlo and C.P. Smyth, J. Phys. Chem., 66, 1105 (1962).
10. J. Crossley, R.I.C. Reviews, 4, 69 (1971).
11. C.P. Smyth, Advan. Mol. Relaxation Processes, 1, 1 (1967-68).
12. J. Crossley, Advan. Mol. Relaxation Processes, 6, 39 (1974).
13. J. Crossley, Advan. Mol. Relaxation Processes, 2, 69 (1970).
14. W. Dannhauser and A. Fleuckinger, Phys. Chem. Liquids, 2, 37 (1970).

15. M.J.C. Van Gemert, G.P. De Loor, P. Bordewijk, A. Quickenden and A. Suggett, Advan. Mol. Relaxation Processes, 5, 301 (1973).
16. A.A. Antony and C.P. Smyth, J. Am. Chem. Soc., 86, 152 (1964).
17. K. Chitoku and K. Higasi, Bull. Chem. Soc. Japan, 39, 2160 (1966).
18. J. Crossley and C.P. Smyth, J. Am. Chem. Soc., 91, 2482 (1969).
19. R.A. Crump and A.H. Price, Trans. Faraday Soc., 66, 92 (1970).

THE STOPPED-FLOW AND TEMPERATURE-JUMP TECHNIQUES - PRINCIPLES AND RECENT ADVANCES

B. H. Robinson

Chemical Laboratory, University of Kent at Canterbury

A. THE STOPPED-FLOW METHOD

Before 1950, reactions occurring in times less than 5s or so were considered by most chemists and biochemists to be 'instantaneous', and many kineticists were content to dismiss such reactions as simply being 'too fast to measure'. This is somewhat surprising in view of the fact that the first fast-reaction technique, the continuous-flow method, had been introduced in 1923 by Hartridge and Roughton[1]. Flow techniques, of which stopped-flow, continuous-flow and accelerated-flow are all examples, extend the classical time-range by three - four orders of magnitude, down to 1 ms or so, the limiting fast time being controlled by the mechanical process of thoroughly mixing the two solutions. Early experimenters used the continuous-flow method, although the technique was extravagent on reagents, because of the difficulty of monitoring rapid concentration changes. However, in the 1930's, with the development of oscilloscopes, the stopped-flow method came to be preferred on the grounds of convenience and reagent economy.

It must be stressed that the principle of the stopped-flow technique is extremely simple, in that it is merely a controlled extension of classical mixing. Details of the construction of stopped-flow instruments and the design parameters controlling performance have often been discussed in recent books and reviews (see Appendix). First-order rate constants up to $10^3 s^{-1}$ can be followed, and with favourable systems, second-order rate constants close to the diffusion-controlled limit ($\simeq 10^{10} dm^3 mol^{-1} s^{-1}$) become accessible.

A particular advantage of the stopped-flow technique is that several types of apparatus are readily available commercially;

Wyn-Jones (ed.), Chemical and Biological Applications of Relaxation Spectrometry, 41–48.

popular instruments are manufactured in the U.S., U.K. and Japan. Prices range from \$3000 - \$10,000, and on numbers sold world-wide, the stopped-flow technique would seem to be the most popular and versatile of the fast reaction methods.

i) Recent Developments in Stopped-Flow Instrumentation

a) Mechanical Unit:- The mechanical unit has not been significantly improved upon over the last decade. The 'dead' time, i.e. the sum of the mixing and flow times to the point of observation, is still of the order of 1 ms. Attempts to extend the time-range down to $\simeq$ 100μs by the use of more sophisticated mixer designs coupled with high fluid-flow velocities have not been widely adopted. For fast mixing, high linear flow velocities are required, and this can lead to cavitation (bubble formation) and vibration effects on stopping the flow. The manifestation of these artefacts on the recorded transient has been thoroughly investigated recently[2]. It appears that the best procedure is to keep the mechanical system relatively simple for optimum results unless the chemical system under study necessitates 'working at the limit'. b) Detection of the concentration change:- One of the great advantages of the stopped-flow technique is that it can readily be used with a variety of detection systems. Ultra-violet and visible spectrophotometric detection are normally available on commercial instruments with the option of fluorescence detection. In the fluorescence mode, very low reagent concentration changes can be monitored, and diffusion-controlled reactions become amenable to study. Also dynamic changes in (bio)polymers with intrinsic or extrinsic (labelled) fluorescent probes can be studied. Light-scattering detection is useful in the study of such processes as micellization, coagulation etc., and the oligomerization of sub-unit enzymes and proteins. An apparatus has recently been described[3] which utilizes a continuous mode He/Ne laser as the light source. Other detection methods include NMR[4], ESR[5], ORD[6], CD[7], IR[8], conductance[9], pH and temperature. A full description of the versatility of the stopped-flow method has been published, in which recent applications are discussed[10]. For many systems, especially when the overall reaction is complex, e.g. enzyme catalysis, it is useful to monitor the same reaction by means of different detection methods, since these are often specific, and can monitor different aspects of the overall scheme.

A useful accessory which has recently become available is a rapid wavelength-scanning accessory for the measurement of the complete spectra of intermediates in fast chemical reactions.[11] This device has been successfully employed in studies of dye binding to micelles by Yasunaga et al[12]. c) Accuracy of the Method: By means of conventional data-recording and analysis methods (oscilloscope for recording etc.), it is possible to obtain a reproducibility of ± 2% or so, which is usually associated with the overall accuracy in the rate constant determination. However, analogue to digital converters have recently become available (Dr. W. Knoche[13] - this volume) which enable the analogue trace to be digitised, the data being stored in a memory prior to direct read-out onto punch-tape for subsequent computer processing. Alternatively, a small computer may be used directly. This new method of data-capture is particularly advantageous when complex transients (non-exponential curves) are analysed. Transient recorders suitable for stopped-flow applications are available from only $1,000, and hence are very competitive with the storage oscilloscopes, which are generally used for semi-permanently recording the transient in analogue form. (For temperature-jump applications, due to the shorter sampling time required, the cost of the transient recorder is considerably higher).

Using a differential stopped-flow technique[14], it has been found that an accuracy equivalent to ± 0.3% can be achieved. The major factor limiting the accuracy appears to be the requirement of accurate volume delivery from the two drive-syringes. To this end, precision-bore gas-tight syringes should be used. It is also necessary to control the temperature of the reagents to ± 0.01K, and this is achieved by direct immersion of most of the mechanical unit, including the mixing-chamber and observation-region, in a precision thermostat.[15]

Rate determinations to this degree of accuracy (in the $1 - 10^{-3}$s region) allow the possibility of measurement of subtle differential kinetic effects such as: i) secondary isotope effects ii) salt effects iii) solvent isotope (H_2O/D_2O) effects and iv) heat capacities of activation.

B. THE TEMPERATURE-JUMP RELAXATION TECHNIQUE

Perturbation relaxation methods have been most elegantly developed and exploited over the past twenty years mainly by Eigen, De Maeyer and co-workers[16]. The temperature-jump technique is only one example of a range of single-step perturbation techniques, others being pressure-jump and electric-field jump. The advantage of the relaxation methods is that the physical process of mixing of reagents to initiate the reaction is eliminated. While the stopped-flow technique is only suitable for systems with $t_{\frac{1}{2}} \geq 1$ ms, the temperature-jump technique can be used for the study of relaxation times (τ) down to 1 μs, and, in some cases, even faster relaxations (τ = 10 - 100ns) can be studied.

The temperature-jump principle (of an 'instantaneous' (~ 1 μs) change in temperature ($\lesssim$ 10K) followed by a slower readjustment of the chemical equilibrium to the new thermodynamic conditions at the higher temperature recorded by monitoring an internal parameter e.g., optical absorbance) is thoroughly discussed in standard texts (see Appendix), as is the evaluation, from the measured relaxation time, of the rate constants characterizing the equilibrium. Coupling of reactions (e.g., by means of a pH indicator) is sometimes a convenient method of monitoring a reaction which has $\Delta H^o \sim 0$.

i) Heating Methods

Of particular interest is a comparison of the various ways of effecting the temperature-jump. The three main methods utilize one of: a) Joule (or resistive) heating, b) Micro-wave heating or c) Laser heating. A comparison of these three methods is shown in Table 1. Joule heating (by capacitor-discharge) has the advantage of being available in commerical instruments. It is necessary to discharge a high-voltage capacitor(C) through a low resistance medium (R) to obtain a fast heating time τ_H, since $\tau_H \approx \frac{1}{2}RC$. Thus the ionic strength is normally adjusted to 0.1M by addition of salt. An ultra-fast temperature-jump apparatus uses a co-axial cable as capacitor[17]. Through impedance matching, a rectangular heating pulse ($\simeq$ 50ns) can be produced.

A conducting medium is not required in the micro-wave version, and reactions in a pure aqueous medium can be studied. (Any polar solvent absorbing at the micro-wave frequency can in principle be used). An early success was the measurement of the rate of recombination of the ions of water[18] by means of a micro-wave temperature-jump instrument equipped with conductimetric detection. Micro-wave instruments equipped with optical detection have since been reported[19]. Since the temperature-jump is small in single-pulse operation, interference and colour filters are used when appropriate to optimise the light intensity in the sample cell and hence the signal-to-noise ratio of the photomultiplier detector.

In the laser temperature-jump method, a solid-state laser is pulsed in either the 'Q'-switched (20ns) or non-Q-switched mode (500μs) to produce monochromatic light. The main problem with this method is to efficiently convert the light pulse (time duration in parentheses) into heat energy in the solvent. With a ruby lasing-rod (694nm emission), it is usual to add dye absorbers to the solvent (e.g., Napthol Green B in H_2O; vanadyl phthalocyanine in aprotic solvents). Unfortunately, these dyes tend to 'bleach' (i.e. lose efficiency) at the high power levels ($\simeq 10^8$W) produced under Q-switched conditions[26]. It is obviously preferable to absorb the laser light directly employing excited vibrational levels of the solvent rather than indirectly through electronic levels of an added absorber. The difficulty is that most solvents do not absorb at 694nm, and water only absorbs to the extent of 5% cm^{-1} at 1060nm, the wavelength corresponding to use of a neodymium rod. However, the problem has been successfully overcome

TABLE 1. TEMPERATURE-JUMP METHODS

Heating Mode	Joule	Micro-wave	Laser
Heating Time	0.1 → 10 μs	1 μs	20 ns/500 μs
Max. Temp. Rise	10K	0.5 K	1K/10K
Solvent Requirements	Conducting $I > 10^{-2}M$ (in H_2O)	Polar	Must absorb at laser emission wavelength. (Dye absorbers can added)
Special Features	i) Commercially available ii) Range of Detection Methods - UV/VIS, Fluorescence, ORD/CD, Light Scattering	i) Low temp. rise ii) Repetitive pulsing possible - hence signal averaging can be used	i) Coupling of techniques to a) Stopped-Flow b) High Pressure ii) Small volume sample cell can be used with focussed beam[27]

recently by employing the stimulated Raman effect in liquid nitrogen to shift the laser wavelength of neodymium from 1060 nm to 1410 nm.[20] This corresponds effectively to total absorption in H_2O and so H_2O/D_2O mixtures must be employed. An alternative method is through modification of the laser cavity, suppressing the 1060 nm emission from neodymium and inducing laser output at 1340 nm (Dr. Giannini - this volume), which is an ideal wavelength for absorption into water and several other common solvents. Another approach is to use a tunable pulsed dye (liquid) laser over the ultra-violet/visible range, but to date power levels achieved are rather low.

A particular advantage of the laser technique is that it is easily coupled to other techniques; in conjunction with a high-pressure vessel operating up to 3 k Bar[21], kinetics in both aqueous and aprotic media [23] can be studied, which enables $\Delta V^{\neq}$ to be determined for fast kinetic processes. This parameter should be of value in the elucidation of reaction mechanisms in solution.

Mention should also be made of the use of pulsed solid-state lasers in a flash-photolysis perturbation mode. Reactions recently studied include the rate of recombination of the ions of water[24] (following vibrational excitation) and the dynamics of the square-planar to octahedral interconversion[25] of Ni(2,3,2-tet). In these systems, the equilibrium is shifted by photo dissociation, but the factors determining the amplitude of the resulting transient are not at present completely understood.

Important additional information can, however, be obtained from the amplitude (A) of relaxation transients, where A is the total change in the monitoring parameter. An amplitude analysis is desirable, and is particularly informative, when the overall reaction is complex. Values of K_i and ΔH_i^o can be resolved to high accuracy for some systems. (See D. Thusius, C.F. Bernasconi - this volume).

C. AREAS OF INTEREST

Just a few of the areas of chemical and biological research where the techniques of stopped-flow (SF) and temperature-jump (TJ) have found extensive application are summarised below.

(i) In the study of elementary reactions: a) Proton-transfer processes in both aqueous (TJ) and aprotic (SF,TJ) solvents. b) Electron-transfer processes. c) Charge-transfer complex and hydrogen-bond formation(TJ). d) Metal-ligand substitution reactions. (Provides indirect information on solvent-exchange kinetics around metal ions) (SF,TJ).

(ii) Cooperative Processes (in Biology). (a) Conformational transitions (Helix → Coil; Double-helix → Coil) (TJ). b) Allosteri in sub-unit enzymes (Dynamics of enzyme regulation) (SF,TJ).

c) Binding of small molecules e.g. dyes, mutogens, carcinogens to macromolecules (DNA). (TJ).

iii) Resolution of individual steps in enzyme-substrate interactions (SF,TJ).

(iv) Nucleic-acid/protein and hapten/antibody interactions (SF,TJ).

(v) Alkali-metal antibiotic interactions (Ion transport through membranes. (TJ).

(vi) Micellization; solubilization; molecular association (SF,TJ).

(vii) Study of factors affecting reactions at or near the diffusion-controlled limit (TJ).

(viii) Characterization of the 'transition-state' for reactions in solution. (From measurement of $\Delta H^{\ddagger}$, $\Delta V^{\ddagger}$, $\Delta S^{\ddagger}$, isotope effects, solvent effects etc.)

Thus, it can be seen that the versatility of these techniques has resulted in their wide application in the elucidation of fundamental problems in chemistry and biology. Should more information be required on instrumentation and applications, the books (reviews) listed in the Appendix will be helpful.

APPENDIX

E.F. Caldin, Fast Reactions in Solution (Blackwells) 1964
M. Eigen and L. De Maeyer in "Technique of Organic Chemistry" Vol. 8, Part II, Eds., Friess, and Weissberger, (Interscience, New York). 1963
K. Kustin (Ed), Fast Reactions (Methods in Enzymology) Vol. XVI (Academic, London and New York). 1969
A.F. Yapel (Jr) and R. Lumry; Methods of Biochemical Analysis, Vol. 20, p 169-350 (Wiley-Interscience) 1970
D.N. Hague, Fast Reactions (Wiley) 1971
S. Claesson (Ed) Fast Reactions and Primary Processes in Chemical Kinetics (Nobel Symposium 5, Interscience) 1967
H. Winkler, Endeavour, Vol. 33, p.73 1974
E.F. Caldin, The Temperature-Jump Method. Chemistry in Britain, Volume 11, p.4. 1975
Z.A. Schelly and E.M. Eyring, J. Chem. Educ., 48, 695 1971
R.M. Reich, Analyt. Chem., 43, 85-97A. 1971.

REFERENCES

1. H. Hartridge and F.J.W. Roughton, Proc. Roy. Soc. A, **104**, 376 (1923).

2. M.M. Wong and Z.A. Schelly, Rev. Sci Instr., 44, 1226 (1973).
3. D. Riesner and H. Buenemann, Proc. Nat. Acad. Sci., 70, 890 (1973).
4. J. Grimaldi, J. Balds, C. McMurrary and B.D. Sykes, J. Amer. Chem. Soc., 94, 764, (1972).
5. J.C. Kertesz and W. Wolf, J. Phys. (E)., Sci. Instr., 6, 1009 (1973).
6. K. Hiromi, S. Duo, S. Itoh and T. Nagamura, J. Biochem., 6, 64 (1968).
7. M. Anson and P.M. Bayley, J. Phys. (E)., Sci Instr. 7, 481 (1974).
8. J.P. Maher, University of Bristol, Personal Communication.
9. P.A. Tregloan and G.S. Laurence, J. Sci. Instr., 42, 869 (1965).
10. M.H. Davies, J.R. Keeffe and B.H. Robinson, Ann. Rep. (Chem. Soc) A, (1974).
11. T. Yasunaga, Personal Communication. R.M. Wightman, R.L. Scott, C.N. Reilley, R.W. Murray and J.W. Burnett, Analyt. Chem. 46, 1492 (1974).
12. N. Tatsumoto, K. Takeda, S. Isshiki and T. Yasunaga, Bull. Chem. Soc., Japan, 47 (2), 289 (1974).
13. Dr. W. Knoche - this volume.
14. E.F. Caldin, K.J.A. Hargreaves and B.H. Robinson, Paper in preparation.
15. E.F. Caldin, A. Queen and J.E. Crooks, J. Phya. (E)., 6, 930 (1972).
16. M. Eigen in "Nobel Symposium - Volume 5" - Interscience (1967).
17. G.W. Hoffmann, Rev. Sci. Instr., 42, 1643 (1971).
18. G. Erth and H. Gerischer, Z. Elekrochem., 65, 629 (1967).
19. E.F. Caldin and J.E. Crooks, J. Sci. Instr., 44, 449 (1967).
20. J.W. Beitz, G.W. Flynn, D.H. Turner and N. Sutin, J. Amer. Chem. Soc., 92, 4130 (1970).
21. E.F. Caldin, M.W. Grant, B.B. Hasinoff and P.A. Tregloan, J. Phys. (E)., 6, 349 (1973).
22. E.F. Caldin and M.W. Grant, J.C.S. Faraday 1, 69, 1648 (1973).
23. J.E. Crooks and B.H. Robinson, Trans. Faraday Soc., 67, 1707 (1971).
24. D.M. Goodall and R.C. Greenhow, Chem. Phys. Lett., 9, 583, (1971).
25. K.J. Ivin, R. Jamison and J.J. McGarvey, J. Amer. Chem. Soc., 94, 1763 (1972). N. Sutin and C. Creutz, J. Amer. Chem. Soc., 95, 7177 (1973).
26. E.F. Caldin, J.E. Crooks and B.H. Robinson, J. Phys. (E)., Sci. Instr., 4, 165 (1971).
27. R. Riglar, A. Jost and L. DeMaeyer, Exp. Cell. Research, 62, 107 (1970).

THE SOLVENT-JUMP RELAXATION METHOD USING STOPPED AND CONTINUOUS FLOW

Z. A. Schelly

Department of Chemistry, University of Georgia,
Athens, Georgia 30602, U.S.A.

SUMMARY. The step perturbation through mixing is described for measuring relaxation times longer than 1 msec. Both transient and steady state detections are possible.

Chemical equilibria can be disturbed by changing external or internal parameters of the system in question. The external parameters usually altered are the temperature T, the pressure P, or the electric field E, and the internal ones are the concentration of solutes and/or the composition of the solvent. A perturbation of the total concentration can be achieved by sudden dilution (concentration-jump) using rapid mixing devices[1,2]. Similarly, the composition of the solvent can be abruptly changed by fast mixing of two equilibrium solutions in different solvents, or one equilibrium solution with another solvent. Since the total solute concentration is not necessarily altered in the experiment, the name solvent-jump is more generally descriptive of the method than the name concentration-jump. This type of experiment was first suggested by Ljunggren and Lamm[3] and has been successfully applied to slow[4] as well as to fast reactions[5,6].

If we consider the chemical equilibrium

$$\nu_1 A_1 + \nu_2 A_2 + \ldots \underset{k_{-1}}{\overset{k_1}{\rightleftharpoons}} \nu_i A_i + \nu_{i+1} A_{i+1} + \ldots \qquad (1)$$

the equilibrium constant K is given by the product of the activities a_i, $K = \prod_i a_i^{\nu_i}$, with stoichiometric mole numbers ν_i negative on the left and positive on the right side of the reaction Since a_i =

Wyn-Jones (ed.), Chemical and Biological Applications of Relaxation Spectrometry, 49–51.

$C_i\gamma_i$, we can write

$$\ln K = \Sigma\nu_i \ln a_i = \Sigma \nu_i \ln C_i + \Sigma \nu_i \ln \gamma_i \tag{2}$$

Dilution of the equilibrium solution with the same solvent has no effect on K, but it changes the concentrations C_i and the activity coefficients γ_i. Mixing with another solvent results in a new medium, and all three (K, C_i and γ_i) may be effected. If the new values of C_i and γ_i do not satisfy the equilibrium condition

$$-A \equiv \Sigma\nu_i\mu_i = \Sigma \nu_i\mu_i^o + RT\,(\Sigma \nu_i \ln C_i + \Sigma \nu_i \ln\gamma_i) = 0 \tag{3}$$

the system will relax to zero affinity, A = 0. μ_i^o are the chemical potentials of the species in a standard state.

In the case of a simple equilibrium $A + B \rightleftarrows C$, the proper dilution ratio n, that will result in a sufficiently small amplitude δC_i, can be estimated[7] from

$$n = (1 + \phi)^2/(1 + \phi/K_o C_A) \tag{4}$$

if activity coefficients are neglected. For a monomer-dimer equilibrium (i.e. A = B), eq. (4) modifies to

$$n = (1 + \phi)^2/(1 + 2\phi/K_o C_A) \tag{5}$$

In both equations ϕ is the tolerated perturbation factor, defined by $\delta C_i = \phi_o C_i/n$ and $\bar{C}_i = {}_oC_i/n + \delta C_i$, where ${}_oC_i$ are the equilibrium concentrations before and $\bar{C}_i$ after mixing. If the equilibrium solution is diluted with another solvent, instead of eq. (4) we have

$$n = K_1(1 + \phi)^2/(K + \phi/{}_oC_A) \tag{6}$$

where K_1 is the new equilibrium constant in the new mixture-medium.

After a sudden n-fold dilution of a one-step equilibrium the concentrations C_i as a function of time are given by

$$C_i = \bar{C}_i\{1 - \exp(-t/\tau)\} + ({}_oC_i/n)\exp(-t/\tau) \tag{7}$$

The relaxation time τ can be obtained from the transient observation of $C_i(t)$ usually in a stopped flow apparatus.

In the non-transient method[8] a continuous flow is set up, and the relaxation time is determined from a single measurement of an integrated partial relaxation amplitude A. The integration is done coaxially with the flow direction along the distance coordinate, that is counted from the mixing jet. The steady state value of A(τ) is given by

$$\frac{A-\bar{A}}{\bar{A}-A_o/n} \quad t_s = \tau\{\exp(-\ell/v\tau)-\exp(-d/v\tau)\} \tag{8}$$

where $\bar{A}$ and A_o are the final (at time infinity) and initial (before dilution) equilibrium absorbances, respectively, with path length s of the optical observation. t_s is the residence time of the solution between the distance coordinates ℓ and d.

The steady state observation of the non-transient method makes the solvent jump method easily adaptable to OR, CD and ORD detection.

Both methods have been extensively used in our laboratory, most recently in the study of the dimerization of rhodamine type laser dyes in aqueous and ethanolic solutions[6].

REFERENCES

1. B. Chance, Q.H. Gibson, R.H. Eisenhardt and K.K. Lonberg-Holm, eds., "*Rapid Mixing and Sampling Techniques in Biochemistry*", Academic Press, New York, 1964.
2. The rate of mixing is limited by cavitation. See: M.M. Wong, Z.A. Schelly, *Rev. Sci. Instr.*, **44**, 1226 (1973); also ref. 1).
3. S. Ljunggren, O. Lamm, *Acta Chem. Scand.*, **12**, 1834 (1958).
4. J.H. Swinehart, G.W. Castellan, *Inorg. Chem.*, **3**, 278 (1964).
5. Z.A. Schelly, R.D. Faring, E.M. Eyring, *J. Phys. Chem.*, **74**, 617 (1970).
6. M.M. Wong, Z.A. Schelly, *J. Phys. Chem.*, **78**, 1891 (1974).
7. D.Y. Chao, Z.A. Schelly, to be published.
8. Z.A. Schelly, M.M. Wong, *Intern. J. Chem. Kin*, **6**, 687 (1974).

A LASER TEMPERATURE JUMP AND PHOTOLYSIS APPARATUS WITH REPETITIVE TUNABLE EXCITATION

I. Giannini

Laboratorio Ricerche di Base, Snamprogetti,
Monterotondo, Roma

ABSTRACT. An apparatus will be described which can be used to study small liquid and solid samples by perturbation techniques both as temperature jump and pulsed photolysis. We use a repetitive pulsed laser source tunable in the visible and near I.R. Precise monitor beam optics and fast electronics coupled to averaging techniques allow the detection of less than 0.1% OD changes with 10 ns time resolution.

INTRODUCTION

The relevance of relaxation techniques in kinetic studies is generally well known and was emphasized in many reviews, papers[1-3]. On the other hand by the recent advances in laser technology, some very powerful and ductile pulsed light sources are now available.

We describe in the following a new instrument for relaxation measurements where the main perturbation source is a tunable and pulsed laser system, which can be utilized alternatively as a temperature jump or a flash photolysis apparatus. As main features, this apparatus joins some different aspects of the existing laser-T-jump [4-7], flash photolysis[8,9] instrument, i.e. the near infrared (I.R.) heating of temperature jump operation[4] to the repetitive and averaging techniques already used by Witt and coworkers for flash photolysis operation[8].

EXPERIMENTAL

A schematic diagram of the apparatus is shown in fig. 1.

Wyn-Jones (ed.), Chemical and Biological Applications of Relaxation Spectrometry, 53–59.

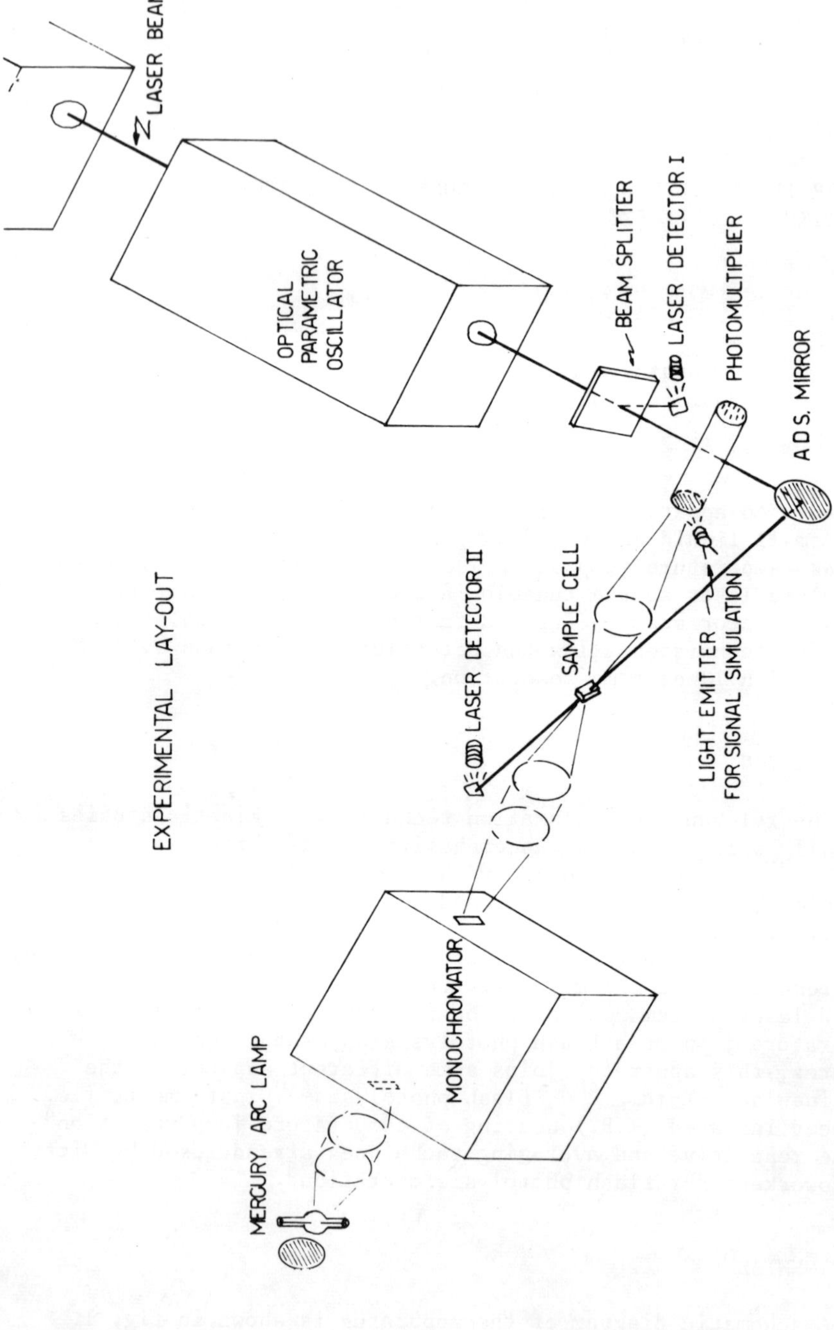

Fig. 1. Schematic diagram of the laser-T-jump apparatus.

The laser system was the Nd:YAG Laser mod. 1000 of Chromatix, which is coupled with a tunable Optical Parametric Oscillator (OPO) (Chromatix mod. 1020). A complete description of the system can be found elsewhere[10]. The wavelength of the output radiation has some fixed powerful lines (the shortest at 973 nm, the longest at 1.36 μ) and a tunable region ranging from 600 nm to ~ 3 μ. The laser pulses have a duration of ~ 100 ns and the repetition rate generally used is 75 Hz. The energy of each pulse varies from 0.1 to 3 mJ. The highest power is obtained at the laser output at 1.32 ÷ 1.36 μ by a high transmission (50%) output mirror. At this wavelength water absorption is good (~ 2 cm^{-1}) and we can obtain temperature rise of 1 - 2°C for each pulse by focusing the laser beam on our sample.

On the area of the cell where the laser hits we have the monochromatic image of a high pressure Hg arc lamp with a very small spot dimension (HBO 100W/2 Osram). The light from the lamp goes through a high aperture quartz optics and a small monochromator (SPEX Minimate 1 : 4 aperture). We can work with a spot image of about 0.1 mm in max. size. By this way the volume of the observed reaction is about 10^{-5} μl, i.e. extremely small. This condition limits the upper observable time by diffusion effects to about 1 ms. The lower observable time is limited to some fraction of the laser pulse width (i.e. some 10^{-8} sec); in fact we can do an accurate waveshape measurement of laser pulse, and reconstruct the signal by computer deconvolution.

The laser pulse is monitored by different detectors: an ultra fast silicon pin-photodiode is used in the visible region (Hp 4220); while in the near I.R., where no fast detectors are available, either Germanium (OAP 12 Philips) or In As (Barnes A 100) photovoltaic detectors are used, coupled to an integrating amplifier device. The resulting response is the integral of the laser pulse, with distortion less than few percent up to 10 ns resolution. The monitor beam crosses the laser beam at about 15°. The cell geometry can be varied: we have alternatively used quartz cell disposed at Brewster angle as in Fig. 1; or 1 mm. diameter capillary with flowing solutions. Accurate thermostating of the cell is obtained in some different heat exchanger.

Temperature gradient between the laser spot and the thermostatic solution is measured by absorption of a coloured indicator (10^{-4} M Phenol Red in 0.1 M Tris-HCl solution, pH 8, was used): the colour change is measured by the monitor beam in the absence of laser and varying the temperature of the bath; the laser effect is then obtained. All the measurements can be made with temperature gradient less than 2°C.

The absorbancy change following the laser pulse is measured by a Philips XP 1113 photomultiplier equipped with a wide band (~ 100 MHz) preamplifier; the photomultiplier output can be switched from anode to the 4th dynode.

A fluorescence emission monitor is also compatible with the

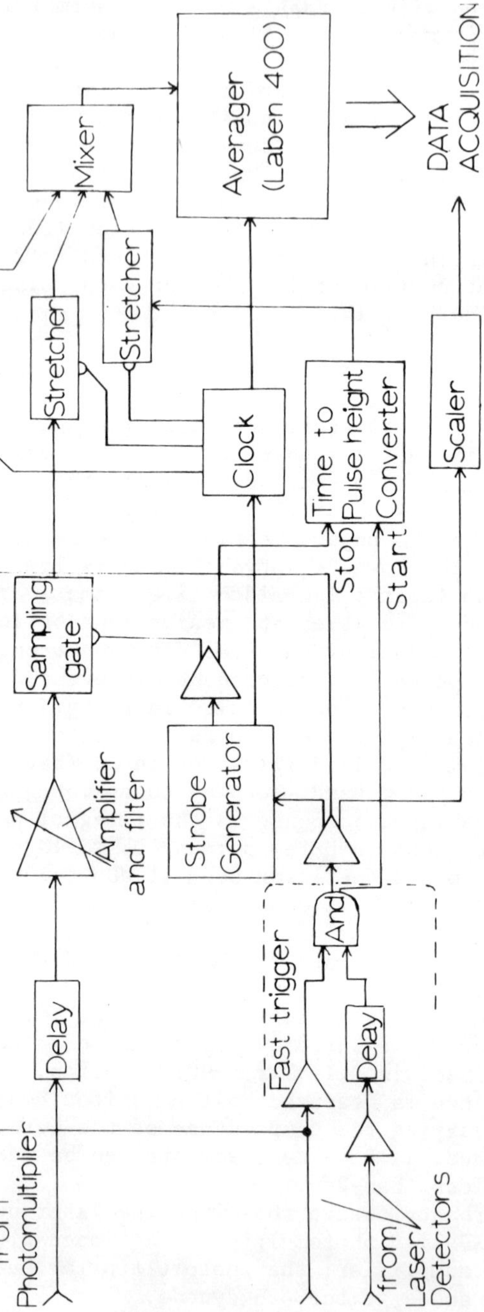

Fig. 2. Electronics block diagram.

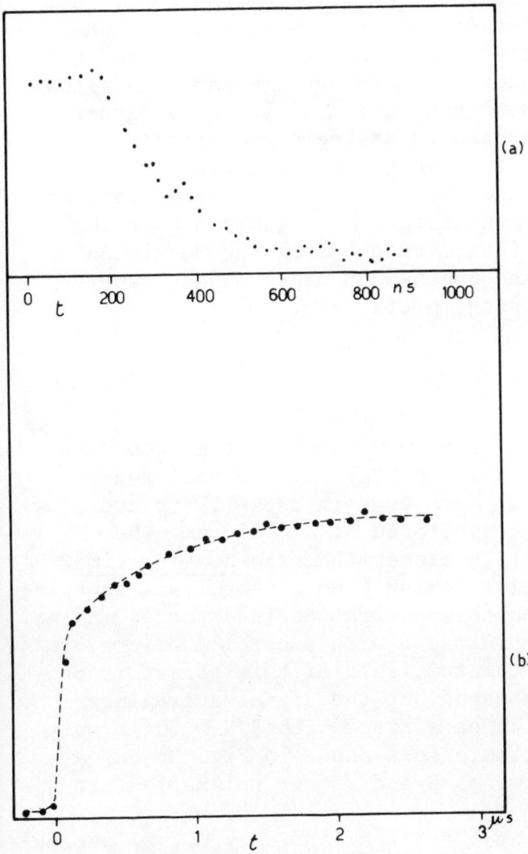

Fig. 3a. T-Jump operation; the laser is working at 1.36 μ with special output mirror. A flowing solution in a capillary is heated 1 2°C per pulse. The sample was a fast reacting mixture of Phenol Red 2.10^{-4}M and Tris HCl 0.1M pH 8 - Absorbance change follows with small delay (∿ 20ns) the heating effect.

b. Flash photolysis operation. A photoisomerization effect is observed in a Carbonic anhydrase azosulfonamide complex. The fast relaxation has τ = 800 ns time. The equilibrium was reached at longer times with two more relaxation effects[11].

actual optics design, and will be used in the near future. The electronic signal processing follow the block diagram of Fig. 2 and is only partially realized with commercial circuitery. Fast trigger is obtained from laser detectors; the signal from photomultiplier is amplified, filtered and then analyzed in the fast sampling gate, the strobe generator commands the sampling signal sequence. The signal, the laser monitor, the time delay generated by the strobe are first memorized in analogic memories (capacitors in "stretcher" circuits) and then sequentially digitalized and memorized in a multichannel analyzer (Laben 400), with averager. We usually can sum over some thounsands of signals due to the high repetition rate of the laser. Output from the multichannel was displayed on the scope and punched on paper tape. Computer processing gives normalized data, plots, etc.

PERFORMANCES

A computer plotted trace of ~ 1000 samples of a temperature jump operated experiment is shown in Fig. 3a. A phenol red 2.10^{-4} M pH 8. in Tris HCl 0.1 M is used in a capillary and flow geometry. The change in pH is monitored at λ = 430 nm. The resulting trace is only a little slower than the heating time.

The Fig. 3b shows the plot coming from a photolysis experiment now under way. A fast photoisomerization is detected in the complex of Human Carbonic Anhydrase B with a chromophoric azosulfonamide (Neoprontosil). The Quantum Yield of this effect is of the order of few part per thousand but the signal averaging system allows the registration of a trace with S/N ~ 20 in only 5 minutes. The first relaxation effect shown in Fig. 3b has a relaxation time of 800 ns (two more and slower relaxation are observed in this system)[11].

The laser was operated at λ = 532 nm and only 0.2 mJ per pulse in order to avoid heating; the monitor was at λ = 405 nm corresponding to the second singlet transition of the dye.

ACKNOWLEDGEMENTS

The realisation of this instrument should have been impossible without the collaboration of R. Colilli. We are indebted to Dr. L. Micheli and P. Grasselli for the essential contributions they have done during the work for their thesis at the University of Rome.

REFERENCES

1. M. Eigen and L. de Maeyer (1964) "Technique of Organic Chem. (S.L. Friess, E.S. Lewis and A. Weissberger eds.) Vol. 8, No.2 p. 895 Wiley Interscience New York.

2. G.G. Hammes, Adv. in Protein Chem., 23, 1 (1968).
3. G.G. Hammes and P.R. Schimmel "The Enzymes" vol. II, p. 39, P.D. Boyer ed. Academic Press, New York (1970).
4. J.U. Beitz, G.W. Flynn, P.H. Turner and N. Sutin, J Am. Chem. Soc., 92, 4130 (1970).
D.H. Turner, G.W. Flynn, N. Sutin and J.U. Beitz, J Am. Chem. Soc., 94, 1554 (1972).
5. E.M. Eyring and B.C. Bennion, Ann. Rev. Phys. Chem., 18, 129, (1967).
6. H. Hoffman; E. Yeager, J. Stuher, Rev. Sci. Instr., 39, 649 (1968).
7. E.F. Caldin, J.E. Crooks and P.H. Robinson, J. Phys. E: Sci. Instr., 4, 165 (1971).
8. H. Ruppel and H.T. Witt, "Methods in Enzimology" vol. 16, "Fast reactions" p. 316, K. Kustin ed. Academic Press, New York (1969).
9. B. Alpert, R. Banerjee and L. Lindquist, Proc. Nat. Acad. Sci. USA, 71, 558 (1974).
10. S.E. Harris, Proc. IEEE, 57, 2096 (1969), R.W. Wallace and S.E. Harris, Appl. Phys. Letters, 15, 111 (1969).
11. I. Giannini and G. Sodini - to be published - see also this volume.

THE PROBLEM OF INHOMOGENEOUS HEATING IN LASER-T-JUMP EXPERIMENTS

Z.A. Schelly

Department of Chemistry, University of Georgia,
Athens, Georgia 30602, U.S.A.

ABSTRACT. Due to exponential extinction of the heating laser pulse in the solution, the chemical relaxation takes place in an anisotropic medium in laser-T-jump experiments. The problem is analysed in terms of relaxation amplitudes as a function of time and position in the cell.

The limitations of the standard Joule-heating temperature-jump method[1], namely a heating time of about 1μsec and the requirement of high ionic strength, can be circumvented by heating through electronic[2,3] or vibrational[4,5,6] excitation of the sample. In the latter case the energy of a single ca. 20 n sec Q-switched pulse from a ruby or neodymium laser is absorbed by the solvent or by an added dye. If the photons of the pulse are not absorbed by the medium sufficiently for heating, the wave length can be shifted, using stimulated raman scattering, to ranges where absorption is stronger[7,8]. In recent studies the raman shifted lines of a Nd:glass laser (λ = 1.06μ) in liquid nitrogen[9] (λ = 1.41μ) and in H_2-gas[10] (λ = 1.89μ) were successfully used for T-jump.

In laser temperature-jump experiments the following assumptions are made: (1) after absorbing the energy of the laser pulse the Boltzmann distribution in all modes is achieved much faster than the chemical relaxation of the system occurs; (2) during the chemical relaxation all components are in the ground state as they were before the perturbation; (3) the perturbation of the system is small enough for linearization of the rate equations; and (4) the chemical relaxation takes place in an isotropic medium.

The validity of the first three assumptions depends on the specific system, as well as on the experimental conditions that

Wyn-Jones (ed.), Chemical and Biological Applications of Relaxation Spectrometry, 61–67.

have to be optimized. In the intense radiation field of the focused or collimated laser pulse non-linear effects, as well as plasma formation, cavitation, dielectric breakdown, etc. may occur[11]. At a power density of 0.4 to 0.9 GW cm^{-2}, for example, transient luminosity (the emission of a continuum with time constants in the 10^2 to 10^3 n sec range) was observed in pure water, and also in optical glass and Plex glass after the irradiation with a 20 n sec neodymium laser pulse[12]. Although the laser-T-jump method is not restricted to systems in the ground state, the knowledge of the state of the components, as well as the dynamics of the processes involved is necessary for the unambiguous specification of the rate constants.

The invalidity of assumption (4) is more serious, since it is inherent in the exponential extinction of the laser pulse, causing inhomogeneous heating of the sample. The incident intensity I_o (in J cm^{-2}) of the laser pulse decreases according to the Bouguer-Lambert-Beer law to $I(x)$

$$I(x) = I_o \exp(-\varepsilon_s . c_s . x) \qquad (1)$$

after passing through a distance x in the solution, if only the solvent s is absorbing. ε_s and c_s are the molar natural extinction coefficient and the molar concentration of the solvent, respectively. Since the temperature change δT along x is given by $\delta T = dI(x)/(C_v dx)$, where C_v is the heat capacity of the solution in J cm^{-3} K^{-1}, a temperature gradient is established in the cell simultaneously with the pulse, expressed as (see also Fig. 1)

$$\delta T(x) = \delta T_o \exp(-\varepsilon_s . c_s . x) \qquad (2)$$

δT_o in eq (2) is the temperature increase at the front surface of the solution.

At this point one has to make an additional assumption, which at the same time represents a limitation for the method: (5) The relaxation times measured must be much faster than the temperature equilibration through diffusion. However, even if condition (5) is met, the rate constants will be different at different points of the cell. If the ΔH of the reaction is large enough for a measurable perturbation, at least one of the activation enthalpies must be

$$|\Delta H_i^{\neq}| \geq |\Delta H| ,$$

since

$$\Delta H = \Delta H_f^{\neq} - \Delta H_r^{\neq}$$

indicating the temperature sensitivity of the rate constant k_i.

There is a possibility of decreasing the temperature difference between the two ends of the solution by back-reflecting (the unabsorbed part of) the pulse into the cell. To find the optional path length x where the function $\delta T(x)$ should be "folded" (see Fig. 1), one has to solve the equation

$$\delta T_o + \delta T(2x) = 2\delta T(x) \tag{3}$$

After substituting (2) into (3) one obtains

$$1 + \exp(-\varepsilon.c.2x) = 2 \exp(-\varepsilon.c.x) \tag{4}$$

and with $y = \exp(-\varepsilon.c.x)$

$$1 + y^2 = 2y \tag{5}$$

The root of (5), y = 1, indicates that eq. (3) has only a trivial solution, x = 0. Thus, one should use the smallest experimentally possible path length x. This result is valid also for experimental arrangements where the laser pulse is split into two beams of equal intensity and then are led from opposite direction into the cell. Of course, for the same reason, the smallest possible path length should be used also in the single passage arrangement. The advantage of the back-reflecting is usually cancelled by difficulties that arise from frequent damage of the reflecting surface. This can be avoided by using a prism as a reflector. In this case, however, no coaxial detection can be applied, the importance of which is pointed out in the following sections.

Figure 2 shows a schematic representation of the observed property P(x,t), which is linearly proportional to the concentration c_i of each chemical component. If optical detection is used P is the absorbance A. For a one-step reaction at time t the instananeous deviation of P from its final equilibrium value is given by

$$\delta P(t) = \delta P_o \exp(-t/\tau), \tag{6}$$

where δP_o is the relaxation amplitude[13,14], and $\tau(x)$ the relaxation time of the process caused by a step-perturbation of the system. The relaxation amplitude can be expressed as

$$\delta P_o = \sum_i \nu_i \frac{\partial P_i}{\partial \bar{c}_i} \left(\frac{\delta \bar{c}_i}{\nu_i}\right) \tag{7}$$

and the virtual change of the equilibrium concentration $\delta \bar{c}_i$ is given by

$$\delta \bar{c}_i = \nu_i \Gamma \delta \ln K \tag{8}$$

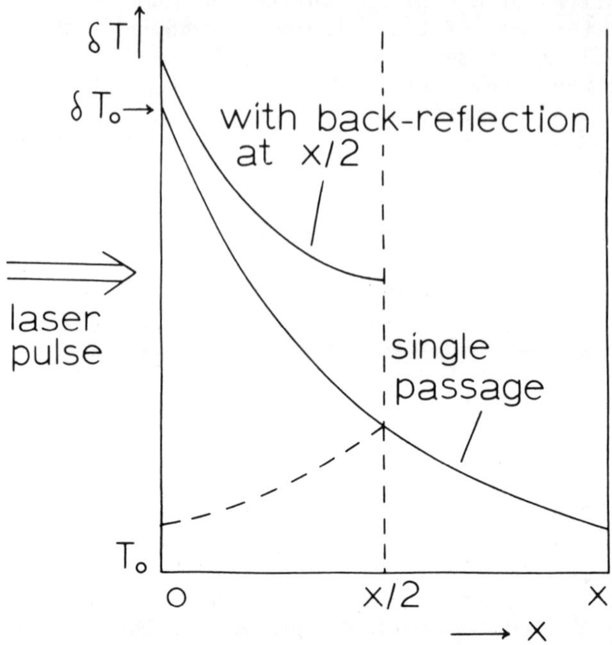

Fig 1. The determination of the optional path length x from equation (1)

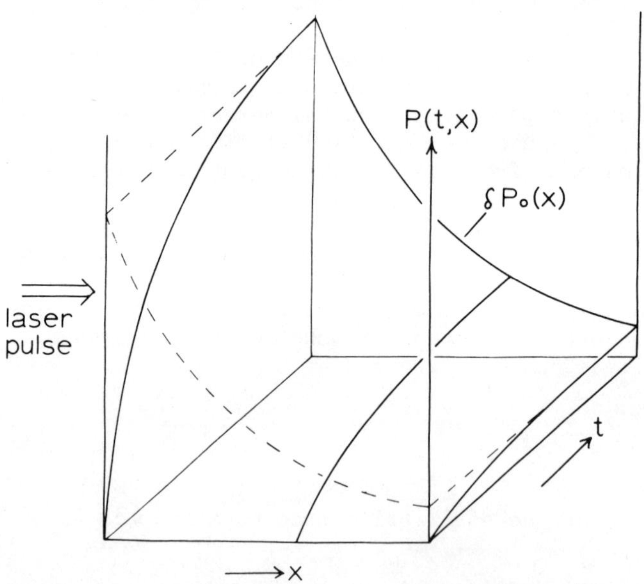

Fig 2. Schematic representation of the observed property P(x, t).

if one neglects volume changes and activity coefficients. In eq. (8) $\Gamma^{-1} = \sum \nu_i^2/\bar{c}_i$, $\delta \ln K = (\Delta H/RT^2)\delta T$ and $K = \prod \bar{c}_i^{\nu_i}$. The stoichiometric coefficients ν_i are defined as negative for reactants and positive for products, reading the reaction equation from left to right. Thus, the explicit form of the function $\delta P_o = f.\delta T$ is

$$\delta P_o = \sum_i \nu_i \frac{\partial P_i}{\partial \bar{c}_i} (\sum_i \nu_i^2/\bar{c}_i)^{-1} (\Delta H/RT^2)\delta T \qquad (9)$$

where $\delta T(x) = T_\infty(x) - T_o$ and $T^2 = T_\infty T_o$, with the subscripts referring to the time before and after the instantaneous perturbation took place. Substituting (2) and (9) into (6) one obtains P(t,x), represented by the surface in Fig. 2

$$P(t,x) = \delta T_o . f. \exp(-\varepsilon_s . c_s . x - t/\tau) \qquad (10)$$

The transient observation of the relaxation yields a signal S(t)

$$S(t) = b\int_0^x \delta P(t,x)dx \qquad (11)$$

where the proportionality constants b depends on instrumental conditions. If $\delta T \ll T$, as it is usually the case, the final overall temperature ϑ (reached after temperature equilibration) can be used for T in eq. (9). ϑ can be calculated from $\vartheta = T_o + \delta T_\infty$, where T_o is the initial overall temperature in the cell, and δT_∞ is obtained from $\int \delta T(x)dx = \delta T_\infty x$, under adiabatic conditions. With the approximation $T^2 = \vartheta^2$ the integration in (11) becomes simpler, nevertheless, the position dependence of $\bar{c}_i(x)$ and of $\tau(x)$ has to be considered explicity. $\bar{c}_i(x)$ can be determined in preliminary equilibrium measurements, and for the relaxation times $\tau(x)$, values have to be computed which are internally consistent with S(t) and with the proposed mechanism.

Errors may be introduced also in the experimental integration of (11) by the detector. Using crossed beam monitoring, where heating pulse and analyzing beam are perpendicular, the "quality" of the integration depends on the uniformity in luminous sensitivity of the detector surface. Additionally, the crossed beam arrangement is very sensitive to interference of signal changes caused by the temperature dependent transient change of the refractive index of the medium. Therefore, coaxial detection is preferred, where the analyzing beam is mixed with the perturbation pulse and thus passes through the solution in the direction of the temperature gradient (Fig. 3).

Ways alternative to the short path length and coaxial detection to minimize the effects of the temperature gradient in the cell are offered by using (1) repetitive, small T-jumps and sampling the output signals[15]; and (2) evanescent photons probing only the first

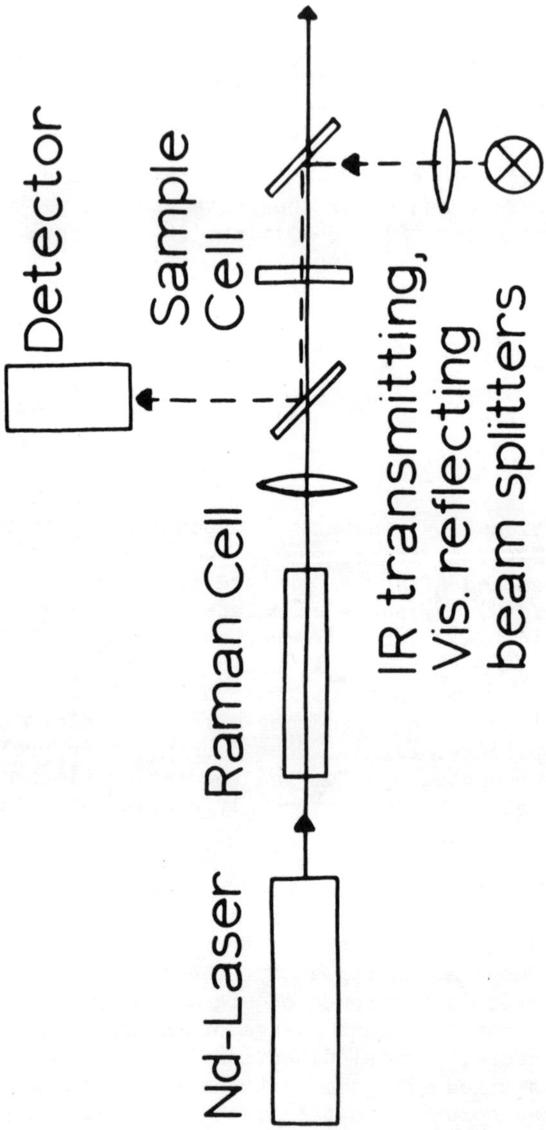

Fig 3. Block diagram of apparatus

liquid surface in the cell. Dr. Giannini[16] and Professor Eyring[17] are presenting reports on these subjects.

Acknowledgement. The author is grateful to the Alexander von Humboldt-Stiftung for a lecturer fellowship and to Professors L. DeMaeyer and M. Eigen for their hospitality at the Max-Planck-Institut fur biophysikalische Chemie in Gottingen.

REFERENCES

1. G.H. Czerlinski and M. Eigen, Z. Elektrochem., 63, 652 (1959).
2. L.S. Nelson and J.L. Lundberg, Nature, 179, 367 (1957).
3. H. Strehlow and S. Kalarickal, Ber. Bunsenges., 70, 139 (1966).
4. W.H. Inskeep, D.L. Jones, W.T. Silfvast and E.M. Eyring, Proc. Nat. Acad. Sci. (USA), 59, 1027 (1958).
5. H. Hoffmann, E. Yeager and J. Stuehr, Rev. Sci. Instrum., 39, 649 (1968).
6. E.F. Caldin, J.E. Crooks and B.H. Robinson, J. Phys., E:Sci. Instr., 4, 165 (1971).
7. V.V. Bochavow, A.Z. Grasyuk, I.G. Zubarev and V.F. Mulikov, Zh. Eksp. Teor. Fiz., 56, 430 (1969).
8. J. Ducuing, G. Jeffrin and J.P. Coffinet, Opt. Commun., 2, 6 (1970).
9. D.H. Turner, G.W. Flynn, N. Sutin and J.V. Beitz, J. Amer. Chem. Soc., 94, 1554 (1972).
10. L. DeMaeyer and S. Ameen, to be published.
11. J.F. Ready, "Effects of High-Power Laser Radiation", Academic Press, New York, N.Y., 1971.
12. Z.A. Schelly, J. Lang and E.M. Eyring, Monatsh. Chem., 104, 1672 (1973).
13. M. Eigen and L. DeMaeyer, "Techniques of Organic Chemistry", Vol. 8, 2nd. ed., Part 2, S.L. Friess, E.S. Lewis and A. Weissberger, Eds., Wiley, New York, N.Y., 1963, p.931.
14. D. Thusius, J. Am. Chem. Soc., 94, 356 (1972).
15. P. Braun, E.P. Kilian and H. Ruppel, Ber. Bunsenges., 72, 1085 (1968).
16. I. Giannini, this volume.
17. E.M. Eyring, this volume.

KINETIC STUDIES OF METAL ION COMPLEXATION AND DYE DIMERIZATION BY LASER ULTRASOUND AND LASER T-JUMP

M.M. Farrow, N. Purdie, A.L. Cummings, W. Hermann, Jr., and Edward M. Eyring,
University of Utah, U.S.A.

ABSTRACT. There are two comparatively new relaxation techniques for measuring nanosecond time scale reaction rates in liquids: the laser Debye-Sears ultrasonic absorption method and the laser Raman temperature jump technique. Advantages and limitations of these techniques are reviewed, and their use in kinetic studies of metal ion complexation and dye dimerization is surveyed.

LASER ULTRASOUND

In 1922, Brillouin[1] predicted and ten years later Debye and Sears[2] and Lucas and Biquard[3] demonstrated that an ultrasonic wave will set up a series of compressions and rarefractions in a medium which will act as a diffraction grating. As the ultrasonic pressure wave moves through the solution, the amount of the change in the index of refraction in the medium caused by the acoustic wave, and thus the efficiency of the 3-dimensional diffraction grating, is proportional to the acoustic intensity.

This phenomenon may be divided into three ultrasonic frequency regions where the interaction arises from differing effects:

The first region, occurring at comparatively low ultrasonic frequencies of the order of hundreds of KHz, is ray bending and may be explained as a focusing of the incident beam into the region of highest refractive index where acoustic intensity is at a maximum. Only a very moderate modulation in intensity and small scattering angle are obtained, which therefore limits its application in chemical kinetics.

The second region occurs in the low MHz ultrasonic region. The effect in this region is known as the Debye-Sears effect

Wyn-Jones (ed.), Chemical and Biological Applications of Relaxation Spectrometry, 69–83.

and is essentially an ultrasonically generated diffraction grating with the pattern of compressions and rarefactions replacing the lines of a grating. This region is characterized by several orders of diffraction (we have observed as many as thirty) depending on the acoustic intensity.

Since the ultrasonic diffraction grating is three dimensional instead of two dimensional, destructive interference arises when the "depth" of the grating or the interaction length becomes too large. For the 514 nm argon-ion laser line and at an ultrasonic frequency of 5MHz this interaction length can be as large as 4 cm, but at 25 MHz it should not exceed about 2 mm. This interference may be overcome by allowing the incident beam to strike the acoustic wave front at the Bragg angle. This identifies the third region of interest which is known as the Bragg reflection region since the effect is analogous to the reflection of x-rays off a crystal plane. Thus, many orders of diffraction are observed in the Debye-Sears region whereas only the first order diffracted beam remains in the Bragg region.

This third region, the Bragg reflection region, is being used to modulate lasers for communications and in information retrieval systems,[4] There is extensive literature available[5] on the deflection of light by sound, but the sound propagation medium generally used is solid due to the losses in liquids at high frequencies.

To the chemist, it is these losses that are of interest as a probe for deducing fast reaction kinetics from the ultrasonic relaxation spectra of solutions. Although optical methods for obtaining ultrasonic absorption data have been used by some,[6] application has been limited[7] to the fairly restricted range of about 10 to 100 MHz with some differential double beam work being done at lower frequencies.[8]

The method most in favour today for ultrasonic absorption spectroscopy in the 5 MHz to 500 MHz range is the pulse method[9] in which an ultrasonic pulse is transmitted from one transducer crystal and received by another matched transducer after passage through the solution. The received signal is then amplified and displayed on an oscilloscope and compared to a signal of the same frequency whose amplitude is known. The separation of the two crystals is varied a known amount and the signal loss (or gain) is a measure of the absorption in the soluti

There are numerous difficulties involved in the pulse method including:

1. Measuring low level RF signals accurately by matching visually two peaks on an oscilloscope.
2. Maintaining exact parallelism of the transducers (to within half an acoustic wavelength) during translation of the transducer.

3. The need for acoustical delay lines to prevent overlapping of the electro-magnetically propagated pulse (in air) from the signal source with the transmitted pulse (through solution).
4. The mechanical system (sample cell volume, micrometer drive etc.) must change for various acoustic frequency ranges. For example, in a not unusual system it is possible that the solution would absorb 100 db/cm at 100 MHz but only 1 db/cm at 10 MHz. Put another way, an acoustic path of 1 km would be necessary to produce a 10 db change at 100 KHz but only 10^{-2} mm at 1 GHz.

In mid-1973, we began using a laser optical ultrasonic system to which we originally assigned the acronym "LDS" (Laser Debye-Sears) because we were then able to work only in the low MHz ultrasonic frequency range. More recently we have been able to extend the useful range of the method to over 345 MHz with a single quartz crystal and associated mechanical system. Not only are the data obtained with this system more precise than those typically available from the pulse technique, but the data can also be accumulated at least ten times as rapidly. Key contributing factors that make this extended frequency range possible are the stabilized argon-ion laser light source and the recognition that a 5 MHz X-cut quartz crystal (half inch diameter) can be driven successfully at very high harmonics (sixty-ninth in our experiments).

The quartz crystal is not a uniformly efficient transducer even on its odd harmonics, and the RF output of the electronic system is not uniformly reproducible over a broad frequency range. For these reasons and since absorption is a function of the rate of change of the (acoustic) field intensity with distance, the information on acoustic absorption must be obtained at each frequency by actually changing the transducer-light beam separation. A plot of log [Photo detection output voltage] vs. separation is linear and of slope $2\alpha_{total}$ where α_{total} is the total absorption of ultrasonic energy in neper cm^{-1} (8.686 db/neper).

In Figure 1 we provide a schematic of the optical path and in Figure 2 a block diagram of the electronics. The thermostatted Teflon sample cell has a ~50 ml liquid volume and a hemispherical bottom to prevent formation of ultrasonic standing waves. In Figure 1, the location of three stepping motors is indicated all of which will eventually be computer driven. As is obvious from the Bragg relation, the angle of laser beam incidence on the sample liquid must be varied continuously with ultrasonic frequency.

To verify system accuracy we first made measurements on pure water. We obtained for H_2O at 25°C over the frequency range

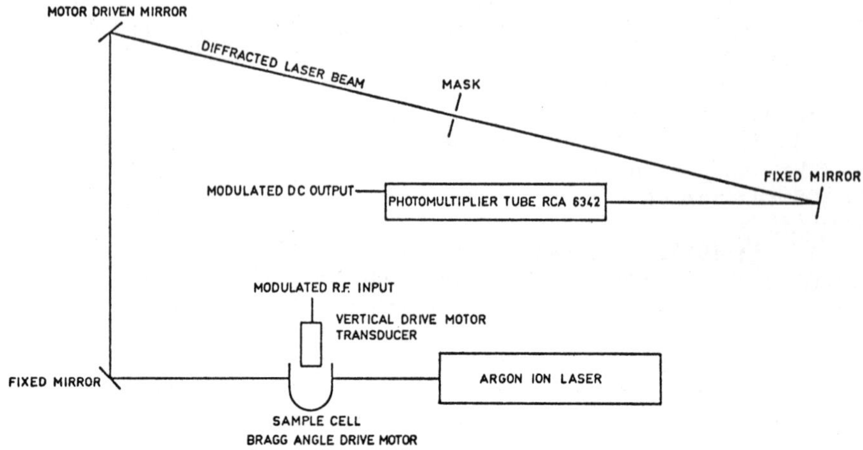

Fig. 1. Schematic of the light path in the laser optical ultrasoni system.

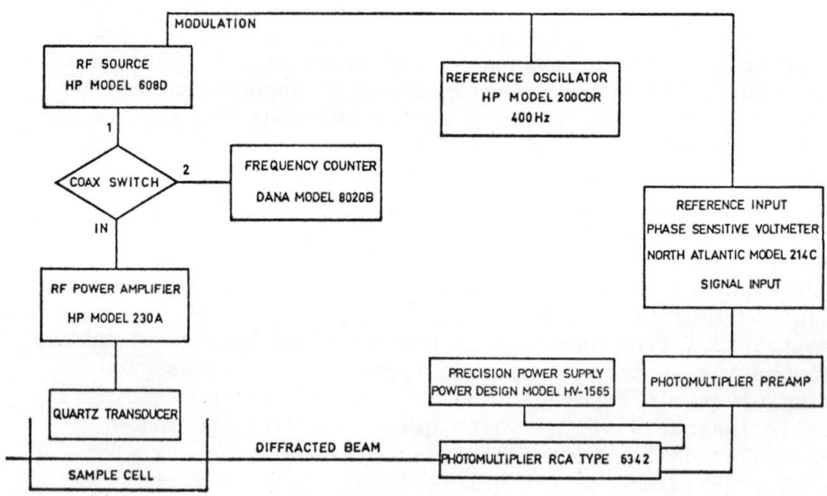

Fig. 2. Electronic block diagram of the laser optical ultrasonic system.

15 to 345 MHz a α/f^2 value of 21.7 X 10^{-17} nepers $cm.^{-1}sec^2$ which is the literature value.[10]

The first chemical system studied was the formation of the aqueous samarium (III) sulfate complex. This was for comparison with the pulse technique since it is well established by this latter method that solutions of $Sm_2(SO_4)_3$ show an absorption maximum in the 30 - 50 MHz range as well as some broadening of the high frequency shoulder.[11]

Ultrasonic absorption spectra over the frequency range 15 MHz to 205 - 245 MHz are shown for two Samarium sulfate solution concentrations (3.52 X 10^{-2}M and 1.17 X 10^{-2}M) in Fig. 3 where the broadening from a single relaxation at frequencies > 1000 MHz is immediately obvious. The open circles are experimental absorptions. The data have been analyzed as two coupled relaxations, and the solid lines in Fig. 3 are the computer simulated absorption curve and the corresponding single relaxations derived from the analysis. The data have been interpreted in terms of a two-step mechanism encompassing the Eigen-Tamm mechanism.[12]

$$Sm^{3+} + SO_4^{2-} \underset{k_{31}}{\overset{k_{13}}{\rightleftharpoons}} Sm(W)SO_4^{+} \qquad (1)$$

$$Sm(W)SO_4^{+} \underset{k_{43}}{\overset{k_{34}}{\rightleftharpoons}} SmSO_4^{+} + W \qquad (2)$$

An exact solution of the rate equations is necessary in this case because of the close coupling of the chemical equilibria. This solution correlates both relaxation times and all four rate constants, according to the relation.

$$(\tau_{1,2})^{-1} = 1/2 \quad \{S \pm (S^2 - 4P)^{1/2}\} \qquad (3)$$

where the positive sign corresponds with τ_1^{-1}, the negative sign with

τ_2^{-1}, $S = (k_{13}\theta + k_{31} + k_{34} + k_{43})$, $P = \{k_{13}\theta\ (k_{34} + k_{43}) + k_{31}\ k_{43}\}$

and $$\theta = \pi_f \left[\{\overline{Sm}^{3+}\} + \{\overline{SO}_4^{2-}\} + \{\overline{SO}_4^{2-}\} \left(\frac{\partial \ln \pi_f}{\partial \ln \beta}\right)_c \right]$$

In the expression for θ, $\pi_f = \gamma_1^{12}$ where γ_1 is the activity coefficient of the univalent ion, the bars refer to equilibrium concentrations, and β and C are the degree of association and analytical salt concentration respectively. Two concentrations and four relaxation frequencies are sufficient to solve eq. (3). If we denote the experimentally determined frequencies by f_1 and f_2 and we know that $\tau_1^{-1} = 2\pi f_1$, and $\tau_2^{-1} = 2\pi f_2$, S and P can be calculated from ultrasonic absorption curves using the equations $S = 2\pi(f_1 + f_2)$ and $P = 4\pi^2 f_1 f_2$. Simultaneous solution of two equations for S and two equations for P gives as a result four rate constants. These results are summarized for the combination of four different solution concentrations in Table 1.

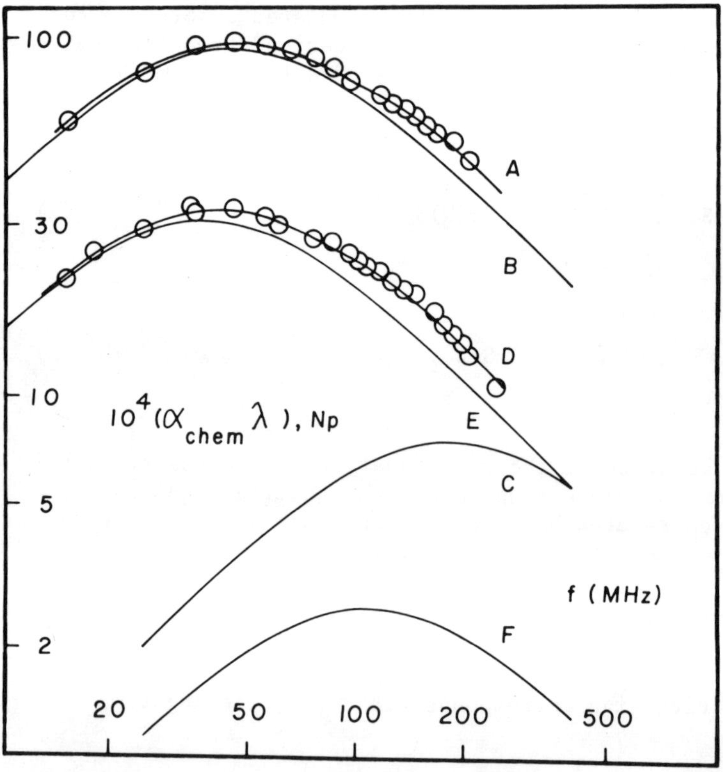

Fig. 3. Excess ultrasonic absorption spectra for samarium sulfate solutions plotted as 10^4 (α_{chem}) neper vs. frequency in MHz. A, B and C are the computer simulated double and composite single relaxation curves for 3.52 X 10^{-2}M $Sm_2(SO_4)_3$. D, E and F are the corresponding curves for 1.17 X 10^{-2}M $Sm_2(SO_4)_3$.

Table I. Rate Constants and Stepwise Equilibrium Constants for $SmSO_4^+$ Formation

	Method of Analysis		
	Eq. (3)[a]	Reidler and Silber[b] (2-step)	Theoretical[c] (2-step)
$k_{13}, M^{-1}sec^{-1}$	1.6×10^{11}		1.3×10^{11}
k_{31}, sec^{-1}	1.3×10^{8}		1.7×10^{8}
K_{13}	1230	365	765
k_{34}, sec^{-1}	2.3×10^{8}	4.1×10^{8}	
k_{43}, sec^{-1}	9.4×10^{7}	3.4×10^{7}	
K_{34}	2.4	12.1	
K_T(kinetic)	4185	4782	
K_T(conduct)	4550	4550	

[a]Error in k_{13} estimated to be approximately ± 30%.

[b]J. Reidler and H.B. Silber, J. Phys. Chem., 77, 1275 (1973).

[c]Calculated from equations of Debye and Eigen, see M. Eigen and L. DeMaeyer, in A. Weissberger (ed.), "Technique of Organic Chemistry", Vol. VIII, Part II, Interscience, New York, (1963.)

A criterion for estimating the internal consistency of the rate data is to calculate the overall association constant for the process using $K_T = K_{13} (1 + K_{34})$, and to compare this with the value obtained conductiometrically.[13]

Also included in Table 1 are results from a previous treatment of the same system in which K_{13} was estimated and in which no attempt was made to fit the data to a double relaxation. The agreement between the constants k_{13} and k_{31} obtained from theory and from eq. (5) is gratifying although we would be more comfortable if the experimental value of k_{13} were smaller than the value calculated from the Debye equation rather than slightly larger.

Coloured sample solutions such as aqueous cerium (IV) sulfate that absorb strongly in the blue-green spectral region can present a problem for this argon-ion laser ultrasonic technique. To carry out a preliminary kinetic study of this sample system it was necessary to use a pumped dye laser as a light source which allowed a shift in the incident beam to ~580 nm which is not absorbed (as strongly). Initial analysis of the cerium (IV) sulfate system indicates the probability of either a double or even a triple relaxation occurring in the frequency range used and no attempt has as yet been made to assign a mechanism.

There appears to be some self-focusing of the laser beam occurring in aqueous solutions of silver (I) thiosulfate and cerium (IV) sulfate which leads to localized heating and thus beam spreading. This has been corrected to some extent by placing a shutter between the laser and the cell and only opening it for the few seconds required to measure the diffracted beam at each position of the transducer.

LASER TEMPERATURE-JUMP

A pulsed laser obviously can effect small temperature jumps in those condensed media with reasonable absorbances at the laser wave length. From the relaxation kineticist's point of view laser heating has several potential advantages over the Joule heating (high voltage capacitive discharge) T-jump relaxation method. Three of these are the feasibility of rapidly heating solvents of low dielectric constant (in which ionic strengths of 0.1 to 1.0M are frequently unattainable), heating very small solution volumes thanks to the ease of focusing of the collimated laser beam, and heating on a time scale of nanoseconds rather than microseconds so that very much faster chemical relaxations are observable.

A variety of laser T-jump studies have been carried out in several laboratories[14] some of which are being reported in this Institute. We will focus our attention here only on the laser Raman T-jump technique devised by Turner et al.[15]

The fundamental output wavelength of a neodymium:glass laser,

1.06 μm, is only weakly absorbed by an aqueous sample solution. When such a laser is "Q-switched" to produce ~20nsec duration pulses and the 1.06 μm laser light is passed through ~25 cm of liquid nitrogen essentially none of the light is absorbed but ~15% of it is Raman shifted to a wavelength of 1.41 μm that is strongly absorbed by water. Our laser Raman T-jump apparatus, closely resembling that of Turner et al.[15], is shown schematically in Fig. 4. The aqueous triiodide equilibrium[15] provides a particularly useful test of the quality of our apparatus because the optical absorbance change associated with a ~3° temperature rise is in this case comparatively small.

There are a number of experimental difficulties with the technique of Turner et al[15] that are worth noting. In laser heating, unlike Joule heating of a liquid, the Beer's Law absorption of energy has an exponential profile that produces significantly higher temperatures near the front surface of the cell than at greater distances through the sample liquid. Thus rate constants determined from measured chemical relaxations in the laser heated cell will be valid for some arduously defined average "final" temperature of the sample liquid. To minimize the impact of this intrinsic difficulty with laser heating Turner[16] used very thin (< 1mm thick) sample cells. The spectrophotometric detection of chemical relaxations is most sensitive when the light absorbance at the detecting wavelength is ~0.8 optical density units. Comparatively few potential sample equilibria are so highly colored at a suitable analyzing wavelength that they will produce this optical density in a ~1 mm optical path length cell. However, it may be possible to markedly lengthen the optical path length in such a cell with thin film technology.

Using optical waveguide techniques[17,18] it should be feasible to probe the color change at the first liquid surface of the laser Raman T-jump cell with evanescent photons.[19] Briefly, the restructured apparatus would consist of an asymmetric-slab waveguide[20] (to carry the spectrophotometric sampling beam) with passive cladding on one side and active cladding on the other. The passive cladding would be the glass or quartz wall of the sample cell on which the 1.41 μm laser light is incident. The active cladding would be the sample solution and more specifically that portion of the solution in which the laser heat is first dissipated. This arrangement is shown in Fig. 5. Three significant improvements in the apparatus of Turner et al.[15] would follow from this modification: 1) the optical path length for determining the chemical relaxation time could be varied from ~1 cm to lengths as short as masking techniques permit, 2) since the penetration of evanescent photons is small (the order of the wavelength of analyzing light) the ΔT of the probed region would be well defined, and 3) the waveguide-coupler system would greatly simplify the crossed 1.41 μm laser light-visible wavelength analyzing light optical alignment.

Other disadvantages of the laser Raman technique are 1) the development of acoustic waves (that produce cavitation) in so thin

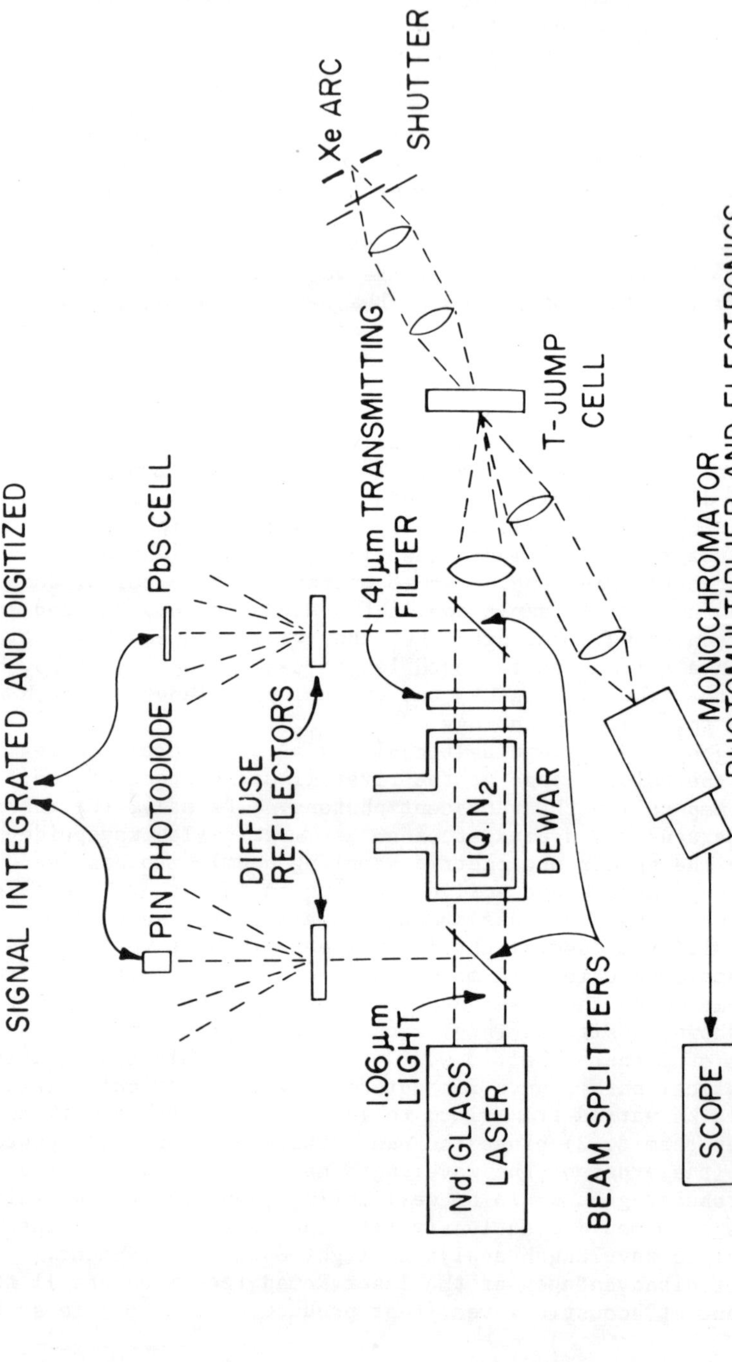

Fig. 4. Schematic of Utah laser Raman temperature-jump apparatus closely resembling that of Turner et al.[15]

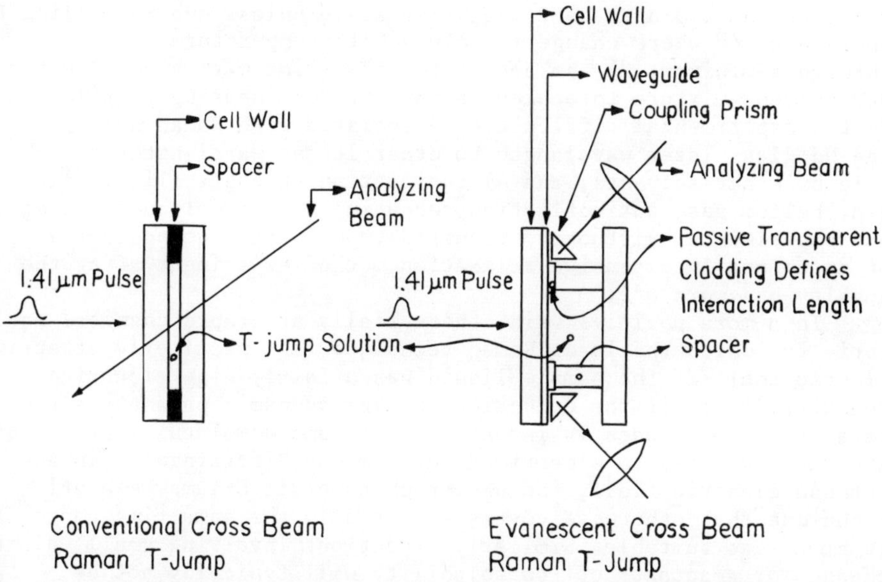

Fig. 5. Schematic comparison of the sample cell of Turner et al. and another with a potentially longer sampling light path.

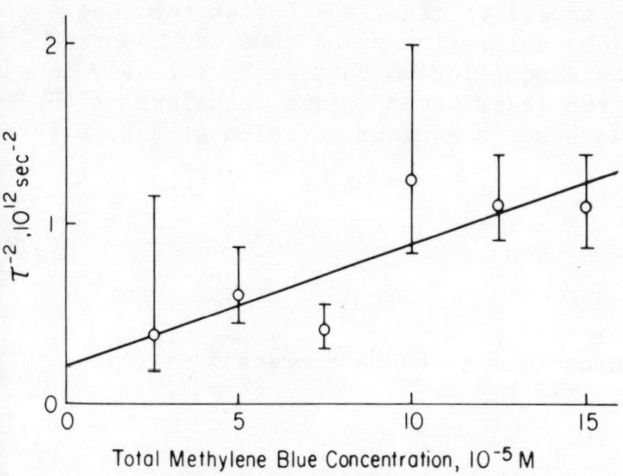

Fig. 6. Plot of laser Raman temperature-jump relaxation method kinetic data for methylene blue dimerization. Data points are a mix of results obtained at glass electrode pH values of 5.8 and 7.3 in a 50/50 water-D_2O mixture at 25°. Vertical error bars denote standard deviation in reciprocal relaxation times. The least squares straight line diagonal is a fit of the average data points at each concentration, denoted by open circles.

a sample cell that essentially limit successful kinetic measurements to the 3.0 nsec to 2 μsec time scale unless one is willing to operate at 4° where change in volume with temperature passes through a minimum. 2) the necessity of working with a 50/50 water: D_2O solvent mixture to achieve a more uniform heating profile, 3) the experimental difficulties associated with Raman shifting the Nd:glass laser wavelength to other longer wavelengths absorbable by other solvents, and 4) the continuing costs (liquid nitrogen, helium gas, antireflection recoating of optical surfaces etc.) and experimental virtuosity required to maintain the productivity of an unusually expensive relaxation method experiment after the novelty has worn off.

In a more positive vein, there really are rapid chemical equilibria for which the laser Raman technique is a peculiarly effective kinetic tool. If the sample liquid has a fairly high electrical conductivity or if the deduction of thermodynamic parameters from relaxation amplitudes is intended or if sample molecules are large and may give rise to electric dichroism and birefringence in an intense electric field, the spectrophotometric E-jump relaxation technique that otherwise covers essentially the same time range wil be much less suitable. Similarly, reactions involving small volume changes or reactants of low solubility will typically not be susceptible to study by acoustic methods that actually can go to much shorter relaxation times. (As noted above it is easy to make ultrasonic absorption measurements at frequencies as high as 350 MH that from $\tau^{-1} = 2\pi f$ correspond to a relaxation time of only 0.45 nsec. This is nearly two orders of magnitude faster than the shortest laser Raman T-jump relaxation times thus far reported.)

An example of a sample equilibrium that we have lately found susceptible to study by the laser Raman T-jump technique is the dimerization of methylene blue in aqueous solution at 25°. Writing the equilibrium as

$$2M \underset{k_{-1}}{\overset{k_1}{\rightleftharpoons}} M_2 \qquad (4)$$

the experimental relaxation time τ can be expressed as

$$\tau^{-2} = k_{-1} + 8k_1 k_{-1} \{M\}_T \qquad (5)$$

where $\{M\}_T$ is the total concentration of methylene blue. Laser Ram data are plotted in this manner in Fig. 6. The resulting rate constants are compared in Table II with similar constants obtained by Turner et al[21] for proflavin dimerization and by Hammes and Hubba for acridine orange dimerization.

The value of the dimerization equilibrium constant for aqueou methylene blue is the quotient, k_1/k_{-1}, of the rate constants. The resulting value, $3.5 \times 10^3 M^{-1}$, agrees well with previously report

Table II. Dye Dimerization Rate and Equilibrium Constants

	k_{-1}, sec^{-1}	k_1, $M^{-1}sec^{-1}$	K_{dimer}, M^{-1}
Proflavin[a] 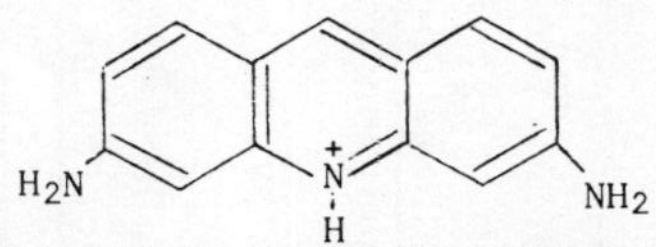	$(2.0 \pm 0.3) \times 10^6$	$(7.9 \pm 1.0) \times 10^8$	$395 \pm 80\ M^{-1}$
Acridine Orange[b] 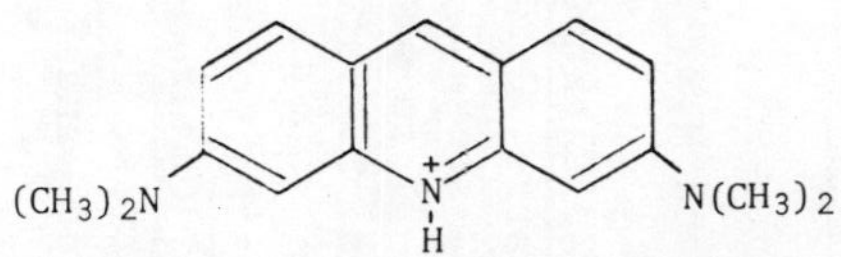	1.8×10^4	2.7×10^8	$1.5 \times 10^4\ M^{-1}$
Methylene Blue 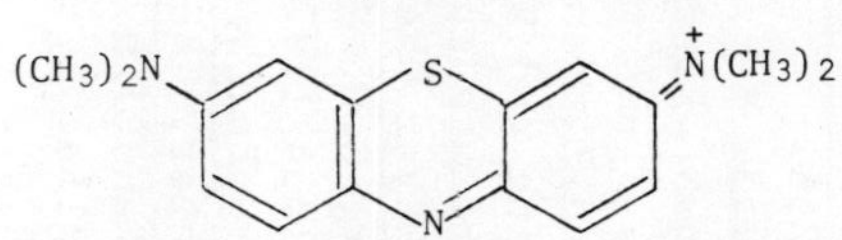	4.8×10^5	1.7×10^9	$3.5 \times 10^3\ M^{-1}$

[a] D.H. Turner et al., Nature, 239, 215 (1972).

[b] G.G. Hammes and C.D. Hubbard, J. Phys. Chem., 70, 1615 (1966).

non-kinetic values[23-25]: 3.6×10^3, 5.9×10^3 and $2.2 \times 10^3 M^{-1}$.

Turner et al[21] suggested that k_{-1} for proflavin was 100 times that for acridine orange because of "the hydrophilic nature of the amino groups in proflavin." Since nitrogen substitution of methylene blue resembles that of acridine orange and yet the methylene blue k_{-1} more closely resembles that of proflavin, it is clear that similar dimerization data for other structurally related dye molecules will be needed before we can accept or reject the Turner hypothesis.

ACKNOWLEDGEMENT. This work was sponsored by Grant AFOSR 73-2444 from the Directorate of Chemical Sciences, Air Force Office of Scientific Research and by a contract from the Office of Naval Research. The Nd:glass laser of the T-jump studies was refurbished under a contract from the Aerospace Research Laboratories, Wright-Patterson AFB.

REFERENCES

1. L. Brillouin, Ann. Physique, 17, 88 (1922).
2. P. Debye and F.W. Sears, Proc. Nat. Acad. Sci. U.S., 18, 409 (1932).
3. R. Lucas and P. Biquard, Compt. Rend., 194, 2132 (1932).
4. R. Adler, IEEE Spectrum, 4, 42, May (1967).
5. T. Nowicki, Electro-Optical Systems Design, 23, February (1974
6. G. Maass,, Ph.D. thesis, University of Goettingen, Goettingen, Germany.
7. J. Steuhr, Techniques of Chemistry, Vol. VI, 3rd- ed., G.G. Hammes, ed., John Wiley and Sons, New York, 1974. p.258.
8. T. Tanaka, Rev. Sci. Instrum., 43, 164 (1972).
9. Ref. 7, p.256.
10. Ref. 7, p.253.
11. Reviewed in N. Purdie and M.M. Farrow, Coord. Chem. Rev., 11, 189 (1973).
12. M. Eigen and K. Tamm, Z. Elektrochem., 66, 93 (1962).
13. F.H. Spedding and S. Jaffe, J. Am. Chem. Soc., 76, 882 (1954).
14. For a literature survey see J.T. Knudtson and E.M. Eyring, Ann. Rev. Phys. Chem., 25 in press.
15. D.H. Turner, G.W. Flynn, N. Sutin, J.V. Beitz, J. Am. Chem. Soc., 94, 1554 (1972).
16. D.H. Turner, Ph.D. Thesis, Columbia University, New York, 197
17. J.E. Midwinter, IEEE J. Quantum Electronics, QE-7, 339-344.
18. J.N. Polky and J.H. Harris, J. Opt. Soc. Am., 62, 1081 (1972)
19. C.K. Carniglia, L. Mandel and K.H. Drexhage, J. Opt. Soc. Am. 62, 479 (1972).
20. K.O. Hill, R.I. MacDonald and A. Watanabe, J. Opt. Soc. Am., 64, 263 (1974).

21. D.H. Turner, G.W. Flynn, S.K. Lundberg, L.D. Faller and N. Sutin, Nature, 239, 215 (1972).
22. G.G. Hammes and C.D. Hubbard, J. Phys. Chem., 70, 1615-1622 (1966).
23. E. Rabinowitch and L.F. Epstein, J. Am. Chem. Soc., 63, 69 (1941).
24. A.K. Ghosh and P. Mukerjee, J. Am. Chem. Soc., 92, 6408 (1970).
25. K. Bergmann and C.T. O'Konski, J. Phys. Chem., 67, 2169 (1963).

THE ELECTRIC FIELD JUMP RELAXATION METHOD

E.M. Eyring

Department of Chemistry, University of Utah, U.S.A.

ABSTRACT. A brief survey of the recent literature on the electric field jump relaxation technique and its uses in chemistry is presented. Applications of this kinetic method at Utah are also mentioned.

The evolution of the electric field jump (E-jump) or dissociation field effect relaxation method from its first conception by Eigen[1] makes an entertaining story to tell but a superfluous one to write. The principal references on the subject[2-4] are detailed and discourage attempts at amplification. At least in part because of unusual expense and experimental difficulties, the reviews[5-12] written since 1967 that consider this technique and the descriptions[13-19] of equipment modifications approach in number those papers[20-39] actually devoted to new chemical results obtained by this method. Strehlow[11] succinctly characterized the disadvantages of the E-jump technique as "high dilution necessary" and "sophisticated equipment". When the technique is applied to polymeric electolytes several competing, observable effects[38-40] can discourage even fairly intrepid experimentalists.

We will restrict ourselves here to representative E-jump studies at Utah with occasional references to related work from other laboratories.

Two versions of Eigen's E-jump relaxation method have been tried at Utah. The first incarnation[14] involved the application of square high voltage waves (up to ~10 μsec long with voltages of ~50 KV and consequent field intensities of ~100KV/cm in a sample cell with an interelectrode distance of 0.5 cm) from a coaxial delay line to an asymmetric Wheatstone bridge. The first E-jump apparatus of this general type had been used by Eigen and

Wyn-Jones (ed.), Chemical and Biological Applications of Relaxation Spectrometry, 85–90.

De Maeyer[42] to determine the now classic rate constant for the recombination of a hydronium ion with hydroxide ion in pure water. At Utah, this conductimetric instrument was used principally to study the kinetics of hydrolysis of aquated metal ions such as $Al^{3+} \underset{}{\overset{k_1}{\rightleftharpoons}} AlOH^{2+}(aq)+H^{+}(aq)$ and $FeOH^{2+}(aq) \overset{k_2}{\rightleftharpoons} Fe(OH)_2^{+}(aq)+H^{+}(aq)$ for which the ion recombination rate constants were found at 25° to be $k_{-1} = 4.4 \cdot 10^9\ M^{-1}sec^{-1}$ and $k_{-2} = 8.0 \cdot 10^9\ M^{-1}sec^{-1}$ respectively. Our interest in these reactions was fostered by their relevance to corrosion processes.[44] Wendt[45] has provided an interesting survey of the metal ion hydrolysis-polymerization kinetics field.

Two advantages of the square wave E-jump technique with conductimetric detection are that a percent change in sample conductance as low as 0.1 is readily detectable and the sample substance can be colorless (and coupled indicator equilibria are unnecessary). Principal disadvantages are the need for extraordinarily pure (low conductivity) solvent and the time consuming balancing of the Wheatstone bridge and the difficult achievement of linearity in the high voltage probes between the bridge and oscilloscope. In spite of these disadvantages, the method is still used occasionally as in the recent determination by Yasunaga et al.[33] of the kinetics of the alpha helix - random coil transition of aqueous poly-L-glutamic acid.

The second and more useful incarnation of the E-jump technique in our laboratory has been a square-wave apparatus with spectrophotometric detection (in the visible).[16,19] The first successful E-jump relaxation method studies performed with spectrophotometric detection were reported by Ilgenfritz.[13] He studied the kinetics of acid dissociation-ion recombination of aqueous sulfonephthalein acid-base indicators. Refinements we have added to his basic experiment include dual triggered spark gaps to obviate the need for long lengths of expensive coaxial cable and a pulsed spectrophotometer light source[19] to improve the signal to noise ratio (see Figure 1). Presently, we are adding a transient digitizer (Tektronix R7912)- minicomputer (Intel 8080) system to speed the translation of detected photomultiplier voltage transients into calculated rate constants. We forsee significant economies in the time of our students when they not only discontinue reading photographs of individual exponential decay curves but can also discover early in the course of a day's experiments that different experimental conditions such as pH or temperature would produce more informative results.

In recent months we have used our spectrophotometric E-jump apparatus to measure the rate of complexation of aqueous samarium (III) ion by murexide ion,[36] the protonation-deprotonation kinetics of several aqueous dinitrophenols, and an interaction between aqueous amylose helices and a polyiodide.

One can reasonably expect the E-jump technique to be frequently used in studying fast inorganic and organic reactions in non-aqueous solvents as well as kinetic properties of solution of polymers in the next few years. Much of this work will need to be done conductimetrically, but the spectrophotometric E-jump technique will

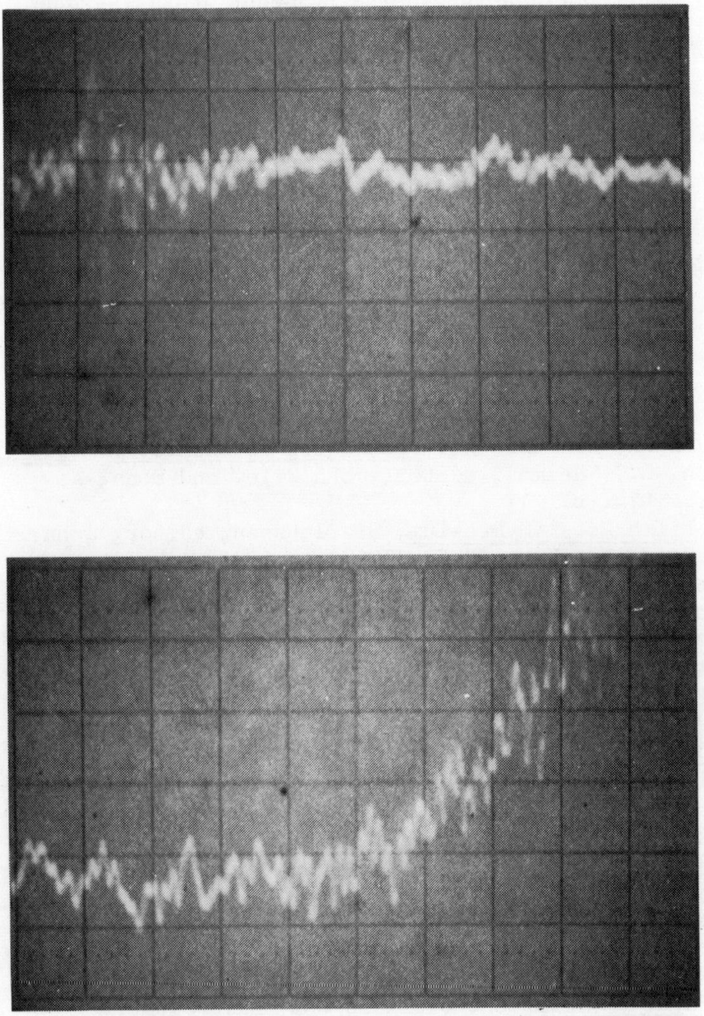

Fig. 1. Enhancement of a spectrophotometric E-jump relaxation curve amplitude by pulsing the xenon arc lamp. Sample solution: 5×10^{-4}M aqueous 2,5-dinitrophenol, pH = 4.3, ; λ = 440 nm. Vertical scale: 10 millivolts per major division. Horizontal scale: 0.1 microseconds per major division. In the experiment recorded in the bottom picture the light intensity for the 3 msec duration of the lamp pulse is 100 times greater than in the experiment recorded in the top photo.

undoubtedly also fill an important niche in these kinetic studies.

ACKNOWLEDGEMENT. This work was sponsored by Grant AFOSR 75-2444 from the Directorate of Chemical Science, Air Force Office of Scientific Research.

REFERENCES

1. M. Eigen, Disc. Faraday Soc., 17, 194 (1954).
2. M. Eigen and L. De Maeyer, Technique of Organic Chemistry, Vol. VIII, Part II, 2nd edition, S.L. Friess, E.S. Lewis and A. Weissberger, editors, Interscience Publishers, New York, N.Y., 1963 p. 895.
3. L.C.M. De Maeyer, Methods in Enzymology, Vol. XVI, Fast Reactions, K. Kustin, editor, Academic Press, New York, N.Y., 1969, p.80.
4. L. De Maeyer and A. Persoons, Techniques of Chemistry, Vol. VI, 3rd edition, G.G. Hammes, editor, John Wiley and Sons, New York, N.Y., 1974, p. 211.
5. M. Eigen, Fifth Nobel Symposium, S. Claesson, editor, John Wiley and Sons, New York, N.Y., 1967, p. 333.
6. E.M. Eyring and D.L. Cole, Fifth Nobel Symposium, S. Claesson, editor, John Wiley and Sons, p. 255.
7. D.N. Hague, Fast Reactions, Wiley-Interscience, New York, N.Y., 1971, p. 41.
8. Z.A. Schelly and E.M. Eyring, J. Chem. Ed., 48, A695 (1971).
9. G. Ilgenfritz, Probes of Structure and Function of Macromolecules and Membranes, Vol. 1, B. Chance, C.-p. Lee and J.K. Blasie, editors, Academic Press, New York, N.Y., 1971, p. 505.
10. J.E. Crooks, MTP International Review of Science, Chemical Kinetics, Physical Chemistry, Series One, Vol. 9, J.C. Polanyi, editor, Butterworths, London, 1972, p. 312.
11. H. Strehlow, Advan. Mol. Relaxation Processes, 2, 235 (1972).
12. H. Winkler, Endeavour, 33, 73 (1974).
13. G. Ilgenfritz, Chemische Relaxation in Starken Electrischen Feldern, Ph.D. Disseration, Georg-August-Universitat, Goettingen, 1966.
14. D.T. Rampton, L.P. Holmes, D.L. Cole, R.P. Jensen and E.M. Eyring, Rev. Sci. Instrum., 38, 1637 (1967).
15. B.R. Staples, D.J. Turner and G. Atkinson, Chem. Instrumentation, 2, 127 (1969).
16. S.L. Olsen, R.L. Silver, L.P. Holmes, J.J. Auborn, P. Warrick, Jr. and E.M. Eyring, Rev. Sci. Instrum., 42, 1247 (1971).
17. H.H. Grunhagen, Biophysik, 10, 347 (1973).
18. H.H. Grunhagen, Messtechnik, 1/74, (1974).
19. S.L. Olsen, L.P. Holmes and E.M. Eyring, Rev. Sci. Instrum., 45, 859 (1974).

20. R. Winkler, Ph.D. Dissertation, Vienna, 1969.
21. D.L. Cole, L.D. Rich, J.D. Owen and E.M. Eyring, Inorg. Chem. 8, 682 (1969).
22. L.D. Rich, D.L. Cole and E.M. Eyring, J. Phys. Chem., 73, 713 (1969).
23. L.P. Holmes, A. Silzars, D.L. Cole, L.D. Rich and E.M. Eyring, J. Phys. Chem., 73, 737 (1969).
24. T.M. Schuster and G. Ilgenfritz, Symmetry and Function of Biological Systems at the Macromolecular Level, Nobel Symposium II, A. Engstrom and B. Strandberg, editors, John Wiley and Sons, New York, N.Y., 1969, p. 181.
25. P. Hemmes, L.D. Rich, D.L. Cole and E.M. Eyring, J. Phys. Chem., 74, (1970).
26. P. Hemmes, L.D. Rich, D.L. Cole and E.M. Eyring, J. Phys. Chem., 75, 929 (1971).
27. J.J. Auborn, P. Warrick, Jr. and E.M. Eyring, Dissociation Field Effect Kinetics of Aqueous Acetic Acid and Bromocresol Green, J. Phys. Chem., 75, 2488 (1971).
28. J.J. Auborn, P. Warrick, Jr. and E.M. Eyring, J. Phys. Chem., 75, 3026 (1971).
29. G. Ilgenfritz and T.M. Schuster, Probes of Structure and Function of Macromolecules and Membranes, Vol. II, Probes of Enzymes and Hemoproteins, B. Chance, T. Yonetani and A.S. Mildvan, editors, Academic Press, New York, N.Y. 1971, p. 229 - 310.
30. G. Ilgenfritz and T.M. Schuster, Probes of Structure and Function of Macromolecules and Membranes, Vol. II, Probes of Enzymes and Hemoproteins, p. 399 - 406.
31. P. Warrick, Jr., J.J. Auborn and E.M. Eyring, J. Phys. Chem. 76, 1184 (1972).
32. R.G. Sandberg, G.H. Henderson, R.D. White and E.M. Eyring, J. Phys. Chem., 76, 4023 (1972).
33. T. Yasunaga, T. Sano, K. Takahaski, H. Takenaka and S. Ito, Chem. Lett. (Japan), 405 (1973).
34. D.J. Lentz, J.E.C. Hutchins and E.M. Eyring, J. Phys. Chem., 78, 1021 (1974).
35. A.P. Persoons, J. Phys. Chem., 78, 1210 (1974).
36. M.M. Farrow, N. Purdie and E.M. Eyring, Inorg. Chem., 13 2024 (1974).
37. H.H. Grunhagen, Ph.D Dissertation, Carolo-Wilhelmina Technical University, Braunschweig, 1974.
38. H. Hirohara, K.J. Ivin and J.J. McGarvey, J. Am. Chem. Soc., 96, 3311 (1974).
39. H. Hirohara, K.J. Ivin, J.J. McGarvey and J. Wilson, J. Am. Chem. Soc., 96, 4435 (1974).
40. M. Eigen and G. Schwarz, Proceedings of an International Symposium held in Trieste, June 1959, B. Presce, editor, Pergamon Press, New York, N.Y., 1962, p. 309.
41. N. Ise, M. Eigen and G. Schwarz, Biopolymers, 1, 343 (1963).
42. E. Fredericq and C. Houssier, Electric Dichroism and Electric Birefringence, Oxford University Press, London, 1973.

43. M. Eigen and L. De Maeyer, Z. Elektrochem., 59, 986 (1955).
44. M. Marek and R.F. Hochman, Corrosion-NACE, 30, 208 (1974).
45. H. Wendt, Chimia, 27, 575 (1973).

PRESSURE-JUMP METHODS

W. Knoche

Max-Planck-Institut fur biophysikalische Chemie,
Gottingen, Germany

1. INTRODUCTION

The progress of most chemical reactions in solution is accompanied by a change in volume. Therefore, according to Le Chatelier's principle, the equilibrium position of a reversible reaction can be shifted by changing the pressure applied to the solution. For the reaction

$$\Sigma \nu_i A_i = 0 \tag{1}$$

where the ν_i are the stoichiometric numbers (positive for the products and negative for the reactants), the volume of reaction is defined as

$$\Delta V = \Sigma \nu_i V_i \tag{2}$$

with V_i the partial molar volume of the species A_i. The equilibrium between reactants and products is determined by the equilibrium constant.

$$K = \Pi \{A_i\}^{\nu_i} \tag{3}$$

At constant temperature, the relation between equilibrium constant and pressure is given by equation

$$\left(\frac{\partial (\ln K)}{\partial p}\right)_T = -\frac{\Delta V}{RT} \tag{4}$$

Wyn-Jones (ed.), Chemical and Biological Applications of Relaxation Spectrometry, 91–102.

In pressure-jump relaxation experiments, the duration of the pressure change is short compared with the relaxation time. Such fast pressure changes occur under adiabatic rather than isothermal conditions, and instead of eq. (4) the following equation has to be applied:

$$\left(\frac{\partial \ln K}{\partial p}\right)_S = \left(\frac{\partial \ln K}{\partial p}\right)_T + \left(\frac{\partial \ln K}{\partial T}\right)_p \left(\frac{\partial T}{\partial p}\right)_S = -\frac{\Delta V}{RT} + \frac{\alpha T.\Delta H}{\rho c_p.RT^2} \qquad (5)$$

α is the coefficient of thermal expansion of the solution, ρ its density, and c_p its specific heat.

The first pressure-jump apparatus was developed by Ljungren and Lamm[1]. Nearly simultaneously Strehlow and Becker[2] developed a different pressure-jump technique which after further improvements has been extensively used in several laboratories. In this version, the autoclave is closed by a metal rupture membrane, the pressure is increased to about 100 atm, and by the rupturing of the membrane a fast pressure drop of 1 atm is realized within less than 0.1 msec. An even faster pressure change occurs within the very thin transition region of a shock wave. This fact was used by Jost[3] in a pressure-jump apparatus applicable to relaxation times between 1 μsec and 1 msec. A similar technique has been proposed by Hoffmann and Yeager[4] which has been used more frequently and will be discussed in Part III.

In most applications of the pressure-jump techniques the progress of reaction has been observed by measuring the electrical conductance[5,6] but also spectrophotometric[7,8] and thermometric[9] readout have been used.

II. PRESSURE-JUMP APPARATUS

The most frequently used pressure-jump technique follows the general plan of Strehlow and Becker[2]. Figure 1 is the schematic diagram of the equipment required. Inside the autoclave (where the pressure can be changed rapidly) two identical conductivity cells (1) are mounted: one is filled with the solution under investigation, and the second one with a solution having the same electrical conductivity but showing no relaxation in the time range studied. The autoclave is closed with a burst diaphragm which ruptures spontaneously at a pressure of about 130 atm. Before the measurement begins, the bridge is tuned at a pressure of 1 atm. The pressure is then increased slowly until the burst diaphragm blows out. The pressure decreases within less than 100 μsec to ambient, and the voltage peak of the piezoelectric capacitor (5) triggers the oscilloscope and the digitizer. The trace on the oscilloscope screen now shows how the investigated solution regains equilibrium at the ambient pressure of 1 atm. Simultaneously, the signal is digitized, whereupon the computer calculates

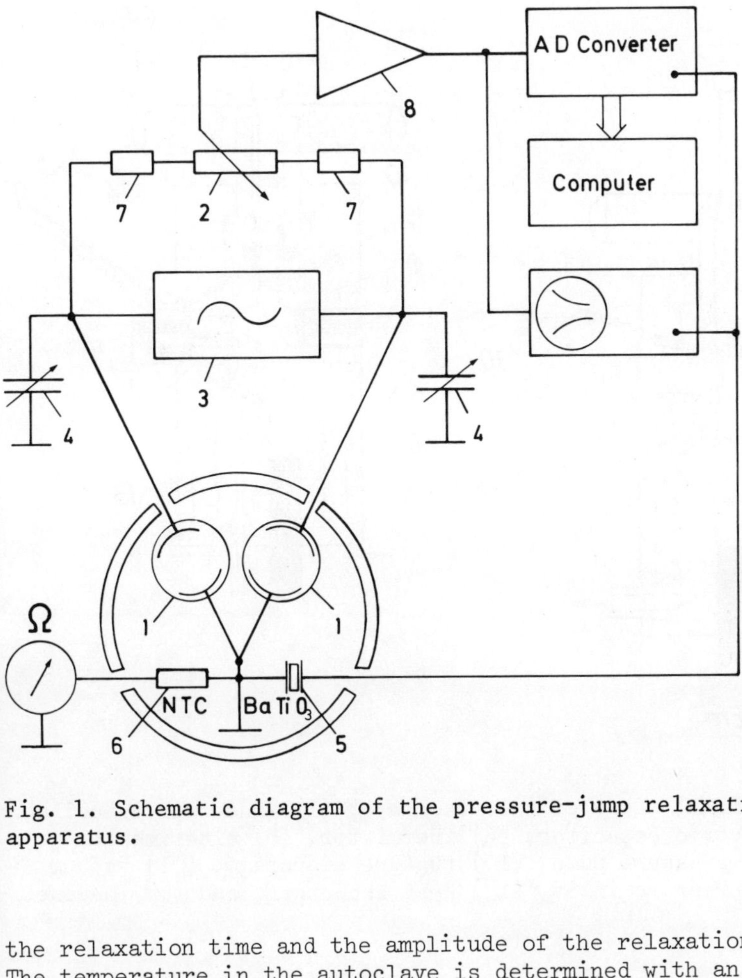

Fig. 1. Schematic diagram of the pressure-jump relaxation apparatus.

the relaxation time and the amplitude of the relaxation effect. The temperature in the autoclave is determined with an accuracy of $\pm$ 0.1°C by measuring the electric resistance of the NTC resistor (6). Details of the digitizing of the signal and of the data processing are discussed in a later contribution of this conference[10].

Figure 2 shows sectional views of an autoclave used in our laboratory[5]. At the pressure increase, energy is stored in the autoclave according to

$$E = \int \beta \, Vp \, dp \tag{6}$$

V is the volume, p the pressure and β the compressibility of the liquid in the pressure chamber. This energy is released at the pressure jump and is partly needed to break the rupture membrane.

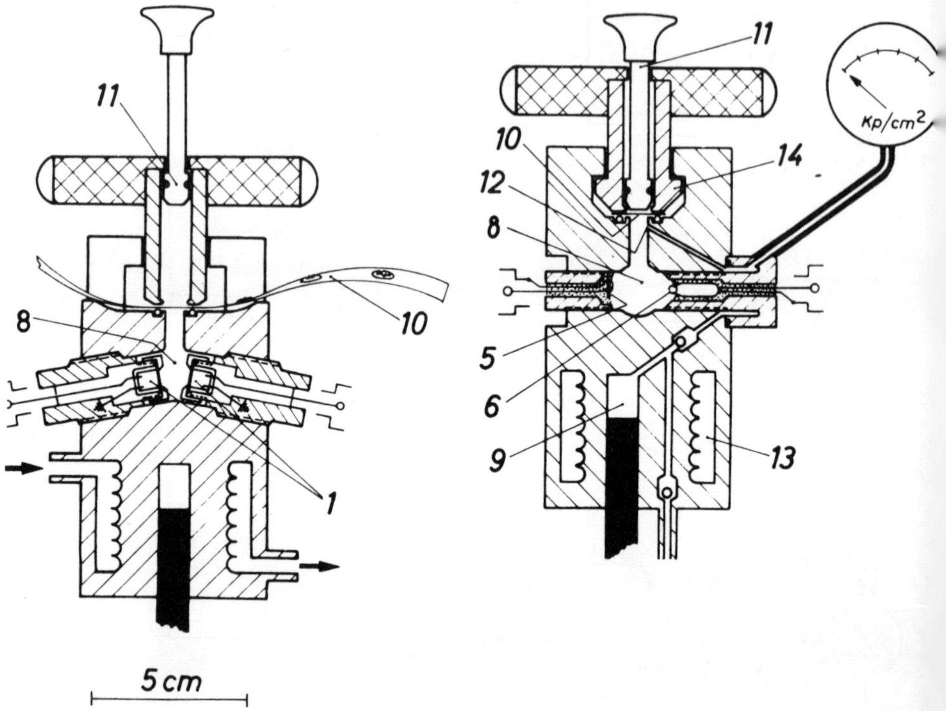

Fig. 2. Sectional views of the autoclave. (1) Conductivity cells, (5) piezoelectric capacitor, (6) thermistor, (8) experimental chamber, (9) pressure pump, (10) rupture diaphragm, (11) vacuum pump , (12) pressure inlet, (13) heat exchanger, and (14) bayonet socket.

However, it also causes the autoclave to oscillate, thus disturbing the determination of the cell resistances. To minimize the energy, the volume of the experimental chamber (8) is reduced as far as possible, and the pressure pump (9) has been constructed as an integrated part of the autoclave to reduce all supply lines to a minimum. Moreover, water is used as the pressure transducing liquid because of its small compressibility. The conductivity cells are mounted with small inclination so that no air bubbles are trapped in the experimental chamber because they cause pressure oscillations after the pressure jump. By the same reason, the pressure transducing water as well as the solutions in the conductivity cells must be carefully degassed. To avoid disturbances by acoustical noise during the pressure jump in the compartment above the rupture diaphragm (10), the pressure is reduced to about 5 Torrs. This is achieved by a simple pump (11). The pressure inlet (12) is close beneath the rupture membrane; thus the incoming water does not stream along the measuring cells and small air bubbles (which, despite all precautions, may be contained in the water) do not reach the experimental chamber.

The temperature of the autoclave is regulated between 0 and 80°C by thermostating the liquid streaming rapidly through the heat exchanger (13). The autoclave is made of bronze and is surrounded by a 5-mm thick cover made of PVC (not shown in Fig. 2) for the thermal insulation. Thus temperature gradients in the autoclave may be neglected. Since the pump is thermostated as well, the temperature of the experimental chamber is not altered by the water pumped into the autoclave. A band of 0.1 mm thick brass is used as bursting membrane. To perform an experiment, only the bayonet socket (14) has to be loosened by half a turn and the band has to be moved forward.

The conductivity cells are permanently mounted on screws and the electrical leads are soldered to BNC sockets. They can easily be removed from the autoclave for filling. The construction of the cells is shown in Fig. 3. They are made of glass with sealed-in platinized platinum electrodes and imbedded in plastics to insulate them from mechanical shock and the vibrations of the autoclave. A thin membrane of PVC covers the solution, and screw-caps close the cells.

If the precautions discussed above are observed, the mechanical disturbances occurring immediately after the pressure jump only cause changes in the relative resistance of the cells which are smaller than $\delta R/R = 10^{-4}$.

The adiabatic pressure jump δp is connected with a temperature change δT according to eq. (5) which yields inserting for water and diluted aqueous solutions at 20°C, $\alpha = 0.21 \times 10^{-3} K^{-1}$, $\rho = 1\ g/cm^3$, $c_p = 1$ cal/g K: $(\partial T/\partial p)_S = 1.45$ K/atm. The disturbances of the measurements by the thermal equilibration after the pressure jump largely cancels by comparing the resistance of the relaxing solution with that of a nonrelaxing solution.

The change in resistance caused by the chemical relaxation may be as small as 10^{-4}. Therefore a relative change of the conductance of 10^{-5} has to be clearly measurable to obtain the relaxation time with some accuracy. This does not require that the resistance itself has to be known with this precision. Bridge arrangements are especially suitable for the measurement of such small differences. Alternating current has to be used to drive the bridge because direct current would cause polarization effects on the electrodes. Since rapid changes have to be followed, the frequency of the current has to be relatively high. A schematic diagram of a bridge is shown in Fig. 1 too. For relaxation times longer than 50 μsec, a frequency of 40 kHz is sufficient. As detection system for a shock-tube apparatus a similar arrangement with a frequency of 2 MHz has been used[4]. The sample and the reference cell (1), the two resistors (7), and the potentiometer (2) are combined in a Wheatstone-bridge arrangment. The bridge is balanced by means of the potentiometer (a 10-turn film potentiometer) and the tunable capacitors (4). The resistance of each of the two resistors (7) (each 10 to 100 pF) is 4.5 times that of the balancing potentiometer, so that its sensitivity is increased 10 times. This is necessary because the balancing has to be performed with a relative accuracy of 10^{-5} in the most sensitive measurements. The bridge is mounted as symmetrically to ground as possible. Nevertheless, the balancing of the bridge is slightly frequency dependent due to differences in the cell conductances and the high frequency behaviour of the electrode capacitances. Therefore, even if the bridge is balanced for the fundamental frequency of the driving voltage, the harmonics still cause a remaining signal. This requires that the current provided by the oscillator is extremely free from harmonics. The oscillator (3) supplies a very stable voltage with an amplitude of 3 V peak to peak. A higher voltage cannot be applied without causing temperature changes in the cells by Joule heating. The output of the oscillator is ungrounded and capacitively symmetrical to ground. Its output impedance is about 10 Ohm and small compared to the resistance of the cells (100 to 10^5 Ohm in all experiments). Thus, a change in the conductivity of the solution caused by the relaxation process does not alter the voltage applied to the bridge. Using cell resistances larger than 10^5 Ohm, the sensitivity of the measurements is seriously reduced by the shunt capacities of the electrodes and the electric leads. The amplifier (8) has an amplification factor of 100. It is tuned to the frequency of the oscillator with a bandwidth (B) of

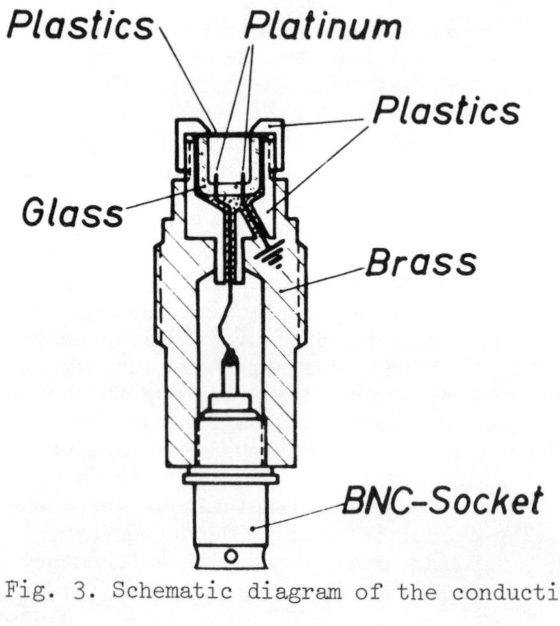

Fig. 3. Schematic diagram of the conductivity cells.

10 kHz. Thereby, the signal to noise ratio (S/N) is improved considerably, as S/N is proportional to $B^{-1/2}$. The response time (T_r) of the amplifier is determined by the bandwidth according to $T_r = (2\pi B)^{-1}$. This response time (for the amplifier used $T_r = 1.6 \times 10^{-5}$ sec) does not seriously distort the measurements, because the equipment is only used for relaxation times longer than 5×10^{-5} sec. The reduction of the bandwidth also eliminates disturbances by 50 (or 60) cycle hum and by the noise from radio stations.

The voltage on the oscilloscope (δU) is related to the change in the resistance (δR) by

$$\frac{\delta U}{U} = \frac{1}{4} F \frac{\delta R}{R} \tag{7}$$

U is the voltage of the oscillator, F the amplification factor of the amplifier, and R the resistance of the cells. For further details of the conductometric recording see ref. 6.

With this relatively simple equipment relaxation times between 50 μsec and 50 sec have been measured. As an example, in Fig. 4 the relaxation of an aqueous solution of $In_2(SO_4)_3$ is shown where two relaxation effects are observed. The method is applicable to all reactions which have a finite reaction volume, i.e., their equilibrium is shifted by changing the pressure, and in which the conductivity of the solution changes during the chemical reaction. It has been applied to study metal-complex formation in aqueous and non-aqueous solution, the hydration of organic compounds, biochemical reactions and the formation of micelles.

With the equipment described above, chemical relaxation can be measured only if the progress of reaction is accompanied by a change in electrical conductivity. To increase the applicability a pressure-jump apparatus with spectrophotometric detection has been developed[8]. Cross sections of this device are shown in Fig. 5. The outer dimensions of the autoclave are chosen such that it can be used with the optical arrangement of the conventional (and commercially available) temperature-jump apparatus of the Goettingen-type. Therefore, for details of the detection method, publications concerning that technique should be consulted. The construction of the autoclave is similar to that described above and can be seen in Fig. 5. Only the following differences should be mentioned. Instead of the conductivity cells two quartz windows are mounted into the autoclave. To increase the pressure, a small pump is connected to inlet. For controlling the temperature in the autoclave, the thermostating of the cell-holder is used. A similar cell with four windows to measure fluorescence or light-scattering is under construction.

With the apparatus shown in Fig. 5 the kinetics of the hydration of pyruvical-ethylester has been studied[11]. This reaction shows relaxation times of several seconds. It cannot be studied by temperature-jump methods because in this time range strong

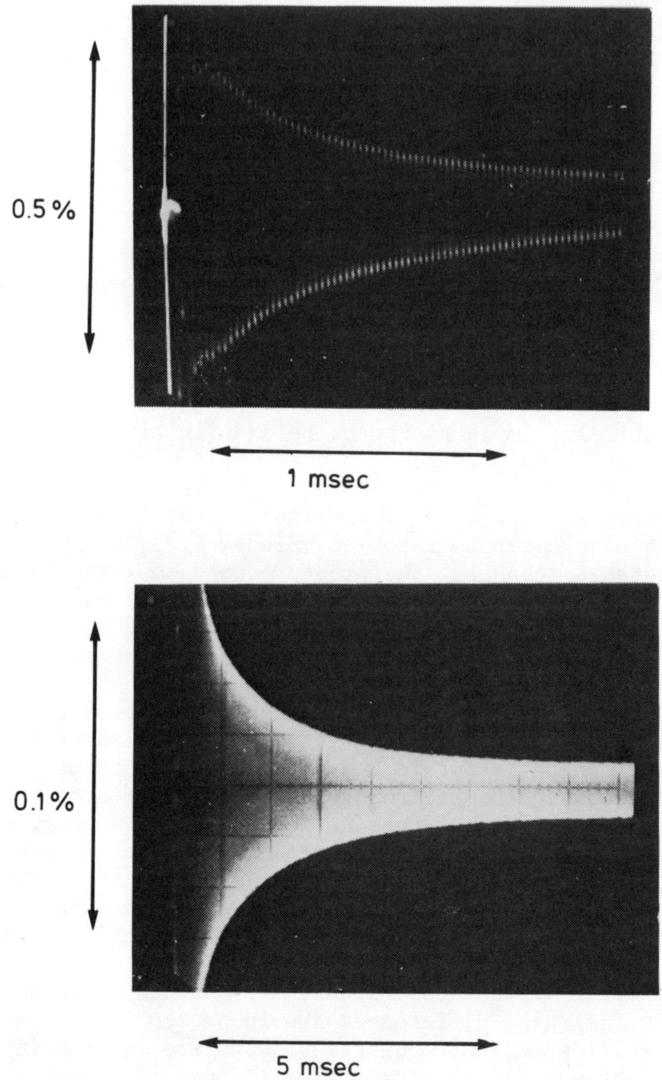

Fig. 4. Relative change of conductivity as function of time for pressure jumps of 130 atm applied to solution of 0.005 M $In_2(SO_4)_3$, pH = 2.82 at 34°C. Two relaxation effects are observed with τ_I = 0.35 msec and τ_{II} = 1.9 msec.

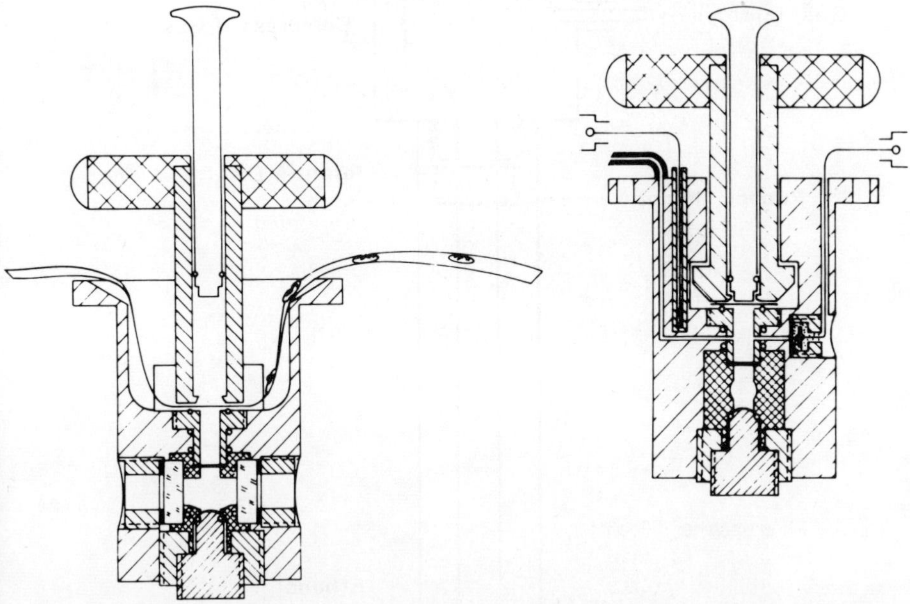

Fig. 5. Sectional views of the autoclave with spectrophotometric detection.

interference occurs with thermal equilibration processes. Since it is a reaction between uncharged particles, conventional pressure-jump technique with conductiometric readout cannot be used either.

III. SHOCK-WAVE APPARATUS

With the techniques described above, a pressure jump can be realized within 50 to 100 µsec. This sets the short-time limit of the relaxation times that can be measured. To investigate faster reactions, the very steep pressure increase in shock waves has been utilized.

A shock-wave apparatus as developed by Hoffmann and Yeager[4] is shown in Fig. 6. It consists of a high pressure and low pressure part separated by a bursting disc. The low-pressure part is a vertically mounted tube with a length of 125 cm and an inner diameter of 1.2 cm. It is filled with ethanol or hexane, whereas

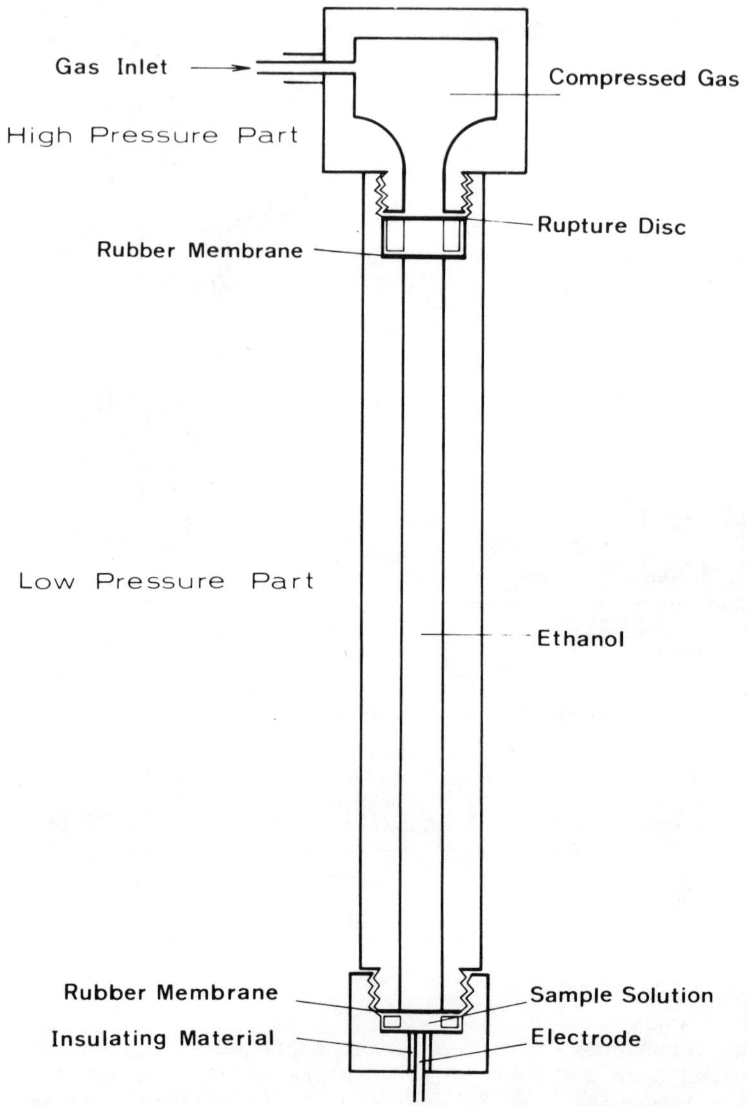

Fig. 6. Shock-wave apparatus developed by Hoffmann and Yeager[4].

the high-pressure chamber contains gas. The sample cell is placed at the lower end of the tube and is covered by a thin rubber membrane. In the upper part the pressure is increased. The rupture disc is inflexible so that the pressure is not transduced to the lower part until at about 150 atm the disc bursts. This generates a pressure wave expanding into the low-pressure section. Since the velocity of sound increases with pressure, the front of the pressure wave sharpens when propagating into the liquid column. The volume of the high-pressure chamber is sufficiently large so that the pressure of the gas is not noticeably reduced by the compression of the liquid.

The distance L over which the wave must travel to achieve a step-like pressure increase at the wave front is obtained by

$$L = u_o^2 \; \frac{\delta t}{\delta p}\left(\left(\frac{\partial u}{\partial p}\right)_T\right)^{-1} = 2\, u_o^5 \; \frac{\delta t}{\delta p}\left(\left(\frac{\partial^2 p}{\partial \rho^2}\right)_S\right)^{-1} \quad (8)$$

where u_o is the velocity of sound and δt the time needed for the increase of the pressure at the upper end of the low-pressure part from 1 atm. to δp. For the described apparatus, the time needed for the bursting of the rupture disc is $\delta t \approx 50$ μsec and the pressure increase is $\delta p \approx 150$ atm. Inserting these values together with those for the sound velocity and the pressure dependence of the compressibility obtainable from the literature[12] it is seen that a length of 125 cm is sufficient to obtain an essentially step-like pressure increase at the lower end of the tube. A rubber membrane is placed close beneath the rupture disc to prevent gas from dissolving in the liquid after the pressure jump.

At the lower end of the tube a second rubber membrane separates the sample solution from the pressure transmitting liquid. The change in concentration is observed conductiometrically. The electrode arrangement is critical, since it determined the time resolution of the apparatus. This is given by the time required for the shock wave to pass through that part of the solution that contributes to the conductance of the cell. With the arrangement shown in Fig. 6, where the conductance is measured between the central electrode and the grounded bottom of the cell, the time resolution is about 1 μsec. For relaxation times longer than 10 μsec, a dual conductance cell was used. The shock wave travels through the sample cell and is reflected at the lower end of the tube. Thereby the pressure is doubled to 300 atm. It remains at this value for the time the pressure front needs to travel back up the liquid column and down again, i.e., for twice the length divided by the sound velocity. Thus, the long-time limit of the apparatus is about 2 msec.

The two main pressure-jump methods as described in this contribution have the advantages of relatively simple construction and easy operation. They have been applied over a wide range to study reactions with relaxation times between 5×10^{-6} and 50 secs.

REFERENCES

1. S. Ljunggren and O. Lamm, Acta Chem. Scand. 12, 1834 (1958).
2. H. Strehlow and M. Becker, Z. Elektrochem. 63, 457, (1959).
3. A. Jost, Ber. Bunsenges. physik. Chem. 70, 1057 (1966).
4. H. Hoffmann and E. Yeager, Rev. Sci. Instr. 39, 1151 (1968).
5. W. Knoche and G. Wiese. Chem. Instr. 5, 91 (1973-74).
6. W. Knoche, Techniques of Chemistry, Vol. 6, Part 2, p. 187, (ed.: G.G. Hammes), Wiley (Interscience), New York, 1973.
7. D.E. Goldsack, R.E. Hurst and J. Love, Analytical Biochem. 28, 273 (1969).
8. W. Knoche and G. Wiese, to be published in Rev. Scient. Instr.
9. J. Helisch and W. Knoche, Ber. Bunsenges. physik. Chem. 75, 951 (1971).
10. W. Knoche, this publication.
11. H.-J. Buschmann and W. Knoche, to be published in Ber. Bunsenges. physik. Chem., 79, (1975).
12. H.E. Eduljee, D.M. Newitt and K.E. Weale, J. Chem. Soc., 3086, 1951.

ON DATA PROCESSING, EVALUATION AND PRECISION OF RELAXATION MEASUREMENTS

Wilhelm Knoche

Max-Planck-Institut fur biophysikalische Chemie,
Gottingen, Germany

INTRODUCTION

Chemical relaxation techniques have been introduced by Eigen and coworkers about twenty years ago. Since that time, the precision of the techniques has been improved considerably and especially the signal-to-noise ratio of the measurements has been enlarged. Due to the higher precision it may turn out that there are two relaxation effects in a system studied, where only a single one was detected with less accurate equipment, or that there is a strong disturbance by an instrumental effect missed in former measurements. To discriminate between a single or a multiple exponential, to eliminate disturbances, and to increase the accuracy of the results also the techniques of evaluating the measurements and of data processing have been developed further.

In this contribution, it is not intended to give a summary over all data-handling systems developed or proposed so far. Only an on-line data processing system is discussed that has been extensively used in our laboratory for the evaluation of jump-relaxation measurements. The equipment includes a small computer for the immediate evaluation of the experiments. It enables us to sample the measurements for further improvement of the signal-to noise ratio and to evaluate time constants and amplitudes of two or more superimposed exponentials.

THE DIGITIZING INTERFACE

In jump-relaxation experiments we obtain a signal of the form

$$y = A \exp(-t/\tau) + C \qquad (1)$$

Wyn-Jones (ed.), Chemical and Biological Applications of Relaxation Spectrometry, 103–111.

or, in the case of superposition of several relaxing effects

$$y = \sum A_i \exp(-t/\tau_i) + C \tag{2}$$

In principle, in the case of Eq. (1) three values of y at different times are sufficient to obtain the three unknowns A, τ, and C. Of course, caused by the limited precision of the measurements, more values of y are needed to get the unknowns with good accuracy. Furthermore, it is not known whether an experiment is described by Eq. (1) or by Eq. (2). To increase the accuracy one may measure a large number of values y at different times t and fit those to Eq. (2). But the general shape of the signal is well known (i.e., the superposition of exponentially decaying curves). Therefore, it is not necessary to accumulate an exceedingly high number of amplitude measurements if the single values of y have high precision. This can be achieved by proper filtering to increase the signal-to-noise ratio. For an exponentially decaying signal, the nearly ideal filtering is the integration of the signal during time intervals Δt with $0.2\tau \leqslant \Delta t \leqslant 0.3\tau$. Thus, high frequency noise (compared to $1/\tau$) is eliminated to a large extent. Low frequency disturbances only change the value of C. Of course, disturbances in the frequency range $1/\tau$ cannot be reduced by filtering since they have the same time constants as the relaxing effect. However, if these disturbances are uncorrelated to the chemical relaxation, they can be eliminated by sampling, i.e. by the superposition of several experiments.

According to these considerations a digital evaluation technique has been developed[1]. A block scheme of the complete experimental setup with a pressure-jump relaxation apparatus is given in Fig. 1. The detection of the change of concentration in the pressure-jump experimental setup is performed by measuring the conductivity in a 40 kHz Wheatstone bridge. Two conductivity cells in the autoclave form two branches of the bridge (one cell is filled with a solution containing the relaxing system, the other one with a non-relaxing solution of about the same resistance to compensate for trivial physical effects). Furthermore, the autoclave contains a piezoelectric $BaTiO_3$-capacitor to trigger the oscilloscope and the digitizer at the pressure jump. The principle of the pressure-jump relaxation technique has been described elsewhere[2].

The signal of the bridge is fed to the input of a storage-oscilloscope which is mainly needed as a monitor to eliminate exceptionally disturbed measurements. Moreover, the vertical amplifier of the oscilloscope is used as a preamplifier providing the same voltage for a signal filling the oscilloscope display independently of the bridge signal. Thus, by changing the sensitivity of the oscilloscope, the signal is scaled to the full input voltage range of the precision fullwave rectifier ($\pm$10V). The rectified signal is fed to the integrator, which was previously reset to 0 V output voltage, and integrated in time intervals Δt ($\sim 0.3\tau$)

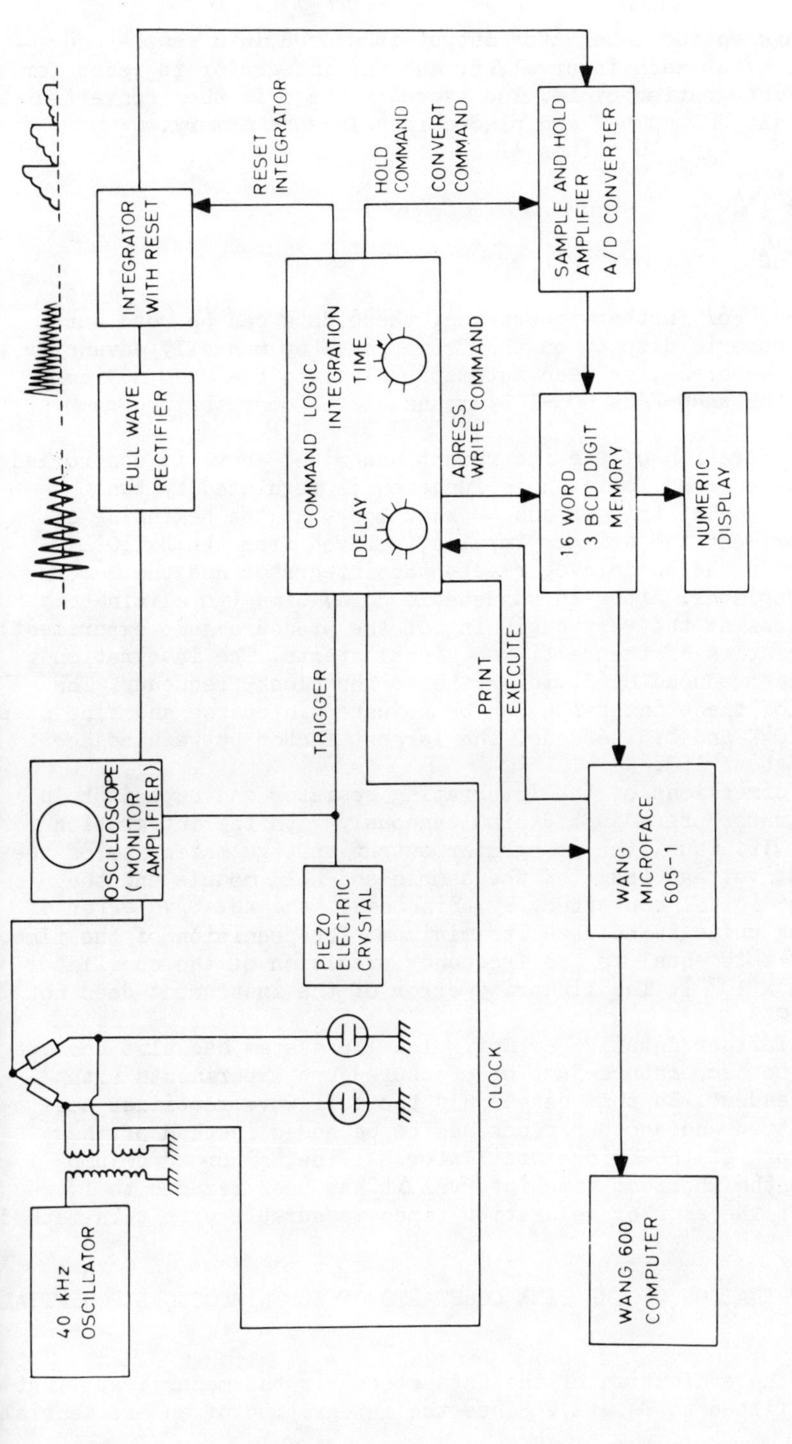

Fig. 1. Block diagram of pressure-jump relaxation apparatus with digitizing interface and computer.

The voltage on the integrator output is stored in a sample-and-hold unit after each interval Δt, and the integrator is reset for the next integration cycle. The stored voltage is then converted to a 3-digit BCD number and placed in a 16-word memory.

In this way, 16 values of

$$\int_{n\Delta t}^{(n+1)\Delta t} y \, dt$$

are stored. For further processing, these data can be read out from the numeric display on the front panel by manually advancing the memory address, or read automatically into the WANG 600 computer via the WANG-FACE 605-1 by means of the computer programme control.

The operation of the instrument described above is controlled by digital command logic whose function is regulated by two external signals, trigger and 40 kHz clock. At the beginning of each experiment the trigger impulse, derived from the $BaTiO_3$-capacitor in the autoclave, resets the integrator and the memory address register. After an adjustable delay time (to eliminate disturbances at the very beginning of the pressure-jump experiment) the integration of the rectified signal starts. The integration time Δt is produced by dividing the 40 kHz clock-frequency. The duration of these intervals can be adjusted in coarse and fine steps between 10^{-4} and 8.5 seconds. The largest factor between adjacent steps is about 1.3.

The dimensions of the integrating resistor and capacitor in the integrator are adjusted simultaneously with the integration intervals Δt. Thus, the integrator output voltage makes use of the full input voltage range of the sample-and-hold module and the analog-to-digital converter. By this means, the relative error of all analog units is held at its minimum. The precision of the time interval Δt is equal to the frequency precision of the oscillator ($\delta f/f \sim 3 \times 10^{-4}$). The linearity error of the instrument does not exceed 0.25%.

For further details see Ref. 1 . The system has also been adjusted to temperature-jump or pressure-jump experiments with optical readout. In that case, only the full wave rectifier has to be omitted and another clock has to be added instead of the 40 kHz clock of the bridge oscillator. At the temperature-jump equipment the shortest time interval Δt has been reduced to 1 μsec because of the shorter relaxation times measurable with this method

THE DETERMINATION OF THE TIME CONSTANTS OF SUPERIMPOSED EXPONENTIALS

For the evaluation of the data stored in the memory, at first they are fitted to Eq. (1). Since the integration of an exponential function yields again an exponential function with the same time constant, the relaxation time obtained is not influenced by the integration in the digitizer. For the fitting of the data to Eq. (1), 12 successive data of the memory are used. If $\Delta t \approx 0.25\tau$, the remaining amplitude at $t = 12\Delta t$ is $y - C < 0.02A$, i.e., nearly no information is lost by using only the data up to $t = 12$ t.

By a least square fit to Eq. (1) using the first eight amplitudes, the time constant τ is evaluated. Then the calculation is repeated starting with the second amplitude and including the ninth amplitude and so on. Thus five time constants, τ_o to τ_4, are obtained. Of course, if the curve is really a single exponential, the five τ_n values should be the same within experimental error.

However, if two (or more) exponentials are superimposed, the τ_n values will change systematically with n. In this case, the measurements have to be fitted to Eq. (2) where C may alter in different experiments. The case with only two overlapping exponential functions will be considered first. We have

$$y = Ae^{-t/\tau_A} + Be^{-t/\tau_B} + C \tag{3}$$

The usual procedure to obtain the two relaxation times by plotting ln y vs. t gives very unsatisfactory results if τ_A/τ_B does not differ much from 1. In fact, if τ_A and τ_B differ less than by a factor of 3, Eq. (2) can be well approximated by a single exponential expression

$$y = A' e^{-t/\tau} + C' \tag{4}$$

with

$$\frac{1}{\tau} \approx \frac{A}{A+B}\frac{1}{\tau_A} + \frac{B}{A+B}\frac{1}{\tau_B} \tag{5}$$

As an example, in Fig. 2 the most favourable case is shown where both exponentials have the same amplitudes (A = B). Even then a high accuracy is required to discriminate between a single and a double exponential and the existence of two exponentials can hardly be recognized from a semilogarithmic plot. Also, if $A/\tau_B \gg 3$ different sets of parameters A, B, τ_A, and τ_B may fit the data nearly equally well. If the amplitudes A and B differ by more than a factor of three, the exponential with the smaller amplitude may be considered as a disturbance and at least for the main exponential a precise value of τ can be obtained as discussed below.

There may be different reasons for exponential disturbances: an obvious case is the superposition of two (or more) chemical relaxations with similar relaxation times. Disturbances may be caused also by heat conducting processes. Evidently in temperature-

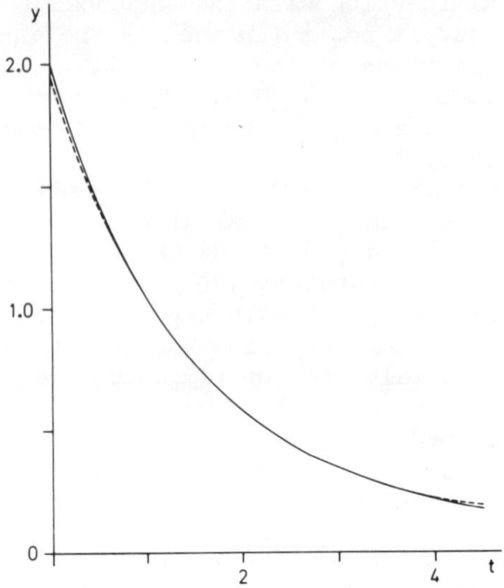

Fig. 2. Plots of $y_1 = e^{-t} + e^{-0.4t}$ and $y_2 = 1.85e^{-0.67t} + 0.1$ (dotted line).

-jump experiments, but also at (adiabatic!) pressure jumps temperature gradients are set up in the measuring cells. The equilibration process occurs exponentially with time. The extent of reaction depends on temperature and also extinction coefficients, conductivities, or other physical quantities by which the relaxation process is measured. Therefore, the thermal equilibration will be detected as an unwanted exponential disturbance of the chemical relaxations. A good compensation method reduces the physical effects and the procedure discussed here may be applicable. Fortunately, the corresponding time constants of temperature equilibration τ_B are known for a given apparatus.

The signal is disturbed too, if the equilibrium is shifted too far by the applied pressure or temperature jump (except for first-order reactions). While this does not occur at pressure jumps with $\delta p < 200$ atm, the situation may be different in temperature-jump experiments as will be demonstrated with a specific example. For a reaction of the type

$$A + B \underset{k_2}{\overset{k_1}{\rightleftharpoons}} C \tag{6}$$

the rate equation is

$$-\frac{d\xi}{dt} = \frac{\xi}{\tau} - k_1\xi^2 \tag{7}$$

where ξ is the instantaneous deviation of the extent of reaction from its equilibrium value, and

$$\frac{1}{\tau} = k_1[c_A + c_B] + k_2 \tag{8}$$

If the temperature jump is small, $k_1\xi^2$ can be neglected compared to ξ/τ and to a good approximation a single exponential with time constant τ is the solution of Eq. (7). However, if $k_1\xi^2$ may not be neglected but still $k_1\xi\tau \lesssim 0.3$, the solution of Eq. (7) is approximately

$$\frac{\xi(t)}{\xi(0)} = (1 + \xi(0)k\tau)e^{-t/\tau} - \xi(0)k\tau e^{-2t/\tau}. \tag{9}$$

$\xi(0)$ is the value of ξ immediately after the temperature jump.

Thus an exponential with time constant $\tau/2$ and smaller amplitude is superimposed. For example, at a temperature jump $\Delta T = 5$ K and an enthalpy of reaction $\Delta H = 10$ kcal/mole the observed "relaxation time" differs up to 6% from the correct value. This error is obtained when $\xi(t)$ is made to fit a single exponential and Eq. (5) is used.

In the case of such exponentially decaying disturbances, there is a systematic change in the five τ_n values calculated at the evaluation of the measurements as discussed above. The slopes and curvatures of the τ_n vs. n curves depend on the ratio of the time constants and on the ratio and relative sign of the amplitudes as shown in Fig. 3. This fact is used to obtain the relaxation time of the main relaxing effect τ_A ("B" refers to the disturbance). As can be seen from figure 3, it is essential to know the curvature of τ_n vs. n because it determines the direction into which the curves have to be extrapolated. If the slope increases or decreases with increasing n, the value of τ_n closest to τ_A is that for n = 0 or for n = 4, respectively. However, a high precision of the measurements and a high accuracy in the τ_n values is necessary to determine the curvature. In the case of high accuracy the correct time constant of the main exponential is given by[3]

$$\tau_A \approx \frac{\tau_1(2\tau_o\tau_2 - \tau_1\tau_2 - \tau_o\tau_1)}{\tau_o\tau_2 - \tau_1^2} \tag{10}$$

According to this equation τ_A is calculated in the computer programme.

When time constant and amplitude of the main exponential are determined, these parameters can be calculated with less accuracy for the smaller effect by fitting $(y - A_1 \exp{-t/\tau_1})$ to Eq. (1). If there are more than two exponentials superimposed, this procedure may be used repetitively.

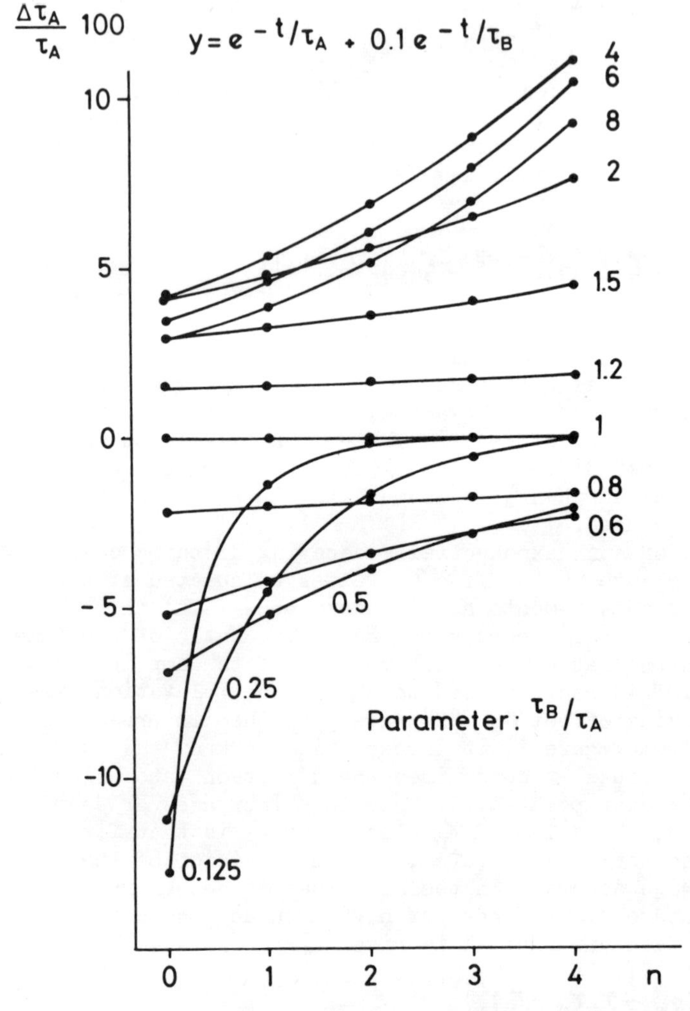

Fig. 3. Calculated τ_n–n curves for disturbed relaxation times. The amplitude of the disturbing exponential is assumed to be 10% of the main amplitude. For other amplitude ratios (including sign), the ordinate of Fig. 2 must be changed accordingly.

This technique is used with a pressure-jump apparatus and the digitizing interface discussed earlier. It is possible to measure relaxation times with an error of about 1% and to determine separately two relaxation times τ_A and τ_B which differ by not more than a factor of 3. If the difference is larger, the relaxation times may be obtained in different experiments with differing time intervals Δt, where one relaxation effect, respectively, causes only a small perturbation. By this method, the experiments shown in ref. 2 , Fig. 4, have been evaluated. However, if more than two relaxation effects differ by a factor of not more than 3 to 4, it is not possible to separate these effects.

To eliminate uncorrelated disturbances (such as noise, hum, mechanical vibrations, or changing light intensity when using optical detection methods) provision is made in the computer programme for adding the data of consecutive pressure-jump experiments. The values of the first amplitude integrals are added up in the first summing register, the values for the second amplitude integrals in the second register and so on. Although the signal/noise ratio increases only with $\sqrt{N}$, with N the number of experiments, the influence of this sampling on the precision of the calculated relaxation times often is surprisingly strong. The reason for this effect can be seen from Fig. 3. In order to obtain the correct value of τ the curvature of τ_n vs. n has to be known. Thus, the noise has to be reduced only so far that this curvature can be seen unambiguously. If the curvature is not known, the best value of τ is given by the average of the τ_n values.

The time needed for the performance of an experiment and for thermal equilibration processes after the jump has to be short if sampling is to be useful. This condition holds for the pressure-jump technique where an experiment can be repeated every 20 seconds (time for temperature equilibration after the adiabatic pressure change), so that sampling may be performed in a reasonable time. In temperature-jump experiments this time is much longer and for effective sampling the circulating of the sample-solution through an heat-exchanger is recommended.

Finally, it should be emphasized that the exact value of a relaxation time is only meaningful, if immediately after the temperature or pressure jump the temperature of the sample is precisely known. For instance, assuming an activation energy of 10 kcal/mole, the rate constant changes by 5% per degree centigrade. That means, if the relaxation time is obtained with an error of ± 1% the temperature has to be known with an accuracy of ± 0.1 K to avoid that this uncertainty contributes to the error in the rate constants.

REFERENCES

1. M. Krizan and H. Strehlow: Chem. Instr. 5, 99 (1973-1974).
2. W. Knoche: "Pressure-Jump Methods", this volume.
3. H. Strehlow and J. Jen: Chem. Instr. 3, 47 (1971).

THE ANALYSIS OF RELAXATION AMPLITUDES

Darwin Thusius

Laboratoire d'Enzymologie Physico-chimique et Moleculaire, Orsay, France

WHY MEASURE RELAXATION AMPLITUDES?

In stepwise chemical relaxation techniques the amplitude of an individual exponential decay may be defined as the value of the induced signal change at zero time, δS^o.

$$\delta S(t) = \delta S^o e^{-t/\tau} \qquad (1)$$

The signal S is proportional to some physical property such as optical absorbance, fluorescence light intensity or conductivity.

Relaxation amplitudes contain unique information on the energetics of a reaction mechanism and on the structure of intermediates. Often this information cannot be derived from static experiments. A case in point is the evaluation of the enthalpy changes and extinction coefficients of a binding reaction followed by an isomerization.

$$A + B \underset{①}{\rightleftarrows} AB \underset{②}{\rightleftarrows} AB^{*} \qquad (I)$$

Calorimetry would only yield an overall reaction enthalpy which is a function of the two elementary reaction heats. Similarly, a spectrophotometric titration would give an apparent extinction coefficient which is a function of the four individual constants, ε_A, ε_B, ε_{AB} and ε_{AB^*}. Although ε_A and ε_B can be measured directly, ε_{AB} and ε_{AB^*} remain undetermined.

Spreading out the overall signal change on the time axis opens the way for an evaluation of elementary thermodynamic functions and physical constants of intermediates. The signal time course following a rapid perturbation of reaction I is

Wyn-Jones (ed.), Chemical and Biological Applications of Relaxation Spectrometry, 113–119.

$$\delta S(t) = \delta S_1^o e^{-t/\tau_1} + \delta S_2^o e^{-t/\tau_2} \quad (2)$$

In principle, relaxation times τ_1 and τ_2 can be used to estimate enthalpy changes by taking differences between activation energies for the forward and reverse rate constants of the two elementary steps. This requires, however, a large number of precise data obtained in a wide concentration and temperature range. Relaxation times are clearly independent of extinction coefficients. On the other hand, relaxation amplitudes provide a straightforward means of determining both types of parameters. We shall see below that the quantities ΔH_1^o, ΔH_2^o, ε_{AB} and ε_{AB^*} can in principle be extracted from the concentration dependences of δS_1^o and δS_2^o with a least-squares curve fitting procedure.

Even in those cases where the evaluation of microscopic parameters from amplitudes is not feasible, it may still be possible to demonstrate that the concentration dependences of the initial signal changes are at least consistent with an assumed mechanism. As amplitudes are functions of different system parameters than relaxation times, they provide a valuable control in model selection. In fact there are several cases where mechanisms proposed on the basis of relaxation times alone have been rejected following an amplitude analysis[1]. In particular, amplitude considerations allow one to comment on relative signal changes. An elegant verification of a reaction mechanism is to rationalize in terms of analytical amplitude expressions the disappearance of one or more decay curves at given reactant concentrations. Finally we recall that one can never "prove" a reaction mechanism; one can only hope to show that all available information is consistent with the same model. For this reason it is important to quantitatively correlate equilibrium data, relaxation times <u>and</u> relaxation amplitudes.

AMPLITUDE ANALYSIS STRATEGY

In spite of the rich information content of relaxation amplitudes, there is a tendency to ignore amplitude data and to concentrate on the interpretation of relaxation times. This preference arises from the fact that the primary objective of a kinetic experiment is in large part the determination of specific rate constants. In addition, relaxation times can reveal subtle mechanistic features which are not reflected in amplitudes. This is particularly true for mechanisms involving intermediates at vanishingly small concentrations. While relaxation time behaviour may indicate the presence of "steady state" intermediates, amplitudes are only sensitive to macroscopic stoichiometry.

A number of authors have presented theoretical treatments

of relaxation amplitudes[2-8]. These discussions have either dealt with restricted families of reactions[3,6,8] or have emphasized the calculation of overall concentration shifts[2,4,5] rather than observed initial signal changes. For the practical exploitation of amplitudes, it is desirable to have a general strategy allowing the experimenter to objectively fit his data to an assumed mechanism with a minimum amount of computational labour.

My approach to the problem involves expressing the observed amplitudes explicitly as linear sums of equilibrium concentrations. The coefficients of the linear equations are functions of elementary thermodynamic functions and "specific signal changes"[7]. The number of terms in the sum is related to the number of elementary steps, or in the limit of widely separated relaxation times, to the relative decay rate. If $\tau_1 << \tau_2 << \ldots\ldots \tau_N$, the ith relaxation amplitude is a sum of i(i+1)/2 terms:

$$
\begin{aligned}
\delta S_i^o = b_{11}X_{11} &+ b_{12}X_{12} + \ldots\ldots + b_{1i}X_{1i} \\
&+ b_{22}X_{22} + \ldots\ldots + b_{2i}X_{2i} \\
&\qquad \vdots \\
&\qquad\qquad + b_{ii}X_{ii}
\end{aligned}
\qquad (3)
$$

where the $X_{\alpha\beta}$ are functions of equilibrium concentrations and the $b_{\alpha\beta}$ are the following constants.

$$
\begin{aligned}
b_{\alpha\alpha} &= \Delta\phi_\alpha \delta \ln K_\alpha, \quad \Delta\phi_\alpha = \Sigma V_{i\alpha}\phi_i \\
b_{\alpha\beta} &= \Delta\phi_\alpha \delta \ln K_\beta + \Delta\phi_\beta \delta \ln K_\alpha
\end{aligned}
\qquad (4)
$$

Here $\Delta\phi_\alpha$, $\delta \ln K_\alpha$ and $V_{i\alpha}$ are respectively the specific signal change of reaction α, the thermodynamic function of reaction α, and the stoichiometric coefficient of species i in reaction α. The specific signal, defined as

$$\phi_i = (\partial S/\partial \bar{C}_i)_{i \neq j} \qquad (5)$$

($\bar{C}_i$ = equilibrium concentration of species i) allows application of the above formalism to any detection system. For example,

absorbance	scattered light
$\phi_i = \ell\varepsilon_i$	$\phi_i = \gamma M_i^2$

where ℓ, ε, γ, and M_i denote optical length, molar extinction coefficient, instrument constant and molecular weight. Similarly, the thermodynamic functions are defined in terms of the perturbation method. For example,

T-jump	P-jump
$\delta \ln K = (\Delta H^{o}/RT^{2})\delta T$	$\delta \ln K = -(\Delta V^{o}/RT)\delta P$

A distinct advantage over previous computational methods is that the $X_{\alpha\beta}$ can be expressed in terms of equilibrium concentrations by a straightforward manipulation of matrix terms derived by an inspection of the reaction mechanism. If the equilibrium concentrations are known, either from static measurements or from an analysis of the relaxation times, the amplitudes may be fit to the assumed mechanism with a linear least-square procedure. A "goodness of fit" test can therefore be applied without any knowledge of the elementary $\Delta\phi_{\alpha}$ and $\delta \ln K_{\alpha}$. In those cases where the equilibrium concentrations are not known, an iterative, non-linear curve fitting procedure may be used.[9]

(A) Practical Application

Now, let us return to reaction I. We assume the limiting case of general interest where the binding step is much faster than the isomerization. We further assume that component B is in large excess. In terms of the above formalism, the amplitudes for this system are:

$$\delta S^{o}_{1}/A^{o} = (\Delta\phi_{1}\delta \ln K_{1})\left[1/((1 + B^{o}K_{1})(1 + (B^{o}K_{1})^{-1}))\right] \quad (7)$$

$$\delta S^{o}_{2}/A^{o} = (\Delta\phi_{1}\delta \ln K_{1})X_{11} + (\Delta\phi_{1}\delta \ln K_{2} + \Delta\phi_{2}\delta \ln K_{1})X_{12} + (\Delta\phi_{2}\delta \ln K_{2})X_{22} \quad (8)$$

where

$$X_{11}=\{(1+K_{ap}B^{o})\left[1+(K_{ap}B^{o})^{-1}\right]\}^{-1}-\{\left[1+(1+K_{23})(K_{ap}B^{o})^{-1}\right](1+K_{ap}B^{o})\}^{-1}$$

$$X_{12} = \{(1 + K_{23}^{-1})(1 + K_{ap}B^{o})\left[1 + (K_{ap}B^{o})^{-1}\right]\}^{-1} \quad (9)$$

$$X_{22} = (K_{ap}B^{o} + K_{23} + 1)\{(1 + K_{23})(1 + K_{23}^{-1})(1 + K_{ap}B^{o})\left[1+(K_{ap}B^{o})^{-1}\right.$$

A^{o} and B^{o} denote analytical concentrations.

Although numerically complex, the right side of equation (8) contains only 3 parameters : K_{ap}, the overall equilibrium constant; B^{o}, the known concentrations of species B; and K_{23}, the isomerization equilibrium constant. The constant K_{ap} can be determined by equilibrium titration, whereas K_{23} can be derived from the concentration dependence of τ_{2}. The coefficients $\Delta\phi_{1}\delta \ln K_{1}$, $(\Delta\phi_{1}\delta \ln K_{2} + \Delta\phi_{2}\delta \ln K_{1})$ and $\Delta\phi_{2}\delta \ln K_{2}$ may then be evaluated with one- and three-term linear least-squares treatments of δS^{o}_{1} and δS^{o}_{2}, respectively. By plotting the data as $\delta S^{o}_{1}/A^{o}$ vs B^{o} one can quickly check the quality of the fit by verifying that the theoretical curves calculated with equations (8) and (9) pass throug the experimental points with no systematic deviations.

If the least-squares coefficients have been determined with sufficient precision, they may be used to calculate $\Delta\phi_1 \delta \ln K_2$ (or $\Delta\phi_2 \delta \ln K_1$), reducing the results to three independent products. Evaluation of the individual $\Delta\phi_\alpha$ and $\delta \ln K_\alpha$, however, requires an additional experimental parameter. In principle the signal change or thermodynamic function of the overall reaction provides this independent information.

We have followed the above curve fitting procedure in a fluorescence temperature jump study of proflavin binding to calf thymus DNA[7]. Earlier, Li and Crothers studied the kinetics of the same reaction with an absorbance temperature jump technique.[10] These authors interpret the relaxation spectrum in terms of scheme I, with proflavin bound to the DNA surface in complex AB, and intercalated between base pairs in complex AB^*. An advantage of fluorescence detection is that one can work with very large DNA/dye ratios (> 100:1), thus reducing proflavin stacking on the polymer.

To a good approximation the fluorescence decay following a rapid thermal perturbation is described by a single exponential, at least in our experimental conditions. The reciprocal relaxation time exhibits a hyperbolic concentration dependence, indicative of a slow isomerization coupled to a rapid binding reaction (Figure 1). The amplitude concentration dependence provides strong independent support of this model (Figure 2). Our results are therefore formally consistent with the intercalation mechanism of Li and Crothers.

Values of the elementary thermodynamic functions and specific signals of this system can yield direct evidence for the proposed structural differences between the two dye-DNA complexes. For example, it has been proposed that proflavin quenching observed in static experiments is due entirely to intercalation.[11] Therefore an amplitude analysis should be consistent with $q_{AB} >> q_{AB^*}$, where q denotes quantum yield. In regard to thermodynamic functions, Li and Crothers estimate from differences between activation energies that $\Delta H_1^o << \Delta H_2^o$, which is in accord with two different binding modes.

We have attempted to extract elementary specific signals and enthalpies from the amplitude regression coefficients, b_{11}, b_{12} and b_{22}. It can be shown, however, that when the equilibrium concentration of species AB^* is much larger than that of AB- which is the case here - an approximately linear relation exists between X_{11} and X_{12}.[7] This results in large uncertainties in the regression coefficients, and frustrates the estimation of underlying mechanistic parameters. On the other hand, computer simulation shows that the analysis in question should be feasible by doubling the present number of experimental points.

Acknowledgment

This work was supported by the Centre National de la Recherche Scientifique.

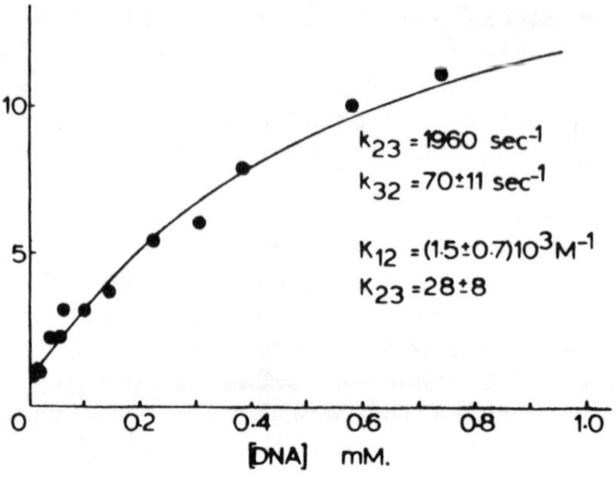

Fig. 1. Relaxation times of the slow kinetic phase observed by fluorescence detection of proflavin binding to DNA. The rate and equilibrium constants were evaluated with non-linear regression assuming the two-step model $A + B \underset{k_{21}}{\overset{k_{12}}{\rightleftharpoons}} AB \underset{k_{32}}{\overset{k_{23}}{\rightleftharpoons}} AB^*$, with $K_{12} = k_{12}/k_{21}$ and $K_{23} = k_{23}/k_{32}$. Conditions: 0.2 M $NaNO_3$, 0.01 M phosphate, pH 7.0, final temperature 13.5°C, excitation wavelength 436 nm.

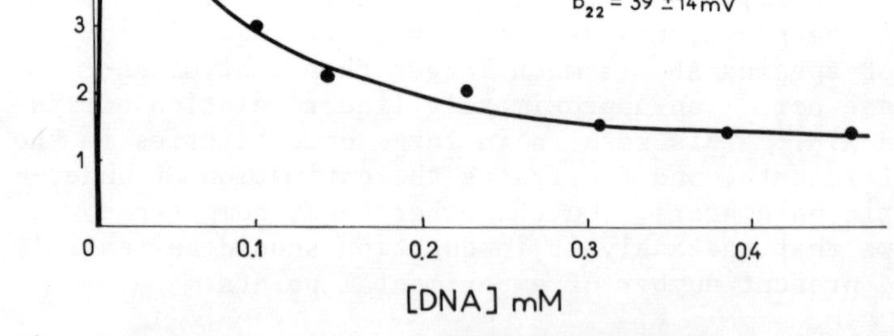

Fig. 2. Fluorescence relaxation amplitudes. The line was calculated with the coefficients b_{ij} estimated with linear regression.

REFERENCES

1. See, for example, T. M. Schuster and G. Ilgenfritz in "Probes of Structure and Function of Macromolecules and Membranes", Chance and Yonetani, eds, Vol II, pp367-383, Academic Press, New York (1971).
2. M. Eigen and L. De Maeyer in "Technique of Organic Chemistry", S. L. Friess, E. S. Lewis and A. Weissberger, Ed., Vol.VIII 2nd ed., Part 2, Wiley, New York, (1963).
3. G. H. Czerlinski, "Chemical Relaxation", Marcel Dekker, New York, N.Y., Chapt. 6. (1966).
4. P. Schimmel, J. Chem. Phys., 54, 4136 (1971).
5. R. White, Ph.D. Thesis, University of Washington.
6. D. Thusius, J. Am. Chem. Soc., 94, 356 (1972).
7. D. Thusius, G. Foucault and F. Guillain in "Dynamic Aspects of Conformational Changes in Biological Macromolecules", Sadron, ed., Reidel, Boston, p.271 (1973).
8. R. Winkler, Doctoral Dissertation, Göttingen, 1969.
9. D. Thusius, Biochimie, 55, 277 (1973).
10. H. J. Li and D. M. Crothers, J. Mol. Biol., 39, 461 (1969).
11. Q. Van Winkle and H. G. Heilweil, J. Phys. Chem. 59, 939 (1955).

NORMAL MODE ANALYSIS OF CHEMICAL REACTIONS

Claude F. Bernasconi

Thimann Laboratories, University of California,
Santa Cruz, California 95064

In a system involving several consecutive reactions which equilibrate at comparable rates there is a strong mutual coupling between the various reactions. In such a situation the relaxation times refer to the rate of equilibrations along the "normal modes" of the system rather than to the equilibration of individual reaction steps, in analogy to the normal frequencies of a vibrating molecule. Depending on the specific situation this coupling may affect the relaxation amplitudes in different ways, sometimes leading to a very small amplitude associated with one or more relaxation effects. As a consequence some relaxation effects may become invisible inducing the experimenter to erroneous conclusions about his system. An understanding of the normal modes of reaction and with it of the factors which influence relaxation amplitudes can help in recognizing such situations and in designing better experiments.

This is illustrated with the two-step system of the general form

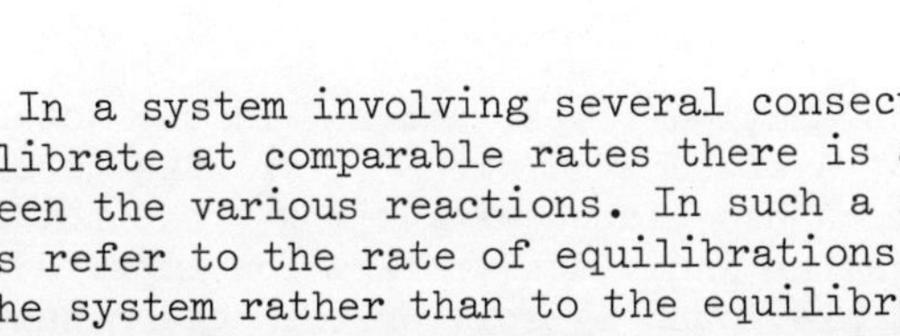

(1)

The linearized rate equations are given by

$$dx_1/dt + a_{11}x_1 + a_{12}x_2 = 0$$
$$dx_2/dt + a_{21}x_1 + a_{22}x_2 = 0 \qquad (2)$$

Wyn-Jones (ed.), Chemical and Biological Applications of Relaxation Spectrometry, 121–132.

where

$$
\begin{aligned}
x_1 &= \Delta A \qquad x_2 = \Delta B \quad (-x_1 - x_2 = \Delta M) \\
a_{11} &= k_1(\{\overline{M}\} + \{\overline{N}\}) + k_{-1} = k_1' + k_{-1} \\
a_{12} &= k_1' \\
a_{21} &= k_2(\{M\} + \{N\}) = k_2' \\
a_{22} &= k_2' + k_{-2}
\end{aligned}
\tag{3}
$$

The solution of the system of equation 2 is of the form

$$
\begin{aligned}
x_1 &= x_1^{o1}\exp(-t/\tau_1) + x_1^{o2}\exp(-t/\tau_2) \\
x_2 &= x_2^{o1}\exp(-t/\tau_1) + x_2^{o2}\exp(-t/\tau_2)
\end{aligned}
\tag{4}
$$

where τ_1 and τ_2 are the relaxation times, and x_1^{o1} is the total change in concentration of A associated with τ_1, x_2^{o1} the total change in concentration of A associated with τ_1, x_1^{o2} the total are usually referred to as the individual relaxation amplitudes.

The relaxation times are found by solving the determinantal equation 5

$$
\begin{vmatrix}
a_{11} - 1/\tau & a_{12} \\
a_{21} & a_{22} - 1/\tau
\end{vmatrix}
= 0
\tag{5}
$$

which leads to

$$
\frac{1}{\tau_{1,2}} = \frac{a_{11} + a_{22}}{2} \pm \sqrt{\left(\frac{a_{11} + a_{22}}{2}\right)^2 + a_{12}a_{21} - a_{11}a_{22}}
\tag{6}
$$

The problem of finding the relaxation amplitudes is more difficult and can be treated in different ways.[1] We use here the approach based on the "normal concentrations" which are directly related to the "normal modes of reaction".[2-6] Papers by Eigen[7], Hammes and Steinfeld[8], Kirschner et al.[9], Kustin and Liu[10] describe some other application of normal mode analysis of reactions. The terminology used here is the same as in the analysis of the vibrational normal modes of molecules because the phenomena are strictly comparable. Thus the two reactions of eq 1 correspond to the stretching vibration of a linear three-atomic molecule with the two normal modes:

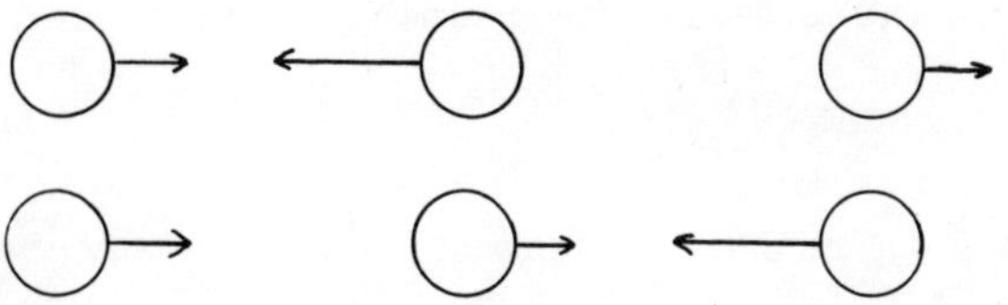

The reciprocal chemical relaxation times correspond to the frequencies of the two vibrational normal modes.

By definition the normal modes are orthogonal, i.e., independent of each other. In the vibrating molecule this feature manifests itself in that the various vibrational modes can be excited independently of each other (e.g. "IR and Raman active" modes in CO_2). The same is true for chemical reactions. In mathematical terms the orthogonality of the normal modes of reactions, or "normal reactions", is expressed as

$$\frac{dy_1}{dt} = -\frac{1}{\tau_1} y_1 \; ; \qquad y_1 = y_1^o \exp(-t/\tau_1) \tag{7}$$

$$\frac{dy_2}{dt} = -\frac{1}{\tau_2} y_2 \; ; \qquad y_2 = y_2^o \exp(-t/\tau_2) \tag{8}$$

where y_1 and y_2 are the normal concentration variables; they are linear combinations of the true concentration variables x_1 and x_2, and hence x_1 and x_2 in turn are linear combinations of y_1 and y_2, eq. 9 and 10.

$$\begin{aligned} y_1 &= m_{11}x_1 + m_{12}x_2 \\ y_2 &= m_{21}x_1 + m_{22}x_2 \end{aligned} \tag{9}$$

$$\begin{aligned} x_1 &= m_{-11}y_1 + m_{-12}y_2 \\ x_2 &= m_{-21}y_1 + m_{-22}y_2 \end{aligned} \tag{10}$$

It is now evident that combining eq. 10 and 8 with 4 leads to $x_1^{o1} = m_{-11}y_1^o$, $x_1^{o2} = m_{-12}y_2^o$, etc., i.e., the amplitudes can be found if we know the m_{-ij}'s and y_1^o, y_2^o. The procedure of finding these quantities is usually cast into the language of matrix algebra where y_1, y_2 and x_1, x_2 are regarded as the components of the vectors $\vec{y}$ and $\vec{x}$.[2-6] The m_{ij}'s and m_{-ij}'s of eq. 9 and 10 are then the elements of the matrices M and M^{-1} respectively

$$\vec{y} = M\vec{x}; \quad \vec{x} = M^{-1}\vec{y} \tag{11}$$

The derivation of M and M^{-1} is rather lengthy and has been treated elsewhere in considerable detail.[6] We just give the results here

$$\begin{aligned} y_1 &= x_1 - \frac{a_{12}}{a_{22} - 1/\tau_1} x_2 \\ y_2 &= - \frac{a_{21}}{a_{11} - 1/\tau_2} x_1 + x_2 \end{aligned} \tag{12}$$

$$\begin{aligned} x_1 &= \left(y_1 + \frac{a_{12}}{a_{22} - 1/\tau_1} y_2 \right) / \mathrm{detM} \\ x_2 &= \left(\frac{a_{21}}{a_{11} - 1/\tau_2} y_1 + y_2 \right) / \mathrm{detM} \end{aligned} \tag{13}$$

$$\mathrm{detM} = 1 - \frac{a_{12}a_{21}}{(a_{22} - 1/\tau_1)(a_{11} - 1/\tau_2)} \tag{14}$$

Note that y_1 and y_2 are not normalized[6] but this is without consequence in the current context.

It is apparent that for the general case the expressions of eq. 12 and 13 become extremely complicated. However in most real systems simplifications are possible because some special conditions prevail. The most common cases are: I. k_1', $k_{-1} \gg k_2'$, k_{-2}; II. k_1', $k_2' \gg k_{-1}$, k_{-2}; III. k_{-1}, $k_{-2} \gg k_1'$, k_2'. They have been discussed in considerable detail elsewhere.[6]

A fourth situation, viz $k_{-2} = k_{-1}$, is of particular interest in the context of normal reactions and is discussed now. A case where this situation is likely to prevail is the reaction of nucleophiles such as hydroxide or alkoxide ions with 1,3,5-trinitrobenzene; in a rapid first step a Meisenheimer complex (M) is formed which reacts further to form a cis (B) or trans (A) 1:2 complex.[11] Other examples have been discussed.[11]

After presenting the mathematical treatment we shall see that simple physical reasoning can provide the same results; at the same time the meaning of the normal reaction will become clear.

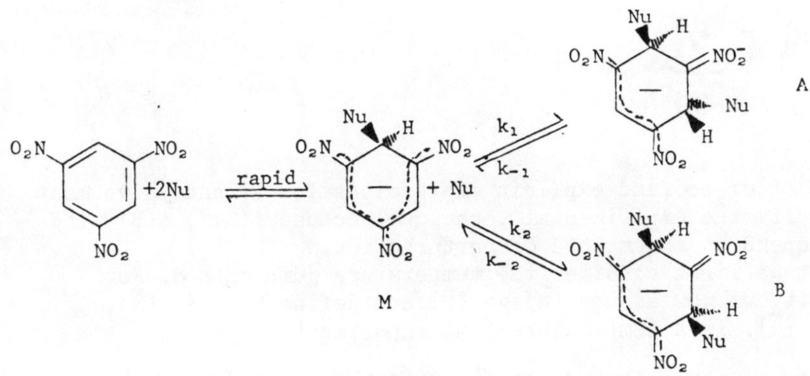

MATHEMATICAL TREATMENT OF CASE $k_{-2} = k_{-1}$

Substituting k_{-1} for k_{-2} in a_{22} (eq. 3) and solving eq. 6 leads to

$$\frac{1}{\tau_1} = k_1' + k_2' + k_{-1} \tag{15}$$

$$\frac{1}{\tau_2} = k_{-1} \tag{16}$$

whereas eq. 12 and 13, expressed for the special situation at t = 0 become ($\Delta\{A\}^o = x_1^o$, $\Delta\{B\}^o = x_2^o$, $\Delta\{M\}^o = -x_1^o - x_2^o$ are the total changes in the concentrations after complete equilibration of the system; $\Delta\{A\}^{o1} = x_1^{o1}$, $\Delta\{A\}^{o2} = x_1^{o2}$, etc. are the changes associated with τ_1 and τ_2 respectively)

$$y_1^o = \Delta\{A\}^o + \Delta\{B\}^o = -\Delta\{M\}^o \tag{17}$$

$$y_2^o = -\frac{k_2}{k_1}\Delta\{A\}^o + \Delta\{B\}^o \tag{18}$$

$$\Delta\{A\}^o = \underbrace{\frac{k_1}{k_1 + k_2} y_1^o}_{\Delta\{A\}^{o1}} - \underbrace{\frac{k_1}{k_1 + k_2} y_2^o}_{\Delta\{A\}^{o2}} \tag{19}$$

$$\Delta\{B\}^o = \underbrace{\frac{k_2}{k_1 + k_2} y_1^o}_{\Delta\{B\}^{o1}} + \underbrace{\frac{k_1}{k_1 + k_2} y_2^o}_{\Delta\{B\}^{o2}} \tag{20}$$

$$\Delta\{M\}^o = \underbrace{-y_1^o}_{\Delta\{M\}^{o1}} \tag{21}$$

In order to find explicit expressions for y_1^o and y_2^o we must now specify the experimental technique because $\Delta\{A\}^o$, $\Delta\{B\}^o$ and $\Delta\{M\}^o$ depend on the method of perturbation.

Let us first consider the temperature-jump method. For simplicity we now assume $\{\bar{N}\} >> \{\bar{M}\}$ and define $K_1' = K_1 \{\bar{N}\}$, $K_2' = K_2 \{\bar{N}\}$. In a temperature-jump experiment we have

$$\{\bar{A}\} + \Delta\{A\}^o = (K_1' + \Delta K_1')(\{\bar{M}\} + \Delta\{M\}^o) \tag{22}$$

where $\Delta K_1' = \Delta K_1 \{\bar{N}\}$ with ΔK_1 being the change in equilibrium constant brought about by the temperature-jump (K_1, $\{\bar{A}\}$ and $\{\bar{M}\}$ refer to the temperature after the jump). For small perturbations ($\Delta K_1/K_1 << 1$) this simplifies to

$$\Delta\{A\}^o = K_1'\Delta\{M\}^o + \Delta K_1'\{\bar{M}\} \tag{23}$$

Similarly eq. 24 holds; combining eq. 23 and 24 leads to

$$\Delta\{B\}^o = K_2'\Delta\{M\}^o + K_2'\{\bar{M}\} \tag{24}$$

eq. 25 through 27 ($C_o = \{A\} + \{B\} + \{M\}$)

$$\Delta\{A\}^o = K_1' \frac{(1 + K_2')\dfrac{\Delta K_1}{K_1} - K_2'\dfrac{\Delta K_2}{K_2}}{(1 + K_1' + K_2')^2} C_o \tag{25}$$

$$\Delta\{B\}^o = K_2' \frac{(1 + K_1')\dfrac{\Delta K_2}{K_2} - K_1'\dfrac{\Delta K_1}{K_1}}{(1 + K_1' + K_2')^2} C_o \tag{26}$$

$$\Delta\{M\}^o = - \frac{K_1'\dfrac{\Delta K_1}{K_1} + K_2'\dfrac{\Delta K_2}{K_2}}{(1 + K_1' + K_2')^2} C_o \tag{27}$$

with

$$\frac{\Delta K_1}{K_1} = \frac{\Delta H_1}{RT^2}\,\Delta T; \quad \frac{\Delta K_2}{K_2} = \frac{\Delta H_1}{RT^2}\,\Delta T \tag{28}$$

where ΔH_1 and ΔH_2 are the reaction enthalpies and ΔT the extent of the temperature-jump.

Combining eq. 25-27 with 17 through 21 finally affords the desired quantities, viz. the individual relaxation amplitudes

$$\Delta\{A\}^{o1} = \frac{k_1}{k_1 + k_2} \cdot \frac{K_1' \frac{\Delta K_1}{K_1} + K_2' \frac{\Delta K_2}{K_2}}{(1 + K_1' + K_2')^2}\, C_o \tag{29}$$

$$\Delta\{B\}^{o1} = \frac{k_2}{k_1}\,\Delta\{A\}^{o1} \tag{30}$$

$$\Delta\{M\}^{o1} = - \frac{K_1' \frac{\Delta K_1}{K_1} + K_2' \frac{\Delta K_2}{K_2}}{(1 + K_1' + K_2')^2}\, C_o \tag{31}$$

$$\Delta\{A\}^{o2} = \frac{K_1'K_2'\left(\frac{\Delta K_1}{K_1} - \frac{\Delta K_2}{K_2}\right)}{(K_1' + K_2')(1 + K_1' + K_2')}\, C_o \tag{32}$$

$$\Delta\{B\}^{o2} = -\Delta\{A\}^{o2} \tag{33}$$

$$\Delta\{M\}^{o2} = 0 \tag{34}$$

The most striking result is of course that $\Delta\{M\}^{o2} = 0$, which means that in a temperature-jump experiment the second relaxation effect induces no change in the concentration of M, i.e. τ_2 would remain undetected if only M is used to monitor chemical relaxation. Closer inspection of eq. 32 and 33 reveals that τ_2 may not induce a significant change in the concentration of A and B either. This is because the situation where $k_{-2} = k_{-1}$ is most likely to prevail when A and B are chemically very similar as the cis and trans isomers of the diadduct of 1,3,5-trinitrobenzene and thus $\Delta H_1 \approx \Delta H_2$, or $\Delta K_1/K_1 - \Delta K_2/K_2 \approx 0$. Hence τ_1 may remain the only detectable relaxation time.

The question arises whether τ_2 may be more easily detected if another perturbation method is used. A possibility would be a stopped-flow experiment, initiated by mixing M with N (with $\{N\} \gg \{M\}$ to assure first-order kinetics). Here $\Delta\{A\}^o$, $\Delta\{B\}^o$ and $\Delta\{M\}^o$ are given by

$$\Delta\{A\}^{o} = -\frac{K_1'}{1 + K_1' + K_2'} C_o \tag{35}$$

$$\Delta\{B\}^{o} = -\frac{K_2'}{1 + K_1' + K_2'} C_o \tag{36}$$

$$\Delta\{M\}^{o} = \frac{K_1' + K_2'}{1 + K_1' + K_2'} C_o \tag{37}$$

In combination with eq. 17 to 21 one obtains

$$\Delta\{A\}^{o1} = -\frac{K_1'}{1 + K_1' + K_2'} C_o \tag{38}$$

$$\Delta\{B\}^{o1} = \frac{k_2}{k_1} \Delta\{A\}^{o1} \tag{39}$$

$$\Delta\{M\}^{o1} = \frac{K_1' + K_2'}{1 + K_1' + K_2'} C_o \tag{40}$$

$$\Delta\{A\}^{o2} = \Delta\{B\}^{o2} = \Delta\{M\}^{o2} = y_2^{o} = 0 \tag{41}$$

It is obvious that with respect to detecting the second relaxation effect the situation is even worse than with the temperature-jump method; τ_2 induces no change in the concentration of any of the species.

A different kind of stopped-flow experiment can be visualized by the following scheme

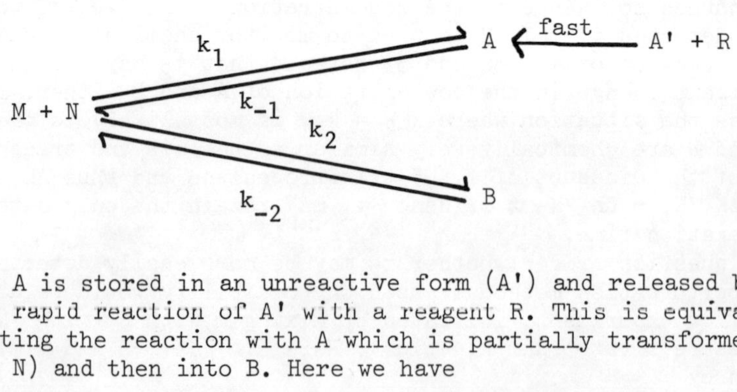

Here A is stored in an unreactive form (A') and released by the very rapid reaction of A' with a reagent R. This is equivalent to starting the reaction with A which is partially transformed into (M + N) and then into B. Here we have

$$\Delta\{A\}^{o} = \frac{1 + K_2'}{1 + K_1' + K_2'} C_o \tag{42}$$

$$\Delta\{B\}^{o} = - \frac{K_2'}{1 + K_1' + K_2'} C_o \tag{43}$$

$$\Delta\{M\}^{o} = - \frac{1}{1 + K_1' + K_2'} C_o \tag{44}$$

and for the individual changes

$$\Delta\{A\}^{o1} = \frac{k_1}{k_1 + k_2} \cdot \frac{1}{1 + K_1' + K_2'} C_o \tag{45}$$

$$\Delta\{B\}^{o1} = \frac{k_2}{k_1} \Delta\{A\}^{o1} \tag{46}$$

$$\Delta\{M\}^{o1} = - \frac{1}{1 + K_1' + K_2'} C_o \tag{47}$$

$$\Delta\{A\}^{o2} = \frac{K_2'}{K_1' + K_2'} C_o \tag{48}$$

$$\Delta\{B\}^{o2} = -\Delta\{A\}^{o2} \tag{49}$$

$$\Delta\{M\}^{o2} = 0 \tag{50}$$

Although $\Delta\{M\}^{o2} = 0$ we see now that $\Delta\{A\}^{o2}$ and $\Delta\{B\}^{o2}$ are quite large, in fact for typical situations (e.g. $K_1' = K_2' = 1$) they may be even larger than $\Delta\{A\}^{o1}$ and $\Delta\{B\}^{o1}$ and thus τ_2 is in principle easily detectable by monitoring A or B. Note however that if for example changes in light absorption are used to monitor the relaxation a wavelength where $\varepsilon_A \neq \varepsilon_B$ must be found. This is because the change in OD associated with τ_2 is given by

$$\Delta OD^{o2} = \ell(\varepsilon_A \, \Delta\{A\}^{o2} + \varepsilon_B \, \Delta\{B\}^{o2} + \varepsilon_B \, \Delta\{M\}^{o2}) \tag{51}$$

Since $\Delta\{M\}^{o2} = 0$ this simplifies to $\Delta OD^{o2} = \ell(\varepsilon_A - \varepsilon_B)\Delta\{A\}^{o2}$.

PHYSICAL INTERPRETATION

The easiest way of understanding the physical meaning of the above results is by deriving $1/\tau_1$, $1/\tau_2$, y_1 and y_2 (eq. 15 to 18) on the basis of physical reasoning and intuition. For this we recall that y_1 and y_2 must obey eq. 7 and 8 respectively.

In view of eq. 15 and 16 we suspect that the first normal reaction, associated with τ_1 and y_1, involves the reaction of M with N while the second normal reaction ($1/\tau_2$, y_2) does not. In fact the first normal mode corresponds to the equilibration between M + N and the sum of A and B, as represented by eq. 52.

$$M + N \underset{k_{-1}}{\overset{k_1 + k_2}{\rightleftharpoons}} (A, B) \tag{52}$$

Setting up the rate equation

$$d\Delta\{M\}/dt = -(k_1' + k_2')\Delta\{M\} + k_{-1}(\Delta\{A\} + \Delta\{B\}) \tag{53}$$

and substituting $-\Delta\{M\}$ for $(\Delta\{A\} + \Delta\{B\})$ immediately affords

$$d\Delta\{M\}/dt = -(k_1' + k_2' + k_{-1})\Delta\{M\} \tag{54}$$

which in fact is of the form of eq. 7 with $1/\tau_1$ and y_1 according to 15 and 17 respectively.

It is to be noted that no assumption was made about the ratio {A}:{B}, i.e. the process of eq. 52 is independent of whether A and B are or are not at equilibrium with each other. Hence we suspect the equilibration between A and B which in turn must be independent of reaction 52 corresponds to the second normal mode, as represented by eq. 55.

$$A \underset{\overleftarrow{k}}{\overset{\overrightarrow{k}}{\rightleftharpoons}} B \tag{55}$$

This can now be confirmed as follows. When equilibrium between A and B has been reached eq. 56 holds; {B} and {A} may

$$\{B\} - \frac{k_2}{k_1}\{A\} = 0 \tag{56}$$

or may not refer to the final equilibrium concentrations ($\{\bar{A}\}$, $\{\bar{B}\}$), depending on whether the first normal reaction has reached equilibrium or not. This can be expressed as

$$\{\bar{B}\} + \Delta\{B\} - \frac{k_2}{k_1}(\{\bar{A}\} + \Delta\{A\}) = 0 \qquad (57)$$

Combining eq. 56 (written for $\{A\} = \{\bar{A}\}$, $\{B\} = \{\bar{B}\}$) with eq. 57 provides

$$\Delta\{B\} - \frac{k_2}{k_1}\Delta\{A\} = 0 \qquad (58)$$

If equilibrium between A and B is <u>not</u> established, eq. 58 is <u>not</u> fulfilled; the difference $\Delta\{B\} - k_2\Delta\{A\}/k_1$ is thus a measure of how far the second normal reaction is from equilibrium. One now easily obtains

$$d(\Delta\{B\} - \frac{k_2}{k_1}\Delta\{A\})/dt = -k_{-1}(\Delta\{B\} - \frac{k_2}{k_1}\Delta\{A\}) \qquad (59)$$

which is of the form of eq. 8 with $1/\tau_2$ and y_2 according to eq. 16 and 18 respectively. (Note that since $1/\tau_2 = k_{-1} = \overrightarrow{k} + \overleftarrow{k}$, and $\overrightarrow{k}/\overleftarrow{k} = k_2/k_1$, it follows that $\overrightarrow{k} = k_{-1}k_2/(k_1 + k_2)$, $\overleftarrow{k} = k_{-1}k_1/(k_1 + k_2)$).

We now appreciate the meaning of the fact that in a stopped-flow experiment initiated by mixing M + N $y_2^o = 0$ and with it $\Delta\{A\}^{o2} = \Delta\{B\}^{o2} = \Delta\{M\}^{o2} = 0$. It is a consequence of A and B being formed directly in their final equilibrium ratio (eq. 58), so that equilibration along the second normal reaction becomes redundant. In the alternative stopped-flow experiment (releasing A by reacting A' with R; thus at $t = 0$ $\{A\} = C_o$, $\{B\} = 0$) A and B are very far from their final equilibrium ratio at the beginning, thus $y_2^o \neq 0$ and τ_2 is very much in evidence.

The small amplitude for τ_2 in a temperature-jump experiment is now also easily understood in physical terms. The equilibrium constant for the second normal reaction, eq. 55, is $K_{AB} = K_2/K_1$ and thus the normal enthalpy of reaction $\Delta H_{AB} = \Delta H_2 - \Delta H_1$. For similar ΔH_1 and ΔH_2 we obtain $\Delta H_{AB} \approx 0$ and thus the equilibrium displacement of reaction 55 induced by a temperature-jump is only very small.

To conclude let us now come back to the analogy with the vibrating linear three-atomic molecule. In the chemical system we always have $\Delta\{M\}^{o2} = 0$ regardless of the method of perturbation. This corresponds to a second vibrational mode (low frequency vibration) where the middle atom is at rest all the time, i.e.

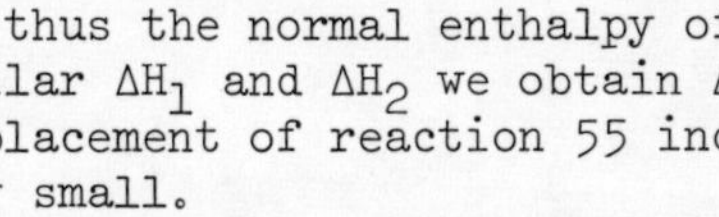

Second Normal Mode

This situation occurs for example in symmetrical molecules like CO_2 or CS_2.

REFERENCES

1. For an alternative treatment see e.g. (a) D. Thusius, J. Amer. Chem. Soc., 94, 356 (1972); (b) D. Thusius, this publication.
2. (a) M. Eigen and L. DeMaeyer in A. Weissberger, "Technique of Organic Chemistry", Vol. VIII, Part 2, Wiley-Interscience, New York, N.Y., 1963, p. 895; (b) M. Eigen and L. DeMaeyer in G. G. Hammes, "Technique of Chemistry", Vol. VI, Wiley-Interscience, New York, N.Y., 1974, p. 63.
3. K. Kustin, D. Shear and D. Kleitman, J. Theor. Biol, 9, 186, (1965).
4. G. Schwarz, Rev. Mod. Phys., 40, 206 (1968).
5. G. G. Hammes and P. R. Schimmel, Enzymes, 2, 67 (1970).
6. C. F. Bernasconi, "Relaxation Kinetics", Academic Press, New York, N.Y., in press.
7. M. Eigen in "Fifth Nobel Symposium", S. Claessen, Ed., Interscience, New York, N.Y., 1967, p. 333.
8. G. G. Hammes and J. I. Steinfeld, J. Amer. Chem. Soc., 84 4639 (1962).
9. K. Kirschner, E. Gallego, I. Schuster and D. Goodall, J. Mol. Biol., 58, 29 (1971).
10. K. Kustin and S. T. Liu, J. Chem. Soc., Dalton Trans., 278 (1973).
11. C. F. Bernasconi and R. G. Bergstrom, J. Amer. Chem. Soc., 96, 2397 (1974).

BRIEF REVIEW OF THE CHEMICAL RELAXATION STUDIES OF MICELLAR EQUILIBRIA

R. Zana

C.R.M., C.N.R.S., 6 rue Boussingault,
67083 Strasbourg Cedex - France

INTRODUCTION

After the pioneering stopped-flow study of Jaycock and Ottewill[1], P-Jump[2-7], T-Jump[7-11], stopped flow[12-14] and ultrasonic absorption[15-27] have been widely used to investigate the kinetics of chemical equilibria in micellar solutions of detergents. Some thirty papers have been published in the past fifteen years with seven of them in 1973[4,5,7,13,17,26,29] and seven more since the beginning of 1974[6,14,25,28,30-32] thereby indicating a growing interest on this subject among workers. Mention should also be made of the studies where the kinetics of micellar equilibria is approached by NMR[29,30,33-36], ESR[37-41] and fluorescence decay[42-46].

The examination of the above papers shows that until 1973, T-Jump and P-Jump experiments consistently revealed the existence of a single relaxation process with a relaxation time in the range 100 μsec-10 msec for fairly dilute micellar solutions. On the other hand ultrasonic absorption indicated the existence of a relaxation process with a relaxation time well below 1 μsec. This method also revealed an additional relaxation process occurring only in micellar solutions at concentration close the second c.m.c. (critical micelle concentration). Very few points of agreement have been reached among workers concerning the origin and characteristics of these relaxation processes. For this reason, it appeared worthwhile at the beginning of this two-day session of the Summer School on the kinetics of micellar processes to present a brief review of:

1) the different types of chemical equilibria to which have been assigned the relaxation processes observed by means of T-Jump, P-Jump, shock tube, stopped flow and ultrasonic absorption, and
2) the various points of disagreement among workers.

Wyn-Jones (ed.), Chemical and Biological Applications of Relaxation Spectrometry, 133–138.

CHEMICAL PROCESSES IN MICELLAR SOLUTIONS

Let us consider a micellar solution of an ionic detergent A, with counterions C. For the sake of simplicity we assume both A and C to be monovalent. Let n and z be the aggregation number and electrical charge of the most stable micelle denoted by A_n^z. The relaxation processes observed with micellar solutions have been assigned to the following equilibria.

1. Counterion association/dissociation to/from micelles (ionization) :

$$A_n^z \rightleftharpoons A_n^{z+1} + C \qquad (1)$$

2. Detergent ion association/dissociation to/from micelles (exchange process)

$$A_n^z \rightleftharpoons A_{n-1}^{z-1} + A \qquad (2)$$

3. Micelle formation/dissolution

$$nA \rightleftharpoons A_n \qquad (3)$$

4. Change of micellar shape and/or size. At the second cmc it is usually assumed that "spherical" micelles become "cylindrical"[47,48]

a - Change of shape at constant aggregation number

$$A_n \rightleftharpoons A'_n \qquad (4a)$$

b - Change of shape and of size

$$A_n + mA \rightleftharpoons A_{n+m} \qquad (4b)$$

Remarks a) For the sake of simplicity equilibria (1) and (2) have been written with only one counterion and detergent ion associatin (or dissociating) to (or from) micelles. These processes may however involve more than one counterion or detergent ion.

b) Equilibria (1) to (4) are chemically coupled. Indeed the dissociation of a counterion may induce that of a detergent ion and vice-versa. On the other hand equilibria (3) and (4b) do not occur in one step but by a series of bimolecular equilibria such as (1) and (2).

POINTS OF DISAGREEMENT AMONG WORKERS

1) The dissociation rate constants determined by means of different methods are very different.

NMR measurements indicate that the exchange between micellize and free detergent molecules is fast on the NMR time scale[33-36].

This result yields a value above 10^4 - 10^5 sec^{-1} for the dissociation rate constant of one detergent molecule with 12 carbon atoms from a micelle[29]. This conclusion agrees with the results of ultrasonic absorption[21,24,25,27] and EPR[37] studies but contradicts those obtained in T-Jump and P-Jump studies which yield rate constants of $10^2 sec^{-1}$ [3-11].

2) In several instances, different results have been reported in studies performed with the same method on a given detergent, by different workers.

(a) - Number of relaxation processes characterizing detergent solutions. In a T-Jump study, Kresheck et al[8] detected only one relaxation process due to micellar equilibria in solutions of dodecylpyridinium iodide (DPI). A similar observation was reported by other workers in T-Jump and P-Jump studies of a number of detergents[3-5,7,9-11]. In contrast, two processes were detected in more recent T-Jump, P-Jump and shock tube studies of micellar solutions of DPI, DPBr and sodium dodecylsulfate (NaDS)[26,28]. The relaxation times τ_1 and τ_2 characterizing the faster and the slower of these two processes, respectively, are in a ratio of 10^{-2} to 10^{-3}. If these last results prove to be correct they would provide a simple explanation for the disagreement among the results of various workers, using different techniques, emphasized in the preceding paragraph.

On the other hand two relaxations processes characterized by the relaxation times τ'_1 and τ_o have been observed by means of ultrasonic absorption for alkali metal salts of alkylsulfates and alkylcarboxylates[20]. Other workers, however, detected only one process, in similar experiments with the same detergents and the same technique[22,23]. Note that the process characterized by τ_o appears only in concentrated micellar solutions at concentration close to the second cmc. On the other hand the process characterized by τ'_1 appears to be identical to the fast process detected by T-Jump and shock tube and characterized above by τ_1 as recently shown with DPI, DPBr, and NaDS[6,28].

(b) - Concentration dependence of τ_o, τ_1 and τ_2. τ_o^{-1} has been found to either increase with the detergent concentration C[20] or be independent of C[26].

τ^{-1} has been found to increase linearly with C at C > cmc[20,24,25] and to show a departure from linearity at C close to the second cmc[20,25]. Other workers[27] also reported the existence of a minimum on the curve τ_1^{-1} vs C at C close to the cmc. Note however that the ultrasonic frequency range investigated was too narrow to allow an accurate determination of the value τ_1^{-1}

Finally, τ_2^{-1} has been found to either increase linearly with C[3,7-11], or show a curvature toward the C-axis[4,5]; or show practically no variation with C[6]. Note that recent works[6,28] have shown that a change from a linear variation to practically no variation of τ_2^{-1} with C can be obtained by carefully purifying the detergent.

(c) - Ionic strength dependence of τ_1 and τ_2. Kresheck et al[8] reported no effect of the ionic strength on τ_2 for LPI solutions between 0.1 and 0.5 M NaCl but more recent works[6,28] showed a very strong dependence on this parameter below 0.1 M added salt. τ_1 does not appear to depend on ionic strength[25]

3. Contradictory and/or conflicting assignments have been made for the observed relaxation processes.

The process observed by means of ultrasonic absorption for micellar solutions at concentration close to the second cmc and characterized by the relaxation time τ_o has been attributed to either reaction (4a)[26] or (4b)[20].

The process characterized by the relaxation time τ_1 has been assigned by some authors to the ionization equilibrium (1)[22,23] and by other workers to the exchange reaction (2)[20,21,24-27].

The process characterized by the relaxation time τ_2 has been assigned to either the exchange reaction[3,7-11,13,14], or the micelle formation-dissolution equilibrium (3)[6,28-31] or reactions between micelles of intermediate size[32].

4. Conflicting models have been proposed for micellar equilibria.

For the sake of simplicity we shall restrict ourselves to dilute micellar solutions, ie, when the process characterized by τ_o is absent. One group of workers attributes τ_1 to the exchange process and τ_2 to the micellization-dissolution equilibrium (reaction (3))[6,28,29,31]. A somewhat similar model is also presented by Nakagawa[30].

In a recent paper Colen[32] assigns the slow process to reactions between micelles of intermediate size, and the fast process to the exchange reaction.

Finally, Kresheck et al[8] assign the slow process to the dissociation of one detergent molecule from the most stable micelle and assume that all other bimolecular reactions (2) to be very fast. This model was later modified by Graber and Zana[21] to take into account the polydispersity of micelles.

It is hoped that the results which will be presented during this session on micelles and the discussions between participants will help clearing up some of the above controversies, and that agreement will be reached on such points as:

- the number of relaxation times characterizing micellar solutions.
- their dependence on parameters such as the detergent concentration, the ionic strength, the paraffinic chain length, etc. ....
- the qualitative assignment of the various relaxation times.

REFERENCES

1. M.J. Jaycock and R.H. Ottewill, Proc. Fourth Int. Cong. Surface Active Substances, Brussels 1964, Section B, paper 8.
2. F.J. Minjlieff and R. Dietmarsch, Nature, 208, 889 (1965).
3. K. Takeda and T. Yasunaga, J. Coll. Interface Sci., 40,
4. T. Janjic and H. Hoffmann, Zeit. Physikal. Chem. N.F., 86, 322, (1973).
5. U. Herrmann and M. Kahlweit, Ber. Bunsenges. Physik. Chem., 77, 1129 (1973).
6. R. Folger, H. Hoffmann and W. Ulricht, Ber. Bunsenges. Physik. Chem., 1974, in press.
7. K. Takeda and T. Yasunaga, J. Coll. Interface Sci., 45, 406 (1973).
8. G. Kresheck, E. Hamori, G. Davenport and H. Scheraga, J. Amer. Chem. Soc., 88, 246 (1966)
9. B. Bennion, L. Tong, L. Holmes and E. Erying, J. Phys. Chem., 73, 3288 (1969).
10. B. Bennion and E. Eyring, J. Coll. Interface Sci., 32, 286 (1970).
11. J. Lang and E. Eyring, J. Polymer Sci., A-2, 10,89 (1972).
12. J. Lang, M. Auborn and E. Eyring, J. Coll. Interface Sci., 41, 484 (1972).
13. T. Yasunaga, K. Takeda and S. Harada, J. Coll. Interface Sci., 42, 457 (1973).
14. K. Takeda, N. Tatsumoto and T. Yasunaga, J. Coll. Interface Sci., 47, 128 (1974).
15. J. McKellar and J. Andreae, Nature, 195, 865 (1962).
16. P. White, D. Moule and G. Benson, Trans. Far. Soc., 54, 1638 (1958).
17. S. Nishikawa and T. Yasunaga, Bull. Chem. Soc. Jap., 46, 1098, (1973).
18. S. Nishikawa, T. Yasunaga and K. Takahashi, Bull. Chem. Soc. Jap., 46, 2992 (1972).
19. R. Zana and J. Lang, C.R. Acad. Sci. (Paris), C 266, 893 (1968) and ibid., C 266, 1347 (1968).
20. E. Graber, J. Lang and R. Zana, Koll. Z.u.Z.Polym., 238, 470 (1970).
21. E. Graber and R. Zana, Koll. Z.u. Z.Polym., 238, 479 (1970).
22. T. Yasunaga, H. Oguri and M. Miura, J. Coll. Inter. Sci., 23, 352 (1967).
23. T. Yasunaga, S. Fujii and M. Miura, J. Coll. Inter. Sci., 30, 399 (1969).
24. P. Sams, E. Wyn-Jones and J. Rassing, Chem. Phys. Lett., 13, 233 (1972).
25. J. Rassing, P. Sams and E. Wyn-Jones, J.C.S. Farad. Trans.II, 69, 180 (1973) and ibid., 70, 1247 (1974).
26. J. Rassing and E. Wyn-Jones, Chem. Phys. Lett., 21, 93 (1973).

27. a) D. Adair, V. Reinsborough, N. Plavac and J. Valleau, Can. J. Chem., 52, 429 (1974).
b) this volume
28. J. Land, C. Tondre, R. Zana, R. Bauer, H. Hoffmann and W. Ulbricht, J. Phys. Chem., 1974, in press.
29. N. Muller, J. Phys. Chem., 76, 3017 (1972) and "Recent Advances in the Chemistry of Micelles" in Reaction Kinetics in Micelles, E. Cordes Ed., Plenum Press, New York 1973, p.1.
30. T. Nakagawa, Colloid Polym. Sci., 252, 56 (1974).
31. E. Aniansson and S. N. Wall, J. Phys. Chem., 78, 1024 (1974).
32. A. Colen, J. Phys. Chem., 78, 1676 (1974).
33. T. Nakagawa and H. Inoue, Proc. Fourth Int. Cong. Active Substances, Brussels 1964, Section B, paper 11.; H. Inoue and T. Nakagawa, J. Phys. Chem., 70, 1108 (1966); T. Nakagawa, H. Inoue, H. Jizomoto and K. Horiuchi, Koll. Z.u.Z.Polym., 229, 159 (1969); T. Nakagawa and K. Tori, ibid., 194, 143 (1964).
34. N. Muller and R. Birkham, J. Phys. Chem., 71, 957 (1971); N. Muller and T. Johnson, ibid, 73, 2042 (1969); N. Muller and F. Platko, ibid, 75, 547 (1971).
35. J. Eriksson and G. Gillberg, Acta Chim. Scand., 20, 2019 (1966)
36. M. Alexandre, C. Fouchet and A. Rigny, J. Chim. Phys., 70 1073 (1973).
37. K. K. Fox, Trans. Farad. Soc., 67, 2802 (1971).
38. N. Atherton and S. Strach, J.C.S. Farad. Trans.II, 68 374, (1972).
39. J. Oakes, J.C.S. Farad. Trans. II, 68, 1464 (1972).
40. T. Nakagawa and H. Jizimoto, Koll. Z.u.Z.Polym., 250,594(1972)
41. J. Brotherus and P. Tormala, Koll. Z.u.Z.Polym., 251,774(1972)
42. R. Dorrance and T. Hunter, J.C.S. Farad. Trans I, 68, 1312(197
43. R. Hautala, N. Schore and N. Turro, J. Amer. Chem. Soc., 95, 5508 (1973).
44. M. Gratzel and J. Thomas, J. Amer. Chem. Soc., 95, 6885 (1973)
45. N. Schore and J. Turro, J. Amer. Chem. Soc., 96, 306 (1974).
46. P. Infelta, M. Gratzel and J. Thomas, J. Phys. Chem., 78, 190, (1974).
47. F. Reiss-Husson and V. Luzzati, J. Phys. Chem., 68, 3504 (1964
48. P. Ekwall and P. Holmberg, Acta Chim. Scand., 19, 445 (1965).

CHEMICAL RELAXATION STUDIES OF MICELLAR EQUILIBRIA

R. Zana

CNRS, Centre De Recherches Sur Les Macromolecules,
6, rue Boussingault, Strasbourg Cedex 67083 - FRANCE

I - ULTRASONIC STUDIES[1-3]

The detergents investigated were alkali metal salts of alkylcarboxylates with 6, 7, 8, 10 and 12 carbon atoms.

At concentration C below the critical micelle concentration (c.m.c.) the excess absorption is close to zero, but rises sharply at C>c.m.c. and goes successively through a maximum and a minimum which are shifted to higher concentration as the ultrasonic frequency f is increased.

The curves excess absorption vs f are characterized by a single relaxation time τ_1, whose reciprocal increases linearly with C. At higher concentration, close to the second c.m.c. a second relaxation process is observed, whose (relaxation time)$^{-1}$ also increases with concentration[2]. This process which is very likely associated in some manner to changes of micellar shape[4,5] will not be discussed any further.

The relaxation process detected in dilute micellar solutions has been assigned to the exchange of detergent ions between the micelles and the surrounding solution, and not to the association/dissociation of counterions (ionization) to/from micelles as was done by other workers[6-8] for the following reasons:

(1) the ionization equilibrium should be characterized by a non linear dependence of τ_1^{-1} on concentration[6-8], in disagreement with the experimental results[3].

(2) the excess absorption associated with ionization process should be mainly dependent on the nature of the counterion but not on the chain length of the detergent ion. In fact, the experimental results show an opposite behaviour[3]. Also for a given detergent ion the plots relaxation frequency vs C extrapolate to the same point at C = 0 for Na^+, K^+ and Cs^+ counterions[3]. Excess absorption have also been found in micellar solutions of

Wyn-Jones (ed.), Chemical and Biological Applications of Relaxation Spectrometry, 139–141.

tetraalkylammonium salts of decanoic acid, although the dissociation of such large ions should only give rise to a very small volume change, and in turn to a negligible excess absorption, owing to the very small electrostriction of these ions[9].

(3) one would expect very different results for two ionic detergents of the same chain length and c.m.c. and of opposite electrical charge, in disagreement with the results for octylammonium chloride and potassium nonanoate, for which practically coincident τ_1^{-1} vs C curves were obtained[9].

(4) the ionization equilibrium assignment leads to a value of 15 cm^3/mol. of sodium ion dissociating from a micelle[6]. Such a value is much larger than the one which can be reasonably expected for such a process, on the basis of the partial molal volume[10] of Na^+ and the contribution of electrostriction to this quantity. On the other hand, Mukerjee[11] and Stigter[12] have shown that the binding of counterions to micelles should give rise to volume changes of less than 2 cm^3/mol., *i.e.* an order of magnitude smaller than the value given above[6].

(5) on the assumption that the excess absorption of micellar solutions is due to the counterion ionization equilibrium, one would expect this absorption to be much smaller than that due to binding of counterions by polyions in solutions of polyelectrolytes Indeed, extensive studies have shown that the larger the polyion charge density, the larger the excess absorption due to counterion binding[13]. On the other hand polyions are usually more densely charged than micelles. However the excess absorption found for micellar solutions is one to two orders of magnitude larger than for polysalt solutions, and thus cannot be due to counterion ionization.

A model has been presented for micellar exchange, derived from that postulated by Kresheck *et al*[14], and including the width of the micellar distribution. The excess absorption was assigned to an exchange of 2m detergent ions between two stable micelles. The calculated values of the volume change for the transfer of one detergent ion from the micelle to the solution were found to be in agreement with the experimental ones when assuming that this exchange was much faster than the equilibrium between stable micelles and monomers, although out model did not explicitely state this assumption[3].

II - T-JUMP, P-JUMP and SHOCK TUBE STUDIES[15]

These studies, performed in collaboration with the group of Dr. H. Hoffmann (University of Erlangen, W.Germany), revealed the presence of two relaxation times in the ratio 10^2 to 10^3 for several ionic detergents: sodium lauryl sulfate, dodecylpyridinium iodide and chloride[15].

The fast relaxation process appears to be identical with that detected by means of ultrasonic absorption.[15,16]

The slow relaxation process is characterized by a relaxation time τ_2, highly sensitive to ionic strength, impurity content of the detergent, temperature, and chain length of the detergent ion. Its concentration dependence is very complex: depending on the nature of the soap and the ionic strength, τ_2^{-1} may increase, decrease go through a maximum and/or a minimum in the concentration range investigated[15]. This process has been attributed to the micellization-dissolution equilibrium, in agreement with the models recently postulated by several workers.[17-19]

REFERENCES

1. R. ZANA and J. LANG, C.R. Acad. Sci. (Paris), C 266, 893 (1968); ibid., C 266, 1347 (1968).
2. E. GRABER, J. LANG and R. ZANA, Koll. Z. u. Z. Polym., 238, 470 (1970).
3. E. GRABER and R. ZANA, ibid., 238, 479 (1970).
4. F. REISS-HUSSON and V. LUZZATI, J. Phys. Chem., 68, 3504, (1964).
5. P. EKWALL and P. HOLMBERG, Acta Chim. Scand., 19, 455, (1965).
6. T. YASUNAGA, H. OGURI and M. MIURA, J. Coll. Interface Sci., 23, 352 (1967).
7. T. YASUNAGA, S. FUJII and M. MIURA, ibid., 30, 399 (1969).
8. LEE KUN MOO, Daehan Hwahak Hwoejee, 17, 72 (1973) in Chem. Abst., 79 4798 (1973).
9. S. YIV, J. LANG and R. ZANA, unpublished results.
10. R. ZANA and E. YEAGER, J. Phys. Chem., 71, 521 and 4241 (1967).
11. P. MUKERJEE, ibid., 66, 943 (1962).
12. D STIGER, ibid., 68, 3603 (1964).
13. R. ZANA and C. TONDRE, this volume.
14. G. KRESHECK, E. HAMORI, G. DAVENPORT and H. SCHERAGA, J. Amer, Chem. Soc., 88, 246 (1966).
15. J. LANG, C. TONDRE, R. ZANA, R. BAUER, H. HOFFMANN, and W. ULBRICHT, J. Phys. Chem., in the press.
16. H. HOFFMANN, this volume.
17. N. MULLER, J. Phys. Chem., 76, 3017 (1972).
18. T. NAKAGAWA, Colloid Polym. Sci., 252, 56 (1974).
19. E. ANIANSSON and S. N. WALL, J. Phys. Chem., 78, 1024 (1974).

KINETIC STUDIES OF THE RAPID REACTIONS IN n-ALKYL SULFATE SOLUTION

Tatsuya Yasunaga, Kunio Takeda, Nobuhide Tatsumoto and Hiromoto Uehara

Department of Chemistry, University of Hiroshima, Japan

The kinetics of micellization have been studied using the various relaxation techniques on several detergent micelles[1-16]. Synthetically speaking, the relaxation times obtained by means of the ultrasonic absorption techniques[1-7] are above two order of magnitude faster than those by means of the temperature-jump or the pressure-jump techniques[8-15]. In the latter case, all of the published papers containing ours[12,13] have reported that the reciprocal relaxation times increase with the total detergent concentration in the concentration range larger than the critical micellar concentration (c.m.c.) except two papers[14,15], which also indicate the apparent concentration dependences of the relaxation times in the lower concentration range. More recently, however, we have obtained the relaxation times independent of the total detergent concentration on sodium dodecyl sulfate (SDS) by means of the pressure-jump technique.

On the other hand, the kinetics on the solubilization of pinacyanol chloride into SDS micelle have been studied by means of the stopped-flow technique[17].

1. MICELLIZATION KINETICS

The SDS was prepared almost according to the method of Dreger et al.[18]. The purity of dodecyl alcohol was above 99% by means of gas chromatography. The c.m.c. was determined to be 8.3 mM at 25°C by the electric conductivity method. The materials below were used as additives. Sodium decyl sulfate (SDeS) was prepared, as same manner as SDS, from decyl alcohol, the purity of which was above 99.8%. The purities of ethyl, butyl, octyl, decyl and dodecyl alcohol are above 99.5, 99.9, 97, 99.8 and 99% respectively. The

Wyn-Jones (ed.), Chemical and Biological Applications of Relaxation Spectrometry, 143–157.

purities of sodium tetradecyl sulfate (STS) and hexyl alcohol are not known.

The pressure-jump apparatus used is as follows. Increasing pressure is applied to the sample cell arrangement until a brass diaphragm bursts at a pressure of 60 atm. Although it bursts at 30 atm in the previous case[12,13], the bursting at 60 atm makes possible to enlarge the measurable concentration range to some extent. After the sudden decrease of pressure, the concentration changes of the species in solution were followed by means of the electric conductivity method.

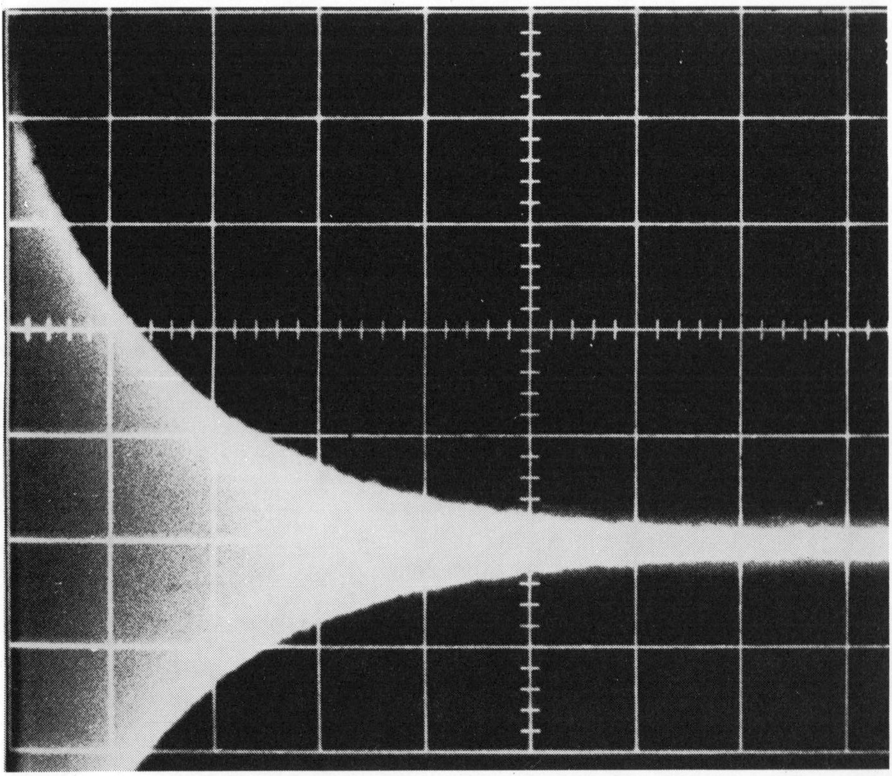

Fig. 1. Typical relaxation effect. Total SDS concentration, C_o, 1.0×10^{-2}M; Sweep, 2 msec/division; Temperature, 15°C.

The relaxation effects were observed only above the c.m.c. in the aqueous SDS solution. A typical one is shown in Fig. 1. The reciprocal relaxation times obtained are plotted against the total SDS concentration at several temperatures in Fig. 2. The vertical lines were drawn through the experimental points. As shown in this figure, the relaxation times are almost independent of the

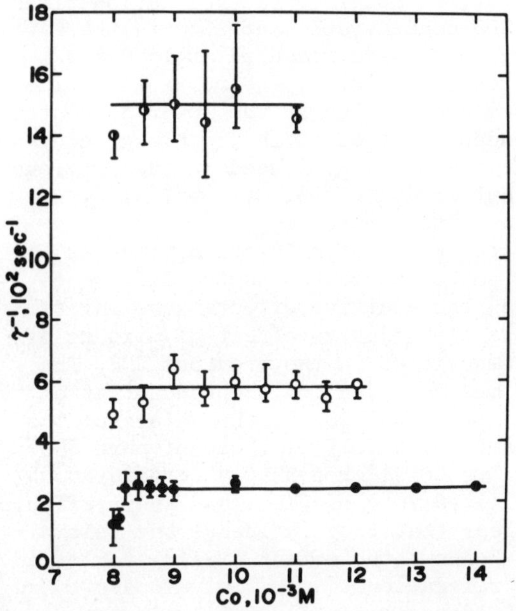

Fig. 2. Plot of reciprocal relaxation times, τ^{-1} vs. total SDS concentration, C_o. ●, 15°C; o, 20°C; ◑, 25°C.

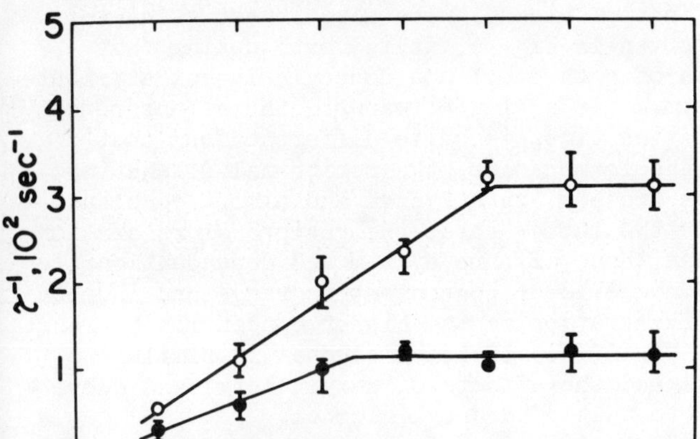

Fig. 3. Plot of reciprocal relaxation times, τ^{-1} vs. total SDS concentration, C_o. ●, 15°C; o, 20°C.

concentration.

On the other hand, however, we obtained the relaxation times which were dependent on the SDS concentration at 20°C in the previous work[12]. In addition, those were measured at 15°C and were shown with the previous ones in Fig. 3. The relaxation times at 15°C are dependent on the concentration in the lower concentration range but are clearly independent of it in the higher concentration range. The micro-analysis of the SDS used in the previous work, gives the following results: C_{10}, 0.9%; C_{12}, 96.8%; C_{14}, 2.0%; water, 0.56%; inorganic salts, trace.

Then, the appearance of two kinds of relaxation times as seen in Figs. 2 and 3, seems to be attributed to the difference of purities of samples. Hence, the additive effects were investigated as follows. Fig. 4 shows the additive effect of mixture of SDeS and STS which are main impurities in the previous SDS. The tendency of the relaxation times dependent on the concentration is rather similar to that of the previous SDS and the values of the reciprocal relaxation times are below half of those of pure SDS system. In the previous step, we considered that a certain amount of impurities would be solubilized into micelles and their effects were negligible, but it is clear that they influence the relaxation times unexpectedly. Fig. 5 shows the effect of addition of a slight amount of SDeS on the relaxation times. The values of reciprocal relaxation times increase gradually until they approximately become to those of pure system and flatten out at a certain concentration. This means that the quantity of additive against that of micelle decreases and above a certain concentration, the quantity of micelle not solubilizing the additive overcomes that of one solubilizing it.

The additive effect of alcohol were also investigated and typical results are given in Fig. 6. It is worth noting that although the addition of both ethyl and dodecyl alcohol attribute to the decrease of c.m.c. in a similar manner, the effect of ethyl alcohol on the relaxation times is quite different from that of dodecyl alcohol. In the former case, the reciprocal relaxation times are larger than those of pure system and almost constant. On the other hand, in the latter case, the reciprocal relaxation times are smaller than those of pure system and dependent on the total concentration in the lower concentration range and independent of the total concentration in the higher concentration range as the case of SDeS. The effect of butyl alcohol is similar to that of ethyl alcohol and the effects of hexyl, octyl and decyl alcohols are similar to that of dodecyl alcohol.

There is the concentration dependence of relaxation times in the narrow concentration range around the c.m.c. in pure SDS system as seen in Fig. 2. This is also considered to be due to the trace amount of impurities in SDS prepared this time. Accordingly, we may conclude that whether the relaxation times are dependent on the total detergent concentration is due to the presence of impurities and then in the completely pure SDS, the

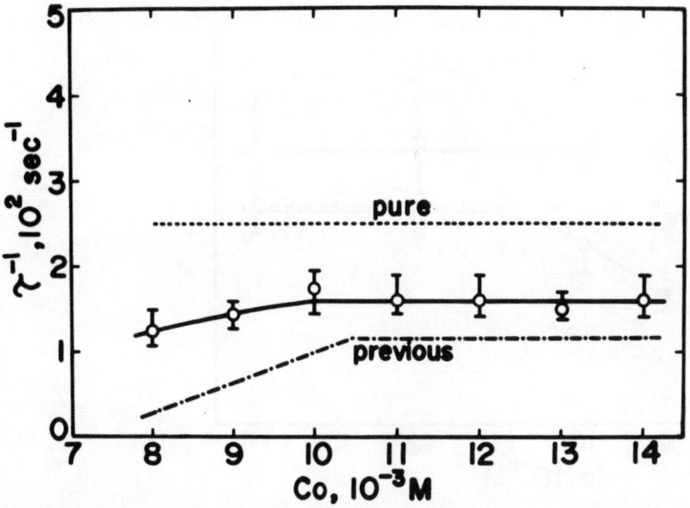

Fig. 4. Plot of reciprocal relaxation times, τ^{-1} vs. total SDS concentration, C_o under the addition of mixture of SDeS (0.9 Wt. %/SDS) and STS (2.0 Wt. %/SDS) at 15°C.

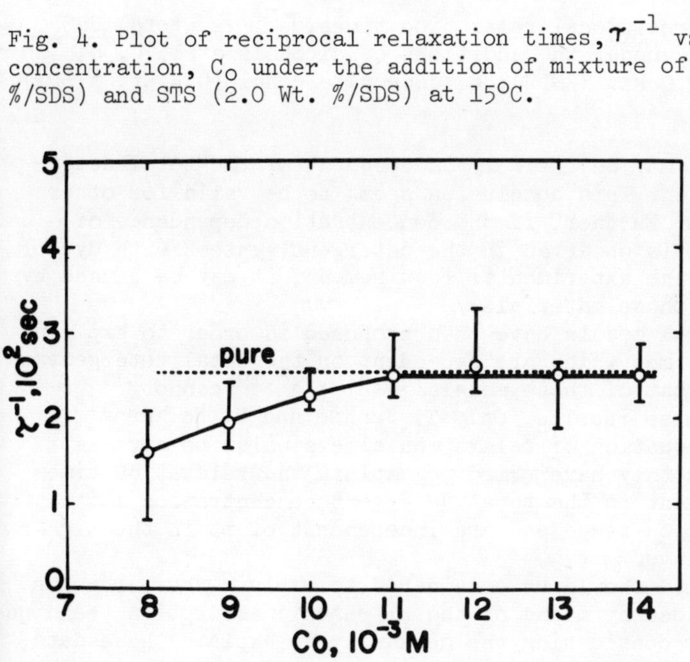

Fig. 5. Plot of reciprocal relaxation times, τ^{-1} vs. total SDS concentration, C_o under the addition of SDeS (0.4 Wt. %/SDS) at 15°C.

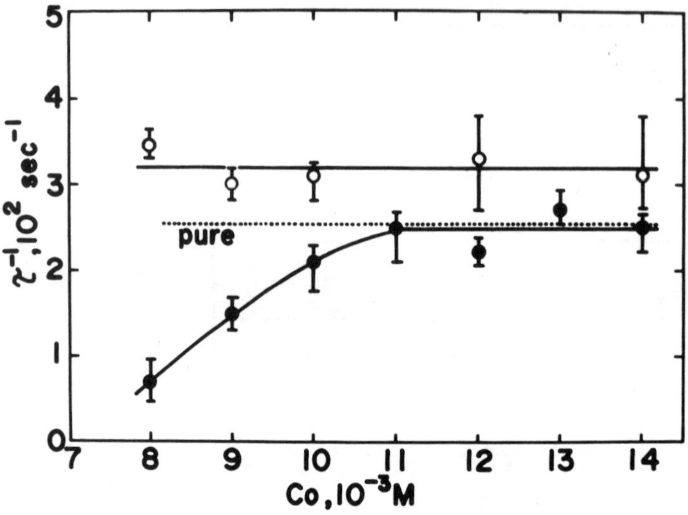

Fig. 6. Plot of reciprocal relaxation times, τ^{-1} vs. total SDS concentration, C_o under the additions of alcohols at 15°C. o, ethyl alcohol (2.0 Wt. %/SDS); •, dodecyl alcohol (0.4 Wt. %/SDS).

relaxation times may be entirely independent of the total detergent concentration. This conclusion seems to be valid for other detergent system. Further, if the concentration dependence of relaxation times is observed in the detergent systems with dye or electrolyte for the experimental requirement, it may be caused by the addition of these materials.

While several models have been proposed in order to explain the relaxation times which are dependent on the total detergent concentration, most of those models[1,3,4,5,7,8,19] cannot be applied to these results. Only T. Janjic and H. Hoffmann[15] have given the equation of relaxation time similar to ours as stated below and they have tried to explain the relaxation times which are dependent on the total detergent concentration in the lower concentration range and are independent of it in the higher concentration range.

The mechanism should be considered to explain these results together with those by means of the ultrasonic absorption technique While we are now considering the mechanism to explain these data, the one we propose is model (1), where A is a detergent monomer n and m are arbitrary number, A_n and A_{n+m} are micelles and k_{12}, k_{21}, k_{23} and k_{32} are overall rate constants. The presumption that the distribution of micelle-size is symmetric, leads to the

$$nA \underset{k_{21}}{\overset{k_{12}}{\rightleftharpoons}} A_n \underset{k_{32}}{\overset{k_{23},\ +mA}{\rightleftharpoons}} A_{n+m} \qquad (1)$$

$$\text{fast } (\tau_{12}) \qquad \text{slow } (\tau_{23})$$

assumption that

$$\{A_n\}_e \simeq \{A_{n+m}\}_e \qquad (2)$$

where the subscript e designates equilibrium concentrations. The rate equations are given for two step overall reactions as follows,

$$\frac{d\{A_n\}}{dt} = k_{12}\{A\}^n - k_{21}\{A_n\} - k_{23}\{A\}^m\{A_n\} + k_{32}\{A_{n+m}\} \qquad (3)$$

$$\frac{d\{A_{n+m}\}}{dt} = k_{23}\{A\}^m\{A_n\} - k_{32}\{A_{n+m}\} \qquad (4)$$

Then, the next equation is derived by the usual calculation,

$$\begin{vmatrix} a_{11} - \tau_{12}^{-1}, & a_{12} \\ a_{21}, & a_{22} - \tau_{23}^{-1} \end{vmatrix} = 0 \qquad (5)$$

where

$$\begin{aligned} a_{11} &= k_{12}n^2\{A\}_e^{n-1} + k_{21} + k_{23}(\{A\}_e^m - nm\,\{A\}_e^{m-1}\{A_n\}_e) \\ a_{12} &= k_{12}n(n+m)\,\{A\}_e^{n-1} - k_{23}m(n+m)\,\{A\}_e^{m-1}\{A_n\}_e - k_{32} \\ a_{21} &= k_{23}nm\{A\}_e^{m-1}\{A_n\}_e - k_{23}\{A\}_e^m \\ a_{22} &= k_{23}m(n+m)\,\{A\}_e^{m-1}\{A_n\}_e + k_{32}\ . \end{aligned} \qquad (6)$$

Further, if $\quad a_{11} >> a_{22}$ (7)

as presumed from Model (1), we obtain

$$\tau_{12}^{-1} = a_{11} \tag{8}$$

$$\tau_{23}^{-1} = a_{22} - \frac{a_{12}}{a_{11}} a_{21} \tag{9}$$

Here, the assumption that

$$n >> m \tag{10}$$

leads to the following equations,

$$\tau_{12}^{-1} = \frac{nk_{21}}{2\{A\}_e} C_o - (\frac{n}{2} - 1)k_{21} \tag{11}$$

$$\tau_{23}^{-1} = k_{32}\left(1 + \frac{n^2K_{12}\{A\}_e^{n-1}}{n^2K_{12}\{A\}_e^{n-1} + 1}\right) \tag{12}$$

where $K_{12} = \dfrac{\{A_n\}_e}{\{A\}_e^n}$. (13)

The second term in parentheses of Eq. 12 becomes to unity slightly above the c.m.c. and then it can be rewritten

$$\tau_{23}^{-1} = 2k_{32} \tag{14}$$

The value of rate constants k_{21} and k_{32} listed in Table 1 were calculated from Eq. 11 and the results of others[2,3,6] and Eq. 14 and Fig. 1, respectively. The activation energy for the slow process was obtained to be 33 kcal/mole from the usual Arrhenius plot.

The value of rate constant k_{32} decreases with the increase of the relative quantity of SDeS or higher alcohols to micelle as mentioned above. It is may be because these additives form the mixed micelles with SDS or they are solubilized into micelles. Such micelles are considered to be more stable than those with no additives and then the rate of the exchange of m detergent monomers to and from micelles becomes slower.

While, the value of k_{32} increases with the increase of

TABLE I. RATE CONSTANTS FOR MICELLIZATION

	k_{21} $(10^5 sec^{-1})$		k_{32} $(10^2 sec^{-1})$		
	25°C	30°C	15°C	20°C	25°C
NaC_8 [a)]	2.9				
SOS [b)]	2.7	4.8			
NaC_{10} [c)]	3.8				
SDS			1.3	2.9	7.5

a) Sodium Caprylate, b) Sodium Octyl Sulfate, c) Sodium Caprate.

additive such as ethyl alcohol. These lower alcohols are highly soluble in water in comparison with the higher alcohols and subsequently they seem to affect the water structure or the state of hydration on the micellar surface. However, if only micelles are affected by the addition of these lower alcohols, the concentration dependence of the relaxation times must be observed as the case of addition of higher alcohols. Nevertheless, the relaxation times are almost independent of the total detergent concentration as shown in Fig. 6. Then the lower alcohols affect not only the water structre or the state of hydration on the micellar surface, but also those on the free detergent monomers. In addition, these alcohols act as the breaker for water structure. As a result of these effects, the monomers and micelles act more actively and then the rate of the exchange of monomers to and from micelle becomes faster.

Although the relaxation times are anticipated to be completely independent of the total detergent concentration in the completely pure SDS system as mentioned above, Eq. 12 indicates the concentration dependence of relaxation time in the vicinity of c.m.c. as far as n is constant. These questions must be added to this model, but we consider that the rationality of this kind of model can be supported experimentally in other systems.

2. SOLUBILIZATION KINETICS

The kinetics of solubilization has been neglected in spite of the importance of the solubilization phenomena. The investigated system is the most simple one that a dye, pinacyanol chloride is solubilized into SDS micelles[17]. The SDS used in this work, is the previous one stated above and a stopped-flow apparatus is described in the other paper[16]. A rapid scanning spectrophotometer, JASCO RSP 10 whose scanning speed is 400 nm/msec, was also combined with the stopped-flow apparatus in order to check the simultaneous absorption changes at all the characteristic bands, α (615 nm), β (570 nm) and γ (480 nm) bands.

When pinacyanol chloride solution is mixed with the solution containing SDS micelles above two-fold c.m.c. (above the c.m.c. after mixing), the absorption increases with time at the α and β bands characterizing micelle solubilizing pinacynaol chloride, and decreases at the γ-band characterizing the salt of dye and detergent molecules, as shown in Fig. 7, which is the typical result by means of a stopped-flow rapid scanning spectroscopy. This tendency of absorption changes was not observed under the other experimental conditions. The absorption at the γ-band increases abruptly and decreases gradually with time. In addition, no intermediate apparently exists except the salt of dye and detergent, because no new absorption band appears in Fig. 7. These facts suggest that the salt of dye and detergent forms first and is solubilized into micelle as expressed below (eqs. (15) and (16)).

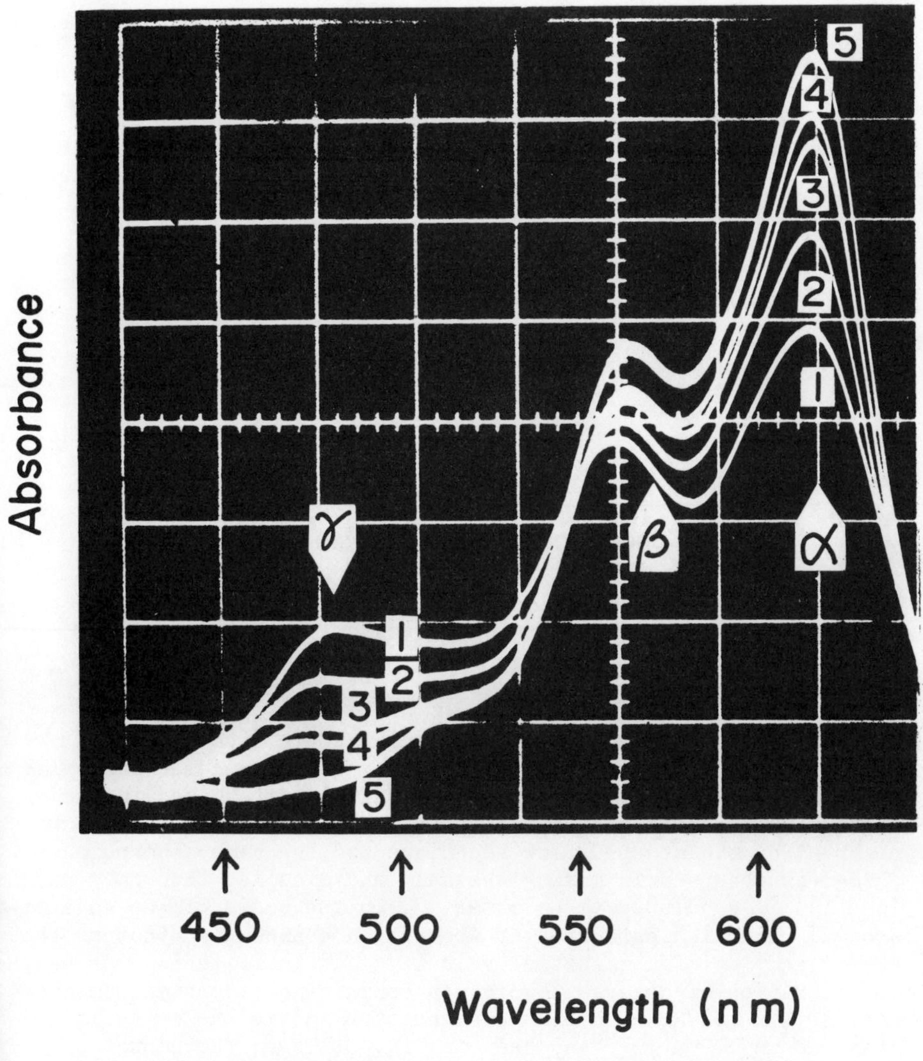

Fig. 7. Change in absorption spectrum with time at 30°C. A solution: 3.2 x 10^{-2}M SDS; B solution: 2 x 10^{-5}M pinacyanol chloride. Time after mixing, 1: 1 msec; 2: 50 msec; 3: 200 msec; 4: 450 msec; 5: 800 msec.

$$D + A \rightleftharpoons DA \quad (15)$$

$$xDA + A_n \rightleftharpoons D_xA_{n+x} \quad (16)$$

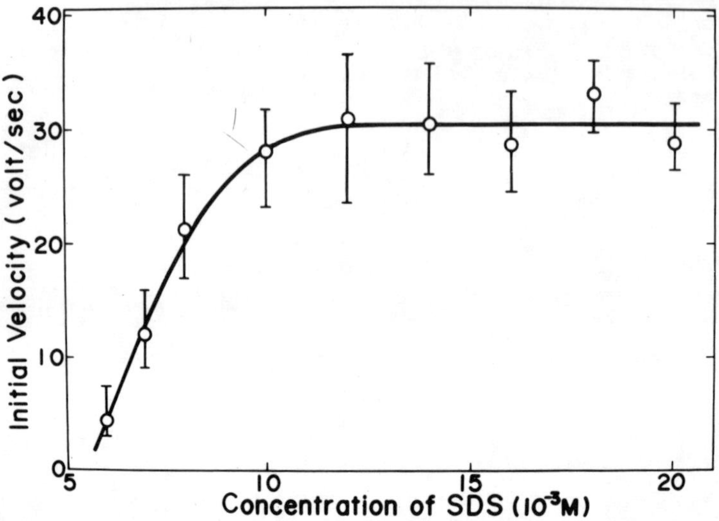

Fig. 8. Plot of initial velocity vs. concentration of SDS after mixing at 50°C. Concentration of pinacyanol chloride after mixing: 1 x 10^{-5}M.

where D is dye molecule, DA is the salt of dye with detergent molecules, x is small integer and D_xA_{n+x} is the micelle solubilizing the salt. The component ratio of salt is determined in our other work[20]. While the rate of salt formation is too fast to be observed by the stopped-flow technique and the same absorption change is observed in mixing the salt and micelle, then only reaction (16) is concluded to be observed. On the basis of the data at t[he] α-band, which indicate greater absorption change than that at the β-band, a plot of the initial velocity of solubilization versus the concentration of detergent solution containing a constant concentration, 1 x 10^{-5}M, of dye, shows that the initial velocity is constant above 1.2 x 10^{-2}M as seen in Fig. 8. Thus, the amount of micelle seems to be in excess of that of dye in the higher concentration range. Further, the ratio of micelle to dye is more than 8 above 1.2 x 10^{-2}M of detergent, and so the number of salt (x) solubilized into one micelle is probably unity. Therefore the reaction (16) can be rewritten as the following pseudo-unimolecular reaction in the high concentration range,

$$DA \underset{k_b}{\overset{k_f}{\rightleftharpoons}} DA_{n+1} \quad \text{(in excess of } A_n\text{)} \tag{17}$$

TABLE II. RATE CONSTANTS FOR SOLUBILIZATION OF PINACYANOL CHLORIDE INTO SDS MICELLE

A solution	B solution	k_f (sec^{-1})			
		30°C	35°C	40°C	50°C
2.0×10^{-2}M SDS	2.0×10^{-5}M Dye	3.6	5.3	10	20
2.4×10^{-2}M SDS					21
2.8×10^{-2}M SDS					20
3.2×10^{-2}M SDS					21
3.6×10^{-2}M SDS					21
4.0×10^{-2}M SDS					20
2.8×10^{-2}M SDS	4.0×10^{-3}M SDS + 2.0×10^{-5}M Dye				21

where k_f and k_b are rate constants. Here, since the absorption changes were observed at the α-band, the next equation is derived in the same manner as that used in Ref. 16,

$$\ln(\ln\Delta E_t/\Delta E_\infty) = -kt + \ln(\ln\Delta E_0/\Delta E_\infty) \quad (18)$$

where ΔE_t is the intensity of the voltage at time, t, ΔE_∞ at infinite time, and ΔE_0, at time 0. The initial part of the absorption change with time in the stopped-flow measurement gives the rate of forward reaction. The values of rate constant, k_f are listed in Table 11 and the activation energy was calculated to be 16 kcal/mole. The agreement between the values of rate constant in two cases as seen in this table, leads to the conclusion that the reactions (15) - (16) are reasonable, i.e., the dye is solubilized into micelle in the state of a salt with a detergent molecule. Since the absorption at the α-band is characteristic of pinacyanol chloride in polar organic solvents, the type of solubilization seems to be the penetration into the palisade layer rather than the incorporation in the hydrocarbon center of micelle. Accordingly, the charged end of dye molecule penetrates into the water layer and the hydrocarbon end is oriented towards the hydrocarbon center as does the detergent molecule in the micelle.

REFERENCES

1. T. Yasunaga, H. Oguri and M. Miura, J. Colloid Interface Sci., 23, 352 (1967).
2. T. Yasunaga, S. Fufii and M. Miura, J. Colloid Interface Sci., 30, 399 (1969).
3. E. Graber and R. Zana, Kolloid-Z. Z. Polym., 238, 470, 479 (1970).
4. P.J. Sams, E. Wyn-Jones and J. Rassing, Chem. Phys. Letters, 13, 233 (1972).
5. J.E. Rassing and E. Wyn-Jones, Chem. Phys. Letters, 21, 93 (1973).
6. J. Rassing, P.J. Sams and E. Wyn-Jones, J. Chem. Soc. Faraday Trans. II, 69, 180 (1973).
7. D.A.W. Adair, V.C. Reinsborough, N. Plavac and J.P. Valleau, Can. J. Chem., 52, (1974).
8. G.C. Kresheck, E. Hamori, G. Davenport and E.M. Scheraga, J. Amer. Chem. Soc., 88, 246 (1966).
9. B.C. Bennion, L.K.J. Tong, L.P. Holmes and E.M. Eyring, J. Phys. Chem., 73, 3288 (1969).
10. B.C. Bennion and E.M. Eyring, J. Colloid Interface Sci., 32, 286 (1970).
11. J. Lang and E.M. Eyring, J. Polym. Sci. Part A-2, 10, 89 (1972).

12. K. Takeda and T. Yasunaga, J. Colloid Interface Sci., 40, 127 (1972).
13. K. Takeda and T. Yasunaga, J. Colloid Interface Sci., 45, 406 (1973).
14. U. Herrmann, and M. Kahlweit, Ber. Bunsenges. Phys. Chem., 77, 1119 (1973).
15. T. Janjic and H. Hoffmann, Z Phys. Chem. (Neue Folge), 86, 322 (1973).
16. T. Yasunaga, K. Takeda and S. Harada, J. Colloid Interface Sci., 42, 457 (1973).
17. K. Takeda, N. Tatsumoto and T. Yasunaga, J. Colloid Interface Sci., 47, 128 (1974).
18. E.E. Dreger, G.I. Keim and G.D. Miles, Ind. Eng. Chem., 36, 610 (1944).
19. T. Nakagawa, Colloid Polym. Sci., 252, 56 (1974).
20. N. Tatsumoto, K. Takeda, S. Isshiki and T. Yasunaga, Bull. Chem. Soc. Japan, 47, 289 (1974).

ULTRASONIC PULSE SPECTROSCOPY IN MICELLAR SOLUTIONS

V.C. Reinsborough

Department of Chemistry, Mount Allison University,
Sackville, New Brunswick, Canada.

The pulse technique despite its relatively high accuracy has limited applicability in studying chemical relaxations in micellar systems. At room temperature, only micelles consisting of 6-8 carbon atoms in monomer length show relaxation frequencies within the usual 10-150 MHz frequency range of the pulse technique. At higher temperatures and in different solvents, it is sometimes possible to shift the relaxation frequency for more voluminous micelles into this range. For example, cetyldimethyl-benzylammonium chloride micelles in molten pyridinium chloride at 155°C gave a concentration-independent relaxation frequency f_c of 42 ± 7 MHz[1].

In aqueous Li, Na and K octylsulphate and hexylsulphate solutions at 40°, the single-valued relaxation frequency was found to be independent of concentration in the dilute micellar range[2]. Generally, the relaxation frequency had been found to increase linearly with the concentration in micellar systems[3-7], so that this finding came as a surprise. It received an easy explanation from an approximate treatment of a model for micelle treatment. Of all the multiple steps in the hierarchy of reactions

$$m + M_{n-1} \rightleftarrows M_n \tag{1}$$

involved in micelle association/disassociation, one was deemed more important than the others (n is the number of monomers, m, and M_n represents an n-mer or micelle containing n monomers). For a single such step, f_c satisfies the equation (2):

$$2\pi f_c = k_d + k_a(\text{c.m.c.}) \tag{2}$$

where c.m.c. denotes the critical micelle concentration.

Since the other studies had generally been concerned with

Wyn-Jones (ed.), Chemical and Biological Applications of Relaxation Spectrometry, 159–162.

the more concentrated micellar range, a thorough investigation over a wider concentration range was undertaken with the alkali metal heptylsulfates from the c.m.c. (around 0.22m) to roughly 1 molal[8]. The values of the amplitude parameter A and the relaxation frequency f_c are depicted in Figure 1 for micellar sodium heptylsulphate solutions at 40°C. The initial decrease in the f_c curve is likely real since the same pattern was observed in all the other alkali metal heptylsulphates even though the experimental error associated with the first few points is undoubtedly greater than with the higher concentration values. However the decrease of about 20% that occurs in the dilute micellar range could be due to secondary effects such as changing solution properties. Thus, it can still be asserted that in the dilute micellar range f_c is relatively concentration independent in marked contrast to the linear increase that commences at 0.35m. This concentration coincides with the peak of the A-curve and is also the second cm.c. as indicated by viscosity, density, and conductivity measurements[9].

Several models to account for the linear increase have already been proposed at this conference but none has met with general approval. No attempt will therefore be made at this stage to interpret this portion of the curve. It is obvious however that investigators who obtain only the f_c values for the concentrated micelle range and extrapolate this curve to the c.m.c. or to zero solute concentration are ignoring the completely different behaviour of the f_c in the dilute micellar range.

The heptyl and octylsulphate data for dilute micellar solutions are open to the criticism that the relaxation frequencies calculated from the equation for a single relaxation[10] lie just below the lowest frequency available to the pulse technique. In other words, only the upper half of the relaxation region is being explored so that it is difficult to eliminate completely the possibility of a multiple relaxation. However, the close fit of the results to single relaxation curves and the systematic behaviour of the relaxation parameters (especially the relative constancy in B) tend to confirm the adequacy of the analysis. Moreover, on moving into the hexylsulphate system, the f_c is shifted into the operating range of the pulse technique and, if anything, the standard deviations increase rather than decrease when absorption data are fitted to the single-relaxation equation. Recently, this work was repeated over a wide concentration range with sodium hexylsulphate solutions doped with varying amounts of sodium chloride and, in all cases, curves similar to Figure 1 were obtained[11].

The relaxation frequency and A-values are shown in Figure 2 for hexylammonium chloride solutions at 25°C[12]. The c.m.c. as indicated by the intersection of the A-curve with zero concentration is 0.89 m in good agreement with the literature value of 0.9M[1] The relaxation frequency varies from 36-76 MHz well within the range of the pulse technique. The two curves of Figure 2 are

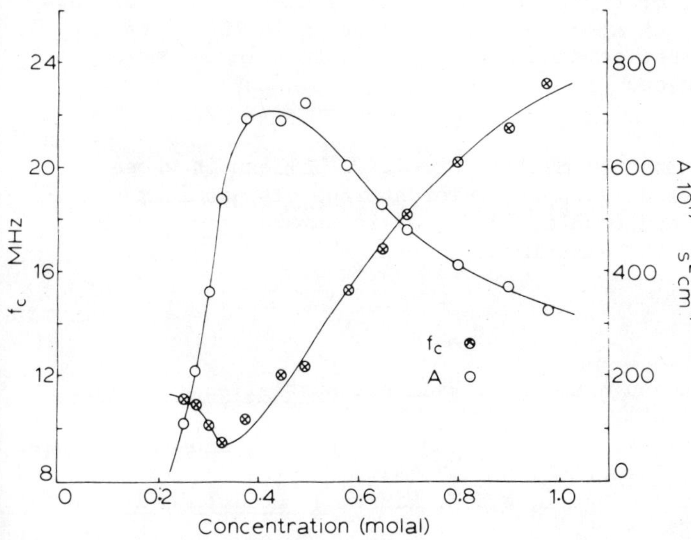

Fig. 1. Relaxational parameters f_c (⊗) and A (○) for sodium heptylsulphate solutions at 40°.

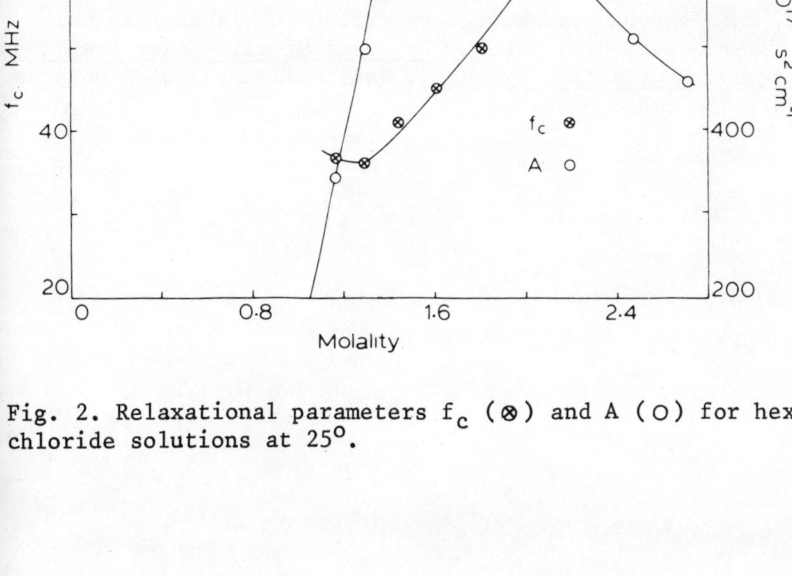

Fig. 2. Relaxational parameters f_c (⊗) and A (○) for hexylammonium chloride solutions at 25°.

similar to those of Figure 1 so that it is clear that, for this cationic surfactant also, micellar formation in the dilute micellar range must be distinguished from that in the more concentrated region.

Acknowledgement must be made to Dr. J. P. Valleau in whose laboratory at the University of Toronto the ultrasonic measurements were made and Douglas Adair, Sally Zamora, and Hayden Trenholm who did the experimental work.

REFERENCES

1. V.C. Reinsborough and J.P. Valleau, Australian J. Chem., 21, 2905 (1968).
2. D.A.W. Adair, V.C. Reinsborough, N. Plavac, and J.P. Valleau, Canadian J. Chem., 52, 429 (1974).
3. T. Yasunaga, H. Oguri, and M. Miura, J. Colloid and Interface Sci., 23, 352 (1967).
4. T. Yasunaga, S. Fujii, and M. Miura, J. Colloid and Interface Sci., 30, 399 (1969).
5. R. Zana and J. Lang, C. R. Acad. Sc. Paris, Ser.C., 266, 893 (1968).
6. E. Graber, J. Lang and R. Zana, Kolloid Z.Z.Polymer, 238, 470 (1970).
7. J. Rassing, P.J. Sams, and E. Wyn-Jones, J. Chem. Soc. Faraday II, 69, 180 (1973).
8. D.A.W. Adair, V.C. Reinsborough, and J.P. Valleau, Canadian J. Chemistry, in press.
9. D.A.W. Adair and V.C. Reinsborough, unpublished results.
10. J. E. Rassing, "Ultrasonic Relaxation Spectrometry", this publication.
11. V.C. Reinsborough and S.K. Zamora, unpublished results.
12. V.C. Reinsborough and H.M. Trenholm, unpublished results.
13. P. Mukerjee and K.J. Mysels "Critical Micelle Concentrations of Aqueous Surfactant Systems", National Bureau of Standards (U.S.) NSRDS-NBS 36, 1971.

KINETICS OF MICELLIZATION FROM ULTRASONIC RELAXATION STUDIES - THE FAST PROCESS DESCRIBED BY THE TWO STATE MODEL.

P.J. Sams, J.E. Rassing and E. Wyn-Jones

Department of Chemistry and Applied Chemistry,
University of Salford, Salford M5 4WT

ABSTRACT. In surfactant solutions at concentrations in excess of the c.m.c. the concentration dependence of both the ultrasonic relaxation times and amplitudes can be adequately described by relaxation equations derived from a two state kinetic model which describes the exchange rate between monomers and micelles in solution. The binding of monomers to the micelle is described in terms of the Langmuir adsorption theory.

This paper traces the progress that we have made in these laboratories in efforts to explain our ultrasonic relaxation data on solutions of pure surfactants containing micelles. Whilst the initial work was in progress Zana et al[1,2] and also Eyring and his collaborators[3,4] reported, respectively some of their ultrasonic and T-jump measurements on surfactant solutions. At that time the general conclusions from our and other relaxation data on solutions containing micelles were (1) a single relaxation process was observed above the c.m.c. and (2) in the majority of cases $1/\tau$ was found to be a linear function of overall concentration. For the micellization process, where aggregates are thought to be formed by a polymerization mechanism involving the step-wise addition of monomers to polymers this behaviour of the relaxation data was surprisingly uncomplicated and simple. As a result, these data prompted us to explore the possibility of developing a relatively simple kinetic model to describe the response in the microscopic behaviour of a surfactant solution containing micelles following perturbation. On the basis of these observations and the description of micelle formation as a pseudo phase separation phenomenon we introduced[5] the "two state model" which basically involves the transitions of monomers from the solution phase to the micelle phase.

Wyn-Jones (ed.), Chemical and Biological Applications of Relaxation Spectrometry, 163–169.

In the equilibrium situation there is, of course, a continuous exchange between the monomers and the micelles. Using the Hartley model for the micelles where the monomers are located on the surface of a sphere this exchange process can be envisaged as a collision between a small object (the monomer) and the large spherical micelle. On this basis we assumed that the rate at which monomers associate with the micelle is proportional to the concentration of free monomers in solution, the concentration of micelles and the cross-section or "size" of the micelle which was expressed as the number of monomers in the micelle. In connection with this latter assumption it can be argued that if the monomers are located on the surface of the spherical micelle then the cross-section becomes proportional to the number of monomers in the micelle because, in this case, the cross-section is proportional to the surface area of the micelle. The rate at which monomers dissociate from the micelle was assumed to be proportional to the concentration of micelles, and the number of monomers in the micelles. This latter situation is analogous to the well known fact that the probability of the dissociation of a monomer from a polymer in which the monomers are equally bonded is proportional to the number of monomers.

This kinetic treatment leads to the rate equation (1),

$$\frac{-d\{A_1\}}{dt} = k_a\{A_1\}\{C - \{A_1\}\} - k_d\{C - \{A_1\}\} \qquad (1)$$

where $\{A_1\}$ and C are respectively the molar concentration of monomers and total surfactant in solution and the k's are the rate constants. The relaxation equation (2)

$$1/\tau = k_a C - k_d \qquad (2)$$

can be derived from equation (1) where τ is the single relaxation time for the mechanism. Equation (2) shows that $1/\tau$ is a linear function of C as shown in Figs. 1 and 2. Furthermore, the intercept on the concentration axis (i.e. when $1/\tau = 0$) is equal to k_d/k_a (= $\{A_1\}$) which in turn is close to the c.m.c. The main assumptions inherent in this mechanism are that (1) the micelles are spherical with the monomers located on the surface of the sphere and (2) that the micelles are much larger than the monomers At concentrations in excess of the c.m.c. these conditions are fulfilled by most surfactants. It is worth pointing out here that recent experiments, including our own (Fig. 1) have shown that at concentrations very close to the c.m.c. there is a curvature in the $1/\tau$ against C plots. These observations clearly define the lower concentration limit associated with assumption (2) since at these concentrations it is possible that the intermediate species such as dimers, trimers etc. are present in substantial amounts.

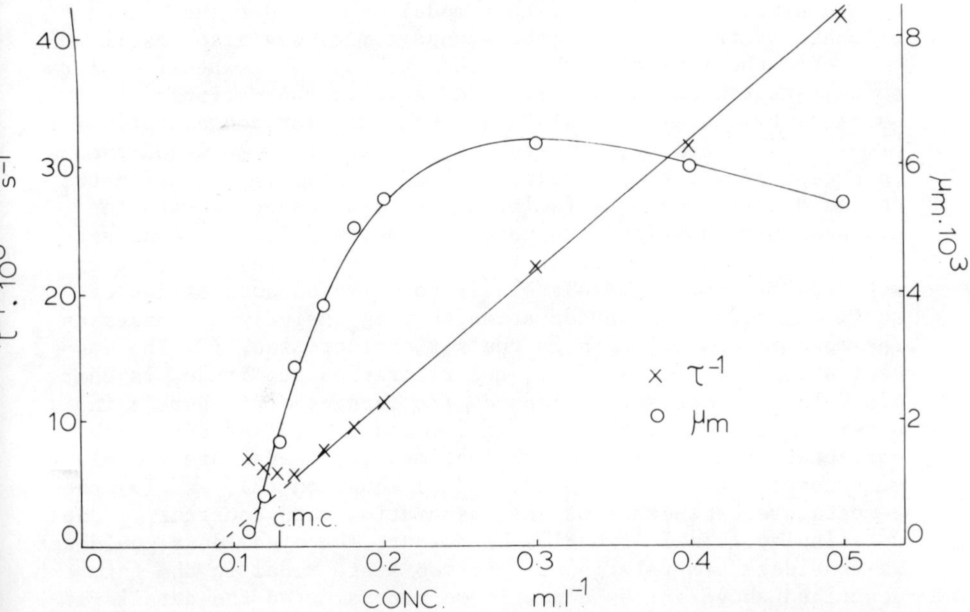

Fig. 1. The concentration dependence of τ and μ_m for sodium octyl sulphate at 25°C.

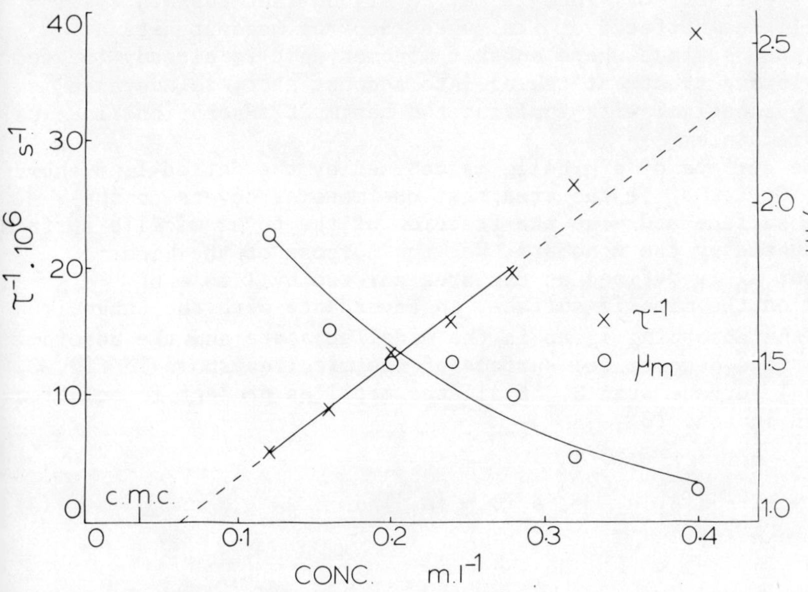

Fig. 2. The concentration dependence of τ and μ_m for sodium decyl sulphate at 25°C.

A simple extension of this model to consider the second relaxation centred around the second c.m.c. was also described[6].

This theoretical work was then followed by extensive ultrasonic measurements on a variety of anionic and cationic surfactants. In all cases[7,8] we found that for concentrations in excess of the c.m.c., $1/\tau$ was a linear function of C and the intercept of $1/\tau = 0$ was always close to the c.m.c. During the course of this work the following aspects connected with the ultrasonic relaxation data were also noticed. (1) At constant concentration of surfactant the relaxation strength decreases with addition of electrolyte. (2) For a homologous series of surfactants the relaxation strength progressively decreases as the monomer chain length in the series increases. (3) The concentration dependence of μ_m, the relaxation amplitude, is complex. In some cases the slope $d\mu_m/dc$ changes sign whereas in others μ_m is constant (see Figs. 1 and 2). (4) At very high surfactant concentration $1/\tau$ sometimes increases more rapidly with concentration than predicted by equation (2), and (5) the temperature dependence of the association rate constant k_a does not display normal Arrhenius behaviour. These new data could not be considered in relation to the two state model in the form described above and as a result we reconsidered the details of the interactions involved in the model. Bearing in mind that the micelle/monomer exchange is analagous to a collision between a small particle and a large sphere it seems plausible to introduce a steric effect since it may be argued that certain collisions will be ineffective, i.e. when the free monomer hits a part of the surface where another monomer unit is already located. The following treatment taking into account steric hinderance is formally identical with applying the Langmuir adsorption kinetics to the mechanism[8].

The surface of a micelle is defined by the dotted line shown in fig. 3. Let a_o be the area that one monomer covers on the micelle surface and α be the fraction of the total micelle surface area covered by the monomers. For the purpose of the kinetic treatment a_o is defined as the area covered by 1 mole of monomer on the micelle surface. In accordance with the Langmuir theory the adsorbing agent is the micellar state and the adsorption surface becomes the surface of the micelles shown in fig. 3. The total surface area S, of all the micelles present in solution is given in eqn. (3)

$$S = \sum_{i=2}^{\infty} i\{A_i\}\frac{a_o}{\alpha} = \frac{a_o}{\alpha}(C - \{A_1\}) \qquad (3)$$

where

$$\sum_{i=2}^{\infty} i\{A_i\} = C - \{A_1\}$$

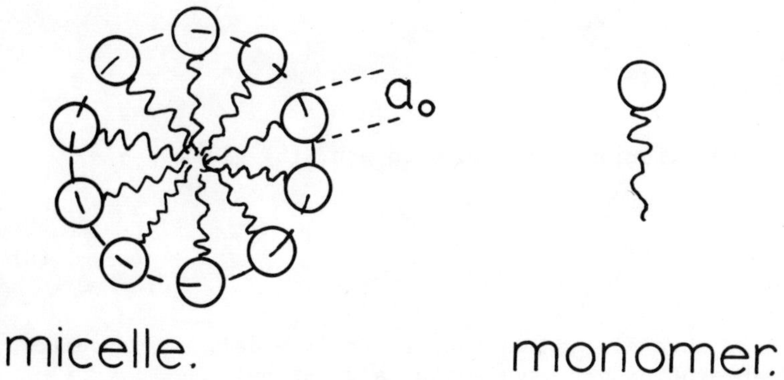

Fig. 3. Model for micelle and monomer.

$\{A_i\}$ denotes the molar concentration of a micelle containing i monomer units. Proceeding in accordance with the kinetic principles of the Langmuir theory, the rate at which monomers disappear from solution is given by

$$-\frac{d\{A_1\}}{dt} = k_1\{A_1\}(1 - \alpha)\ S - k_{-1}\alpha S \qquad (4)$$

where k_1 is the rate constant associated with monomers condensing on the surface of the micelles and k_{-1} is their escaping rate constant.

From eqn (3) and (4) the relaxation time associated with the mechanism now becomes:

$$\frac{1}{\tau} = k_1 a_o \ \frac{(1 - \alpha)}{\alpha}\ C - k_{-1}a_o \qquad (5)$$

If we compare the new relaxation eqn. (5) with that found in the original two state model, eqn. (2), more details concerning the transition probabilities are revealed. The apparent rate constants k_a and k_d are now related to the true rate constants k_1 and k_{-1} by the following equations

$$k_a = k_1 a_o \ \frac{(1 - \alpha)}{\alpha} \qquad (6)$$

and

$$k_d = k_{-1}\, a_o \tag{7}$$

Applying the equilibrium conditions to eqn. (5) gives

$$\frac{k_{-1}\alpha}{k_1(1-\alpha)} = \{A_1\} \approx \text{c.m.c.} \tag{8}$$

In addition we have also shown that for this model, μ_m, the maximum sound absorption per wavelength is given by:

$$\mu_m = \text{const.}\ (\Delta V^o)^2 \left\{\frac{1}{\{A_1\}} + \frac{2}{\{C - \{A_1\}\}}\right\}^{-1} \tag{9}$$

where ΔV^o is the volume difference between monomer and micelle. Using equations (5) - (9) we were able to quantitatively explain the experimental observations (1) - (5) listed above[8]. In addition this model has also been used successfully in connection with the ultrasonic relaxation data on mixed micelles and the solubilization of n-octanol by ionic surfactants[9]. It is also worth pointing out that during the course of this work we realized that the relaxation spectrum of a surfactant solution was characterized by at least two relaxation times which were well separated on the time scale. As we have indicated, the model that is described here is applicable only to the fast process involving the exchange of monomers from the bulk solution to the micelles. It is clearly indicated elsewhere in this volume that the slower process is associated with the dissolution of the micelles.

It has been found that equation (2) could also be derived from the stepwise polymerization scheme (10) by

$$A_1 + A_{i-1} \rightleftharpoons A_i \qquad 2 \leqslant i \leqslant n \tag{10}$$

neglecting the dimerization step. On the basis that this is an unreasonable assumption some authors have used this result to criticize the two state model[10,11]. As we have since shown, the arguments used by these authors can be misleading and often lead to erroneous conclusions. For instance the same relaxation equation can also be derived from scheme (10) with the more reasonable assumption that all intermediate species are in a steady state.[12] At present, however, it seems that both approximations described above are incorrect as it is now evident that the micellization process is characterised by two well separated relaxation times. There is no doubt, however, that within the limits of the concentration range described above the fast relaxation process can be adequately described by the two state model described here. This has been clearly demonstrated by taking into account the concentration dependence of both the ultrasonic relaxation times and amplitudes. Further comments on the fast process are also reported in the following paper.[13]

ACKNOWLEDGEMENT. We thank the S.R.C. for a research grant (E.W-J) and a maintenance award (P.J.S.)(through the C.A.P.S. scheme with Unilevers Ltd., Port Sunlight, Cheshire).

REFERENCES

1. E. Graber, J. Lang and R. Zana, Kolloid Z.Z. Polymere, 238, 470 (1970).
2. E. Graber and R. Zana, Kolloid Z.Z. Polymere, 238, 479 (1970).
3. B.C. Bennion, L.K. Tong, L.P. Holmes and E.M. Eyring, J. Phys. Chem., 73, 3288 (1969).
4. B.C. Bennion and E.M. Eyring, J. Colloid Interface Sci., 32, 286 (1970).
5. P.J. Sams, E. Wyn-Jones and J. Rassing, Chem. Phys. Letters, 13, 233 (1972).
6. J. Rassing and E. Wyn-Jones, Chem. Phys. Letters, 21, 93 (1973).
7. P.J. Sams, E. Wyn-Jones and J. Rassing, J.C.S. Faraday II, 69, 180 (1973).
8. J. Rassing, P.J. Sams and E. Wyn-Jones, J.C.S. Faraday II; 70, 1247 (1974).
9. P.J. Sams, J.E. Rassing and E. Wyn-Jones, Adv. Mol. Relax. Processes, 6, 255 (1975).
10. D.A.W. Adair, V.C. Reinsborough, N. Plavac and J.P. Valleau Canadian J. Chem., 52, 929 (1974).
11. U. Herrman and M. Kahlweit, Ber. Bunseges physik Chem., 77, 1119 (1973).
12. J. Rassing and E. Wyn-Jones, Ber. Bunseges physik Chem., 78, 651 (1974).
13. M. Almgren, this publication.

COMMENTS ON THE FAST RELAXATION PROCESS OBSERVED IN THE STUDY OF MICELLIZATION

Mats Almgren

Department of Physical Chemistry, Chalmers University of Technology and University of Gothenburg, Gothenburg, Sweden

The slower of the two relaxation processes observed in relaxation measurements on micellar solutions is connected with a change in the number of micelles by a step-wise dissolution or built up of micelles. This process seems to be adequately described by the theory of Aniansson and Wall[1]. During the fast process the number of micelles remains constant. The size distribution changes through the uptake (or release) of monomers from (to) the aqueous phase. For this initial process, the theory of Aniansson and Wall predicts a spectrum of concentration independent relaxation times. The analysis has, however, not been driven into sufficient detail for a prediction of the rate of change of the parameters actually measured in various relaxation experiments. At present, it is therefore not possible to decide if the theory is compatible with the experimentally observed linear relation of τ^{-1} on the total concentration well above c.m.c. The "two state model" proposed by Rassing et al[2] is in accord with this finding.

The following discussion will be focused on the change in the number of freee monomers and will, therefore, be relevant for relaxation methods in which the observed relaxation is primarily determined by this change. These include ultrasonic as well as jump methods with conductance, absorbance, or fluorescence read out, but not methods in which the change of light scattering intensity is followed.

The starting point is the multistep scheme

$$A_1 + A_1 \underset{1/2\, k_2^-}{\overset{1/2\, k_2^+}{\rightleftharpoons}} A_2$$

Wyn-Jones (ed.), *Chemical and Biological Applications of Relaxation Spectrometry*, 171–174.

$$A_1 + A_{s-1} \underset{k_s^-}{\overset{k_s^+}{\rightleftharpoons}} A_s$$

where, for convenience the rate constants of the first reversible step has been denoted $1/2\ k_2^+$ and $1/2\ k_2^-$. The rate of change of the monomer concentration becomes

$$\frac{dA_1}{dt} = -\sum_{S=2}^{\infty} k_s^+ A_{s-1} A_1 + \sum_{S=2}^{\infty} k_s^- A_s$$

If we further assume that dimers, trimers etc., if present at appreciable concentrations are very rapidly equilibrated with the monomers, the terms associated with these species may be removed. All terms associated with the very rare intermediate micelles may also be neglected. We then obtain

$$\frac{dA_1}{dt} = -\sum_{S=S_2} k_s^+ A_{s-1} A_1 + \sum_{S=S_2} k_s^- A_s$$

where S_2-1 is the aggregation number of first micelle of appreciable concentration. Introducing the excess variables, $\Delta A_s = A_s - \bar{A}_s$, we obtain after linearization

$$\frac{d\Delta A_1}{dt} = -\sum_{S=S_2}^{\infty} k_s^+ \bar{A}_{s-1} \Delta A_1 - \sum_{S=S_2}^{\infty} k_s^+ \Delta A_{s-1} \bar{A}_1 + \sum_{S=S_2}^{\infty} k_s^- \Delta A_s$$

The relative excess variables are defined by $\xi_s = \Delta A_s / \bar{A}_s$. Introducing these, and utilizing the equilibrium conditions, $k_s^+ \bar{A}_{s-1} \bar{A}_1 = k_s^- \bar{A}_s$, one obtains

$$\frac{d\Delta A_1}{dt} = -\Delta A_1 \sum_{S=S_2}^{\infty} k_s^+ \bar{A}_{s-1} + \sum_{S=S_2}^{\infty} k_s^- \bar{A}_s (\xi_s - \xi_{s-1}) \qquad (1)$$

If the second term on the R.H.S. of this equation is negligible, the relaxation of ΔA_1, should be exponential and the relaxation time given by

$$\tau^{-1} = \sum_{S=S_2}^{\infty} k_s^+ \bar{A}_{s-1} = k^+ \sum_{S=S_2}^{\infty} \bar{A}_{s-1} = k^+ (C_{tot} - \bar{A}_1)\, \bar{n}^{-1}$$

where k^+ denotes the number average of k_s^+, and $\bar{n}$ the mean aggregation number. Assuming further that the size distribution is symmetrical, so that the maximum of the distribution coincides with $\bar{n}$, and that $k^+ \approx k_{\bar{n}}^+$, one obtains the condition for maximum ($\bar{A}_n \approx \bar{A}_{n-1}$) together with the equilibrium condition $A_1 = k_{\bar{n}}^- / k_{\bar{n}}^+$. The equation for τ^{-1} now becomes equivalent with the expression of Rassing et al[2].

$$\tau^{-1} = \frac{k_n^+}{\bar{n}} \; C_{tot} \; - \frac{k_{\bar{n}}^-}{\bar{n}} \tag{2}$$

The omission of the last term of Eqn. (1) thus leads to the same result as the "two state mechanism". This is easily understandable since the omitted term is related to the kinetic difference between micelles of different sizes, whereas the "two state mechanism" assumes every bound monomer to be equivalent with all other independently of the size of the particular micelle in which it happens to reside.

In the treatment of Aniansson and Wall the pseudo-stationary distribution during the slow process is characterized by $\xi_s = s\xi_1 - RJ$ for s values representing the abundant micelles. Introducion of this expression into Eqn. (1) shows that the fast relaxation results in a state where

$$\frac{d\Delta A_1}{dt} = 0$$

due to that the two terms of the right hand number of Eqn. (1) cancel each other. The fast relaxation may therefore not be truly exponential. However, Eqn. (2) should probably still be a good approximation if the following conditions were fullfilled:

$$|\Delta A_1^o| \gg |\Delta A_1^*| \tag{I}$$

$$\sum k_s^+ \bar{A}_{s-1} \; |\Delta A_1^o| \gg \sum k_s^- \bar{A}_s \, |\xi_s^o - \xi_{s-1}^o| \tag{II}$$

Superscript o denotes the initial state and superscript * a hypothetical state in which the fast relaxation has gone to completion and the slow relaxation is still negligible.

A detailed analysis of the consequences of these conditions is outside the scope of this discussion. A few points seem clear, however.

1. Condition II is likely to be best fullfilled when the ξ_s^o-values are small. Since a certain change in A_1 will correspond to the greatest values of ξ_s^o when the total concentration is just above

the c.m.c., deviations from Eqn. (2) should be expected near c.m.c.

2. According to condition I most of the change in A_1 appears during the fast process. Thus, under conditions such that Eqn. (2) applies only the fast relaxation process should be detected by methods sensitive to the change in the number of free monomers. Light-scattering, on the other hand, should be expected to detect only the slow process in which the number of micelles is changed.

From the second point follows, in particular, that in those cases where two relaxation processes have been observed, with comparable amplitudes, the relaxation time of the fast process should not follow Eqn. (2). The fast process may nevertheless appear exponential.

Finally, in the interpretation of experimental data the importance of using models and equations that are clearly stated and rest upon a firm theoretical base should be emphasised. A much more penetrating analysis of the range of validity of Eqn. (2) and the consequences of conditions I and II are therefore required.

REFERENCES

1. E.A.G. Aniansson and S.N. Wall, J. Phys. Chem., **78**, 1024 (1974).
2. J. Rassing, P.J. Sams and E. Wyn-Jones, J.C.S. Faraday II, **70**, 1247 (1974).

ON THE KINETICS OF THE FORMATION OF NaDS MICELLES

U. Herrmann and M. Kahlweit

Max-Planck-Institut fur biophysikalische Chemie
(Karl-Friedrich-Bonhoeffer-Institut), Göttingen

EXPERIMENTS. The system investigated was H_2O-Sodiumdodecylsulfate (NaDS). The solubility of NaDS in H_2O (2° - 21^oC) has been measured by Hutchinson et al[1], and the critical micelle concentration (c.m.c.) (10° - 55^oC) by Goddard and Benson[2]. Their results are shown in Fig. 1. To complete this "phase diagram", we have measured the c.m.c. below the Krafft point ($8.9 \cdot 10^{-6}$ mol cm^{-3}; 282.1 K). For this purpose, solutions up to $16 \cdot 10^{-6}$ mol cm^{-3} NaDS were prepared at room temperature, transferred into a conductivity cell and then cooled down to temperatures as low as 2.5^oC. Like Goddard and Benson[2] we could represent the results by two straight lines, the intersection of which was taken as the c.m.c. at that temperature. These values are shown in Fig. 1 as full circles. The supercooled solutions were quite stable, and precipitation of the solid phase could only be initiated by adding a few crystals. This indicates that the molecules in the micelles are in a liquid like state, so that the micelles as such do not act as nuclei for the solid phase.

The variation of the c.m.c. with pressure has been measured by Hamann[3]. He found that up to 10^3 atm the c.m.c. increases with pressure. In agreement with density measurements of other authors, this indicates that the partial molar volume of NaDS is larger in the micellar than in the dispersed state. The measurements further show, that the slope of the straight line, which represents the electrical conductivity above the c.m.c. also increases with pressure. Since one assumes that the concentration of monomers remains constant above the c.m.c., this indicates that the product of effective charge and mobility of the micelles increases with pressure. We assume that this increase is caused by the decreasing association between micelles and counter-ions. This process sets Na^+ ions free, the hydration shells of which add to the decrease

Wyn-Jones (ed.), Chemical and Biological Applications of Relaxation Spectrometry, 175–180.

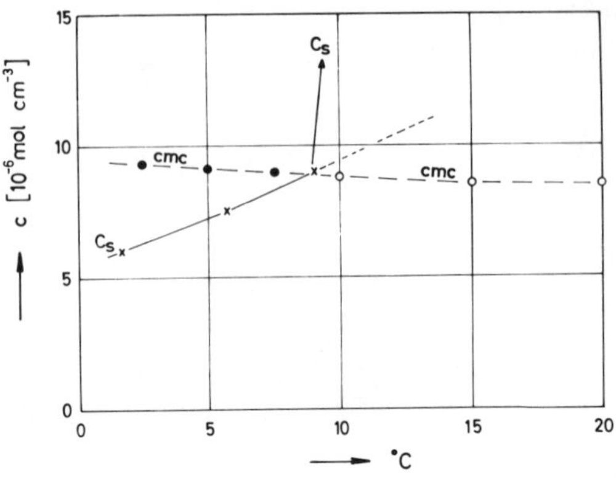

Fig. 1 "Phase diagram" of the system H_2O-NaDs. x x x Solubility[1] o o o c.m.c.[2] ● ● ● c.m.c. (this paper).

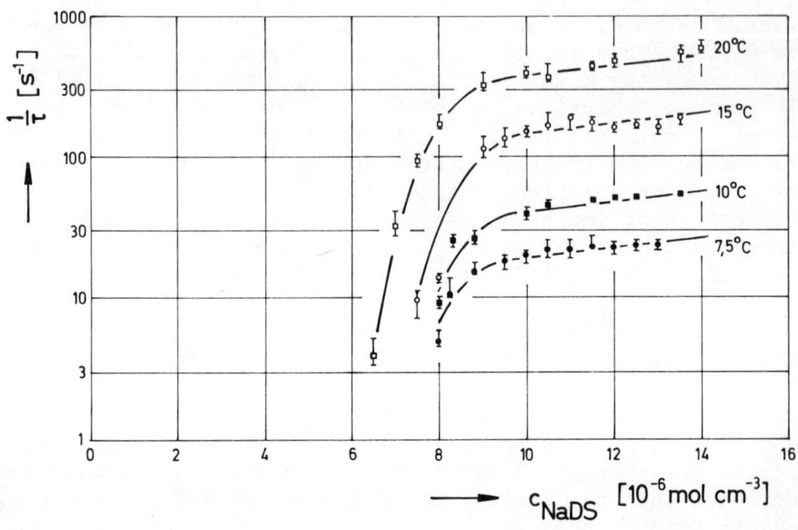

Fig. 2. Reciprocal of the relaxation time $1/\tau$ versus total concentration of NaDS c, with temperature as parameter.

of the volume of the system.

If one sets for $c > c^*$

$$N_1 = N_A c^* \quad \text{and} \quad N_m = (c-c^*)/m \tag{1}$$

and for the number of the dissociated cations

$$N_+ = N_1 + \alpha \cdot m . N_m ; \; 0 < \alpha \leq 1 \tag{2}$$

and further for the charge $|z_1| = z_+ = 1$ and $|z_m| = \alpha . m$, then it follows for the electrical conductivity of the solution

$$K \approx F\{c^*(u_+ + u_1) + \alpha(c - c^*)(u_+ + u_m)\} \tag{3}$$

where the u_i denote the corresponding mobilities. Since u_+ may be found in the literature, one may thus obtain the product $\alpha \cdot u_m$ from

$$\partial K/\partial c \approx F \cdot \alpha(u_+ + u_m); \quad c > c^* \tag{4}$$

The mobility of the monomers u_1, on the other hand, may be obtained from the slope below c^*.

The dependence of the c.m.c. on pressure suggests that one may apply a p-jump-method for the relaxation experiments. This method has already been used by Takeda and Yasunaga[4] at 20°C. The T-jump-method used by Eyring and coworkers[5] at 35°C has the disadvantage that one has to add electrolyte to the solution which may change the relevant properties of the system. Further measurements were carried out by Yasunaga, Takeda and Harada[6] at 34°, 42° and 51°C, using the stopped-flow-method.

In our p-jump-apparatus, which has been described elsewhere[7], pressure up to about 100 atm is applied to the system until a metal membrane bursts. As the system approaches equilibrium at 1 atm, the concentration of the monomers decreases, as does the electrical conductivity. The latter may thus be taken as indicator for the extent of the reaction. In each case, the oscilloscope trace could be represented by a single time constant τ.

We then measured τ as a function of the total NaDS concentration c at six temperatures between 7.5° and 20°C. Below 7.5°C the supercooled solutions became too unstable with respect to the disturbances caused by the pressure jump, while above 20°C the reaction became too fast for the resolution of our apparatus.

In Fig. 2 we have plotted $1/\tau$ versus c with T as parameter. As one can see, the results can hardly be represented by straight lines. This is in contradiction to the findings of the authors cited above[4-6]. Instead, in our measurements the relaxation process becomes detectable at concentrations already well below the c.m.c. There $1/\tau$ increases rather steeply, to flatten off in a convex

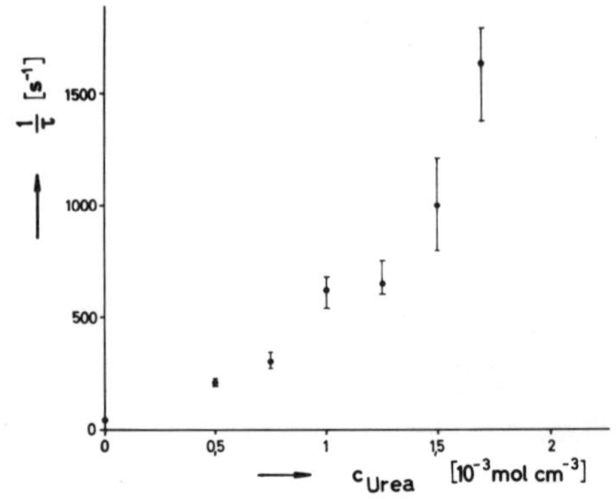

Fig. 3. $1/\tau$ versus concentration of Urea; $c_{NaDS} = 10.8.10^{-6}$ mol cm^{3}; $10^{o}C$.

curve. We are therefore not able to confirm the prediction of a linear dependence of $1/\tau$ on c.

The second goal of this investigation was to study the influence of the water structure on the kinetics of the formation of micelles. This influence is demonstrated by the increase of the c.m.c. in the presence of structure breakers like Urea[8]. We have, therefore, measured the c.m.c. as well as τ (both at $10^{o}C$; the latter at $c_{NaDS} = 10.8.10^{-6}$ mol cm^{-3}) as functions of the Urea concentration. The c.m.c. was again determined by conductivity measurements. Since Urea increases the viscosity η of the solution, we have multiplied K by the corresponding η[9]. It then turns out, that not only the c.m.c. increases with increasing Urea concentration, but also the slope of $(K.\eta)$ above the c.m.c., while below the c.m.c., the slope remains almost unaffected. Urea thus has the same effects as pressure: it accelerates the dissociation reaction of the micelles in relation to the association reaction, and it increases the product of effective charge and mobility of the micelles.

The curves of $1/\tau$ versus c, with the Urea concentration as parameter, have the same shape as those shown in Fig. 2 with, however, considerably higher values of $1/\tau$ (Fig. 3). This accelerating effect was even more pronounced with other substances: In a 10^{-2} m NaDS solution, 10^{-1} m Biuret increases $1/\tau$ by a factor of about 2, and 10^{-1} m 2-Pyridol by even a factor of about 10 (at 10° A variation of the pH between $3 \leq$ pH ≤ 11, on the other hand, had no effect on τ.

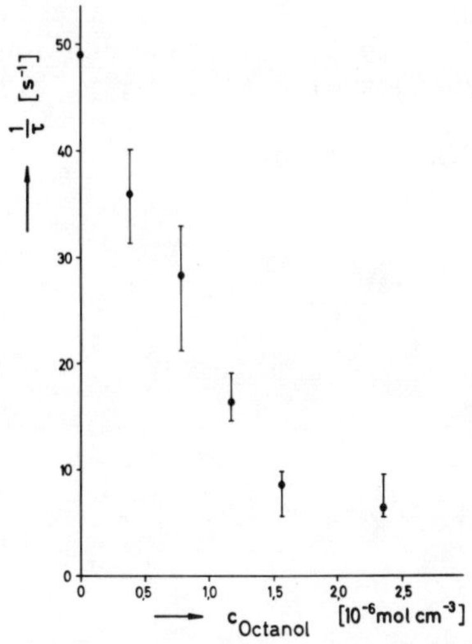

Fig. 4. $1/\tau$ versus concentration of Octanol; $c_{NaDS} = 10.8.10^{-6}$ mol cm^{-3} 10°C.

In view of the applications of surfactants, we have further measured the effect of a solubilization on the relaxation process. For this purpose, increasing amounts of 1-Octanol were added to a solution of $10.8.10^{-6}$ mol cm^{-3} NaDS. The result is shown in Fig. 4: As the Octanol concentration increases to a quarter of the NaDS concentration, $1/\tau$ decreases by one order of magnitude.

At this time, we cannot present the rate constants of the micellisation process, because the earlier developed theories do not agree with our results, however, we try to apply the recently published theory of Aniansson and Wall[10] to our experimental data.

An extended investigation of the reported findings is in progress.

REFERENCES

1. E. Hutchinson, K.E. Manchester and L. Winslow, J. Phys. Chem. 58, 1124 (1954).
2. E.D. Goddard and G.C. Benson, Canadian J. Chem. 35, 986 (1957).
3. S.D. Hamann, J. Phys. Chem. 66, 1359 (1962).
4. K. Takeda and T. Yasunaga, J. Colloid Interf. Sci. 40, 127 (1972).
5. B.C. Bennion, L.K.J. Tong, L.P. Holmes and E.M. Eyring J. Phys. Chem. 73, 3288 (1969).
6. T. Yasunaga, K. Takeda and S. Harada, J. Colloid Interf. Sci. 42, 457 (1973).
7. W. Knoche and G. Wiese, Chem. Inst. (in press).
8. J. Piercy, M.N. Jones and G. Ibbotson, J. Colloid Interf. Sci. 37, 165 (1971).

9. The viscosity data were taken from: H.M. Chadwell and B. Asnes, J. Am. Chem. Soc. **52**, 3507 (1930); the reproduction of these data in the Landolt-Bornstein tables (II, 5a; p.381) is incorrect.
10. E.A.G. Aniansson, M. Almgren and S.N. Wall this volume.

MECHANISM OF MICELLE FORMATION OF SODIUMDODECYLSULFATE (SDS)

H. Hoffmann,

Institute for Physical Chemistry I, University of Erlangen-Nurnberg, Erlangen, Egerlandstrasse 3

1. INTRODUCTION

Chemical relaxation techniques[1] and stopped flow methods[2] have been used to investigate the kinetics of micelle formation. Two relaxation times have been reported for Sodium dodecylsulfate (SDS). At first, Yasunaga, Oguri and Miura[3] observed a fast relaxation process around one μsec by the sound absorption technique. Later measurements with the pressure jump technique revealed a second relaxation effect in the msec time range whose reciprocal value increased linearly with the total concentration of SDS[4]. A similar behaviour had been observed for solutions of SDS with added electrolyte in temperature jump measurements[5]. The linear rise of $1/\tau$ with c_o was considered to be in accord with the existing theory on the kinetics of micelle formation[6].

Contrary to these results, however, recent measurements by Hoffmann and Janjic[7] and by Kahlweit and Hermann[8] could not confirm the original observations by Yasunaga et al., even though the new measurements had been carried out with the same technique and under the same experimental conditions. The new measurements showed that $1/\tau$ increased linearly with c_o only in a very narrow concentration range and became independent of c_o at $c_o > 1.5$ c.m.c. While the last two investigations agreed at least qualitatively on the trend of the relaxation times as a function of c_o, the quantitative agreement left much to be desired. This discrepancy of the data on supposedly pure tenside solutions raised the question of the role of impurities that might have been present in some or all of the reported measurements.

Since the new data could not be reconciled with the existing theory on micelle formation it was of particular interest to reinvestigate the system in order to find out the reason for the

Wyn-Jones (ed.), Chemical and Biological Applications of Relaxation Spectrometry, 181–193.

varying behaviour. In case the linear rise of $1/\tau$ with c_o was not to be confirmed, the generally accepted theory would have to be abandoned.

While the controversial data on SDS were related to the slow relaxation time and which therefore was re-examined, the fast process was also restudied in the new investigation. It is clear that a comprehensive theory of the kinetics of micelle formation must be able to explain both processes and should not be based on only part of the available information.

2. EXPERIMENTAL AND RESULTS

The measurements were carried out with pressure jump[9] and a shock-wave-apparatus[10] which have been previously described. The tensides used for the measurements were a gift from Henkel and Cie GmbH, Dusseldorf. They have been prepared from pure compounds and recrystallized several times from alcohol. The c.m.c. values were determined from conductance measurements (Λ plotted the square of the concentration c_o). The values agreed with data from literature[11]. Surface tension measurements did not show a minimum at the c.m.c. what is usually taken as conclusive evidence for the purity of the compounds.

The $1/\tau$ for solutions of SDS at various temperatures are plotted in Fig. 1[12]. The data for the slow relaxation time from this investigation turned out to be different from all previous measurements as is shown in Fig. 2, where $1/\tau_2$ from this investigation is compared with the older measurements in a plot of $1/\tau_2$ against the total concentration c_o. The new $1/\tau_2$ values are always larger than the old ones, the linear region in the $1/\tau$ against c_o plot has practically disappeared and there are no effects below the c.m.c. The $1/\tau$ values are nearly constant over the measured concentration range, the barely visible maximum in the curves seems to be real[13]. Unpublished data on other systems indicate the same tendency. The maximum becomes more marked when the measurements are extended to larger concentrations[13].

As seen from Fig. 2, in the older measurements relaxation effects were observed below the c.m.c., that means in a concentration region where theoretically no micelles should exist. This is a clear indication that some impurities were present in the older investigations. Several surfactants were therefore added to the pure solutions in order to test whether additives could indeed cause such drastic changes in the τ-values as the observed ones. As additives, compounds were chosen that could have got into the soaps by the production process like Dodecanol or homologs of SDS. The experiment showed that even in minor concentrations of only 1-2% of the main soap, the added compounds indeed caused large shifts of the relaxation times τ. The fast relaxation time τ_1 remained unaffected by the additives. A similar observation has been made recently[1]. The $1/\tau$ values are plotted in Fig. 3.

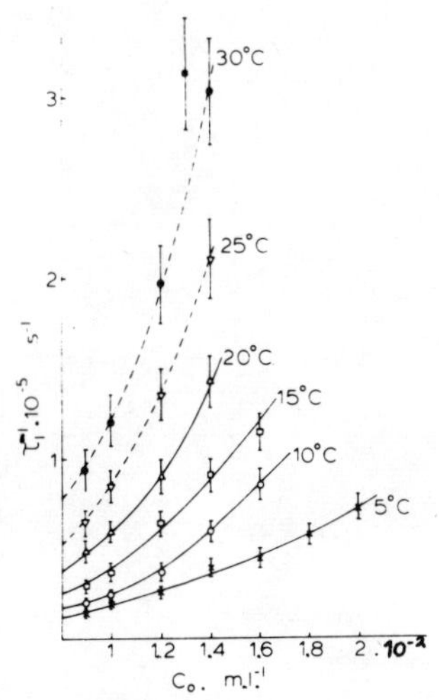

Fig. 1a. Plot of reciprocal relaxation times τ_1 at different temperatures against total concentration of SDS.

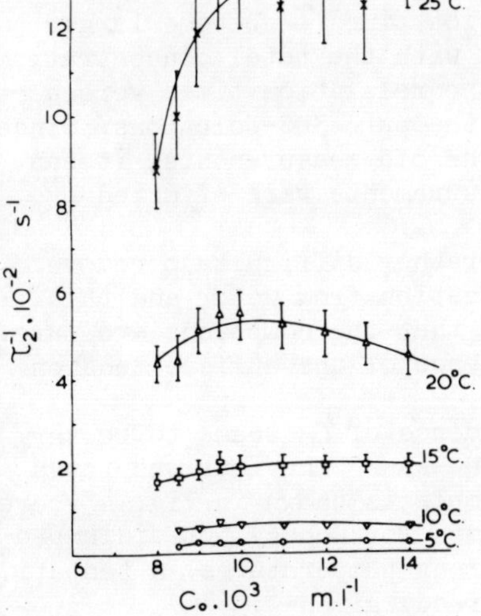

Fig. 1b. Plot of reciprocal relaxation times τ_2 at different temperatures against total concentration of SDS.

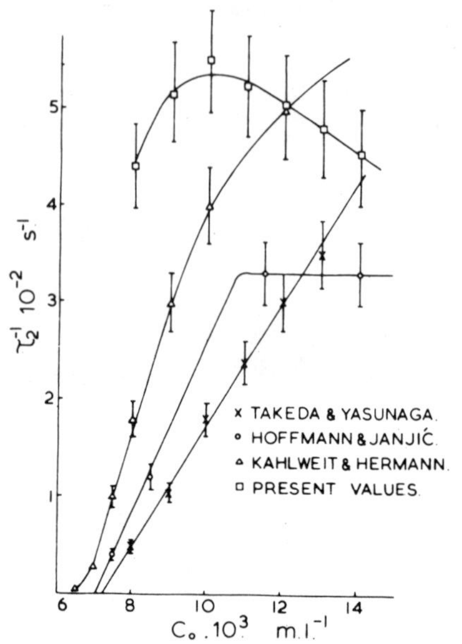

Fig. 2. Plot of reciprocal relaxation times τ_2 from different investigations against total concentration of SDS at 20°C.

In all studied cases the relaxation times τ_2 became larger and the $1/\tau$ values increased linearly with the total concentration right above the c.m.c. Furthermore, the relaxation times were already detectable below the c.m.c. of the pure SDS-solutions. Since all these criteria were observed at the old measurements, it can certainly be concluded that these measurements were effected by unknown impurities.

It is known that impurities are rather difficult to remove from tenside solutions by recrystallization from water and the correct c.m.c. very often is no proof that the compounds are pure. A much more stringent test is the absence of the surface tension minimum at the c.m.c.

The observed concentration dependence of τ_2 seems to be very similar for other ionic detergents. The same behaviour was found for many more systems[14]. An other example is shown in Fig. 4 where the observed reciprocal relaxation times for Dodecylpyridiniumbromide are plotted against c_o for different temperatures. A tentative explanation for this behaviour was already given[13].

The only other known example that shows a completely different concentration dependence in pure solutions that do not contain excess salt is Dodecylpyridiniumiodide for which a more than linear

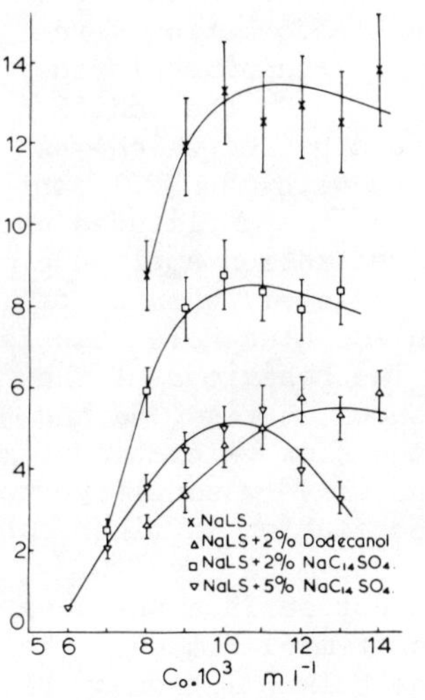

Fig. 3. Plot of reciprocal relaxation times τ_2 at 25°C against total concentration of SDS with different additives.

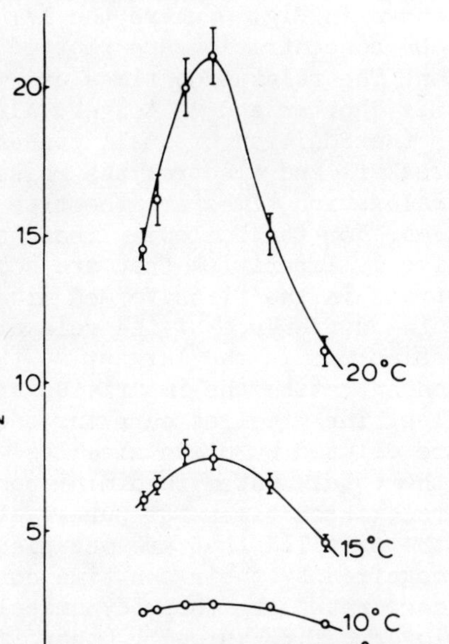

Fig. 4. Plot of reciprocal relaxation times τ_2 at different temperatures against total concentration of DPyBr.

rise of $1/\tau_2$ with c_o was observed[13].

Further interesting results that are very informative for understanding the detailed kinetics of micelle formation were obtained when small amounts of DPCl (Dodecylpyridiniumchloride) were added to solutions of SDS above the c.m.c. On the addition of 0.1 - 1% DPCl a third relaxation time τ_3 appeared which was slower than τ_2 and whose amplitude increased with the DPCl content. Again, as in the case of the other additives, the amplitudes and values of τ_1 remained unaltered while the relaxation amplitude of τ_2 decreased drastically and disappeared completely when c_o of SDS was 1.10^{-2}M and the DPCl concentration amounted more than 1% of c_o. The τ_2-values changed only little. The behaviour of the relaxation times and their amplitudes is shown in Figs. 5a and 5b. The experiments clearly demonstrate that the slow relaxation time that is observed in pure micellar solutions is very sensitive to other additives that can be present in the solution in minor concentrations and which can be incorporated into micelles.

It is likely that molecules which are not surface active are not so effective in changing the relaxation times. But even they can influence the relaxation times from the bulk solution by changing the structure of the solvent as was shown recently by Hermann and Kahlweit[8]. Much larger concentrations however are needed in such a case.

Addition of supporting electrolyte also effects the relaxation times as has been reported and is shown in Fig. 6 where the $1/\tau$ values for both times for a given SDS concentration are plotted against the added NaCl-concentration. The relaxation times are effected in opposite ways: τ_1 becomes shorter and τ_2 longer. All these changes of τ_2 that occur with the addition of small concentrations of additives, structure breakers and electrolytes raise the question whether the reported relaxation times are the true time constants of really pure systems. For the following reasons, the question can be answered positively. Impurities that are solubilized by micelles tend to concentrate in the first formed micelles above the c.m.c. Consequently, the difference of the relaxation times in the pure and not so pure solutions is the largest at the c.m.c. With increasing detergent concentration the impurities are partitioned to more and more micelles. The micelles carrying one or more molecules of an impurity are diluted by clean micelles and the effect of the impurity on the total system is diminished. The relaxation times approach the relaxation times for pure systems When DPCl is used as an impurity, the micelles that are occupied by even a single DP^+ cation are recognized by their own time constant. Micelles of SDS with DP^+ incorporated are thermodynamically more stable as indicated by their lower c.m.c.-value[12]. Other impurities that might be present in solution should therefore preferentially solubilize in these more stable micelles. Previously occupied micelles should be vacated. As seen from Fig. 5 the relaxation time τ_2 that is believed to be due to the clean micelles is very little effected by the addition of DPCl. If anything, the

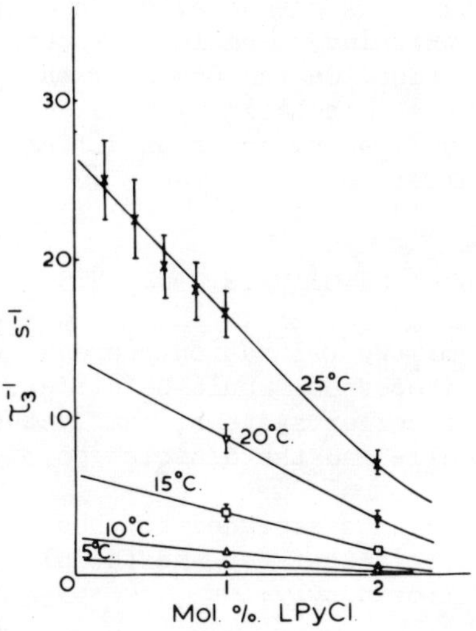

Fig. 5a. Plot of reciprocal relaxation times τ_3 against the concentration of added DPyCl. c_o of SDS = $0.95.10^{-2}$ m/l.

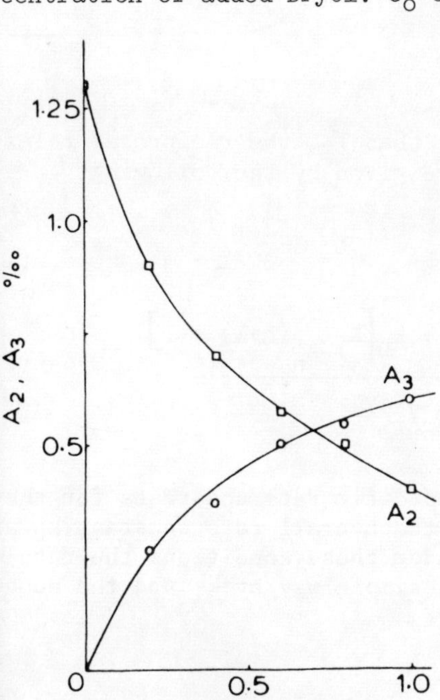

Fig. 5b. Plot of the amplitudes of the relaxation times τ_2 (A_2) and τ_3 (A_3) against the concentration of DPCl in mol % of the SDS-concentration of $c_o = 0.95.10^{-2}$ mol/l.

relaxation times τ_2 are slightly longer with DPCl than without. Altogether these observations can be taken as strong evidence that the relaxation times measured in this investigation for pure SDS solutions were unaffected by impurities. On the basis of the new data the old mechanism that was first postulated by Kresheck et al.[6] must be abandoned. A similar conclusion, based on different information, was put forward by Muller[15].

3. INTERPRETATION OF THE DATA AND EVALUATION OF RATE CONSTANTS

Without having definite evidence of two relaxation processes, Muller[15] postulated recently that two processes should be differentiated in relaxation experiments on micellar systems, namely the dissociation of a monomer from the micelle and the dissolution of a complete micelle.

The observed two relaxation times can be assigned to these two processes, the fast one to the dissociation step and the slow one to the dissolution step, in the following way:

$$(n+1)A \underset{k_D}{\overset{k_F}{\rightleftharpoons}} An + A \underset{k_{-1}}{\overset{k_1}{\rightleftharpoons}} An + 1$$

$$\tau_2 \qquad\qquad \tau_1$$

For the case that τ_1 is much faster than τ_2, the reciprocal relaxation times for such a mechanism are given by the following equations:

$$\frac{1}{\tau_1} = k_{-1} + k_1 (c_A + c_{A_n}) \tag{1}$$

$$\frac{1}{\tau_2} = \frac{nk_F c_A^{n-1}\left[K_1 c_A(n+1) + n\right] + k_D\left[K_1 c_{A_n}(n+1) + 1\right]}{K_1(c_A + c_{A_n}) + 1} \tag{2}$$

Making these assumptions that the specific rate constants for the Multistep dissolution process with the overall rate constant k_D ar the same, Muller could show that under these conditions the rate constant k_D could be expressed in a simple way by k_1 and the numbe of the steps n. The result is:

$$1/\tau_2 = 2\ k_{-1}/n^2$$

Aniansson and Wall[15] very recently gave a more realistic and mathematically detailed description of the rate constants k_{-1} and k_D.

This theory predicts a maximum of $1/\tau_2$ with increasing concentration which is in agreement with the new data.

It is also noteworthy that Nakagawa[17] recognized the need for a second relaxation time to describe the relaxation phenomena in micelle solution even though he was only aware of one relaxation process and he assumed that the other relaxation effect could not be detected experimentally. For a more detailed discussion see Ref. 13. While the various theoretical models may lead to somewhat different conclusions and are based on different theoretical assumptions, it is still noteworthy that all models predict the existence of two relaxation times.

An explanation for the increase of $1/\tau_1$ with c_o has recently been put forward by Rassing and Wyn-Jones[1]. In their treatment of the data, the effect of the charge of the micelles is completely neglected. The present data indicate however that the ionic strength has a definite effect on the relaxation times, as is to be expected for a reaction between highly charged particles[13]. But the ionic strength of the solution does not remain constant when c_o is raised. The change of the ionic strength is not only going to effect the rate constant but also the surface charge per area and also the average number of monomers in the micelles. All these secondary changes should in turn effect the rate constant. It is therefore likely that many factors contribute to the increase of $1/\tau_1$ with c_o.

The inadequacy of the theory can also be shown on the experimental data. The rise of $1/\tau$ close to the c.m.c. is not linear with c_o and the extrapolation of the $1/\tau$ against c_o plot does not cut the abcissa at the c.m.c. as required by theory.

A detailed discussion of existing models will not be given in this contribution, see however Zana's contribution[20]. For the purpose of this contribution it seems sufficient to extract meaningful rate constants from the experimental data. The safest way to do this seems to be to extrapolate the equations (1) and (2) to the c.m.c. For this concentration the expressions simplify to

$$(1/\tau_1)\ \text{c.m.c.} = 2\, k_{-1}$$

$$(1/\tau_2)\ \text{c.m.c.} = k_D/2$$

These values are given in the table for different temperatures. The meaning of these constants is clear from the preceeding discussion.

REMARKS ON THE INCORPORATION OF DPCl INTO SDS MICELLES

No detailed theory exists so far on the incorporation of impurities, like dye molecules or surface active agents of opposite charge into micelles. Such measurements are usually carried out

TABLE 1. VALUES OF $(1/\tau)_{c.m.c.}$ FOR SDS AT DIFFERENT TEMPERATURES

°C 5		10		15		20		25	
$\frac{1}{\tau_1}\{s^{-1}\}$	$\frac{1}{\tau_2}\{s^{-1}\}$	$\frac{1}{\tau_1}\{s^{-1}\}$	$\frac{1}{\tau_2}\{s^{-1}\}$	$\frac{1}{\tau_1}\{s^{-1}\}$	$\frac{1}{\tau_2}\{s^{-1}\}$	$\frac{1}{\tau_1}\{s^{-1}\}$	$\frac{1}{\tau_2}\{s^{-1}\}$	$\frac{1}{\tau_1}\{s^{-1}\}$	$\frac{1}{\tau_2}\{s^{-1}\}$
14000	23	19000	55	26000	170	38000	450	-	900

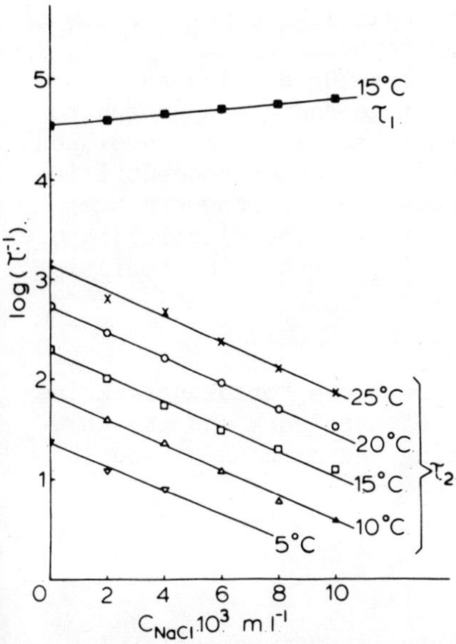

Fig. 6. Plot of the logarithms of the reciprocal relaxation times τ_1 and τ_2 for a total concentration of SDS of $c_o = 1.10^{-2}$ mol/l against added concentration of NaCl.

with dye molecules[18] because of the ease with which such measurements can be made. Very often the absorption of the dye charges and the incorporation can easily be followed by spectroscopy. As is shown in this investigation, the measurements can also be carried out using relaxation techniques and oppositely charged soap molecules instead of the dye. It is usually assumed that the dye molecule penetrates the palisade layer of the micelle in order to get into the micelle. The long time constants that are observed are attributed to the difficulty of the dye molecule to penetrate the palisade layer. Such explanations stand in contradiction to the observation that like-soap molecules are associated into micelles with fast reactions having rate constants that approach the diffusion controlled limits[13]. Judged from these observations, it is likely that the incorporation proceeds by the following mechanism: If a dye or a surface active tenside ion of opposite charge than the micelle is added to a micelle containing solution it immediately forms an ion pair with the readily available monomer ions. This step has already been suggested by Yasunaga and coworkers[19] on the basis of stopped flow measurements. Such a process is likely because the monomer concentration is usually much larger than the micelle concentration. Even in the case that the rate constants for both reactions are of similar size the chance for the dimer formation is much larger than for the association of the dye with the micelle. On the same argument that salt like dimer has a better chance of reacting with further monomer molecules than with a micelle. By these preliminary steps the dye ion surrounds itself with tenside ions of opposite charge and the approach to micelles is made difficult. On the other hand monomers have been used up by these preliminary steps and the monomer concentration has fallen below the c.m.c. As a consequence new monomers must be formed from existing micelles. In this way the dissolution of existing micelles becomes rate determining for the incorporation of dye molecules into micelles.

ACKNOWLEDGEMENTS. I wish to thank the Deutsche Forschungsgemeinschaft for a research grant in support of this work and the Co. Henkel & Cie., Dusseldorf, for providing the compounds used in this investigation.

REFERENCES

1. For a complete summary of available literature see: J. Rassing, P.J. Sams and E. Wyn-Jones, J. Chem. Soc. Farad. Transactions II, (1974) 70, 1247
E. Graber, J. Lang, R. Zana, Kolloid-Z. Polym. 238, 470, 479 (1970).
2. T. Yasunaga, K. Takeda and S. Harado, J. Colloid Interface Sci (1973), 42, 457.
3. T. Yasunaga, H. Oguri and M. Miura, J. Colloid Interface Sci 23, 352 (1967).
4. K. Takeda, T. Yasunaga, J. Colloid Interface Sci., 40, 127 (1972).
5. B.C. Bennion, E.M. Eyring, J. Colloid Interface Sci., 32, 286 (1970).
6. G.C. Kresheck, E. Hamore, G. Davenport, H.A. Scheraga, J. Amer. Chem. Soc., 88, 246 (1966).
7. T. Janjic, H. Hoffmann, Z. Physik. Chem., 86, 322 (1973).
8. U. Hermann, M . Kahlweit, Ber. Bunsenges. Phys. Chem. 77, 1119 (1973).
9. H. Hoffmann, J. Stoehr, E. Yeager in "Chemical Physics of Ionic Solutions", B.E. Conway and R.G. Barradas, editors, J. Wiley and Sons, N.Y., pp255-280 (1966).
10. G. Platz and H. Hoffmann, Ber. Bunsenges. Phys. Chem., 76, 491 (1972).
11. Critical Micelle Concentrations of Aqueous Surfactant Systems ed. P. Mukerjic and K.J. Mysels.
12. for details see:
R. Folger, H. Hoffmann, W. Ulbricht, Ber. Bunsenges Phys. Chem 78 (in press).

13. J. Lamy, C. Tondre, R. Zana, R. Bauer, H. Hoffmann, W. Ulbricht J. Phys. Chem. (in press).
14. H. Hoffmann, W. Ulbricht, unpublished data.
15. N. Muller, J. Phys. Chem. 76, 3017 (1972).
16. E. Anianssen and S. Wall, J. Phys. Chem. 78, 1024 (1974).
17. T. Nakagawa, Colloid and Polymer Sci. 252, 56 (1974).
18. B. Robinson, this publication
19. K. Takeda, N. Tatsumoto, T. Yasunaga, J. Colloid Interface Sci. (In press).
20. R. Zana, this publication.

KINETIC STUDIES OF THE DISSOCIATION OF NONIONIC DETERGENT MICELLES

J. Lang

C.N.R.S., Centre De Recherches Sur Les Macromolecules,
6 rue Boussingault, 67083 Strasbourg Cedex - France

For several years relaxation methods have been applied to the kinetic studies of micellar equilibria. Most of these studies deal with ionic detergents.[1-13] Two kinetic studies however have been devoted to nonionic detergents; a temperature-jump[14] (T-jump) and a stopped-flow[15] study of aqueous solutions of octylphenyl polyoxyethylene alcohol, also called p-1,1,3,3 tetramethylbutylphenyl polyoxyethylene alcohol,

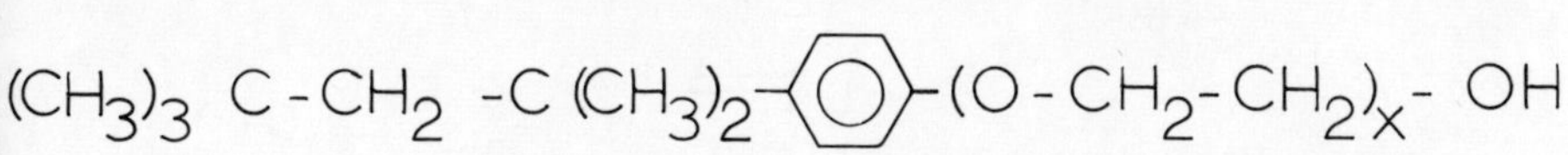

This detergent has been chosen for the different spectroscopic properties of the monomer in the micellar and in the aqueous phase.[16] Thus it has been possible to follow with a spectrophotometric detection any chemical relaxation resulting from the displacement of the equilibrium between micelles and free detergent molecules in the water phase under a sudden temperature rise (T-jump) or a sudden dilution (stopped-flow) of the solution.

In the T-jump experiment two detergents have been studied corresponding to x = 16 and x = 30, i.e. containing, respectively, 16 and 30 moles of ethylene oxide units per mole of detergent. These two detergents will be designated by OPE 16 and OPE 30, respectively.

In the stopped-flow experiments only the detergent having the longest hydrophilic chain, namely OPE 30, has been studied.

The critical micelle concentrations c.m.c. of these two detergents were determined with a Cary 14 spectrophotometer by noting the optical density change which accompanies micellization[16] at 276nm. The c.m.c.'s were found[14] equal to 0.042 and 0.15 g/dl

Wyn-Jones (ed.), Chemical and Biological Applications of Relaxation Spectrometry, 195–200.

for OPE 16 and OPE 30, respectively. Although these two detergents are polydisperse in polyoxyethylene chain length, molecular weights of 910 for OPE 16 and 1526 for OPE 30 were adopted. Thus, the value of the c.m.c. is 4.6×10^{-4}M for OPE 16 and 9.8×10^{-4}M for OPE 30.

I - TEMPERATURE JUMP EXPERIMENTS

The T-jump experiments were carried out at 290nm, on solutions containing 0.15 M KCl and over a wide range of concentrati of detergent in order to determine the relaxation time as a functior of the detergent concentration. However relaxations were observed only for concentrations between 0.037 and 0.11 g/dl for OPE 16 and between 0.09 and 0.3 g/dl for OPE 30. Each solution showed a rapid change in optical density with a relaxation time less than $\sim 10^{-5}$ sec., which corresponds approximately to the heating time of the solution and which has consequently not been studied. This fast step is followed by a slow absorption change which correspond to a single relaxation time τ. τ varies from 400 msec (at 0.037 g/dl to 28.5 msec (at 0.11 g/dl) for OPE 16 and from 66 msec (at 0.09 g/dl) to 1.3 msec (at 0.3 g/dl) for OPE 30. The variation of τ^{-1} vs. the detergent concentration is linear above the c.m.c. both for OPE 16 and OPE 30. Below the c.m.c. this variat is not linear and the relaxations observed can be attributed to the formation of premicellar aggregates which are known to precede the formation of micelles for this type of detergent.[17] The linear dependence of τ^{-1} on the detergent concentration above the c.m.c. h been interpreted with the model proposed by Kresheck et al.[4] These authors assume that the formation of micelles is due to bimolecular reactions of the type

$$\begin{array}{l} A + A \rightleftharpoons A_2 \\ A_{n-1} + A \underset{k_{n,n-1}}{\overset{k_{n-1,n}}{\rightleftharpoons}} A_n \end{array} \qquad (1)$$

where A represents the free detergent molecules and A_i micelles with i detergent molecules. They assume also that reactions of the type

$$A_j + A_k \rightleftharpoons A_n$$

are negligible. This model leads to the following expression for the reciprocal relaxation time:

$$\tau^{-1} = -k_{n,n-1}\ (n-1) + nk_{n,n-1}\ a/[\bar{A}]$$

where a is the total detergent concentration and $[\bar{A}]$ the c.m.c. Knowing the aggregation number n, $k_{n,n-1}$ can then be obtained from the slope of the variation of τ^{-1} vs a. Although n is not available in the literature for OPE 16 and OPE 30, different data for nonionic detergents[14] suggest that n = 48 for OPE 16 and n = 11 for OPE 30. Values of $k_{n,n-1}$ equal to 0.40 ± 0.05 and 73 ± 5 sec^{-1} at 24.8° were thus found for OPE 16 and OPE 30 respectively, showing an increase of $k_{n,n-1}$ with the length of the hydrophilic part of the detergent molecule.

II - STOPPED - FLOW EXPERIMENTS

These kinetic experiments have been undertaken in order to try to test the assumption made by Kresheck et al. that the rate determining step in a micelle dissociation is the release of the first detergent molecule from the micelle, which is characterized by the dissociation rate constant $k_{n,n-1}$. The following reaction

$$A_n \xrightarrow{k_{n,n-1}} A_{n-1} + A \qquad (2)$$

is characterized, according to the value of $k_{n,n-1}$ found by T-jump, by a relaxation time equal to 137 msec, i.e. slow enough to be detected in a stopped-flow experiment. In reaction (2) A_n and A_{n-1}, represent micelles with respectively n and n-1 surfactant molecules, and A the free surfactant molecules.

Each stopped-flow experiment was conducted by mixing the detergent solution at concentration C_o above the c.m.c. with the solvent. The ratio of mixing was generally 1 to 1 in volume. The dissociation of the micelles was followed spectrophotometrically at 276 nm. Variations in absorbance were observed only for concentration C_o between the c.m.c. and three times the c.m.c. Since the dissociation of the micelles involves, beside reaction (2), many other steps, and even the complete dissociation of the micelles in some cases, the method used for the interpretation of the data was to assume that the slope at the origin of the curve absorbance vs time can be associated with the dissociation of the first molecule of detergent from the micelle, i.e., to reaction (2). It must be pointed out that this assumption seems reasonable since the amplitude of the variation of absorbance associated to reaction (2) as the micelle dissociates is probably comparable to the absorbance change due to the dissociation of a second, third, etc... molecule of detergent from the micelles. With this assumption the value of $k_{n,n-1}$ has been found equal to 510 sec^{-1} for OPE 30 at 25°[15], i.e. a value much higher than the value obtained by T-jump, 73 sec^{-1} [12].

III - DISCUSSION

The discrepancy between the values obtained for $k_{n,n-1}$ by T-jump and stopped-flow seems to indicate that the relaxation times measured by T-jump are not directly associated to the release of the first detergent molecule from the micelle. Although the present result, obtained with OPE 30, cannot be extended to the other detergents studied by T-jump, it must be pointed out that for the dodecylpyridinium diiodide for instance, a very fast conductivity change occurring in less than 5 msec and due to the dissociation of a large part of the micelles, has been observed by Jaycock and Ottewill[2] in their stopped flow experiment. This result compared to the result of Kresheck et al[4] who have determined by T-jump a value of 50 sec^{-1} for $k_{n,n-1}$ which corresponds to a relaxation time equal to 20 msec using reaction (2) seems to indicate that for dodecylpyridinium iodide also the relaxation times measured by T-jump are not directly associated with the release of the first detergent molecule from the micelle.

Several other experiments carried out with ionic detergents by ultrasonic absorption,[13,18-20] NMR[21-24] and EPR[25] have given rate constants also much faster than the rate constants obtained for the same detergents by T-jump or P-jump. Several theoretical attempts have been made[26-29] in order to give a more precise interpretation of the different kinetic results which indicate that micellar equilibriae are characterized by at least two relaxation times, a fast relaxation time which is associated with the following reaction

$$A_n \rightleftharpoons A_{n-1} + A \qquad (3)$$

and a slower relaxation time associated with micelle formation-dissolution equilibrium (4).

$$A_n \rightleftharpoons nA \qquad (4)$$

Indeed recent experimental results show that some detergent systems are characterized by two relaxation times[30,31]. Therefore it may well be that some of the T-jump results which were interpreted in terms of exchange reactions similar to reaction (3) should have been related in fact to micelle formation-dissolution reactions such as reaction (4). In the case of OPE 30 it seems, from the comparison of the results obtained by T-jump and stopped-flow, that the T-jump experiments do not give the rate constant of dissociation associated to reaction (3) as it should according to model of Kresheck et al. However more results are necessary in order to assign precisely the relaxation times observed for OPE 30 by this last technique.

REFERENCES

1. M. Czerniawski, Roczniki Chem., 39, 1469 (1965); Chem. Abst., 64, 13430 h (1966).
2. M. J. Jaycock and R. H. Ottewill (1967), in "Chemistry Physics and Application of Surface Active Substances, Proc. 4th Intern. Congr. Surface Active Substances, Brussels, 7-12 September, 1964". J. Th. G. Overbeek, Ed., Vol. 2, Gordo[n] and Breach, New York, pp. 545-553; Chem. Abst. 71, 129040 h (1

3. P. P. Mijnlieff and R. Ditmarsch, Nature, 208, 889 (1965).
4. G. C. Kresheck, E. Hamori, G. Davenport and H. A. Scheraga, J. Amer. Chem. Soc., 88, 246 (1966).
5. T. Yasunaga, H. Oguri and M. Miura, J. Colloid Interface Sci., 23, 352 (1967).
6. T. Yasunaga, S. Fujii and M. Niura, J. Colloid Interface Sci., 30, 399 (1969).
7. B. C. Bennion, L. K. J. Tong, L. P. Holmes and E. M. Eyring, J. Phys. Chem., 73, 3288 (1969).
8. B. C. Bennion, Ph.D. Thesis, University of Utah, (1969).
9. B. C. Bennion and E. M. Eyring, J. Colloid Interface Sci., 32, 286 (1970).
10. R. Zana and J. Lang, Compt. Rend. Acad. Sci. (Paris), C 266, 893 (1968).
11. R. Zana and J. Lang, Compt. Rend. Acad. Sci. (Paris), C 266, 1347 (1968).
12. E. Graber, J. Lang and R. Zana, Kolloid-Z.Z. Polym., 238, 470 (1970).
13. E. Graber and R. Zana, Kolloid-Z.Z. Polym., 238, 479 (1970).
14. J. Lang and E. M. Eyring, J. Polym. Sci., Part A-2, 10, 89, (1972).
15. J. Lang, J. J. Auborn and E. M. Eyring, J. Colloid Interface Sci., 41, 484 (1972).
16. W. B. Gratzer and G. H. Beaven, J. Phys. Chem., 73, 2270, (1969); S. Ikeda and G. D. Fasman, J. Polym. Sci., Part A-1, 8, 991 (1970).
17. P. Mukerjee, Advan. Colloid Interface Sci., 1, 241 (1967).
18. P. Sama, E. Wyn-Jones and J. Rassing, Chem. Phys. Lett., 13, 233, (1972).
19. J. Rassing, P. Sams and E. Wyn-Jones, J.C.S. Farad. Trans. II, 69, 180 (1973); ibid., 70, 1247 (1974).
20. D. Adair, V. Reinsborough, N. Plavac and J. Valleau, Can. J. Chem., 52, 429 (1974).
21. T. Nakagawa and H. Inoue, Proc. Fourth Int. Congr. Active Substances, Brussels 1964, Section B, paper 11; H. Inoue and T. Nakagawa, J. Phys. Chem., 70, 1108 (1966); T. Nakagawa, H. Inoue, H. Jizomoto and K. Horiuchi, Kolloid Z.Z Polym., 229, 159 (1969); T. Makagawa and K. Tori, ibid, 194, 143 (1964).
22. N. Muller and T. Johnson, J. Phys. Chem., 73, 2042 (1969); N. Muller and R. Birkham, ibid, 71, 957 (1971); N. Muller and F. Platko, ibid., 75, 547 (1971).
23. J. Erikson and G. Gillberg, Acta Chim. Scand., 20, 2019 (1966).
24. M. Alexandre, C. Fourchet and A. Rigny, J. Chim. Phys., 70, 1073 (1973).
25. K. K. Fox, Trans. Farad. Soc., 67, 2802 (1971).
26. N. Muller, J. Phys. Chem., 76, 3017 (1972); N. Muller, "Recent Advances in the Chemistry of Micelles" in Reaction Kinetics in Micelles, E. Cordes Ed., Plenum Press, New York, 1973, p.1.

27. T. Nakagawa, Colloid. Polym. Sci., 252, 56 (1974).
28. E. Aniansson and S. N. Wall, J. Phys. Chem., 78, 1024 (1974).
29. A. Colen, J. Phys. Chem., 78, 1676 (1974).
30. R. Folger, H. Hoffmann and W. Ulbricht, Ber. Bunsenges. Physik. Chem., 1974, in press.
31. J. Lang, C. Tondre, R. Zana, R. Bauer, H. Hoffmann and W. Ulbricht, J. Phys. Chem., 1974, in press.

DYNAMICS OF SMALL MOLECULE - MICELLE INTERACTIONS

Brian H. Robinson, Neal C. White, Cecilia Mateo,
Katherine J. Timmins and Alan James.
Chemical Laboratory, University of Kent at
Canterbury, U.K.

INTRODUCTION

Only a few direct studies on the kinetics of solubilization by micelles have been carried out, notably the solubilization of large particles (membranes) by Eyring et al.[1], and of the dye pinacyanol chloride by Yasunaga et al.[2]. However E.S.R.[3] and N.M.R.[4] methods have been used recently to study the dynamics of solubilizate motion. From the initial studies, it seems likely that particles and molecular aggregates are probably solubilized by a fragmentation - induced micelle formation process (since effects are observed below the critical micelle concentration (c.m.c.) of the surfactant) whereas small molecules can often be absorbed into an already existing micelle by an adsorption/absorption process, although an alternative induced - micellization mechanism may operate for certain dyes.

In this paper, a stopped-flow kinetic study of the incorporation of acridine (and related) dyes into anionic micelles formed by sodium n-alkyl sulphates ($C_nH_{2n+1}\ SO_4^-$) is reported. This system has been studied previously by Eyring et al.[5]. The binding of these dyes to macromolecules (e.g. polypeptides, polymers, DNA and membranes) has been much studied over the past few years primarily because they are useful model systems for the study of carcinogenic and mutagenic activity. As a result of this effort, there is now a large amount of spectral evidence available for the interaction of acridine dyes with various types of surface structure, which is useful in the analysis of dye - micelle interactions. The work presented here has applications in the analysis of (i) Micellar catalysis, (ii) Mode of action of carcinogens and (iii) Micellar structure.

Wyn-Jones (ed.), Chemical and Biological Applications of Relaxation Spectrometry, 201–210.

EXPERIMENTAL

The acridine-type dyes used are shown in figure 1.

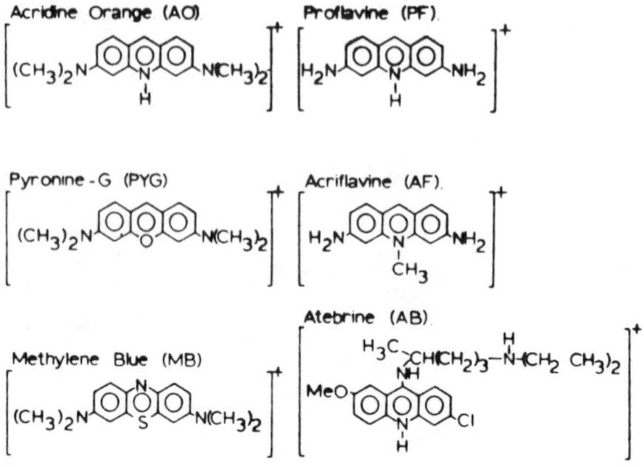

Fig. 1. Acridine dyes employed

Most experiments were performed at pH $\sim$ 7, when the dyes carry a delocalized unit positive charge. Visible spectrophotometric and fluorimetric measurements were made in order to characterize the dye-detergent interaction both above and below the c.m.c. Stopped-flow experiments were performed on a small - volume device equipped with absorbance, light scattering and fluorescence detection.

SPECTRAL CHARACTERIZATION OF THE DYE-SURFACTANT INTERACTIONS

a) Visible spectrophotometry

From the extinction profile in figure 2A for the acridine orange (AO) - sodium dodecyl sulphate (NaLS) system, two distinct types of interaction are apparent. For a fixed low total concentration of dye ($\simeq 10^{-5}$M) and surfactant concentration $\ll$ c.m.c., spectral changes (hypsochromic shift and hypsochromism at the monomer peak wavelength) suggest that dye molecules become stacked in a planar fashion (like playing cards) as the surfactant concentration is increased below the c.m.c. In general, the extent of aggregation increases with both the tendency of the dye to stack in free solution (as measured by K_D in Table 2) and the chain-length of the surfactant. This behaviour seems to indicate a mutually-induced aggregation process involving both dye and surfactant anion. Alternative possibilities are (i) a simple dye-surfactant ion-pair (dye salt) is formed, the coulombic interaction being further stabilized by the hydrophobic nature of both cation and anion, and (ii) the dye induces micelle formation below

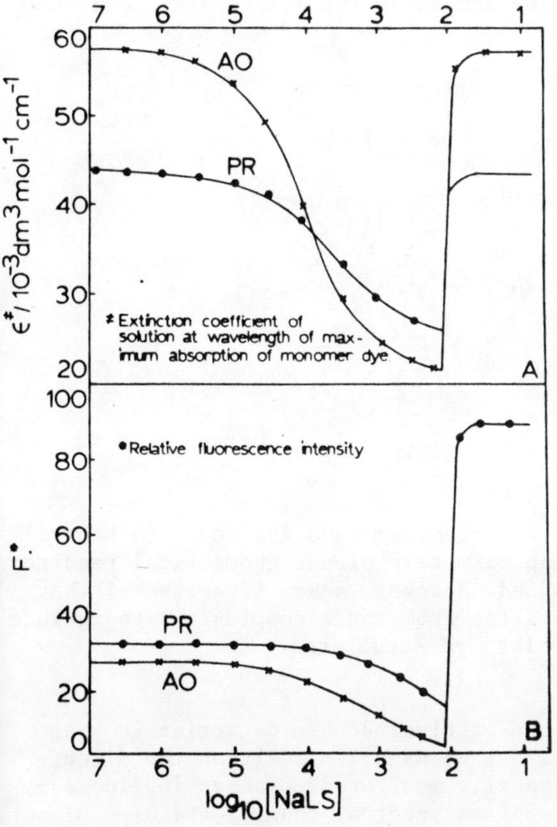

Fig. 2. Spectral profiles for dye-sodium dodecyl sulphate (NaLS) interactions.

the normal c.m.c., as measured in dye-free surfactant solutions, leading to the formation of an 'impure' micelle containing the solubilized dye. However, neither of these alternatives are consistent with the observed spectra, and the dependence of the spectra on surfactant concentration for the different dyes.

In the region of and above the c.m.c. of the surfactant, the spectral changes indicate that a species similar to free monomer dye is rather abruptly restored as the predominant dye species in solution. The slight wavelength shift to the red (≃5 nm) and the increased extinction coefficient suggest that the dye is interacting with the micelle, most probably in an absorbed state in the hydrocarbon region of the micelle. (The spectrum closely resembles that for acridine orange bound at an intercalated site in DNA, and dye dissolved in hydrocarbon solvents).

A highly schematic view of the two types of postulated dye-surfactant interactions below and above the c.m.c. are shown in figure 3.

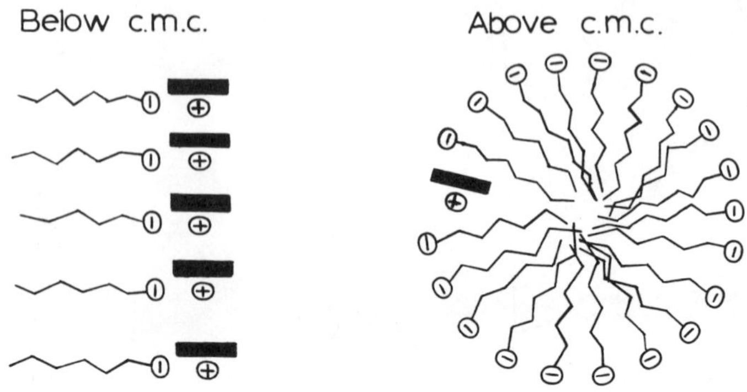

Fig. 3. Postulated dye surfactant interactions.

Other acridine dye-anionic detergent systems are found to behave in a similar way, if the approximately planar geometrical requirements of the dye are satisfied. However, there is evidence[6] that it would be unwise to generalise from these conclusions to include substrate molecules with bulky 3-D structures.

b) Fluorimetry: The advantage of fluorescence detection is that lower concentrations of dye can be used; in addition the direct absorption process is more easily monitored. Changes in fluorescence largely parallel absorption spectral changes (figure 2B). Addition of low concentrations of surfactant $\ll$c.m.c. results in a quenching of fluorescence, consistent again with the formation of "playing-card" dye stacks. However, above the c.m.c., the fluorescence intensity increases dramatically, the relative fluorescence intensity being greater on solubilization than that of the free dye in aqueous solution. These general effects are observed for all the dyes studied. If the c.m.c. is taken as the mid-point of the transition from the (essentially) stacked dye to the solubilized monomer dye, then the c.m.c. values agree to $\pm$5% with those obtained by conductiometric measurements (in the absence of dye). The c.m.c. as generally defined has no sound theoretical basis[7]; the value taken for the c.m.c. is dependent on the experimental method used for its measurement and the method of analysis. Undoubtedly, some dyes give misleading values for the c.m.c. (e.g. crystal violet), but acridine dyes seem to be sensitive to the appearance of micelles in the solution. For the latter dyes, it was found that c.m.c. values obtained by fluorescence and absorbance methods are identical, and independent of dye concentration for c.m.c./[Dye] $>$ 100.

KINETICS OF DYE SOLUBILIZATION (ABSORPTION)

In order to study the dye-micelle interaction, the method used was to mix a low concentration of dye ($\simeq 2 \times 10^{-5}M \rightarrow 5 \times 10^{-6}M$) with an equal volume of a micellar solution at a surfactant concentration < 2 c.m.c., such that after mixing and on completion of the reaction, all the dye was solubilized by the micelle. As the final concentration of micelle is increased above the c.m.c., it is found that the rate of reaction increases until it becomes independent of both dye and micelle concentration. Under these limiting conditions, good first-order plots are obtained, which suggests a simple kinetic scheme.

The overall rate dependence on dye and micelle concentration is consistent with interaction mechanism of the type shown in Scheme 1.

SCHEME 1

$$D + M_S \underset{}{\overset{K_{12}}{\rightleftharpoons}} \underset{ADS}{DM_S} \underset{k_{32}}{\overset{k_{23}}{\rightleftharpoons}} \underset{ABS}{D^*M_S}$$

where M_S - micelle surface binding site, DM_S - represents an adsorbed dye on the micelle surface (weak binding site) and D^*M_S - represents an absorbed dye-dye interacting with hydrocarbon groups in the micelle (strong binding site).

However, when only small amounts of micelle are present, the kinetics are more complex, as shown in Scheme 2, since competition reactions become significant.

SCHEME 2

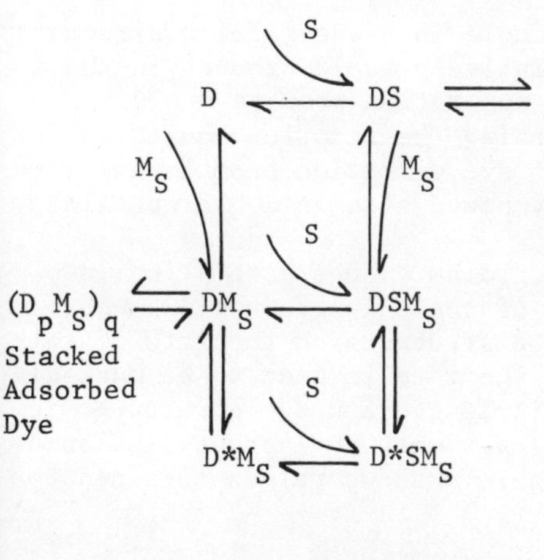

DYE-SURFACTANT AGGREGATE
(Size of aggregate is time - dependent, leading to suspension)

When n = 1, this species corresponds to an induced micelle. S = surfactant molecule $\simeq$ c.m.c.

When $[M_S]_o \gg [D]_o$, (subscript o indicates initial concentrations), two conditions are satisfied. These are: i) The competition reaction $D + S \rightleftharpoons DS$ (and subsequent reactions) is minimised, and ii) Stacking of adsorbed dye on the surface of the micelle is prevented.

Therefore, under these conditions, Scheme 2 reduces to Scheme 1, and since the adsorption process is expected to be fast (k_{12} = diffusion-controlled):

$$k_{obs} = \frac{K_{12} \cdot k_{23}[M_S]}{1 + K_{12}[M_S]} + k_{32} \qquad (1)$$

Under the experimental conditions:

$$[M_S] = \alpha(C_A^o - \text{c.m.c.}) \qquad (2)$$

where α is the fraction of the surface sites not neutralized by sodium counter-ion binding. Also $[M_S] \gg K_{12}^{-1}$ and $k_{23} \gg k_{32}$. Hence $k_{obs} \sim k_{23}$ (i.e. k_{obs} is independent of $[M_S]$ and $[D]$ as observed experimentally).

An experimental estimate of K_{12} can be made from data obtained when the more complete equation (1) is applicable, and is $5 \times 10^3 \rightarrow 1 \times 10^4$ dm^3 mol^{-1} for binding to a charged surface with sulphate head groups. This contrasts with $K \sim 300$ dm^3 mol^{-1} for acridine orange binding to a linear polyglutamate chain (non-cooperative binding)[8] and $K \sim 1$ dm^3 mol^{-1} for dye binding to a single small oppositely charged ion.[9]

Alternative mechanisms for the limiting situation, e.g. induced micellization by surfactant forming around the dye, are inconsistent with a detailed analysis.

KINETIC RESULTS

The absorption of a dye molecule into the micelle structure (measured by $k_{obs} = k_{23}$) is a relatively rapid process in the millisecond time range. A simple adsorption process would be expected to be much faster than this. The reaction monitored is thought to be essentially that of dye diffusion from a weak surface bound site to a more tightly-bound site in a hydrophobic region in the micelle interior.

Table 1 presents data for acridine orange absorption into sodium n-alkyl sulphate micelles of chain length n = 10 → 16. The effect of perturbations in dye structure on the rate of solubilization is shown in Table 2. The rate is seen to be markedly dependent on both surfactant chain-length and dye structure. Table 3 shows that the rate increases with an increase in ionic strength of the medium (NaCl). Table 4 shows values obtained for

Table 1. EFFECT OF CHAIN-LENGTH ON THE RATE OF SOLUBILIZATION OF AO. SURFACTANT = $C_nH_{2n+1}SO_4^-Na^+$

		k_{obs}/s^{-1} (±5%)	
n	T/K	A	F
10	298.2	180	185
12	298.2	47.6	48.3
14	298.2	11.9	9.5
16	316.2	13.1	9.7

A = Absorbance Detection B = Fluorescence Detection

Table 2. EFFECT OF DYE STRUCTURE ON THE RATE OF SOLUBILIZATION INTO NaLS MICELLES (298.2K)

DYE	k_{obs}/s^{-1}	K_D/dm^3mol^{-1}
AF	>500	5.0×10^2
PF	>500	5.0×10^2
AO	47.6	1.0×10^4
MB	47.5	4.5×10^3
PYC	47.2	8.3×10^2
AB	1.4	$\approx 1 \times 10^2$

K_D = Dimerization constant of free dye in aqueous solution.

Table 3. EFFECT OF ADDITIVES ON THE RATE OF SOLUBILIZATION OF AO INTO NaLS MICELLES.

Additive	T/K	k_{obs}/s^{-1}
-	288.2	27.0
0.025M NaCl	288.2	30.4
0.05M NaCl	288.2	41.8
0.10M NaCl	288.2	50.6
-	298.2	47.6
5% $C_{14}H_{29}SO_4^-Na^+$	298.2	41.6
5% $C_{12}H_{25}OH$	298.2	37.6

Table 4. RATE OF SOLUBILIZATION OF SPECIES OF THE TYPE AOR (WHERE R = C_nH_{2n+1}) INTO NaLS MICELLES (298.2K)

n	k_{obs}/s^{-1}	$10^{-4}K_D/dm^3mol^{-1}$
2	37.9	0.95 ± 0.05
4	18.2	0.91 ± 0.08
6	3.4	1.07 ± 0.13
8	0.48	1.48 ± 0.15
12	4.8×10^{-3}	2.91 ± 0.24
16	1.0×10^{-4}	21 ± 13

acridine orange derivatives of general formula (I),

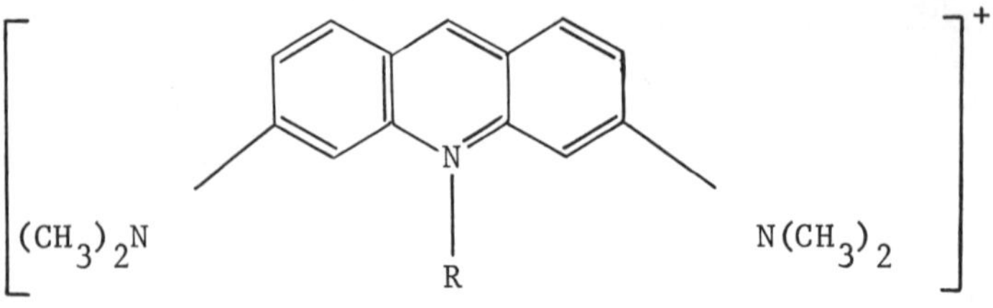

(I)

where R = C_nH_{2n+1}, and n takes values from 0⟶16. The effect of possible additives and impurities is also shown in Table 3.

DISCUSSION OF THE RESULTS

As the chain-length of the surfactant is increased, the c.m.c. decreases. In addition, it is thought that the micelle aggregation-number increases and the head-groups become more compact at the surface. The kinetic measurements are consistent with this last point, the penetration of the dye into the micelle becoming more difficult as the hydrocarbon chain-length increases.

From Table 2, the important result emerges that dyes with the same geometrical shape (see figure 1) are solubilized at identical rates. Thus a major factor hindering absorption is a steric one. There is no correlation of the observed rates with the stacking tendency of the dye (K_D), which is an approximate measure of the lifetime of the dimer unit in solution and reflects in part the charge distribution in the dye.

The increase in the rate of solubilization with increasing ionic strength may tentatively be associated with the likelihood that charge neutralization of the detergent head-groups will reduce the charge retardation effect on the diffusing positively-charged dye.

Table 4 shows that increasing the hydrocarbon moiety of species of type(I)results in a reduction of the rate of solubilization. This is difficult to explain if the dye-induced micelle mechanism is assumed. The stacking tendency increases with hydrocarbon chain-length for n > 4, and gives a measure of the free energy of association of hydrocarbon $-CH_2-$ groups as a function of n.

Addition of small quantities of methanol (Table 3) results in rate enhancement but the c.m.c. is not significantly affected. Similar small effects are observed on addition of the possible micelle (NaLS) impurities sodium tetradecyl sulphate and dodecanol However, the rate of micelle dissociation (as measured by step-perturbation relaxation techniques[10]) is drastically affected.

It can be seen from the results obtained that the stopped-flow kinetic method described in this paper provides the possibility of a subtle probe for micelle-surface structure.

OTHER EXPERIMENTS

Further insight into the nature of dye-micelle and dye-surfactant interactions can be obtained by a range of stopped-flow experiments[2], in which the final equilibrium position is reached from different starting points (figure 4). In an ionic strength-jump mode, the rate of formation of micelles can be followed, using the dye as a probe of micelle formation. An intriguing experiment is to take a dye/micellar solution (>c.m.c.) and mix (dilute) with an equal volume of water, (a concentration-jump experiment resulting in dissolution of the micelle). The acridine dye concentration changes in a none first-order manner over a time period of several seconds for SDS, the rate being dependent on the nature of the dye. The interpretation is not clear at present and further work is in progress; however, it seems most unlikely that micelle breakdown is being measured directly since this is known by other methods to be a rapid process in the ms. time range.

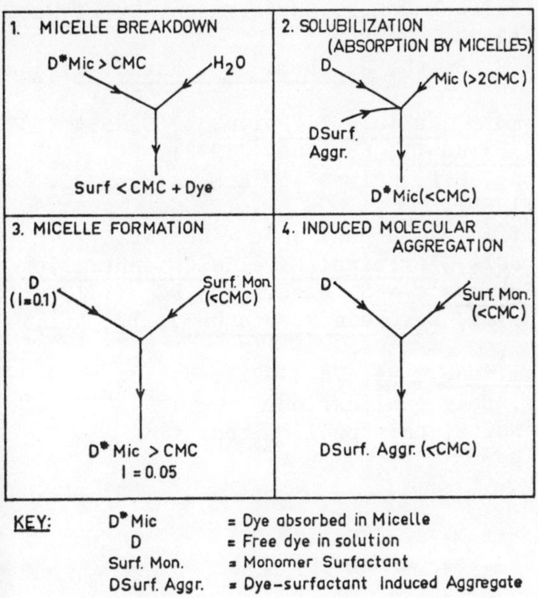

Fig. 4. Stopped-Flow Combinations

The interaction of other chromophores - e.g. cyanine and triphenylmethane dyes - with alkyl sulphate micelles has also been studied. It is interesting that the absorption spectra of some dyes appear to be sensitive to an interaction with surfactant at a concentration in the region of but definitely lower than the c.m.c. For example pinacyanol chloride, crystal violet and the modified surfactant-like acridine orange chromophores (I) exhibit a spectral change at ~ c.m.c./2. This could be associated with the build-up of the intermediate pre-micelle aggregates which were introduced in the development of the cooperative model for micellization[11]. Thus, it is possible that chromophoric (and fluorimetric) probes of different structural types are differently sensitive to the presence of the various aggregated surfactant species in detergent solutions.

ACKNOWLEDGEMENTS

We thank the S.R.C. for awards to NCW and AJ, and for grants towards the purchase of equipment associated with this work.

REFERENCES

1. G.L. Choules, R.G. Sandberg, M. Steggall and E.M. Eyring, Biochemistry, **12**, 4544 (1973).
2. K. Takeda, N. Tatsumoto and T. Yasunaga, J. Coll. Int. Sci., **47**, (1974).
3. A.S. Waggoner, O.H. Griffith and C.R. Christensen, Proc. Nat. Acad. Sci., **57**, 1198 (1967).
   K.K. Fox, I.D. Robb and R. Smith, J. Chem. Soc. Farad. I, **68** 445 (1972).
   T. Nakagawa and H. Tizomoto, Kolloid Z. Polym., **250**, 594 (1972)
4. J. Oakes, J. Chem. Soc. Farad II, **68**, 1464 (1972).
5. B.C. Bennion, L.K.J. Tong, L.P. Holmes and E.M. Eyring, J. Phys. Chem., **73**, 3288 (1969).
6. B.H. Robinson and K.J. Timmins, Unpublished results.
7. P. Mukerjee and K.J. Mysels, 'Critical Micelle Concentrations of Aqueous Surfactant Systems', NBS, Washington DC, 1971, p.1.
8. G. Schwarz and W. Balthasar, European J. Biochem., **12**, 461, (1970).
9. P. Hemmes, J. Amer. Chem. Soc., **94**, 74 (1972).
10. H. Hoffmann and R. Zana, this publication.
11. B.H. Robinson and N.C. White, this publicaiton.

STOPPED FLOW AND TEMPERATURE JUMP STUDIES OF COLLOIDAL SUSPENSIONS

J.D. Owen, Mary Steggall and Edward M. Eyring

Department of Chemistry, University of Utah, U.S.A.

ABSTRACT. Time dependent changes in the intensity of light scattered by colloidal particles can provide clues to the mechanism of surfactant micelle dissociation, erythrocyte solubilization by surfactants, and non-electrolyte transport through biological membranes. Stopped flow and temperature jump kinetic studies of this general type are described.

Four years ago when temperature jump and stopped flow kinetic studies were still being conducted on surfactant micelles in our laboratory,[1-5] the features of the problem that particularly interested us were not the ones on which recent investigators have profitably focused. Although most of our rate measurements were made at or near the critical micelle concentration, we were aware from the work of Kresheck *et al.*[6] and from a few of our own unpublished measurements that at higher surfactant concentrations plots of τ^{-1} versus concentration were not linear over an extended concentration range. However, globular micelles are supplanted by cylindrical micelles of much higher aggregation number in more concentrated surfactant solutions,[7] and this complication led us to suppose that a theoretical treatment of the concentrated surfactant kinetics would be very difficult. The nonlinearity of the dependence of τ^{-1} on surfactant concentration has, as is evident from lectures presented in this Institute and recent publications,[8-11] become a clue in resolving the apparent discrepancy[12] between slow relaxations observed by techniques such as P-jump, T-jump and stopped flow and faster time scale observations made on the same surfactant solutions by NMR, esr and ultrasonic spectroscopy.

Even the more dilute solutions near the critical micelle

Wyn-Jones (ed.), Chemical and Biological Applications of Relaxation Spectrometry, 211–214.

concentration that we chose to work with provided difficulties that still merit investigation. For instance, at or below room temperature, the surfactants dodecylammonium chloride (DAC) and sodium dodecyl sulfonate (SDSn) in moderate concentrations frequently precipitated from aqueous solution, and the rapid temperature jump used in the kinetic experiments also occasionally provided enough of a shock to precipitate SDSn from the aqueous 0.1M NaCl medium. We also unaccountably failed in attempts to measure relaxation times for sodium lauryl sulfate, SDSn, and DAC at temperatures below $\sim 25^{o}$ in spite of apparently favourable enthalpies of micellization. However, the most intriguing experimental difficulty encountered in surfactant studies of this type is sample purity. Hoffmann and Herrmann have both reported in this conference that for very pure sodium dodecyl sulfate τ^{-1} is essentially independent of surfactant concentration above the critical micelle concentration (an explicit contradiction of our earlier work[1]) but becomes linearly dependent on concentration when plausible contaminants are deliberately added. It does not appear possible to account for this discrepancy in terms of differences in ionic strength. It would be surprising but interesting if similar discrepancies emerged from a repetition of earlier rate studies[2,5] of DAC and SDSn using ultrapure surfactants.

In view of the increasing awareness of the roles cell membranes play in both health maintenance and many diseases, including cancer, there is a great deal of current interest in the eluicidation of the structure of these membranes. We have approached this problem by looking at rates of solubilization of Acholeplasma laidlawii membranes with several chemical agents and also at the effects of drugs on permeability characteristics of red blood cell membranes. Our light scattering stopped flow apparatus has been our principal experimental tool in these studies. Reisner and Buenemann[13] have incidentally reported an interesting refinement of this type of apparatus.

Attempts to separate and characterize biological membrane constituents by selective solubilization in such detergents as sodium dodecyl sulfate have succeeded primarily in showing that the membrane is very heterogeneous.[14-16] We first studied the kinetics of Acholeplasma laidlawii membrane solubilization in sodium dodecyl sulfate (SDS).[17] In a dissociated form, the SDS anions most probably interact with the membrane by inserting their hydrophilic tails into the lipid interior of the membrane. It appears there is also some form of membrane-SDS micelle interaction since initial solubilization rates were SDS- concentration dependent both below and above the critical micelle concentration. We next studied the solubilization of mycoplasma membranes in lithium diiodosalicylate (LIS).[18] Marchesi and Andrews[19] had reported the usefulness of LIS in isolating a glyco-protein fraction from red blood cell membranes. The nature of the LIS molecule suggested that it might be capable of breaking some of the non-covalent bonds that were resistant to SDS attack. The kinetic data for solubilization of mycoplasma membranes by LIS showed a surprising temperature dependence. From 16-30°, the solubilization rate decreased with increasing temperature, followed by an increase in rate from 33-40°. The activation parameters, along with viscosity and ultracentrifugation data suggested that the rate-determining step in solubilization may be thermally-caused formation of a "crack" in the membrane which is subsequently coated with LIS. Preliminary experiments on LIS solubilization of red blood cell membranes have not shown this unusual rate-temperature dependence, so it is possible that the mycoplasma system is not representative of biological membranes generally in this regard. Further kinetic studies with LIS and taurocholic acid solvents are in progress.

Another approach to determining the nature of membrane constituents is to study the effects of certain drugs on permeability of membranes to H_2O and small non-electrolytes. Rapid mixing of an isosmolar suspension of red cells with a hyperosmolar non-electrolyte solution in the stopped-flow spectrophotometer causes the cells to initially shrink due to outflow of water. This is followed by swelling of the cells as solute enters, with concomitant water influx. These cell volume changes can be monitored by the rate of change in intensity of 90° scattered light and have been used to determine reflection (σ) and permeability (ω) coefficients. Most of our work is involved with direct measurements of permeability-related phenomena. The rate of volume change at the minimum volume point was utilized by Sha'afi _et al._[20] to measure red cell nonelectrolyte permeability coefficients, while the maximum swelling rate just after the minimum volume is attained has been used to measure liposome relative non-electrolyte permeability.[21,22] Our equipment is best suited for the maximum slope method which we first used to measure the effect of phloretin (a well-known inhibitor of hexone transport in red cells) on small non-electrolyte pathways in red cells.[23] Our recent phloretin work confirms that at high concentrations, the drug decreases the permeability of urea[24] and other hydrophilic molecules while it increases the permeability of lipophilic solutes.[25] We also found that phloretin shows a concentration-dependent bimodal effect on hydrophilic solutes[26], i.e., permeability is increased at concentration < 0.1mM and is decreased at higher phloretin concentrations. It is interesting to note that phloretin exhibits a similar bimodal concentration effect on K^+-dependent phosphatase activity in rat brain.[27] (K^+-dependent phosphatase is the terminal hydrophilic enzyme of ATPase activity after a Na^+-dependent phosphorylation.)

We have more recently used the stopped-flow-light-scattering technique to look at effects on permeability of p-chloromercuribenzene sulfonic acid (PCMBS), a sulfhydryl group modifier, and 4-acetamido-4'-isothiocyano-stilbene-2,2'-disulfonic acid (SITS), which binds specifically to amino groups. It is hoped that studies of these and other chemical modifiers will lead to information concerning the various membrane surface moieties which regulate solute movement across red cell membranes.

ACKNOWLEDGEMENT. This work was supported by the National Institute of Arthritis, Metabolism and Digestive Diseases Grant AM 06231-12 and by Biomedical Science Support Grant RR 07092 from the National Institute of Health.

REFERENCES

1. B.C. Bennion, L.K.J. Tong, L.P. Holmes and E.M. Eyring, "J. Phys. Chem., 73, 3288 (1969).
2. B.C. Bennion and E.M. Eyring, "J. Colloid and Interface Sci., 32, 286 (1970).
3. J. Lang and E.M. Eyring, J. Polymer Sci.:A-2, 10, 89 (1972).
4. J. Lang, J.J. Auborn and E.M. Eyring, J. Colloid and Interface Sci., 41, 484 (1972).
5. B.C. Bennion, "Ph.D. Thesis, University of Utah, 1969.
6. G. C. Kresheck, E. Hamori, G. Davenport and H.A. Scheraga, J. Am. Chem. Soc., 88, 246 (1966).
7. C. Tanford, J. Phys. Chem., 76, 3020 (1972).
8. T. Nakagawa, Colloid and Polymer Sci., 252, 56 (1974).
9. E.A.G. Aniansson and S.N. Wall, J. Phys. Chem., 78, 1024 (1974)
10. A. H. Colen, J. Phys. Chem., 78, 1676, (1974).
11. J. Lang, C. Tondre, R. Zana, R. Bauer, H. Hoffmann and W. Ulbricht, J. Phys. Chem., submitted 1974.
12. N. Muller, J. Phys. Chem., 76, 2017 (1972).
13. D. Riesner and H. Buenemann, Proc. Nat. Acad. Sci. U.S.A., 70, 890 (1973).
14. J.A. Reynolds and C. Tanford, J. Biol. Chem., 245, 5161 (1970).
15. F.H. Kirkpatrick, S.E. Gordesky and G.V. Marinetti, Biochim. Biophys. Acta, 345, 154 (1974).
16. B. Ballou, G. Sundharadas and M.L. Bach, Science,185, 531 (1974).
17. J.J. Auborn, E.M. Eyring and G.L. Choules, Proc. Nat. Acad. Sci. U.S.A., 68, 1996 (1971).
18. G.L. Choules, R.G. Sandberg, M. Steggall and E.M. Eyring, Biochem., 12, 4544 (1973).
19. V.T. Marchesi and E.P. Andrews, Science, 174, 1247 (1971).
20. R.I. Sha'afi, G.T. Rich, D.C. McKulecky and A.K. Solomon, J. Gen. Physiol., 55, 427 (1970).
21. B.E. Cohen and A.D. Bangham, Nature, 236, 173 (1972).
22. M.W. Hill and B.E. Cohen, Biochim. Biophys. Acta, 290, 403, (1972).
23. P.G. LeFevre and J.K. Marshall, J. Biol. Chem., 234, 3022, (1959).
24. R.I. Macey and R.E.L. Farmer, Biochim. Biophys. Acta, 211, 104 (1970).
25. J.D. Owen and A.K. Solomon, Biochim. Biophys. Acta, 290, 414 (1972).
26. J.D. Owen, M. Steggall and E.M. Eyring, J. Membrane Biol. in press.
27. J.D. Robinson, Mol. Pharmacol., 5, 584 (1969).

AN ATTEMPT TO MEASURE THE MONOMER-MICELLE EXCHANGE FREQUENCY USING ELECTRON PARAMAGNETIC RESONANCE

Kay K. Fox

Unilever Research, Port Sunlight Laboratory, Wirral, Cheshire, G.B.

Electron Paramagnetic Resonance (henceforth EPR) has been used to study paramagnetic probes solubilised in micellar systems[1-11] and paramagnetic counter-ions in micellar systems[12-15]. The earlier work with nitroxide probes in micellar systems[1-8] gave no direct information concerning monomer-micelle exchange frequencies, but was often mis-interpreted as showing that the monomer-micelle exchange frequency was fast on the EPR timescale (i.e. greater than 10^9 sec^{-1}). Later work showed that the frequency with which a probe molecule entered or left a micelle was slow on the EPR timescale[9-11] (i.e. less than 10^7 sec^{-1}).

Experiments with a series of cationic paramagnetic surfactants (1)[16-17] were carried out in an attempt to measure the frequency with which a

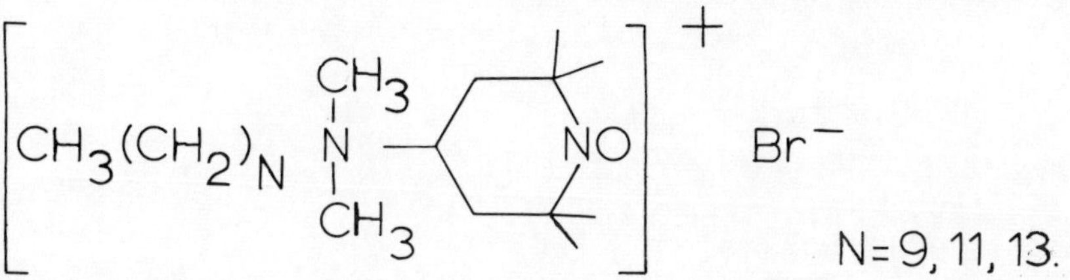

(1)

surfactant monomer enters a micelle. A micelle composed entirely of these paramagnetic monomers would be expected to have an exchange-narrowing spectrum consisting of one broad line, due to the nitroxide headgroups interacting with each other more than 10^8 times per second[18]. A nitroxide surfactant monomer in aqueous

Wyn-Jones (ed.), Chemical and Biological Applications of Relaxation Spectrometry, 215–221.

solution below the c.m.c. would be expected to show the normal 3-lined nitroxide spectrum in the absence of exchange. Above the c.m.c., a superimposed spectrum of 3 sharp lines due to monomeric nitroxide surfactant and one broad line due to micellar surfactant would indicate slow exchange of surfactant between monomeric and micellar states, while observation of one broad line only would indicate fast exchange. The former was observed[16,17] as can be seen from Fig. 1 for the N = 11 compound. Thus this experiment established that the frequency with which this particular cationic nitroxide surfactant exchanged between micellar and aqueous environments was slow on the EPR timescale.

Fig. 1 shows that, as the temperature of the nitroxide-surfactant solution above the c.m.c. increased, the broad micellar line sharpened (exchange narrowing due to increased frequency of collisions between head groups within the micelle) and the sharp monomer lines broadened. It was possible to show[16,17] that the broadening of the monomer lines was greater than the broadening expected for monomer-monomer collisions in solution (Heisenberg spin exchange[19]), and the residual broadening was assigned to monomer-micelle exchange. The monomer-micelle frequencies obtained increased from 10^5 sec^{-1} at 25°C to 10^6 sec^{-1} at 70°C, corresponding to an enthalpy of activation for the micellisation process of about 10 kcal/mole[16,17].

Further experiments with cationic nitroxide surfactants with varying chain lengths produced results which were not compatible with the analysis given above[18]. The unreconcilable data obtained from the addition of structure-making and structure-breaking inorganic salts and/or diamagnetic surfactants to the paramagnetic surfactant systems indicated that either the Heisenberg spin exchange

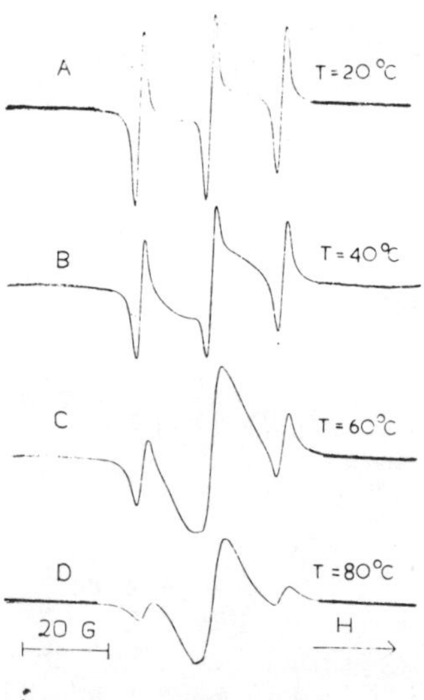

Fig. 1. Changes in 0.05M C_{12} TABNÓ Spectra with Temperature.

between monomeric surfactant molecules had not been correctly allowed for, or that some species other than monomers and micelles was responsible for the observed effects.

The linewidth increase due to Heisenberg spin exchange can be shown to be proportional to T/η, where T is the temperature and η is the viscosity of the solution[19,20]. Deviations from linearity in the direction of a smaller linewidth increase than that expected at higher T/η are sometimes observed, due to the failure of all radical collisions to result in electron exchange (the "weak exchange" condition)[19]. The $M_I = +1$, 0 and -1 nitroxide lines should broaden equally up to a certain value[20], about 4.0 gauss for the system studied here[17], after which the $M_I = \pm 1$ lines broaden more rapidly than the $M_I = 0$ line[20]. Figures 2, 3 and 4 show the variation in linewidth vs T/η for compounds I(N =11), I(N = 9) and I(N = 0) at certain specified concentrations. (I(N=11) is at twice its c.m.c., I(N = 9) is at one half of its c.m .c., while I(N = 0) does not form micelles). The deviation from linearity in Fig. 2 and the appearance of equal broadening for the $M_I = \pm 1$ lines at lower total linewidth than is compatible with Heisenberg spin exchange indicate that other types of exchange are present in the system. Fig. 3 shows that the similar broadening of the $M_I = \pm 1$ nitroxide lines is present in compound I(N = 9) below the c.m.c., while Fig. 4 shows that both this relative linewidth effect and total plot curvatures are present to some extent with the cationic nitroxide headgroup alone.

Zimmerman and Britten[21] have developed equations for the exchange between two sites, one broad and one narrow, both centering on the same frequency ω_o, which take into account the relative populations of the two sites. Woessner[22] has extended this to the case of different resonance frequencies for the two sites. Using these equations, it is possible to show[17] that the linewidth effects observed in Figures 2-4 could be due to a high concentration of monomer exchanging with a low concentration of a species which gave a broad, micelle-like signal. A nitroxide dimer, in which the nitroxide group of one molecule and the quaternary nitrogen of the other molecule were in close proximity, might be expected to have this type of EPR signal.

A search for some sort of non-micellar association in these systems was accordingly undertaken. Fig. 5 shows a plot of the monomer concentration of compound I(N = 11), as measured by EPR intensity methods, versus the total amount of compound I(N = 11) in the solution. The measurements were made at 25°C; the dotted line and the line of X's above the c.m.c. represent different ways of correcting for the micellar intensity underlying each peak. It can be seen that a gradual loss of monomer is evident at concentrations above half the c.m.c. Figure 6 shows that this loss of intensity is also observed with the headgroup alone, thus reducing the probability that the data shown in Fig. 5 can be explained by a diffuse c.m.c. Figure 7 shows that, for the N = 9 compound, the amount of non-micellar aggregate decreases as the

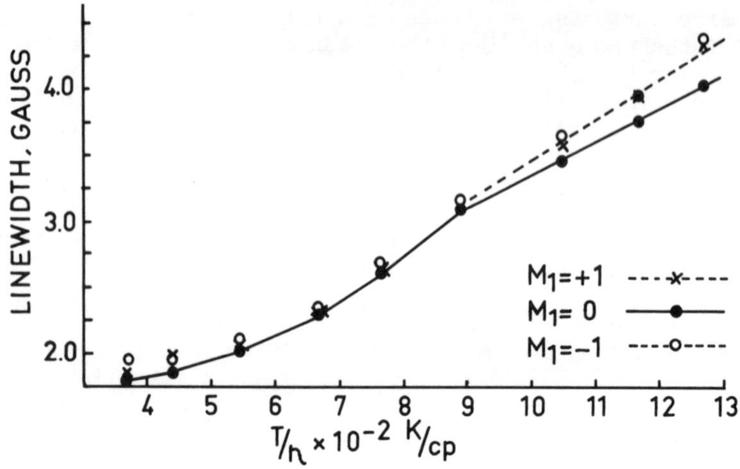

Fig. 2. Linewidth vs. T/η for 1.47 x 10^{-2}M C_{12}TABNȮ.

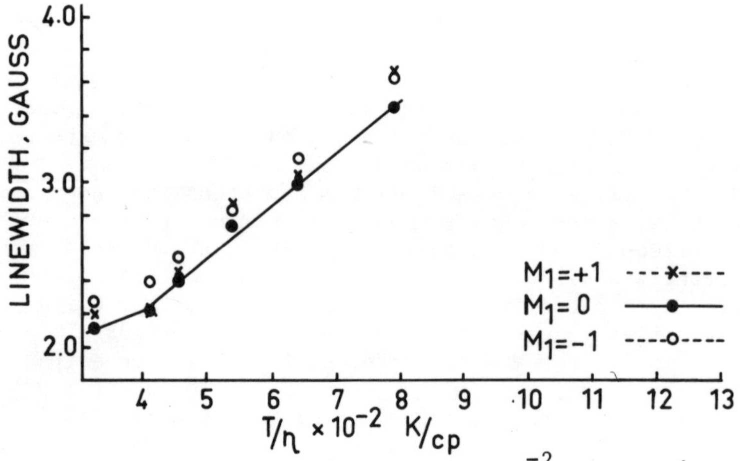

Fig. 3. Linewidth vs. T/η for 2.0 x 10^{-2}M C_{10}TABNȮ

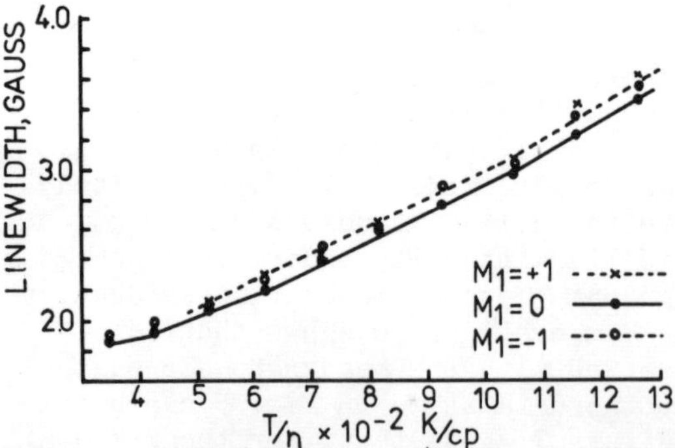

Fig. 4. Linewidth vs. T/η for 8.34 x 10^{-3}M C_1TABNȮ.

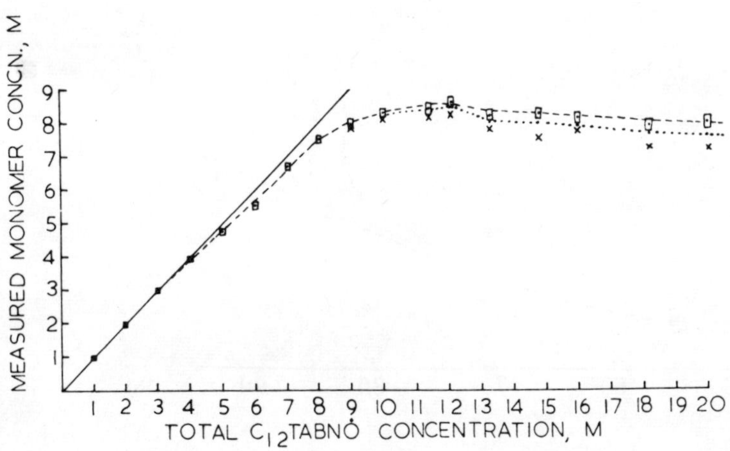

Fig. 5. Variation in the monomer concentration of C_{12}TABNȮ with total C_{12}TABNȮ concentration.

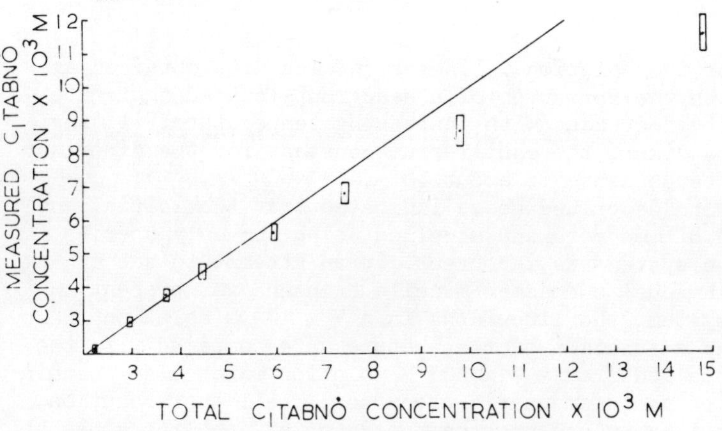

Fig. 6. Association of C_1TABNȮ in solution.

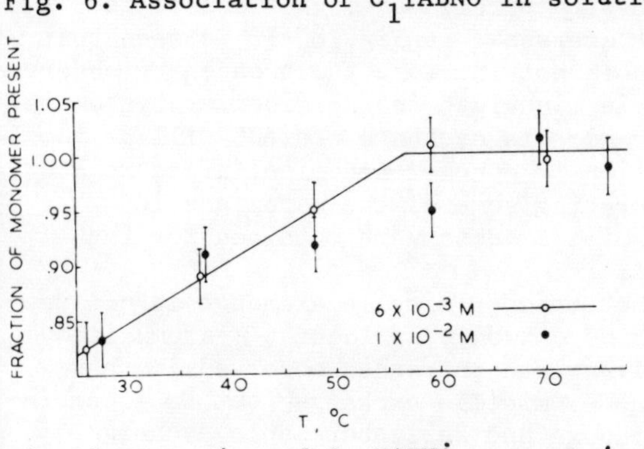

Fig. 7. Fraction of C_{10}TABNȮ present in monomer form vs. temperature.

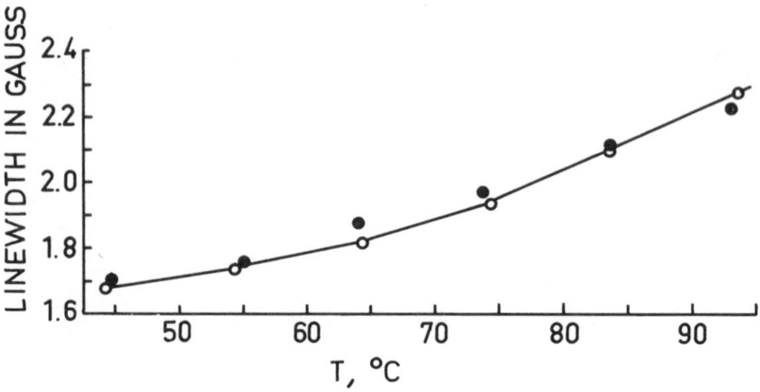

Fig. 8. Linewidth vs. temperature for the $M_I = 0$ line of: • $7 \times 10^{-3}M$ C_{12} TABNȮ ; —o— $7 \times 10^{-3}M$ C_{12} TABNȮ, $8 \times 10^{-3}M$ C_{12} TABNH.

temperature of the solution is increased. The data obtained are consistent with the formation of a headgroup-induced aggregate whose stability decreases with increasing temperature; if the aggregate is a dimer, the equilibrium constant for the dimerization at room temperature is about 10^{17}.

The results described above indicated that the initial analysis in terms of monomers and micelles being the only species present in the system was incorrect. In an attempt to see if any line-broadening due to monomer-micelle exchange was present in the I(N=11) system, the linewidths in a $7 \times 10^{-3}M$ solution (below the c.m.c.) as a function of temperature were compared with the linewidths obtained from a $7 \times 10^{-3}M$ solution which also contained $8 \times 10^{-3}M$ of the diamagnetic precursor. The latter solution contained micelles, whose spectra consisted of one line somewhat broader than that obtained with purely paramagnetic micelles. The results of this experiment are shown in Fig. 8. It is seen that the presence of micelles does not increase the broadening observed - if anything, a decrease in linewidth in the micellar system is observed. Thus any monomer-micelle exchange present will be too small to offset the reduction in monomer-non-micellar-aggregate exchange caused by incorporating some of the paramagnetic material in the micelles. Similar data were obtained for the I(N=9) and I(N=13) systems.

These results show that monomer-micelle exchange is not detectable by EPR methods in the cationic surfactant systems I(N=9), I(N=11) and I(N=13). The linewidth increases observed in these systems are not due to monomer-micelle exchange, but to a combination of monomer-dimer exchange and Heisenberg spin exchange. However, it has been possible to show that in the three surfactant systems studied the monomer-micelle exchange frequency is slow on the EPR timescale. Thus EPR is one of the many techniques which yield an upper limit to the monomer-micelle exchange frequency, but are unable to measure that frequency accurately.

REFERENCES

1. O.H. Griffith and A.S. Waggoner, Accounts Chem. Res., 2, 17 (1969).
2. A.S. Waggoner, O.H. Griffith and C.R. Christensen, Proc. Nat. Acad. Sci., US, 57 (5), 1198 (1967).
3. A.S. Waggoner, A.D. Keith and P. H. Griffith, J. Phys. Chem., 72 (12) 4129 (1968).
4. A.S. Waggoner, T.J. Kingzett, S. Rottschaefer and O.H. Griffith, Chem. Phys. Lipids, 3, 245 (1969).
5. G.P. Rabold, J. Polymer Sci., Part A-1, 1, 1187 (1969).
6. M.G. Goldfield, V.K. Kolkover, E.G. Rosentzev and V.I. Suskina, Kolloid-Z, und Z Polymere, 243, 62 (1971).
7. M.J. Povich, J.A. Mann and A. Kawamoto, J. Colloid and Interface Sci., 41 (1) 145 (1972).
8. C.P. Lee, H. Drott, B. Johansson, T. Yonetaric and N. Chance, Probes of Structure and Function of Macromolecules and Membranes, Proc. Collog. 249 (1971).
9. S. Ohnishi, T.J.R. Cyr and H. Fukushima, Bull. Chem. Soc. Japan, 43, 673 (1970).
10. J. Oakes, Nature, 231, 38 (1971).
11. N.M. Atherton and S.J. Strach, J. Chem. Soc., Faraday Trans II, 68, 374 (1972).
12. N. Miura, A. Hasegawa and Y. Michihara, Bull. Chem. Soc. Japan, 41, (2), 534 (1968).
13. A. Hasegawa, Y. Michihara and M. Miura, Bull. Chem. Soc. Japan, 43, (10), 3116 (1970).
14. E.E. Zaev, Kolloidyni Zhurnal, 34 (2), 304 (1972).
15. J. Oakes, J. Chem. Soc. Faraday Trans.I, 69, 1321 (1973).
16. K.K. Fox, Trans. Faraday Soc., 67, 2802 (1971).
17. K.K. Fox, Ph.D Thesis, University of Leicester, 1974.
18. A. Carrington and A.D. McLachlan, Introduction to Magnetic Resonance, Harper and Row, New York, 1967. Chapter 12.
19. M.P. Eastman, R.G. Kooser, M.R. Das and J.F. Freed, J. Chem. Phys. 51, 2690 (1969).
20. W. Plachy and D. Kivelson, J. Chem. Phys. 47, 3312 (1967).
21. J.R. Zimmerman and W.E. Brittin, J. Phys. Chem., 61, 1328 (1957).
22. D.E. Woessner, J. Chem. Phys., 35, 41 (1961).

KINETICS OF STEP-WISE MICELLE ASSOCIATION AND DISSOCIATION

E.A.G. Aniansson and S.N. Wall

Department of Chemistry, University of Gothenburg,
Fack, S-402 20 Gothenburg, Sweden

ABSTRACT. The kinetic equations have been given a form which suggests an analogy with diffusion in a tube of varying cross section and which clearly identifies the rate-limiting quantities as the products of the disintegration rate constants for the particular micellar sizes and the corresponding equilibrium concentrations. It is rendered highly probable that the rate of the over all disintegration of the micelles is limited by steps pertaining to the rare intermediate micelles and an expression for the time constant of this process is deduced. The inverse of the time constant for the population rearrangement process among the abundant micelles is also derived and shown to be linearly increasing with overall concentration. The degree of association in micelles of surfactant solutions is typically of the order of 100. The treatment of their formation and dissociation has so far been based on rather simplified assumptions[1,2,3]. The following is a short presentation of an attempt[4] at a more general treatment.

DIFFUSION ANALOGY

It is very plausible that the process is a step-wise one:

$$A_1 + A_1 \underset{k_2^-}{\overset{k_2^+}{\rightleftharpoons}} A_2,\ A_1 + A_2 \underset{k_3^-}{\overset{k_3^+}{\rightleftharpoons}} A_3,\ \ldots\ A_1 + A_{s-1} \underset{k_s^-}{\overset{k_s^+}{\rightleftharpoons}} A_s,\ \ldots\ldots \qquad (1)$$

where A_1 denotes the surfactant monomer, A_2 the dimer, etc. In the case of ionic surfactants A_1 denotes the surfactant ion. It is

Wyn-Jones (ed.), Chemical and Biological Applications of Relaxation Spectrometry, 223–238.

probable that in the vast majority of cases the counterions easily adjust to the association and dissociation of the surfactant ions and hence need not be explicitly considered in the kinetic equations. The k_s^+ and k_s^- may, however, be functions of the ionic strength.

We shall look at the process as a flow in aggregation space, J_s denoting the number of particles of aggregation s-1 that by addition of one monomer was transferred to s minus those that by loosing one monomer are transferred from s to s-1. Introducing the relative deviation from equilibrium at s,

$$\xi_s = \frac{A_s - \bar{A}_s}{\bar{A}_s} \tag{1}$$

where A_s now denotes the number of particles at s and $\bar{A}_s$ its equilibrium value, and using the fact that $k_s^+ \bar{A}_1 \bar{A}_{s-1} = k_s^- \bar{A}_s$ one finds that

$$J_s = - k_s^- \bar{A}_s (\xi_s - \xi_{s-1} - \xi_1 - \xi_1 \xi_{s-1}) \tag{2}$$

At sufficiently low deviations from equilibrium, such as are obtained in the relaxation experiments mostly used in experiments of this kind, the last term within brackets may temporarily be neglected. (2) is now reminiscent of an equation for diffusion in a tube of varying cross section k_s^- being, analogous to the diffusion constant, $\bar{A}_s$ to the cross sectional area, $\xi_s - \xi_{s-1}$ to the gradient of temperature and ξ_1 playing the less familiar role of a space-independent driving force.

The ordinary kinetic equations take the form

$$\bar{A}_s \frac{\partial \xi_s}{\partial t} = - (J_{s+1} - J_s) \tag{3}$$

clearly exhibiting the analogy with the continuity equation and confirming the analogy of $\bar{A}_s$ with a cross sectional area. $J_{s+1} - J_s$ corresponds to the divergence of the flow[5]. From these equations we find that the rate limiting quantities are the products $k_s^- \bar{A}_s$ and not the k_s^-, as one might be inclined to think at the outset. In a solution where the size distribution of the micelles is not very broad the dependence of ln $\bar{A}_s$ on s is something like that in figure 1[7,8,9].

The micelle population is more or less sharply peaked (cf. ref. 4, p. 1028) around a most probable value, in this example 100, and several orders of magnitude (10^{-3} or 10^{-4}) lower in the intermediate region.

A hint towards this s-dependence of the k_s^- may be obtained fr

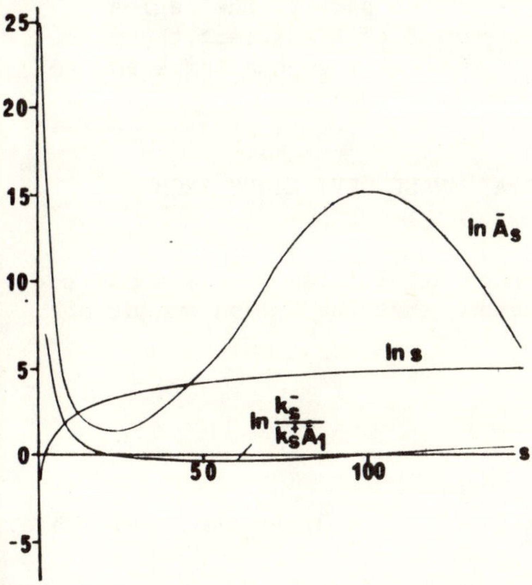

Fig. 1. The s-dependence of the k_s^-, see eqn. (4).

$$\ln \frac{k_s^-}{k_s^+ \bar{A}_1} = -\ln \frac{\bar{A}_s}{\bar{A}_{s-1}} \cong -\frac{d \ln \bar{A}_s}{ds} \tag{4}$$

also shown in figure 1. Although the micelle distribution in this example is not very broad, about 10% half-width at $e^{-1/2}$ of the peak value, (see further below) the ratio

$$\frac{k_s^-}{k_s^+ \bar{A}_1}$$

changes very little over most of the range. It is then plausible that the variations of k_s^- and k_s^+ themselves are not very much larger. If the whole variation is ascribed to k_s^-, these constants will be smaller in the intermediate region than at its ends. A more sharply peaked micelle distribution would indicate even lower relative values of k_s^- below the peak. A recent theory[5] also predicts a similar but even less strong s-dependence of k_s^-.

It is then a plausible assumption indeed that the products $k_s^- A_s$ are very much smaller in the intermediate region than in the regions of abundant micelles and of monomers plus oligomers. Returning to the analogy with diffusion the situation is reminiscent of two thick ends of a tube connected by a very thin tube. One end would correspond to the monomers and the oligomers, the other to the abundant micelles, and the very thin part to the region between. After an initial, short period of adjustment the main process would be a pseudo-stationary flow from one thick end to the other through the thin intermediate tube[5].

OVERALL DISSOLUTION OF MICELLES AT SMALL DEVIATIONS FROM EQUILIBRIUM.

The three parts of the s-space are denoted by $1 \leqslant s \leqslant s_1$, $s_1 < s \leqslant s_2$, $s_2 < s$. We shall assume that the excess amount of material in the second region

$$\sum_{s>s_1}^{\leqslant s_2} s\xi_s \bar{A}_s$$

is negligible and that the term $\xi_1 \xi_{s-1}$ in (2) is also negligible. Then in the thick ends

$$\xi_s - \xi_{s-1} - \xi_1 = \frac{J_s}{\bar{k}_s^- \bar{A}_s} \cong 0 \tag{5}$$

and in the thin part

$$\xi_s - \xi_{s-1} - \xi_1 = \frac{J}{\bar{k}_s^- \bar{A}_s} , \tag{6}$$

where J is the practically s-independent[5] value of the flow during the pseudostationary phase considered.

Summing equation (5) from s = 2 to $s \leqslant s_1$ gives

$$\xi_s = s\,\xi_1, \, (s \leqslant s_1) \tag{7}$$

Further summations of (6) over $s_1 < s \leqslant s_2$ and (5) from $s_2 + 1$ to s give

$$\xi_s = s\xi_1 - RJ, (s > s_2) \tag{8a}$$

where

$$R = \sum_{s>s_1}^{\leqslant s_2} \frac{1}{\bar{k}_s^- \bar{A}_s} \tag{8b}$$

Insertion of (7) and (8) in the expression for material balance,

$$\sum_{s\geq 1} s\xi_s \bar{A}_s = 0 \tag{9}$$

gives

$$J = \frac{\overline{n_1^2}\, c_1 + \overline{n_3^2}\, c_3}{R\, c_3\, \bar{n}_3}\, \xi_1. \tag{10}$$

Using (10) one also finds that the excess number of micelles,

$$\sum_{s>s_2} \xi_s \bar{A}_s,$$

is related to the relative excess number of monomers, ξ_1, by

$$\sum_{s>s_2} \xi_s \bar{A}_s = \frac{\bar{n}_1^2\, c_1 + \sigma^2\, c_3}{\bar{n}_3}\, \xi_1, \tag{11}$$

where

$$c_1 = \sum_{s=1}^{s_1} \bar{A}_s, \quad \bar{n}_1^2 = \frac{\sum_{s=1}^{s_1} s^2 \bar{A}_s}{\sum_{s=1}^{s_1} \bar{A}_s}, \quad \sigma^2 = \bar{n}_3^2 - \bar{n}_3^2,$$

$$c_3 = \sum_{s>s_2} A_s, \quad \bar{n}_3^2 = \frac{\sum_{s>s_2} s^2 \bar{A}_s}{\sum_{s>s_2} \bar{A}_s}, \quad \bar{n}_3 = \frac{\sum_{s>s_2} s \bar{A}_s}{\sum_{s>s_2} \bar{A}_s} \tag{12}$$

The c:s are the equilibrium numbers of aggregates in parts 1 and 3, the n^2:s are the mean square aggregation numbers in the same two parts, σ^2 is the variance of the size distribution of the abundant micelles, and $\bar{n}_3$ the mean aggregation number of the latter.

Inserting (10) and (11) into the differential equation

$$\frac{d(\sum_{s>s_2} \xi_s \bar{A}_s)}{dt} = J \tag{13a}$$

gives

$$\frac{d\xi_1}{dt} = -\frac{1}{\tau}\xi_1, \tag{13b}$$

with the familiar solution

$$\xi_1 = \xi_1(0)\, e^{-\frac{t}{\tau}} \tag{14}$$

where

$$\frac{1}{\tau} = \frac{1}{Rc_3}\, \frac{\overline{n_1^2}\, c_1 + \overline{n_3^2}\, c_3}{\overline{n_1^2}\, c_1 + \sigma^2 c_3} \tag{15}$$

The main features of this expression with regard to its concentration dependence are as follows.

Above the critical micelle concentration (c.m.c.) $\overline{n_1^2}$ and c_1 are practically constant. This also holds for $\overline{n_3^2}$ if σ is not much larger than we have assumed, and for σ if the distribution of abundant micelles is not excessively skew[4].

Eqn. (15) can be simplified somewhat since $\overline{n_1^2} \simeq 1$ in most cases and $\overline{n_3^2} = \bar{n}_3^2 \simeq O(10^4)$. Then $\overline{n_1^2}\, c_1 \simeq \bar{A}_1$ = c.m.c. at $A_{tot}>$ c.m.c. Further, if the micelle distribution is not excessively skew, $\bar{n}_3\, c_3 \simeq A_{tot} - \bar{A}_1 \equiv \bar{A}_{exc}$. Except at concentrations very close to c.m.c. there will hold

$$\overline{n_3^2}\, c_3 \cong \bar{n}_3^2\, c_3 \cong \bar{n}_3\, \bar{A}_{exc} \gg \overline{n_1^2}\, c_1 \cong \bar{A}_1$$

so that (15) takes the form

$$\frac{1}{\tau} \cong \frac{\bar{n}_3^2}{R}\, \frac{1}{\bar{A}_1 + \frac{\sigma^2}{\bar{n}_3}\bar{A}_{exc}} \tag{15a}$$

The second factor will be represented by a hyperbola of the form

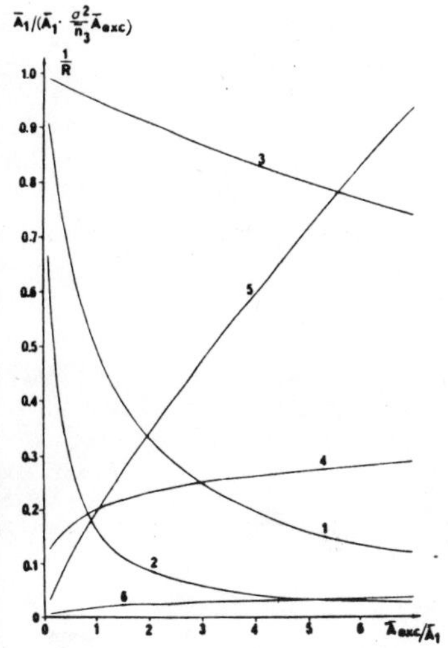

Fig. 2. $\bar{A}_1/(\bar{A}_1 + \frac{\sigma^2}{\bar{n}_3}\bar{A}_{exc})$ versus $\bar{A}_{exc}/\bar{A}_1$ for 1) $\sigma^2/\bar{n}_3 = 1$, 2) $\sigma^2/\bar{n}_3 = 5$, 3) $\sigma^2/\bar{n}_3 = 0.05$. $\frac{1}{R}$ versus $\bar{A}_{exc}/\bar{A}_1$ for the cases 4) a short narrow passage situated around s = 20, 5) a short narrow passage situated around s = 80 and 6) a long narrow passage between s = 20 and 2 = 80.

of curves 1, 2 and 3 in figure 2. The larger the value of $\sigma^2/\bar{n}_3$ is, the faster this factor will decrease towards zero increasing $\bar{A}_{exc}$. The concentration dependence of R will depend on the s-values of the largest terms in the sum (8b) defining R. Since the k_s^- probably vary much less with s than $\bar{A}_s$ it is approximately the concentration dependence of the smallest of the latter that determines the concentration dependence of R.

The scheme (1) is equivalent to the mass action law:

$$\bar{A}_s = K_s \bar{A}_1^s \; ; \; K_s = \prod_{r=2}^{s} \frac{k_r^+}{k_r^-} \tag{16a}$$

Introducing

$$\bar{A}_1 = \left(\frac{\bar{A}_{\bar{n}_3}}{K_{\bar{n}_3}}\right)^{\frac{1}{\bar{n}_3}} \tag{16b}$$

into (16) one obtains

$$\bar{A}_s \propto (\bar{A}_{\bar{n}_3})^{\frac{s}{\bar{n}_3}} \tag{16c}$$

Under the assumptions made above

$$\bar{A}_{\bar{n}_3} \propto \bar{A}_{exc} \tag{16d}$$

so that

$$\bar{A}_s \propto (\bar{A}_{exc})^{\frac{s}{\bar{n}_3}} \tag{16e}$$

In figure 3a is shown the relative increase of $\bar{A}_s$ as a function of s for various values of $\bar{A}_{exc}$, $\bar{A}_1$ having been taken as a natural unit of $\bar{A}_{exc}$. The "tube", then, widens with $\bar{A}_{exc}$, and much faster at large s-values than at small ones. A_2 is practically constant whereas for $s = 0.9\, \bar{n}_3\, \bar{A}_s$ is almost proportional to $\bar{A}_{exc}$, figure 3b.

Curves 4, 5 and 6 in figure 2 show the concentration dependence of R^{-1} for the cases where the "narrow section" is located at s = 20, s = 80 and, finally, extends from $s \cong 20$ to $s \cong 80$. When multiplied with the second factor of (15a) for various values of $\sigma^2/\bar{n}_3$ a very large variety of curve shapes may result, figure 4.

In figure 5 the expression (15a) is compared with two early experimental results using gaussian forms of the size distribution of the abundant micelles and of the s-dependence of $1/k_s^- \bar{A}_s$ around the "narrow section".

Experimental results presented at this symposium show that the almost linear increase of $1/\tau$ with $\bar{A}_{exc}$ found in the earlier experiments is due to strongly surface active impurities. The new results strongly resemble the forms given in figure 5 indicating that larger values of $\sigma^2/\bar{n}_3$ and "narrow sections" extending down to or even located at low s-values are prevalent.

The results for DPI seem to remain an exception to this trend, a fact that may be due to the exceptional affinity between the monomer end-groups and the counter ions, charge transfer complexes.

LARGE DEVIATIONS FROM EQUILIBRIUM

The above treatment holds only when $|\bar{n}_3\, \xi_1| \ll 1$ or when the "narrow section" extends all the way from very low to very high s-values.

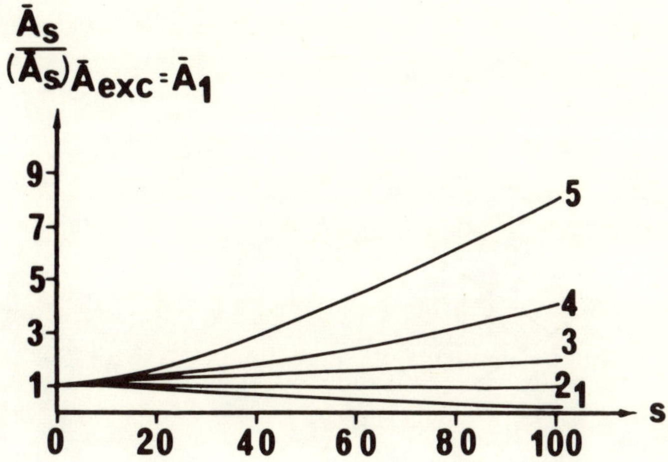

Fig. 3a. $\bar{A}_s/(\bar{A}_s)_{\bar{A}_{exc}} = \bar{A}_1$ versus s for 1) $\bar{A}_{exc} = 0.25\,\bar{A}_1$, 2) $\bar{A}_{exc} = \bar{A}_1$, 3) $\bar{A}_{exc} = 2\bar{A}_1$, 4) $\bar{A}_{exc} = 4\bar{A}_1$, 5) $\bar{A}_{exc} = 8\bar{A}_1$.

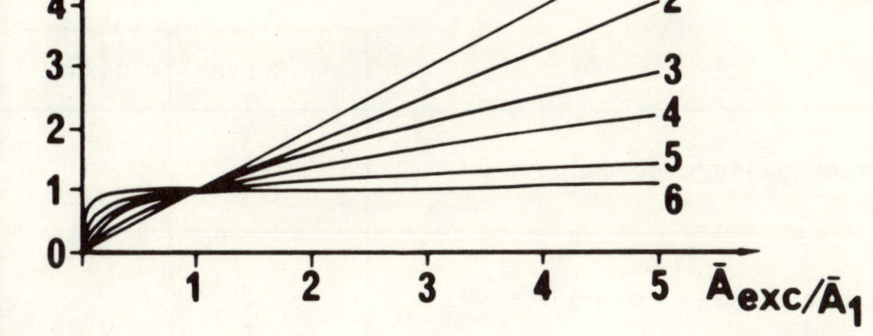

Fig. 3b. $\bar{A}_s/(\bar{A}_s)_{\bar{A}_{exc}} = \bar{A}_1$ versus $\bar{A}_{exc}/\bar{A}_1$ for 1) $s/\bar{n}_3 = 1$, 2) $s/\bar{n}_3 = 0.9$, 3) $s/\bar{n}_3 = 2/3$, 4) $s/\bar{n}_3 = 1/2$, 5) $s/\bar{n}_3 = 1/4$, 6) $s/\bar{n}_3 = 1/20$.

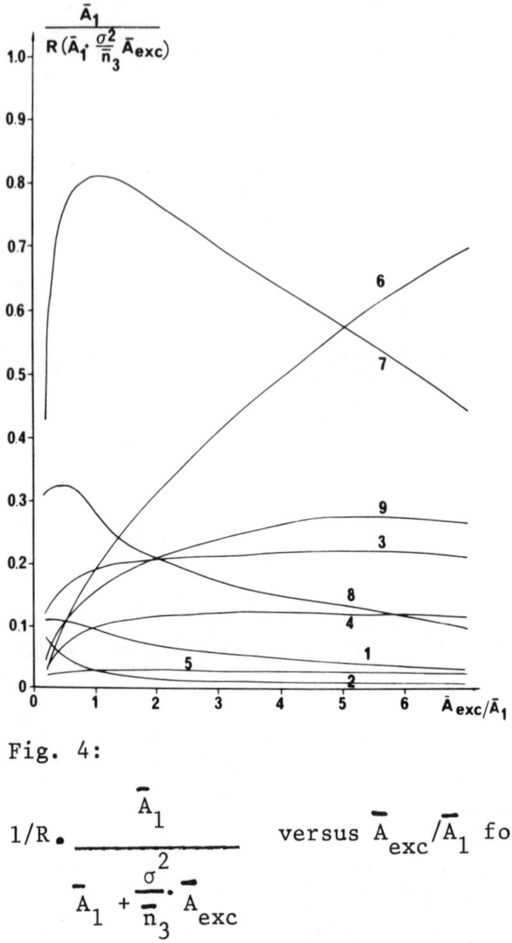

Fig. 4:

$$1/R \cdot \frac{\bar{A}_1}{\bar{A}_1 + \frac{\sigma^2}{\bar{n}_3} \cdot \bar{A}_{exc}}$$ versus $\bar{A}_{exc}/\bar{A}_1$ for the following cases:

	Curve	$\sigma^2/\bar{n}_3$
A short narrow passage around s = 20	1 2 3	1 5 0.05
A short narrow passage around s = 80	4 5 6	1 5 0.05
A long narrow passage between s = 20 and s = 80	7 8 9	1 5 0.05

The values in several of the cases are multiplied with 0.1 or 10.

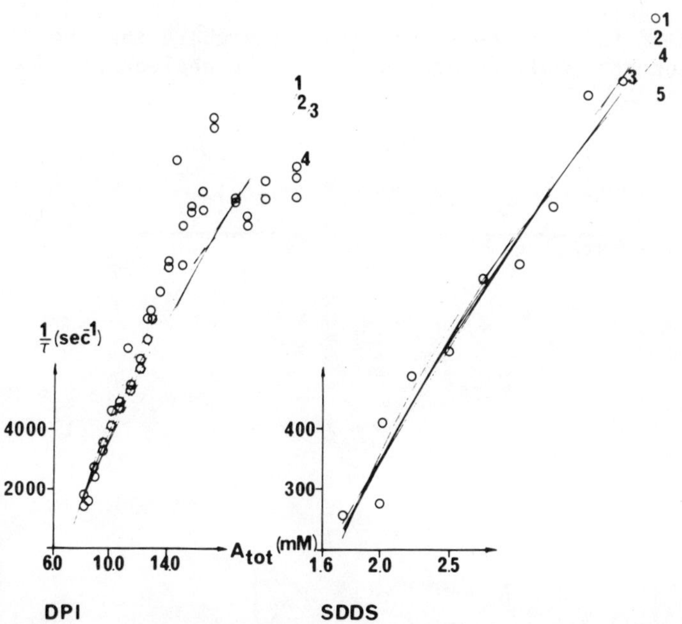

Fig. 5. Experimental results[1,8], (o) and calculations (full lines) based on equation (15). The c.m.c. is defined by $(\partial c_3/\partial \bar{A}_{tot})_{\bar{A}_1=\bar{A}_1'} = 1/2\bar{n}_3'$, a prime indicating quantities at c.m.c. (At larger values of $\bar{A}_{tot}$ the derivative is close to $1/\bar{n}_3$, and at smaller about zero). The assumptions are: $\bar{n}_1 = 1$, $\bar{A}_s' \propto \exp\{-(s - \bar{n}_3')^2/2\sigma^2\}$ for $s > s_2$, $\bar{A}_s/\bar{A}_s' = (\bar{A}_1/\bar{A}_1')^s$, and $1/k_s^- \bar{A}_s \propto \exp\{-(s-s_r')^2/2\sigma_r^2\}$ for $s_1 < s < s_2$. Summations have been replaced by integrations. For given values of $\bar{n}_3'$, σ, s_r' and σ_r those values of R' and $\bar{A}_1'$ were sought which minimize the sum of relative deviations between experimental and calculated values. Abbreviations in table below: DPI = Dodecyl pyridinium iodide. SDDS = Sodium dodecyl sulphate.

Substance	DPI[1]				SDDS[8]				
Curve No. / Parameters	1	2	3	4	1	2	3	4	5
$\bar{A}_1'$ (mM)	5.5	5.6	6.2	6.5	1.15	1.25	1.25	1.30	1.45
$\bar{n}_3'$	87	87	87	87	95	95	95	95	95
σ	1	2	4	4	0	3	1	3	5
s_r'	80	80	80	60	90	90	80	85	80
σ_r	1	5	2	2	3	0	1	7	10
R' 10^6	38	44	66	212	0.88	1.23	1.82	1.84	3.76

An extension of the treatment above to cases where the ξ:s are still small but not small enough to permit the neglect of $\xi_1 \xi_{s-1}$ in (2) yields

$$-\frac{1}{\xi_1}\frac{d\xi_1}{dt} = \frac{1}{\tau(\xi_1)} = \frac{1}{R(\xi_1)\, c_3}\;\frac{\bar{n}_1^2 c_1 + \bar{n}_3^2 c_3 + f(\xi_1)\, c_3}{\bar{n}_1^2 c_1 + \sigma^2 c_3} \tag{17a}$$

where

$$R(\xi_1) = \sum_{s>s_1}^{\leq s_2} \frac{(1+\xi_1)^{\bar{n}_3 - s}}{k_s^- \bar{A}_s} \tag{17b}$$

and

$$f(\xi_1) = \frac{1}{\xi_1}\left[(1+\xi_1)^{\bar{n}_3} - (1+\bar{n}_3\xi_1)\right] c_3 \tag{17c}$$

A closer examination of these expressions reveals that $\tau(\xi_1)$ will remain close to $\tau(0)$ if the narrow section is very long. In cases where the parameters, essentially σ and $\bar{n}_3$ of the second factor of (17a) are known (see further below) $R(\xi_1)$ would be obtainable from determinations of $d\ln|\xi_1|/dt$. A "deconvolution" of the sum (17a) would then give at least an approximate dependence of $1/k_s A_s$ on s in the narrow section.

REARRANGEMENT AMONG THE ABUNDANT MICELLES

Equations (2) and (3) with the differences $\xi_s - \xi_{s-1}$ replaced by $\partial\xi(s)/\partial s$ and $J_{s+1} - J_s$ replaced by $\partial J(s)/\partial s$ can also be used for approximate solutions in cases where ξ_s, k_s^- and $\bar{A}_s$ do not vary too rapidly with s. (3) then takes the form

$$\bar{A}(s)\cdot\frac{\partial\xi(s)}{\partial t} = \frac{\partial}{\partial s}\left[k^-(s)\,\bar{A}(s)\left\{\frac{\partial\xi(s,t)}{\partial s} - \xi_1(t)(1+\xi(s,t))\right\}\right] \tag{18}$$

The reason for writing $\xi_1(t)$ instead of $\xi(1,t)$ will presently become apparent. Approximating $\bar{A}(s)$ by a gaussian distribution

$$\bar{A}(s) = A_o \exp.\ (-(s-\bar{n}_3)^2/2\sigma^2) \tag{19}$$

taking $k_s^- = k^-$, independent of s over the region $\bar{n}_3 - \sigma \lesssim s \lesssim \bar{n}_3 + \sigma$, and introducing

$$z = \frac{s-\bar{n}_3}{\sqrt{2}\,\sigma} \tag{20}$$

equation (18) will take the form

$$\frac{1}{k^-}\frac{\partial \xi(z,t)}{\partial t} = \frac{1}{2\sigma^2}\left(\frac{\partial^2 \xi(z,t)}{\partial z^2} - 2z\frac{\partial \xi(z,t)}{\partial z}\right) +$$

$$\frac{1}{\sqrt{2}\,\sigma}\,\xi_1(t)\left(2z + 2z\xi(z,t) - \frac{\partial \xi(z,t)}{\partial z}\right) \tag{21}$$

Expanding

$$\xi(z,t) = \sum_{n=0}^{\infty} c_n(t)\,H_n(z) \tag{22}$$

where H_n is the Hermite polynominal[15] of order n obeying

$$H_n'' - 2z\,H_n' + 2n\,H_n = 0 \tag{23a}$$

$$\int_{+\infty}^{-\infty} H_n H_m e^{-z^2}\,dz = 2^n n!\sqrt{\pi}\;\delta_{nm} \tag{23b}$$

$$H_n' = 2n\,H_{n-1} \tag{23c}$$

and

$$2z\,H_n = H_{n+1} = 2n\,H_{n-1} \tag{23d}$$

will transform (21) into

$$\Sigma\frac{1}{k^-}\frac{d\ln}{dt}H_n = -\frac{1}{\sigma^2}\Sigma\, n\, c_n H_n + \frac{1}{\sqrt{2}\sigma}\xi_1(t)\,H_1 + \frac{1}{\sqrt{2}\sigma}\xi_1(t)\Sigma c_n H_{n+1} \tag{24}$$

The material balance gives

$$\frac{1}{\sqrt{2}\,\sigma}\,\xi_1(t) = -\frac{1}{\sqrt{2}\,\sigma}\,\frac{\bar{A}_{exc}}{\bar{A}_1}\,c_o(t) - \frac{1}{\bar{n}_3}\,\frac{\bar{A}_{exc}}{\bar{A}_1}\,c_1(t) \tag{25}$$

Since $\xi_1(t)$ or its equivalent is the quantity measured in practice we only need to obtain $c_o(t)$ and $c_1(t)$.

Inserting (25) into (24) and equalizing the coefficients on both sides for each H_n one finds after solving the first order differential equations

$$c_o(t) = (c_o(0)) \tag{26a}$$

and

$$c_1(t) = \left(c_1(0) + \frac{\sigma}{\sqrt{2}}\,\frac{a}{1 + \frac{\sigma^2}{\bar{n}_3}a}\right) e^{-\frac{t}{\tau}} - \frac{\sigma}{\sqrt{2}}\,\frac{a}{1 + \frac{\sigma^2}{\bar{n}_3}a} \tag{26b}$$

where

$$\frac{1}{\tau} = \frac{k^-}{\sigma^2} + \frac{k^-}{\bar{n}_3}\,a\,(1 + c_o) \tag{26c}$$

and

$$a = \frac{\bar{A}_{exc}}{\bar{A}_1} \tag{26d}$$

Inverting (22) one finds

$$c_n(0) = (2^n\,n!\,\sqrt{\pi}\,)^{-1} \int_{-\infty}^{+\infty} \xi(z,0)H_n(z)e^{-z^2}\,dz \tag{27}$$

The extension of the latter integral to $-\infty$ despite the definition (20) is permissible when σ is not very large since the integrand through $\exp(-z^2)$ decreases very rapidly for large values of $|z|$.

Generalization to cases where k_s^- cannot be taken independent of s in the region of abundant micelles or where the size distribution of the latter deviates essentially from a gaussian form is possible via perturbation calculations[16].

Since the deviation from equilibrium of the number of micelles, $\int \xi(s,t)\,\bar{A}(s)ds$, is given by $\sqrt{2\pi}\,\sigma\,A_o c_o(t)$ as is found by inserting (22) into this integral, eq. (26a) expresses the fact that the total number of micelles is constant during this process. If the time constant for the net dissolution of micelles, given in (15), is much longer than that for the rearrangement process, given by (26c), then the present treatment is justified. Inserting plausible values of the quantities involved into these expressions one finds that the two time constants are of different orders of magnitude.

Solving the equations for the higher coefficients c_n, $n > 1$, shorter time constants given by

$$\frac{1}{\tau} = \frac{n\,k^-}{\sigma^2} \tag{28}$$

will enter the expressions. They will, however, not be seen in relaxation experiments unless a quantity more sensitive to the details of the micellar size distribution than ξ_1 or its equivalents are measured. At small deviations from equilibrium c_o will be negligible compared to unity. In such cases eq. (26c) displays a linear concentration dependence from the slope of which $k^-/\bar{n}_3$ is obtained. When $\bar{n}_3$ is known from scattering or thermodynamic measurements the slope will directly give k^-. From the intercept at $a = 0$, k^-/σ^2 is obtained and so from these measurements alone both k^- and σ would, for the first time, be obtainable.

DISCUSSION AND CONCLUSION

In their pioneering work[1] on DPI Sheraga and collaborators proposed a step-wise model for the disintegration (and formation) of micelles. To explain the (almost) linear relation between $1/\tau$ and A_{tot} they assumed that the micelles were strictly monodisperse and that the rate-limiting step was that when the very first monomer left the micelle. With these assumptions one obtains

$$\frac{1}{\tau} = k_n^- \,(1 + n\,a) \tag{29}$$

a result to which also (15) reduces when one puts $R = 1/k_n^- \bar{A}_n$, $C_3 = \bar{A}_n$, $\bar{n}_1^2 = 1$, $c_1 = \bar{A}_1$ $\bar{n}_3^2 = n^2$ and $\sigma^2 = 0$. It is interesting to note this agreement since the assumptions leading to (29) are in contradiction to those leading to (15) in that the rate limiting step is here associated with a species of non-negligible concentration.

It is also interesting that the discrepancy between the low values of k_n^- deduced with (16) and the higher values indicated by resonance results[13] are removed in (15) where $1/k_n^-$ is replaced by $R\,c_3$. In the latter $c_3/\bar{A}_s$ can be very large, allowing much larger values of k_s^-. The somewhat unexpected temperature dependence of k_n^- found[1] using (16) may also find an explanation in the full temperature dependence of $R\,c_3$.

In conclusion it may be said that if the treatment above is

essentially correct (and the new results presented at this symposium would seem to point in that direction) the fast rearrangement process among the abundant micelles would yield both $k^- = k^-_{\bar{n}_3}$ and σ i.e. the two, after $\bar{n}_3$, most important quantities in this micellar size region. The absolute value and temperature dependence of k^- will, in conjunction with theory[6], yield a most valuable insight into the energetics and dynamics of micelles.

For the slow process experiments would yield the smallest products $k_s^- \bar{A}_s$ among those of the rare intermediate micelles. They will be of great value as experimental checks on existing[6,7,8,9] and future theories of micelle formation. The separate, experimental determination of k_s^- and $\bar{A}_s$ in the same region probably has to await a further advance in techniques.

REFERENCES

1. G.C. Kresheck, E. Hamori, G. Davenport and H.A. Scheraga, J. Am. Chem. Soc., 88, 246 (1966).
2. E. Graber, J. Lang, R. Zana, Kolloid-Z. Z. Polym., 238, 470 (1970).
3. J. Rassing, P.J. Sams and E. Wyn-Jones, J. Chem. Soc. Faraday Trans., II, 69, 180 (1973).
4. G. Aniansson and S. Wall, J. Phys. Chem., 78, 1024 (1974).
5. G. Aniansson, this volume.
6. G. Aniansson, M. Almgren and S. Wall, this volume.
7. C.A.J. Hoeve and Benson, J. Phys. Chem., 61, 1149 (1957).
8. D.C. Poland and H.A. Scheraga, J. Phys. Chem., 69, 2431 (1965).
9. D.C. Poland and H.A. Scheraga, J. Colloid and Interface Sci., 21, 273 (1966).
10. G. Aniansson and S. Wall, to be published.
11. B.C. Bennion, L.K.J. Tong, L.P. Holmes and E.M. Eyring, J. Phys. Chem. 73, 3288 (1969).
12. See forthcoming thesis by S. Wall.
13. N. Muller, J. Phys. Chem., 76, 3017 (1972).
14. S. Wall, W. Karlsson and G. Aniansson, unpublished results.
15. G. Arfken, "Mathematical Methods for Physicists", Academic Press, New York, N.Y., 1966.
16. Forthcoming publication.

THE RATE OF MICELLE DISSOCIATION

E.A.G. Aniansson, M. Almgren and S.N. Wall

Department of Physical Chemistry, Chalmers University of Technology and The University of Gothenburg, Fack, S-402 20 Goteborg 5, Sweden

ABSTRACT. The rates with which a monomer leaves and enters a micelle of given aggregation number has been calculated with a generalized verion of Kramers' theory for friction-controlled reactions. The results are compatible with the results from NMR, EPR and ultrasonic measurements.

The rate of micelle dissociation has been measured with several techniques during the last few years[1]. It is generally agreed that the process is a step-wise one in which one monomer at a time leaves the micelle or (microreversibility) enters it. Since the number of steps generally is of the order of 100 rather special and simplifying assumptions were initially made in order to relate the rate constants of the individual steps with the overall rate. Recently a more general attack became possible by regarding the process as a generalized flow in aggregation space[2]. The result of this was a clear identification of the rate-determining quantities in the overall process. In the complete absence of any theory for the values of individual rate constants plausibility arguments had to be relied upon in pursuing the calculation of the overall rate.

There is, therefore, in this field an urgent need for even the most approximate theory giving order-of-magnitude estimates of the individual rate constants. The following is an attempt in this direction which we believe has not only attained this goal but reached considerably beyond it. In this endeavour we were, curiously enough, guided by a parallel attempt to apply the abovementioned approach to the fundamental reaction process ("absolute reaction rate theory") by treating it as a sequence of a large number of consecutive steps. The result of this parallel attempt[3] was a simplified derivation and generalized version of Kramers'

Wyn-Jones (ed.), Chemical and Biological Applications of Relaxation Spectrometry, 239–244.

too little known theory of "friction-controlled" reaction processes.

In the following we shall explicitly treat the case where a long chain monomer leaves (or enters) a spherical micelle, but the results are easily extended to cylindrical or plate-like micelles as well as to interfacial monolayers.

In using the results of ref. (3) we have to consider the free energies of the sequence or sequences of configurations that lead from the intact micelle consisting of $\underline{s}$ monomers, to the configu-ations corresponding to a micelle consisting of $\underline{s}-1$ monomers plus a free monomer. In doing so we can limit ourselves to those configurations where the outgoing monomer intersects the micelle surface only once. As argued below the other configurations are sufficiently rare to be neglected. Taking, then, as the "reaction coordinate" the length ρ of the hydrophobic tail that at the moment is outside the micelle surface, we have to explore the region of ρ - values where the free energy $\underline{G}(\rho)$ of ref. (3) is maximal.

For this purpose it is convenient to start from the equilibrium statistical mechanical theory for micelles, in particular that of Hoeve and Benson[4], and to introduce into it the suitable modifications. These will be:

1. Since one of the monomers is singled out as the outgoing monomer, OM, the number of "identical particles" decreases to $\underline{s}-1$.
2. The surface area $\underline{A}^{(s)}$ of the micelle decreases since the volume of the hydrocarbon core will decrease with an amount equal to the volume of the protruding part of the OM. Of this area the tail of the OM will occupy an area $\underline{a}_1$ which is different from the area occupied by each end group of the remaining monomers, $\underline{a}_o$.
3. The configurational freedom, $\gamma(\rho)$ of the OM will in general be somewhat different from that of the remaining monomers, $\gamma(o)$.
4. The free energy of the system will increase by $\varepsilon(\rho)$, the hydrophobic binding energy of a hydrocarbon tail of length ρ. This function has been found to be roughly linear with ρ and to be roughly equal to $\underline{nkT}$, where $\underline{n}$ is the number of buried CH_2-groups, and as usual $\underline{k}$ and $\underline{T}$ are the Boltzmann constant and absolute temperature, respectively.

For ionic monomers there will be a decrease in free energy equ to the product of the monomer charge and the drop in Gouy potential from the surface to the average position of the OM end group. The change in the "Coulomb free energy" of the remaining monomers will also have to be accounted for to the extent it is not taken care of through the $\underline{a}_o$ - value used.

The change in the moments of inertia of the system is completely negligible for all but the very smallest aggregates, the oligomers. For these the existing statistical mechanical theory is not applicable, so they have to be excluded anyhow from the present treatment. It will, though, be possible to deduce into what direction the deviations will go and something about their order of magnitude.

In order not to burden the equations more than necessary the electrostatic effects will not be included below. The following will then pertain to non-ionic and zwitterionic surfactants as well as to ionic surfactants at such ionic strengths of the solution that the extent of the electric double layer is much shorter than the length of the monomer. Due to the very small partial volume changes during micelle dissociation pressure-volume effects will also be negligible at ordinary pressure and, therefore, dropped here. The inclusion of these effects is fairly straightforward. This will be done in a forthcoming fuller presentation of the whole subject.

Since the total volume of the solution does not influence the rate constants its value will be taken as unity.

Using the same notations as in ref. (3) we can now write:

$$G' = -kT \ln Q_s - kT \ln s, \tag{1}$$

$$G_m^* = -kT \ln Q_{s-1} - kT \ln Q_1 - kT \ln A^{(s-1)} + \Delta, \tag{2}$$

and

$$G_{m+}^* = G_m^* - \Delta, \tag{3}$$

where $\underline{Q}_s$ is the partition function of the micelle with aggregation number $\underline{s}$, $\underline{s} = 1$ denotes the free monomer, $\underline{A}^{(s-1)}$ is the surface area of the micelle of size s-1, $\underline{G}_m^*+$ is the value of $\underline{G}^*$ when the monomer is free from the micelle and Δ is the drop in free energy when the outer tip of the tail leaves the micelle surface.

In the description of Hoeve and Benson[4] Δ will be given by

$$\Delta = kT \ln \frac{A^{(s-1)}}{(A^{(s-1)}-(s-1)\,a_o-a_1)} \cdot \frac{(A^{(a-1)} - (s-1)\,a_o)^{s-1}}{(A^{(s-1)}-(s-1)a_o-a_1)^{s-1}} \cdot \frac{\gamma_{free}}{\gamma_m} \tag{4}$$

where γ_m and γ_{free} are the configurational freedoms of the monomer at the maximum and in the free state, respectively. The first factor in the argument of the logarithm reflects the gain in freedom of the tip. The second describes the gain in freedom of the end groups of the remaining monomers and the third measures the change in freedom for internal bendings and overall rotations of the monomer.

A slight change in the definitions of Hoeve and Benson has been made so that $\underline{Q}_s$ is related to equilibrium concentration in the following way:

$$\ln \bar{A}_s = \ln Q_s + s\lambda\,,\quad s = 1, 2, 3, \ldots, \tag{5}$$

where λ is a Lagrange multiplier independent of s but dependent of the total concentration and $\bar{A}_s$ is the equilibrium concentration of micelles with aggregation number s.

For the outgoing rate constant we obtain according to eqn. (28) of ref. (3)

$$k^{+}_{out} = \frac{D_m}{l_b} \exp\left[-(G^{*}_{m} - G')/kT\right] = \frac{D}{l_b} \cdot \frac{Q_{s-1}\,Q_1}{Q_s} \cdot \frac{A^{(s-1)}}{s}\,\alpha \tag{6}$$

where α is defined by $\Delta = -\,kT\alpha$.

In order to apply the same eqn. (28) to the ingoing rate constant we now define the initial state as that where the free particle is within the distance δr from the micelle surface. G' now becomes $(G^{*}_{m+} - kT \ln \delta r)$, so that

$$k^{\neq}_{in} = \frac{D_m}{l_b}\,\frac{\alpha}{\delta r} \tag{7}$$

These two rate constants are related to the ordinary rate constants k_s^- and k_s^+ defined by

$$A_{s-1} + A_1 \underset{k_s^-}{\overset{k_s^+}{\rightleftharpoons}} A_s \tag{8}$$

via the equations

$$k^{+}_{s} = k^{\neq}_{int}\, A^{(s-1)}\delta r \tag{9}$$

and

$$k^{-}_{s} = s\, k^{\neq}_{out}, \tag{10}$$

so that finally

$$k^{+}_{s} = \frac{D_m}{l_b}\,\alpha\, A^{(s-1)} \tag{11}$$

and

$$k_s^- = \frac{D_m}{l_b} \frac{Q_{s-1} Q_1}{Q_s} \propto A^{(s-1)} \qquad (12)$$

Because of (5) the latter can also be written

$$k_s^- = \frac{D_m}{l_b} \frac{\bar{A}_{s-1} \bar{A}_1}{\bar{A}_s} \propto A^{(s-1)} \qquad (13)$$

from which it is easily checked that the rate constants fulfil the equilibrium condition

$$k_s^+ \bar{A}_{s-1} \bar{A}_1 = k_s^- \bar{A}_s \qquad (14)$$

Thus it is seen that in order to calculate the rate constants we need the diffusion constant, D_m, the length of the barrier, l_b, and the "freedom" expression $\propto A^{(s-1)}$. But we do not need the complete expressions for the partition functions if we can determine experimentally the equilibrium concentration of the micelles of all pertinent sizes, including, according to ref. (2), the rare intermediate ones for the determination of which there exists, at present, no experimental methods.

A close examination[5] shows that for D_m we can with good approximation take the diffusion constant of the monomer in the bulk solution. We can also easily obtain l_b with good approximation by realizing that the free energy, G^*, is a close to linear function of ρ. Now, as will also be seen from a close examination, there holds for the larger micelles,

$$G(1) - G(0) \simeq \varepsilon(1) \qquad (15)$$

so that from eqn. (24) of ref. (3) it follows that

$$l_b = \frac{kT}{\varepsilon(1)} \quad 1. \qquad (16)$$

It can also be shown[5] that for the larger micelles the remaining factors of the expression (12) for k_s^- to a good approximation is given by $l_o^{-1} \exp(-\varepsilon(1)/kT)$, where l_o is equal to Hoeve and Benson's ε, the "freedom of motion" of the monomer end groups normal to the micelle surface. When major electrostatic effects are absent, as assumed here, l_o will be nearly equal to l_b.

For these cases, then, we obtain the simple approximation

$$\frac{1}{s} k_s^- = \frac{D_m}{l_b^2} \exp(-\varepsilon(1)/kT) \qquad (17)$$

Using values appropriate to a monomer containing twelve carbon atoms in an unbranched chain one finds values of $\underline{k}_s^-$ of the order of 10^6 second^{-1}. This is 10^4 times larger than the values inferred from the experiments and model in ref. (1a) pertaining to dodecyl pyridinium iodide in 0.5 molar KCl solutions. In this model only one micelle size of non-vanishing equilibrium concentration was assumed to exist and further the first step in the disintegration sequence was assumed to be rate-determining.

The higher value given by (17) is on the other hand well compatible with the results of the theory of ref. (2) and with the results from NMR[1c] and ultrasonic measurements[1b].

The maximum in free energy occurs in the end of the outgoing process when at most one CH_3 or CH_2 group is left inside the micelle surface. If this group is not the end group of the monomer the configuration must be very strongly bent. Since hydrocarbons of lengths up to 16 C-atoms are known to be little curled[6] in water, this configuration should be rather improbable. Further, the surface area occupied by the outgoing tail would in any case be much larger than $\underline{a}_o$ so that its effect on the crowding would lower α much enough to make this configuration of negligible importance.

For the smallest aggregates, the oligomers, the approximation (17) will not hold. In these aggregates the larger part of the hydrocarbon tails will still be exposed to the surrounding water solution. The values of hydrophobic binding energies to be used instead of $\varepsilon(\underline{1})$ would be only a fraction of the latter value. This would lead to $\underline{k}_s^-$ - values 10^2 or 10^3 times larger than those given by (17).

REFERENCES

1. a. G.C. Kresheck, E. Hamori, G. Davenport and H.A. Scheraga, J. Am. Chem. Soc., **88**, 246 (1966).
   b. J. Rassing, P.J. Sams and E. Wyn-Jones, JCS Faraday II, **70**, 1247 (1974).
   c. For further references to experimental works, see (2).
2. E.A.G. Aniansson and S.N. Wall, J. Phys. Chem., **78**, 1024 (1974).
3. E.A.G. Aniannson, Chem. Phys. Lett., in press.
4. C.A.J. Hoeve and G.C. Benson, J. Phys. Chem., **61**, 1149 (1957).
5. To be published.
6. N. Muller, in: Reaction Kinetics in Micelles, ed. E. Cordes (Plenum Press, New York)(1973).

ON THE RATE OF MANY-STEP PROCESSES

E.A.G. Aniansson

Department of Physical Chemistry, University of Gothenburg, S-40220, Gothenburg, Sweden

ABSTRACT. The kinetic equations for a sequence of mono-molecular transformations are given a form which suggests a very close analogy with diffusion in a tube with varying cross-section and clearly indicates the rate limiting quantities. The overall rate of transformation is deduced for the case when the sequence contains a set of rare intermediate species. The result is further used for a simple derivation of a generalized version of Kramers theory for friction controlled reactions.

In the treatment of the step-wise micelle dissociation[1] involving a number of steps of the order of 100 it was found very useful to regard the process as a flow in aggregation space. Introducing the relative deviation from equilibrium values of concentrations of the reacting species and using equilibrium relations between the rate constants and the equilibrium concentrations it was possible to cast the equations in forms which were closely similar to flow processes such as diffusion and heat conduction. For a sequence of unimolecular processes the analogy is very simple and direct and, as will be shown, leads easily to a generalized version of Kramers' not too well-known theory for friction-controlled reactions[2]. The concepts involved and the just mentioned generalized version proved to be very helpful in developing an a priori theory[3] for the rate constants of the individual steps in the micelle dissociation (the escape of an individual long-chain monomer from the micelle) and since it is most probably applicable to a large class of transformations of loose (strongly entropy-controlled) systems it should be worthy of a separate presentation.

Denoting by A_s (s = ... 1, 2, 3, ...) the s:th species of

Wyn-Jones (ed.), Chemical and Biological Applications of Relaxation Spectrometry, 245–253.

the systems in a chain of transformations

$$A_{s-1} \underset{k_s^-}{\overset{k_s^+}{\rightleftharpoons}} A_s \tag{1}$$

and introducing

$$\eta_s = \frac{A_s}{\bar{A}_s} , \tag{2}$$

where A_s also denotes the concentration of the species and $\bar{A}_s$ denotes its equilibrium value one finds for the flow J_s, the net number of molecules that pass per unit time from s-1 to s,

$$J_s = k_s^+ A_{s-1} - k_s^- A_s = k_s^+ \bar{A}_{s-1} \eta_{s-1} - k_s^- \bar{A}_s \eta_s \tag{3}$$

Or, using that

$$0 = \bar{J}_s = k_s^+ \bar{A}_{s-1} - k_s^- \bar{A}_s \tag{4}$$

the flow equation takes the form

$$J_s = - k_s^- \bar{A}_s (\eta_s - \eta_{s-1}) \tag{5}$$

The quantity within brackets is clearly the equivalent of the spatial derivative of an intensity quantity, e.g. a concentration, whereas the analogous meaning of $k_s^- \bar{A}_s$ can only be seen via the continuity equation

$$\bar{A}_s \frac{\partial \eta_s}{\partial t} + J_{s+1} - J_s = 0 \tag{6}$$

Comparing these expressions with those for the pseudo-one dimensional diffusion in a tube of slowly varying cross section

$$J(x) = - D(x) A(x) \frac{\partial c}{\partial x} \tag{7}$$

and

$$A(x) \frac{\partial c}{\partial t} + \frac{\partial J}{\partial x} = 0 \tag{8}$$

the analogues of k_s^- and $\bar{A}_s$ are obviously the diffusion constant D(x) and the cross sectional area A(x) respectively.

From (5) it is obvious that the rate-limiting quantities are the products $k_s^- \bar{A}_s$ and not k_s^- alone as one would, perhaps, be inclined to think at the outset.

The equations (5) and (6) being linear the general solution for each η_s is a linear combination of exponential decay functions characterized by a set of relaxation time constants (assuming no introduction or removal of matter during the process, i.e. no "sources" or "sinks"). As such they have been thoroughly treated[4] but explicit solutions are in general limited to the cases with very low numbers of steps. Equation (5) and (6) should, however, be quite useful either through approximating them by the continuous counterparts or by using the latter as guides in seeking good starting points in iterative procedures.

OVERALL RATE OF TRANSFORMATION VIA A SET OF RARE INTERMEDIATE SYSTEMS.

The application that will concern us here is, however, still another one. We shall look at the case when $k_s^- \bar{A}_s$ is very much smaller for a set of intermediate s - values than for those on either side of these and in particular when this is due to the smallness of the $\bar{A}_s$ - values in this set. The situation is then analogous to diffusion of a substance from a thick end of a tube, through a very thin connection, to the other thick end. It is then immediately clear that if the intermediate section is sufficiently thin (and not too short) the characteristic time for the flow from one of the thick ends to the other will be much longer than the characteristic time for the equalization of the concentrations within each of the thick ends. For a quantitative criterium, see below.

After an initial short period of adjustment the process will be a pseudo stationary flow in which J_s is to a very good approximation independent of s in the intermediate region and decreasing to zero towards the two ends of the tube. In case this is not intuitively quite clear it can be seen by dividing (6) by $\bar{A}_s$ and considering that, due to the assumed smallness of $\bar{A}_s$, even very small differences between J_{s+1} and J_s would cause η_s to "blow-up" contrary to the pseudo-stationarity considered and to the general nivellation characteristics of diffusion processes.

During the pseudo-stationary phase the problem is further characterized by practically s - independent η_s - values in each of the thick ends as said above. This can also be seen by dividing (5) by $k_s^- \bar{A}_s$ and taking into account the comparative largeness of $\bar{A}_s$ in this region and the fact that J_s is, if anything, smaller in the end parts than in the intermediate region.

The calculation of the flow-rate J from one thick end to the other is now straightforward. Denoting quantities pertaining to the thick ends with ' and " respectively, and denoting the ends of the thin passage by s_1 and s_2

$$\eta'' - \eta' = \eta_{s_2} - \eta_{s_1} = \sum_{s=s_1-1}^{s_2} (\eta_s - \eta_{s-1}) = \sum_{s_1+1}^{s_2} - \frac{J_s}{\bar{k}_s \bar{A}_s} =$$

$$= -J \sum_{s_1+1}^{s_2} \frac{1}{\bar{k}_s \bar{A}_s} = - J R \tag{9}$$

where

$$R = \sum_{s_1+1}^{s_2} \frac{1}{\bar{k}_s \bar{A}_s} \tag{10}$$

is the total flow resistance of the thin passage, a sum of series resistances $1/\bar{k}_s \bar{A}_s$. (The $\bar{k}_s \bar{A}_s$ are, of course, conductances).

Letting now c' and c" denote the instantaneous number of systems on either side of the "narrow passage" we have the following relations

$$c' = \sum_{s \leq s_1} A_s = \sum_{s \leq s_1} \eta_s \bar{A}_s = \eta' \sum_{s \leq s_1} \bar{A}_s = \eta' \bar{c}' \tag{11}$$

$$c'' = \eta'' \bar{c}'' \tag{12}$$

where bars indicate equilibrium values, and since the amount of material in the thin passage will be negligible,

$$c' + c'' = \bar{c}' + \bar{c}'' \tag{13}$$

Using now the relations (9) - (13) in the solution of the rate equation

$$- \frac{dc'}{dt} = J \tag{14}$$

we find the relaxation-type result

$$c'(t) - \bar{c}' = (c'(t{=}0) - \bar{c}') \exp(-t/\tau) \tag{15}$$

where τ, the relaxation time, is

$$\tau = \frac{R}{\frac{1}{\bar{c}'} + \frac{1}{\bar{c}''}} \tag{16}$$

That the relaxation time should be proportional to the total flow resistance is immediately clear and so is the implication that it should increase with the "volumes" $\bar{c}'$ and $\bar{c}''$ of the "thick ends". The full solution of (5) and (6) will give in addition to the relaxation time (16) a number of much smaller ones describing the initial adjustment process preceding the pseudo-stationary final phase described by (15) and (16). The characteristic time for diffusional equalization within one of the three sections of the "tube" is, on the other hand, independent of the "thickness" of the section and of the order of the length squared divided by the diffusion constant. From this a quantitative criterion for the validity of the above treatment is easily obtained.

Of more general interest is, however, the rate constant $k^{\neq}$ that describes the rate when $c'' = 0$. J then takes the form

$$J = \frac{1}{R\bar{c}'} c' \tag{14'}$$

so that (14) gives

$$k^{\neq} = -\frac{1}{c'}\frac{dc'}{dt} = 1/R\bar{c}' = \left[\sum_{s=s_1+1}^{s_1} \frac{\bar{c}'}{\bar{k}_s \bar{A}_s}\right]^{-1} \tag{17}$$

Application to the Elementary Reaction Process.

In order to apply this result to an elementary reaction characterized as usual by a reaction coordinate ρ we divide the latter into small intervals of length λ_s and treat the particles within λ_s as a separate species A_s. If, now, there is an intermediate region of the reaction coordinate where $\bar{A}_s$, the equilibrium number of particles per interval λ_s, is very much smaller than on either side of it equation (17) would apply. This, however, should be the general case since otherwise the definition of reactants vis-à-vis products would be a very delicate one indeed. On the contrary, the general chemical definition of species with regard to domains of the reaction coordinates is very crude and in practice very clear.

It remains, then, to calculate $\bar{k}_s$ and $\bar{A}_s$. The first one, expressing the fraction of particles A_s that per unit time go from s to s-1, will be

$$k_s^- = \frac{\bar{v}}{2\,\lambda_s} \tag{18}$$

where

$$\bar{v} = \sqrt{\frac{2kT}{m}} \tag{19}$$

m is the characteristic mass for the motion along the reaction coordinate, and k and T are the Boltzmann constant and absolute temperature, respectively.

$\bar{A}_s$ can in the usual way be expressed in state sums or, equivalently, in a free energy G_s which will be of the form with respect to λ_s

$$G_s = G_s^* - kT \ln \lambda_s \tag{20}$$

if G_s' changes little compared to kT over the distance λ_s of the reaction coordinate.

Thus

$$\bar{A}_s = A_o \exp(-G_s/kT) = A_o \lambda_s \exp(-G_s^*/kT) \tag{21}$$

where A_o is a constant.

Although

$$k_s^- \bar{A}_s = \frac{\bar{v}}{2} A_o \exp(-G_s^*/kT) \tag{22}$$

is independent of λ_s it is necessary to establish the physical significance of the latter quantity since otherwise the sum in (17) would be indeterminate. The root of this is seen to be in the fact that, on the one hand, λ_s has to be large enough so that on the average a particle after travelling this distance has undergone a sufficiently strong collision that its speed and direction of motion along the reaction coordinate is statistically independent of the neighbouring intervals it came from. Otherwise (3) will not hold but particles coming from the left will have a propensity for going on to the right and vice-versa. The number of particles that per unit of time leave interval λ_s will not only be dependent on A_s but also on A_{s-1} and A_{s+1}.

On the other hand λ_s must not be so large that the particle on the average undergoes several "memory-obliterating" collisions since then (18) is again wrong and has to be exchanged for a random walk expression containing λ_s^{-2}. So, λ_s has to be chosen equal to "the mean free path" along the reaction coordinate.

We can now evaluate (17) and obtain after a quite appropriate replacement of summation by integration

$$k = \frac{\bar{v}}{2\,l_b} \exp(-G_m/kT) \tag{23}$$

where

$$l_b = \int_{\rho_1}^{\rho_2} \exp(-(G_m - G(\rho)/kT))\,d\rho \tag{24}$$

G_m is the maximum value of G_s (eq. (20)) considered as a continuous function $G(\rho)$, and ρ_1, ρ_2 are the values of the reaction coordinate corresponding to s_1 and s_2. Since the integrand of (24) is a very rapidly decreasing function on both sides of the maximum their exact values are immaterial.

Finally,

$$\Delta G_m = G_m - G' \tag{25}$$

where

$$\exp(-G'/kT) = \sum_{s \leqslant s_1} \exp(-G_s/kT) \tag{26}$$

It is found, using the corresponding partition functions, that G' is the ordinary free energy of the reactants.

l_b can be considered as "the width" of the free energy barrier at temperature T. The pre-exponential factor in (23) is simply the frequency with which an equivalent particle oscillates between two walls at a distance l_b apart. The exponential gives the ratio at equilibrium between the number of systems in the mean free path interval λ_m at the top of the free energy barrier and the number of systems in the reactant state.

(23) can be transformed into two other useful forms. The first one is obtained by introducing

$$D_m = \bar{k}_m \lambda_m^2 = \frac{\bar{v}\,\lambda_m}{2} \tag{27}$$

the corresponding diffusion constant along the reaction coordinate at the value of the latter where the free energy (20) is maximal:

$$k^{\neq} = \frac{D_m}{l_b^2} \exp\left(-\left(G^*_m - kT \ln l_b - G'\right)/kT\right) \tag{28}$$

The exponential now gives the fraction of systems at the top of the free energy assuming equilibrium conditions, the barrier taken flat and with width l_b, whereas the pre-exponential factor is the inverse of the characteristic time for diffusion over the length l_b. In this form the essential factors are clearly recognizable and intuitively easily understood.

The second is obtained assuming a parabolic free energy maximum and introducing a new half-width of the barrier, σ, defined by $l_b/\sqrt{\pi}$. It is such that

$$G(\rho_m \pm \sigma) = G_m - kT \tag{29}$$

This results in:

$$k^{\neq} = \frac{\lambda_m}{\sigma} \frac{kT}{h} \exp(-\Delta G^{\neq}/kT) \tag{30}$$

where h is the Planck's constant and $\Delta G^{\neq}$ is almost the conventional activation free energy[5], i.e. lacking both the coordinate and the momentum contributions to the partition function of the "activated state". The difference is that here the "transition state" is taken at the maximum of the free energy (20) whereas the conventional one is taken at the saddle point of Born-Oppenheimer energy surface for the nuclear motions.

In the last form the difference from the "transition-state theory" is very clearly brought forth through the factor λ_m/σ. When the mean free path of the system at the barrier maximum is less than the half-width of the same the rate is correspondingly reduced below the TST - value. This factor could be visualized as a "reflexion coefficient", the collisions of the complex causing a certain fraction to turn back after having passed the barrier top. But there is also another contributing factor. From the analogy with diffusion through a thin passage it is clear that the virtual concentration η is exactly one half of η'. Thus, there is also a deviation from the equilibrium value of 50%.

For λ_m larger than σ (30) would give a larger value than the TST. But this is where the assumption underlying (20) breaks down. If the free energy barrier were potential-energy-dominated the acceleration of the system over the mean free path would render (19) erroneous and the definition (20) uncertain. It is striking though, that the border case $\lambda_m = \sigma$, although fortuitously perhaps, renders exactly the TST-value.

When specialized to a system with no other degrees of freedom than that along the reaction coordinate the results above agree with those of Kramers[2] who treated the case of a particle moving in one dimension acted upon by a conservative force potential and by a surrounding viscous medium. In the present treatment a very wide class of "friction-controlled" transformations are covered including those where the barrier is due to an entropy minimum rather than to an energy maximum.

Kramers' choice of an extremely simple system allowed on the other hand a derivation from more fundamental first principles and a treatment of the region $\lambda_m \gg \sigma$ giving the perhaps strongest single argument in favour of the transition state theory of Henry Eyring.

REFERENCES

1. E.A.G. Aniansson and S.N. Wall, J. Phys. Chem., 78, 1024 (1974).
2. H.A. Kramers, Physica, VII, 284 (1940).
3. E.A.G. Aniansson, M. Almgren and S.N. Wall, Chem. Phys. Letters to be published.
4. H.J.G. Hayman, Trans. Faraday Soc., 65, 2918 (1969), 66, 1402 (1970) and 67, 3240 (1971).
5. S. Glasstone, K.J. Laidler and H. Eyring, The Theory of Rate Processes, McGraw-Hill, (1941).

[illegible]

One hypostatizes directly without [illegible] other systems [illegible]
than the [illegible] the [illegible] the problem [illegible]
[illegible] those [illegible] the [illegible] parties [illegible]
[illegible] imposed upon by [illegible]
[illegible]
[illegible]
[illegible] finding [illegible]
[illegible]
[illegible]
[illegible]
[illegible]
[illegible]

[illegible]

[illegible]

[illegible]

[illegible]

A COOPERATIVE MODEL FOR MICELLE FORMATION

Brian H. Robinson and Neal C. White

Chemical Laboratory, University of Kent at Canterbury

INTRODUCTION.

The thermodynamics of micelle formation have been investigated for several decades[1], but because micelles are labile structures, stabilized by relatively weak hydrophobic and dispersive forces, only in the past few years has it become possible to carry out kinetic studies on the rates of formation and breakdown of micelles, following developments in instrumentation for the measurement of fast reaction rates in solution.

A thermodynamic analysis of micellization based on a multiple-equilibrium scheme of step-wise aggregation has been widely employed[2] and, because it can be readily applied in a kinetic analysis, this approach, first used by Kresheck et al[3], has been generally employed so far by experimentalists interested in the kinetics of micelle formation (and dissolution). An alternative analysis, based on the Langmuir Adsorption Isotherm, has been developed recently by Rassing and Wyn-Jones[4], and an interesting novel approach has been presented by Aniansson and Wall[5].

SIMPLE STEP WISE AGGREGATION SCHEMES

There are two schemes in common use for describing step-wise aggregation. These are essentially limiting cases of a more general scheme: i) All-or-none Micellization: This scheme assumes that micelle formation can be described by a process involving the step-wise addition of monomer to a growing aggregate A_n ($n = 2 \rightarrow N-1$) to form finally a micelle, A_N, of aggregation number N.

Wyn-Jones (ed.), Chemical and Biological Applications of Relaxation Spectrometry, 255–263.

$$\text{i.e.:}\quad A_1 + A_1 \overset{K_D}{\rightleftharpoons} A_2 \qquad \text{Nucleation}$$

$$A_2 + A_1 \overset{K_G}{\rightleftharpoons} A_3$$

$$\vdots$$

$$A_{N-2} + A_1 \overset{K_G}{\rightleftharpoons} A_{N-1} \qquad \text{Growth of Intermediate Aggregates}$$

(Nucleation and Growth of Intermediate Aggregates: Fast)

$$A_{N-1} + A_1 \underset{k_{N,N-1}}{\rightleftharpoons} A_N \qquad \text{Micelle Formation} \qquad \text{Slow}$$

The equilibria $A_n + A_1 \rightleftharpoons A_{n+1}$ ($n = 1 \rightarrow N-2$) are all assumed to be established rapidly, the rate-limiting step in the micelle-breakdown process being the loss of the first monomer from the micelle, controlled by rate constant $k_{N,N-1}$.

Two further simplifying assumptions are made in the interpretation of the kinetics:

a) At the critical micelle concentration (c.m.c.)

$$C_1 = C_A^o >> C_2 \;....\; C_{N-1}$$

where C_n = molar concentration of n-mer, and C_A^o = total surfactant concentration expressed as monomer.

b) Above the c.m.c.

$$C_N, C_1 >> C_2 \;....\; C_{N-1} \text{ and } C_1 = \text{c.m.c.}$$

It is therefore necessary that: $K_{12} << K_{n-1,n} (n=3 \rightarrow N-1) << K_{N-1,N}$

Thus, it can be seen that the overall scheme is made up of steps representing nucleation, growth and micelle formation. For the assumptions regarding intermediate concentrations to hold, it is necessary to view the simple Kresheck model as representative of a very highly cooperative aggregation process, i.e. $K_G = K_{23}$etc.>> K_{12}. If q is defined as K_G/K_D, then q (cooperativity parameter)$\rightarrow \infty$

ii) Non-Cooperative Growth of Aggregates:

For $A_n + A_1 \rightleftharpoons A_{n+1}$

$$K_{n,n+1} = C_{n+1}/C_n.C_1 \quad \text{(neglecting activity coefficients)}$$

$$= K_{12} \text{ (for } n = 2 \rightarrow \infty)$$

For this association scheme, the thermodynamic (and some kinetic) implications have already been discussed[6,7,8]. It is important to realise that this aggregation scheme cannot produce an abrupt change in some observable parameter (e.g. conductivity) which is a characteristic of surfactant systems that exhibit a c.m.c. For this model, the build-up of aggregates will occur gradually as C_A^o is increased, with $C_{n+1} < C_n$, except at $C_A^o \rightarrow \infty$, when $C_{n+1} \rightarrow C_n$.

COOPERATIVE MODEL FOR MICELLE FORMATION

In this article, a general cooperative model for micellization and its implications are discussed, based on extensions of the models described in the previous section.

For many surfactant systems in aqueous media, it is observed experimentally that the transition between monomer and micellar surfactant occurs over a narrow concentration range. Nevertheless, it is in this finite 'transition' region that clues regarding the energetics of micellization are likely to be found. Therefore, it seems useful to attempt to establish the presence or otherwise of 'pre-micellar' aggregates in the region close to the c.m.c. These species appear in the step-wise aggregation scheme during the growth process and would be precursors of the apparently spherical micelles which are thought to be the dominant species for most surfactants at concentrations up to a factor of ten greater than the c.m.c. In the argument developed in this paper, we distinguish the pre-micelle aggregates A_n from the true micelles A_N'. Other workers may prefer to think of these structures as micelles (1)-polydisperse and micelle (2)-monodisperse. It is interesting to note that in the 1930's, there was vigorous debate as to whether micellar species in water should be regarded as spherical (Hartley) or spherical plus Lamellar (McBain). The shape of micelles has been discussed recently in an important paper by Tanford[9], who suggests that micelles are energetically most stable in an ellipsoidal shape). Clearly, it is useful to be able to characterize all types of aggregate formed in surfactant solution.

In the cooperative model, to account for most of the available experimental information, the three stages - nucleation, growth and micelle formation - are retained, but since it seems unlikely on free energy arguments that the pre-micelle aggregates are direct precursors (e.g. mini-spherical micelles) of a more or less monodisperse micelle, the final stage in the aggregation scheme has been shown as a conformational (phase) transition.

Thus we have:

$$A_1 + A_1 \underset{}{\overset{K_D}{\rightleftharpoons}} A_2 \qquad \text{Dimerization (Nucleation)}$$

$$\left.\begin{array}{l} A_2 + A_1 \overset{K_G}{\rightleftharpoons} A_3 \\ \vdots \\ A_{N-1} + A_1 \overset{K_G}{\rightleftharpoons} A_N \end{array}\right\} \quad \text{Growth of Lamellar Aggregates}$$

$$\underset{\text{Lamellar}}{A_N} \overset{Y_N}{\rightleftharpoons} \underset{\text{Spherical}}{A_N'} \qquad \text{Conformational Switch}$$

Although the thermodynamic model does not require a molecular visualization of the intermediate states, it is helpful in understanding the source and magnitude of the cooperativity effects. It is possible that the growth process can be identified (on energetic grounds) with the build-up of a relatively rigid layer of Lamellar (laminar) aggregates. Then as these aggregates reach a critical size N, with increasing C_A^o, we postulate a cooperative conformational switch (analogous to a helix-coil transition[10] or a liquid-crystalline mesophase transition[11]) from a Lamellar aggregate to a spherical micelle of aggregation number N. This is a possible justification for the monodispersity (or narrow range of micelle aggregate sizes) which is often found. (Clearly, if polydispersity is indicated, this transition can be allowed over a range of n values with equilibrium constants Y_n).

The concept of cooperativity is therefore introduced at two levels:

(i) In the build-up of pre-micelle aggregates, all growth equilibrium constants are assumed equal, assuming only nearest-neighbour stacking interactions and neglecting longer-range coulombic forces (This charge effect could obviously be allowed for in a more refin[ed] model, along the lines developed previously for the stacking of planar dyes[8]).

Then: $K_{n,n+1}\ (n = 2 \rightarrow N-1) = K_G$

If we again define $q = K_G/K_D$, then for positive but finite cooperativity, $\infty > q > 1$. For a non-cooperative process, $q = 1$, and for an anti-cooperative process, $1 > q > 0$. (When $q = 0$, only dimers are formed). For real systems, it is expected that K_D will represent a weak stacking process with $1 < K_D < 100$, and that K_D will increase by 15-30% for each additional $-CH_2-$ group in the hydrocarbon moiety. For surfactant systems in water, it is likely that $K_G > K_D$, since a larger area of hydrocarbon-water interface will be eliminated in the growth process. However, it is not expe[c]ted that $\Delta G_G^o > 2\Delta G_D^o$; hence q can take values up to K_D, when K_G = K

(ii) It is likely that the conformational change, when permitted, will also be cooperative, since a series of weak energy links stabilize the two aggregate forms. The value of Y_N can in principle take any value from $0 \rightarrow \infty$.

MATHEMATICAL TREATMENT FOR THE COOPERATIVE MODEL

For nucleation: $A_1 + A_1 \overset{K_D}{\rightleftharpoons} A_2$

where $K_D = C_2/C_1^2 = K_G/q$ (neglecting activity coefficients)

If $q > 1$, then lamellar two-dimensional plate-like aggregates are preferred over linear (1D) aggregates.

For growth: $A_{n-1} + A_1 \overset{K_G}{\rightleftharpoons} A_n$ $(n = 3 \rightarrow N$ (or $3 \rightarrow \infty))$

and we assume $K_G \neq f(n)$, neglecting long-range charge effects.

For micelle formation: $A_N \overset{Y_N}{\rightleftharpoons} A'_N$

To simplify the analysis, it is convenient to introduce a dimensionless parameter $S = K_G C_1$. Then, by application of the Mass Action Law, it can readily be shown that:

$$C_A^o = S/K_G + (S/K_G q)(1 - S)^{-2}(1 - S) - S^{N-1}N(1 - S)^{-1} -1$$
$$+ (SNYS^{N-1})/(K_G q)$$

$$C_1 = S/K_G;$$

$$\sum_2^N nC_n = (S/K_G q)((1-S)^{-2}(1-S) - S^{N-1}N(1-S)^{-1}-1);$$

$$NC'_N = (SNYS^{N-1})/(K_G q).$$

The above equations now permit the monomer concentration $-C_1$, the concentration of intermediate Lamellar aggregates $-\sum_2^N nC_n$, and the concentration of spherical micelle $-NC'_N$, to be computed for any total concentration $-C_A^o$.

IMPLICATIONS OF THE MODEL

a) Thermodynamic: For the situation where cooperativity is very strong:

$$q \to \infty \text{ and } Y \to \infty$$

$$\text{Then } C_A^o = (S/K_G)\,(1 + NYS^{N-1}q^{-1})$$

$$= (S/K_G) \quad (\text{for } C_A^o < \text{c.m.c.})$$

i.e. $C_1 = C_A^o$, until spherical micelles begin to form, when $S = 1$.

$$\text{c.m.c.} = K_G^{-1}$$

This limiting behaviour is shown in figure 1. Below A, monomer is the only species present in solution, but for $C_A^o > A$, $dC_1/dC_A^o = 0$ and $NdC_N/dC_A^o = 1$. At B, $C_A^o = 2K_G^{-1}$ and $C_1 = NC_N'$. Thus a possible quantitative definition of the c.m.c. is at B, and this definition is adopted for the cooperative model since it is more readily applicable for systems exhibiting finite cooperativity. Tanford[9] has recently pointed out that the c.m.c. cannot be defined uniquely from experiment.

Figure 2 shows that the transition region becomes less sharp as q decreases. For real systems, values of q of 10-100 are expected (for C_{10}-C_{12} surfactants), since the dimerization constant is also likely to be of this order.

The profiles in figure 3 give a reasonably good representation of the c.m.c. for sodium lauryl sulphate at I = 0.1M in water. It can be seen that in the transition region, the pre-micelle aggregates are likely to be the dominant species present, and that the c.m.c. concept could be rather misleading. In particular the assumption usually made in kinetic analyses, that $C_1 = C_A^o$ at the c.m.c., is in error when realistic values of q are assumed.

The following additional conclusions emerge from a detailed analysis:

(i) The c.m.c. is not very sensitive to large changes in Y_N.

(ii) Pre-micelle aggregates are favoured for systems with low q and large N. This indicates that such species are favoured for short-chain hydrocarbon (C_8) surfactants where the stacking tendency is reduced.

(iii) When $q > 10^3$, pre-micelle aggregates are not formed in significant amounts. This would be the case for long chain surfactants.

(iv) The c.m.c. increases proportionally as N is increased.

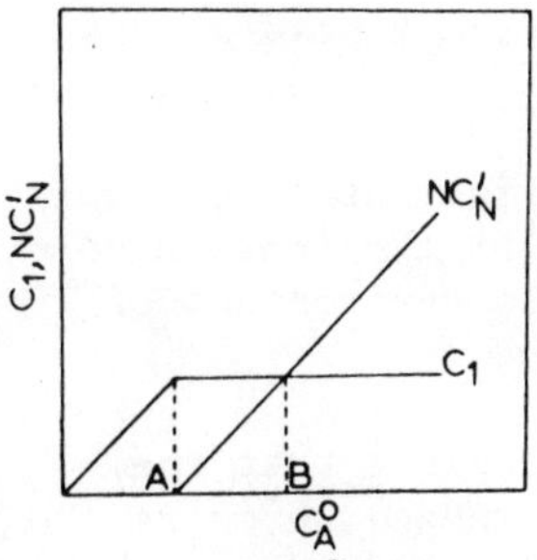

Fig. 1. Dependence of C_1 and NC'_N on C_A^o for the case of infinite cooperativity (i.e. $q \rightarrow \infty$).

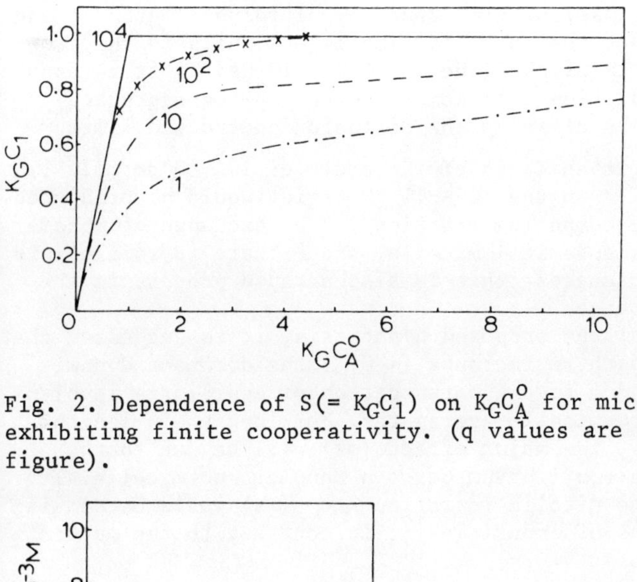

Fig. 2. Dependence of $S(= K_G C_1)$ on $K_G C_A^o$ for micellar systems exhibiting finite cooperativity. (q values are indicated on the figure).

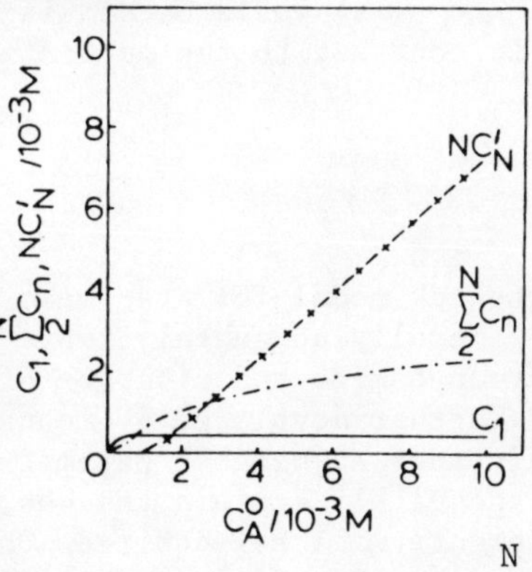

Fig. 3. Dependence of C_1, $\sum_2^N C_n$ and NC'_N on C_A^o. $Y_N = 10^3$, $N = 80$, $K_D = q = 50$.

(v) For surfactants with different head groups, some estimation can be made of relative charge effects of different head groups in the aggregation process.

(vi) Increase in ionic strength (for ionic surfactants) is likely to increase $q(K_G)$ and hence reduce the c.m.c. by the ions in solution exhibiting a charge-screening role in the environment of the growing aggregate.

b) Kinetic: On the cooperative model, it is possible that the slow relaxation time (in the millisecond time range) which is observed by large perturbation techniques (e.g. stopped-flow, temperature-jump and pressure-jump) may be associated in some way with the conformational change, whereas the faster relaxation time (in the microsecond time range), is associated with the perturbation of the growth equilibrium system. Relaxation times have been measured in the microsecond time range by ultrasonic methods, and this perhaps suggests that the lifetime of end units in the growing lamellar stack is of the order of $0.1 \rightarrow 10$ µs. If it is assumed that the rate of addition of monomer to the growing aggregate ($k_{n-1,n}$ $n = 2 \rightarrow N$) is close to the diffusion-controlled limit ($\simeq 10^9 dm^3 mol^{-1} s^{-1}$), then K_G is of the order of 10^2-$10^4 dm^3 mol^{-1}$, and c.m.c.'s ($\simeq K_G^{-1}$) in the 10^{-2}-10^{-4}M region would be predicted. However, a possible competing reaction is the exchange of monomer surfactant with the spherical micelle, since there is some experimental evidence to suggest this is also a rapid process in the micro-second time range.

On the basis of the proposed hypothesis, it is suggested that for short chain-length surfactants ($< C_{10}$) the dominant dynamic effect (µs) may be due to the perturbation of equilibria involving polydisperse pre-micelle aggregates. For long chain-length surfactants ($> C_{12}$), the major effect (ms) will be due to the perturbation of the equilibrium between monomer and micelle. It is expected that the micelle formation step (ms) would be sensitive to small amounts of impurities[12], in contrast to the nucleation and growth steps.

CONCLUSION

It is apparent that the simple Kresheck model for step-wise aggregation to form micelles does not generally adequately represent data recently obtained on the thermodynamics and kinetics of micellization. The scheme is therefore further developed by means of a general cooperative model with only four adjustable parameters - K,q,N and Y_N. It may be that to adequately explain the behaviour of real systems, further refinements will be required. On the approach used, such refinements, e.g. K_G, $Y = f(n)$, are easily incorporated into the model. The calculations, which can be accommodated on a desk computer, are comparatively trivial.

The detailed thermodynamic and kinetic implications of the model will be considered in a separate paper.

APPENDIX.

Some other papers which comment on the significance of the c.m.c. and the cooperative nature of micellization are:

1. P. Mukerjee and K.J. Mysels, "Critical Micelle Concentrations of Aqueous Surfactant Systems", N.B.S., Washington D.C., 1971, p.1.
2. G.S. Hartley, "Aqueous Solution of Paraffin-Chain Salts", Hermann, Paris 1936.
3. K.J. Mysels, J. Colloid Sci., 1955, 10, 507.
4. P. Mukerjee, J. Phys. Chem., 1972, 76, 565.
5. C. Tanford, 'The Hydrophobic Effect', Wiley, New York, 1973.
6. C. Tanford, J. Phys. Chem., 1974, 78, 2469.

REFERENCES

1. E.W. Anacker, Cationic Surfactants, ed. E. Jungermann, (Marcel Dekker, Inc., New York, 1970).
2. J.M. Corkin, J.F. Goodman, T. Walker and T. Wyer, Proc. Roy. Soc., A312, 243 (1969).
3. G.C. Kresheck, E. Hamori, G. Davenport and H.A. Scheraga, J. Amer. Chem. Soc., 88, 246 (1966).
4. J. Rassing, P.J. Sams and E. Wyn-Jones, J.C.S. Faraday 11, 70, 1247 (1974).
5. E. Aniansson and S. Wall, J. Phys. Chem., 78, 1024 (1974).
6. P.J. Sams, E. Wyn-Jones and J. Rassing, Chem. Phys. Letters, 13, 233 (1972).
   J. Rassing, Acta Chem. Scand., 25, 1418 (1971).
7. B.H. Robinson, A. Lofflar and G. Schwarz, J.C.S. Faraday 1 69, 56 (1973).
   B.H. Robinson, A. Seelig-Loffler and G. Schwarz, J.C.S. Faraday 1, in press.
8. D. Thusius, "Mechanism of Glutamate Dehydrogenase Self-Assembly", this publication.
9. C. Tanford, J. Phys. Chem., 78, 2469 (1974).
10. G. Schwarz and T. Seelig, Biopolymers, 6, 1263 (1968).
11. T.J. Tsong, Proc. Nat. Acad. Sci., 71, 2684 (1974).
12. J. Lang, C. Tondre, R. Zana, R. Baner, H. Hoffmann and W. Ulbricht, Reported at N.A.T.O. Summer School on Chemical and Biological Applications of Relaxation Spectroscopy, Salford, 1974.

ON THE MECHANISM OF LIGAND SUBSTITUTION IN WEAK METAL COMPLEXES IN AQUEOUS SOLUTIONS

Wilhelm Knoche

Max-Planck-Institut für biophysikalische Chemie,
Göttingen, Germany

INTRODUCTION

Weak metal complexes may be defined as those that are formed mainly by electrostatic interaction between the metal ion and the ligands. Thus, the contribution of covalent bonding and of ligand-field stabilization may be neglected. For these complexes, Bjerrum[1] introduced the concept of ion pairs and calculated the stability constant K of the equilibrium free ions ⇌ ion pair assuming spherical ions and neglecting dipole interactions. After further improvements of the theory Fuoss[2] proposed the following expression for this constant

$$K = \frac{4\pi N a^3}{3000}\, e^b \qquad (1)$$

with

$$b = \frac{z_+ |z_-| e_o^2}{\varepsilon k T a}$$

a is the center-to-center distance of closest approach of the ions, z_+ and z_- are their valencies, N is the Avogadro number, e_o the elementary charge, ε the dielectric constant, k the Boltzmann constant, and T the temperature.

In these calculations the solvent is assumed to be a continuum with invariable dielectric constant. This assumption is very crude if the size of the solvent molecules is comparable to or even larger than that of the ions. In this case, instead of using a permanent DK, the dipole moment and the polarizability of the solvent molecules have to be considered in the calculation of the electrostatic interaction between the ions. Therefore, the constants calculated with Eq. (1) should be unreliable. However, this is difficult to

Wyn-Jones (ed.), Chemical and Biological Applications of Relaxation Spectrometry, 265–275.

show experimentally, since a) the expression of K_a contains the parameter a which is not known exactly, and b) it is difficult to determine the stability constants of weak complexes, as can be seen by comparing the results obtained by different authors for the same system.

On the basis of chemical relaxation experiments Eigen has shown that for the weak complexes discussed here distinguishable metal complexes exist which have the same stoichiometric composition but different types of solvation. He proposed a mechanism for the complex formation which is now largely accepted - free solvated metal ions M^{m+} and ligands L^{l-} attract each other by Coulombic interaction The "ion pair" is a complex MSSL (S standing for solvent) where the ions are only separated by the solvent molecules of their inner-solvation spheres. In the "outer-sphere complex" (MSL) a monomolecular solvent layer separates the ions. Finally, by the direct contact of the ions, the "inner-sphere complex" ML is formed. The expressions "outer sphere complex" and "inner-sphere complex" refer to the substitution of a solvent molecule in the outer- or inner-coordination sphere of the metal ion. Thus, the reaction scheme for the complex formation can be written as

$$M^{m+} + L^{l-} \rightleftharpoons MSSL^{(m-1)+} \rightleftharpoons MSL^{(m-1)+} \rightleftharpoons ML^{(m-1)+} \tag{2}$$

Of course, all species are solvated; e.g., for octahedrally coordinated metal ions six solvent molecules are bound to M^{m+}.

For weak metal complexes in aqueous solutions a direct exchange between two ligands X and Y according to

$$MX + Y \rightleftharpoons MY + X \tag{3}$$

has not been observed if both X and Y are not solvent molecules. Therefore, this exchange occurs as a two-step reaction

$$MX + Y + H_2O \rightleftharpoons MH_2O + Y + X \rightleftharpoons MY + X + H_2O \tag{4}$$

and only two types of ligand exchange have to be considered, the replacement of a water molecule by another ligand and the reverse reaction. In the reaction schemes (2) and (4) the solvent molecules that are not exchanged at the reaction (5 in the case of octahedrally coordinated complexes) have been omitted.

FORMATION OF THE INNER SPHERE COMPLEX

For many complexes the ligand exchange, i.e., the transition outer-sphere → inner-sphere complex, has been studied in detail. It has been found that the rate of the forward reaction depends a) strongly on the nature of the metal ion, b) is relatively independent of the ligand, and c) is of the same order of magnitude as the rate of exchange of a water molecule in the first coordination shell of the cation as measured with NMR techniques. The most simp explanation of these facts seems to be the assumption of an SN1 mechanism: The water molecule leaves the solvation shell thereby reducing the coordination number by one. This rate determining step is followed by the fast formation of the inner-sphere complex which restores the coordination number to its normal value. However, a quantitative estimation of the energy of activation for such a pure SN1 mechanism leads to values which are far higher than the experimentally observed ones at least for small polyvalent cations, as will be shown below.

For the weak metal complexes discussed here the rate constants for the reaction outer-sphere → inner-sphere complex are summarised in Table 1. It is restricted to ions with complete d-electron shells

Table 1
Approximate Rate Constants for the Replacement of a Water Molecule by Another Ligand in the Inner-Coordination Sphere of Metal Ions, Aqueous Solution, 25°C

M	k [sec^{-1}]	ref.
Be^{2+}	200	[3]
Mg^{2+}	1×10^5	[4]
Sc^{3+}	5×10^7	[4]
Y^{3+}	10^7	[4]
La^{3+}	10^8	[4]
Lanthanides	10^7 -10^8	[4]
Al^{3+}	2.5	[5]
Ga^{3+}	20	[5]
In^{3+}	500	[6]

ligand-field stabilization does not occur. Alkaline ions do not react with monodental ligands. For those ions and also for the heavier earth-alkalines Ca^{2+}, Sr^{2+}, Ba^{2+} complex formation has only been studied with polydental ligands forming chelates. However, these reactions involve more steps than are shown in Eq. (2) and will not be discussed here. As a general rule it can be seen from Table I that the rate constant increases when the radius of the metal ion increases and when its charge decreases. That means, the reaction is faster when the electrostatic interaction between the central ion and the water molecules is smaller. Quantitative results should not be expected by this comparison, since the ions are differently coordinated: Be^{2+} tetrahedrally; Mg^{2+}. Al^{3+}, Ga^{3+} and In^{3+} octahedrally; for Sc^{3+}, Y^{3+}, La^{3+}, and the lanthanides coordination numbers between 7 and 10 are reported.

Reactions scheme (2) indicates that the equilibrium between metal complexes and free ions is determined by three stability constants, and that three relaxation times should be expected. In this contribution, only the last step $MSL^{(m-1)+} \rightleftharpoons ML^{(m-1)+}$ is discussed where the results are not much influenced by the existence of the complex $MSSL^{(m-1)+}$. Moreover, in the case of $BeSO_4$-complexes discussed below, only two different complexes (ML and MSL) have been observed experimentally. Thus, for the sake of simplicity in the following discussion, this complex is neglected yielding the reaction scheme

$$M^{m+} + L^{l-} \underset{k_{21}}{\overset{k_{12}}{\rightleftharpoons}} MSL^{(m-1)+} \underset{k_{32}}{\overset{k_{23}}{\rightleftharpoons}} ML^{(m-1)+} \tag{5}$$

Here the equilibrium is determined by the two constants

$$K_o \equiv \frac{k_{12}}{k_{21}} = \frac{c_2 f_2}{c_+ f_+ c_- f_-} \tag{6}$$

and

$$K_i \equiv \frac{k_{23}}{k_{32}} = \frac{c_3}{c_2} \tag{7}$$

"+" stands for M^{m+}, "−" for L^{l-}, "2" for the outer-sphere, and "3" for the inner-sphere complex. The assumption is made that the activity coefficients depend only on the charge of the ions and on the ionic strength, therefore f_2 and f_3 are assumed to be equal. By thermodynamic methods only the overall association constant

$$K_a = \frac{(c_2+c_3)f_2}{c_+ f_+ c_- f_-} = K_o\ (1+K_i) \tag{8}$$

can be obtained.

For Eq. (5) the relaxation times are given by

$$\frac{1}{\tau_I} = k_{21}\beta_2 + k_{12}\frac{f_+f_-}{f_2}(c_+\beta_- + c_-\beta_+) \tag{9}$$

and

$$\frac{1}{\tau_{II}} = k_{32} + k_{23}\cdot\frac{K_o f_+ f_-(c_+\beta_- + c_-\beta_+)}{f_2\beta_2 + K_o f_+ f_-(c_+\beta_- + c_-\beta_+)} \tag{10}$$

with

$$\beta_i = 1 + \frac{d \ln f_i}{d \ln c_i} \tag{11}$$

τ_{II} refers to the formation of the inner-sphere complex. The rate constants k_{32} and k_{23} as well as the stability constants K_o and K_i can be obtained from τ_{II} only, if its dependence on concentration can be measured precisely. $K_o(1+K_i)$ can be compared with K_a supplied by thermodynamic techniques to check the consistency of the measurements. In the case of weak complexes, the accuracy of $K_o(1+K_i)$ can well compete with the value obtained by thermodynamic methods.[1] The activation energy and also the enthalpies of reaction can be calculated from the temperature dependence of the relaxation times.

FORMATION OF THE Be SO_4-COMPEXES[3]

Beryllium sulphate in aqueous solution was chosen to study the details of complex formation, since τ_{II} depends on concentration and can be measured with high precision of ± 1.5% using the pressure-jump relaxation techniques[7,8]. Moreover, for the tetrahedrally coordinated Be^{2+} it is easier for an octahedrally solvated ion to calculate an electrostatic model that takes into consideration the specific interaction between ions and solvent molecules. For this model, the activation energy for the outer-sphere $\rightleftharpoons$ inner-sphere transition is calculated and compared with the experimentally obtained value.

Fig. 1 shows the results of the kinetic measurements. At constant concentration of beryllium sulphate the relaxation time does not depend on the pH of the solution as has been checked by adding small amounts of $HClO_4$. At each temperature the values of $1/\tau$ have been fitted to Eq. (10) which in the case of $BeSO_4$ ($z_+ = -z_-, c_+ = c_-$) reads

$$\frac{1}{\tau_{II}} = k_{32} + k_{23}\frac{2f_\pm^2 K_o \alpha c_o}{1 + 2f_\pm^2 K_o d c_o} \tag{12}$$

α is the degree of dissociation and c_o the analytical concentration of

$BeSO_4$. Since the association constant K_a is unknown too, Eq. (12) contains three unknowns: k_{32}, $K_i = k_{23}/k_{32}$, and K_o. α is given by $K_a = K_o(1 + K_i) = (1 - \alpha)/(\alpha^2 c_o f_\pm^2)$. The activity coefficient has been calculated by the semi-empirical formula of Davies[9].

$$- \log f_\pm = 0.5 z_+ z_- \{\sqrt{I}\,(1+\sqrt{I})^{-1} - 0.3I\} \qquad (13)$$

By a least square fit the values summarized in Table II are obtained

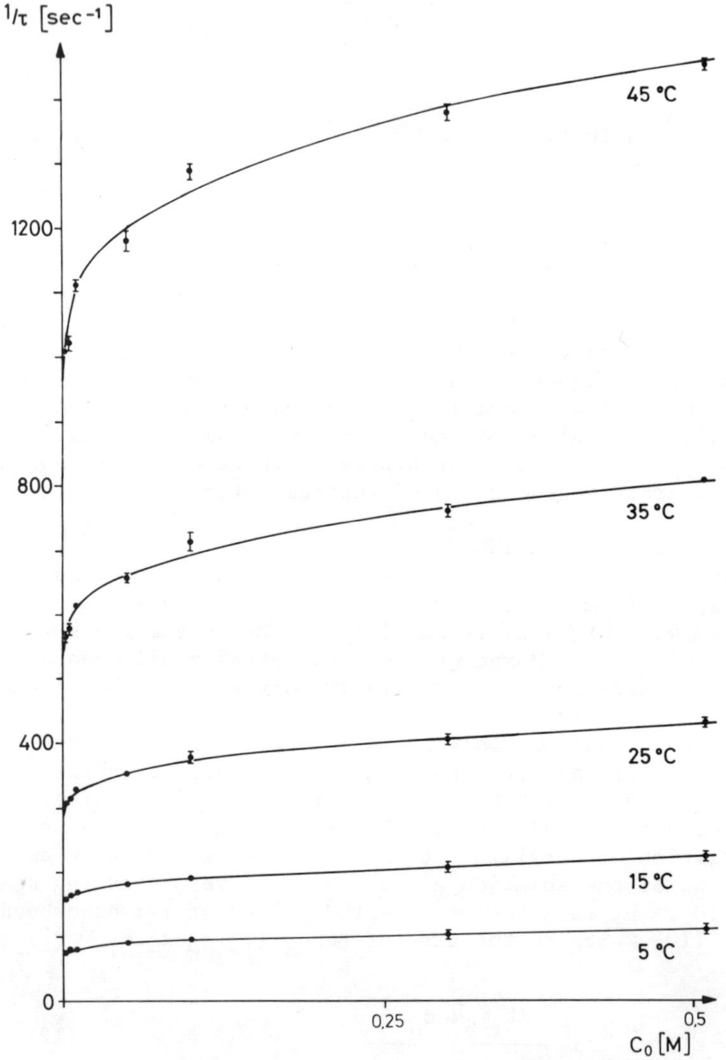

Fig 1. Dependence of the relaxation time on the concentration of l lium sulphate at differenct temperatures. The curves are drawn as the best fit to Eq. (12).

Table II
Kinetic and Thermodynamic Constants of the Formation of the $BeSO_4$ Complexes at 25°C

k_{23}	$= (185 \pm 25)\ sec^{-1}$
k_{32}	$= (285 \pm 10)\ sec^{-1}$
K_o	$= (100 \pm 20)\ M^{-1}$
K_i	$= (0.65 \pm 0.1)$
E_A^{23}	$= (12.0 \pm 1.0)\ kcal\ mole^{-1}$
$\Delta S_i^{\ddagger}$	$= (-9.6 \pm 2)\ cal\ K^{-1}\ mole^{-1}$
ΔH_o	$= (1.4 \pm 0.8)\ kcal\ mole^{-1}$
ΔH_i	$= (0.9 \pm 0.3)\ kcal\ mole^{-1}$
ΔS_o	$= (14.0 \pm 0.3)\ cal\ K^{-1}\ mole^{-1}$
ΔS_i	$= (2.2 \pm 1.5)\ cal\ K^{-1}\ mole^{-1}$
k_{12}	$= (4 \pm 2) \times 10^{10}\ sec^{-1}\ M^{-1}$
k_{21}	$= (4 \pm 2) \times 10^{8}\ sec^{-1}$
ΔV_o	$= (10 \pm 3)\ cm^3\ mole^{-1}$
ΔV_i	$= (3 \pm 2)\ cm^3\ mole^{-1}$

The fast reaction free ions $\rightleftharpoons$ outer-sphere complex has been studied by ultrasonic measurements[10]. The results fit a single relaxation (i.e., the complex MSSL has not been observed). From its relaxation time τ_I, the constants k_{12}, k_{21}, and the reaction volume ΔV_o have been calculated using the value of K_o obtained from τ_{II}. The reaction volume ΔV_i for the reaction outer-sphere $\rightleftharpoons$ inner-sphere complex is supplied by the amplitude of the slow relaxation effect when ΔV_o is known. For further details see ref. 3.

To check the results, K_a has been determined spectrophotometrically yielding a value $K_a = (145 \pm 25)\ M^{-1}$ which agrees well with $K_o(1+Ki) = (165 \pm 30)\ M^{-1}$ obtained by kinetic measurements. Since in the outer-sphere complex the ions (SO_4^{2-} and the solvated Be^{2+}) are large compared to the solvent molecules, K_o may be calculated with some degree of confidence using the Fuoss formula (1). With a distance of closest approach a = 4.6 Å (see below we get K_o (Fuoss) = 125 M^{-1}. Taking into account the crude approximations made at the derivation of this formula, the agreement is surprisingly good.

From the rate constant k_{23} and its temperature dependence the activation energy E_A^{23} = 12 kcal/mole and the activation entropy $\Delta S^{\neq}_{23}$ = -9.6 cal K^{-1} $mole^{-1}$ have been obtained. These values inform of the mechanism of the reaction. A simple electrostatic model has been developed [11] to calculate the activation energy. In this model the ions and water molecules are assumed to be hard spheres with the radii $r(Be^{2+})$ = 0.34 Å, $r(SO_4^{2-})$ = 2.91 Å, $r(H_2O)$ = 1.48 Å, the polarizibilities $\alpha(Be^{2+})$ = 0, $\alpha(SO_4^{2-})$ = 5.8 $Å^3$, $\alpha(H_2O)$ = 1.48 $Å^3$, and with the permanent dipole moment $\mu_o(H_2O)$ = 1.85 Debye. The electrostatic interaction between the ions (Coulomb attraction), between ions and permanent as well as induced dipoles, and between the dipoles have been taken into account. The calculation has to be performed iteratively because the induced dipole moments depend on the local electric field strength which itself depends on the dipole moments. With this model the free enthalpy (electrostatic energy) is determined unambiguously by the distance between the ions and by the arrangement of the water molecules. For details of the calculation see ref. 11. If the assumption of an SN 1-mechanism is made for the transition outer-sphere ⇌ inner-sphere complex, one water molecule has to leave the inner-solvation sphere of the tetrahedrally coordinated Be^{2+} ion before the SO_4^{2-} ion occupies this place. For this process, the smallest activation energy is obtained by assuming a triangular activated complex. Even then, the activation energy is (according this electrostatic model) approximately E_A = 90 kcal/mole. This value is much too large compared to the experimentally determined value E_A = 12 kcal/mole. Therefore, another mechanism is proposed for this process which is shown in Fig. 2. Beryllium ion is tetrahedrally surrounded by four water molecules. Sulphate ion is in contact with three of them, thus forming the outer-sphere complex. A fourth water molecule of the solvation shell of the anion on the opposite

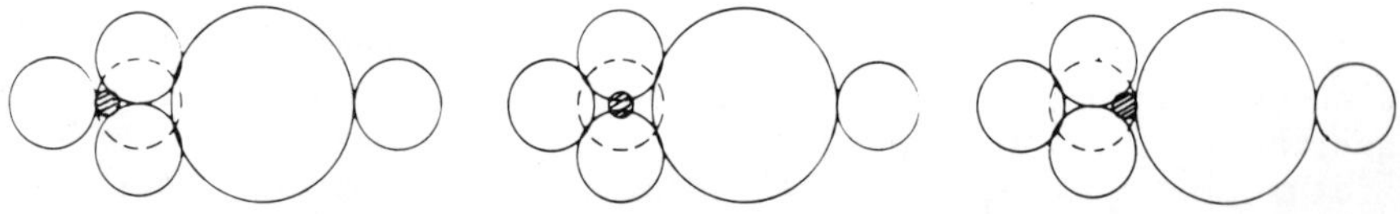

Fig. 2 Complex formation for beryllium sulphate (schematically).

side is also taken into account. The influence of other water molecules being in contact with the sulphate ion largely cancels. Formation of the inner-sphere complex preceeds by the approach of the ions. For each distance ion-ion the water molecules take positions

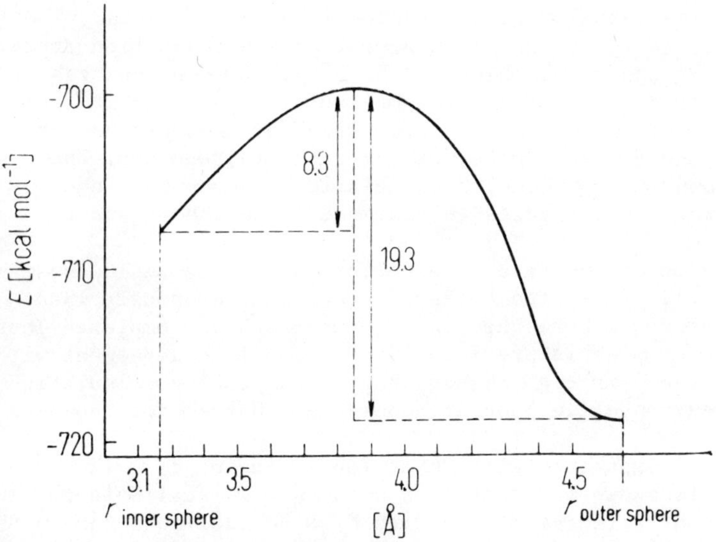

Fig. 3. Calculated electrostatic energy profile for the reaction $BeSO_4 \rightleftharpoons BEOH_2SO_4$.

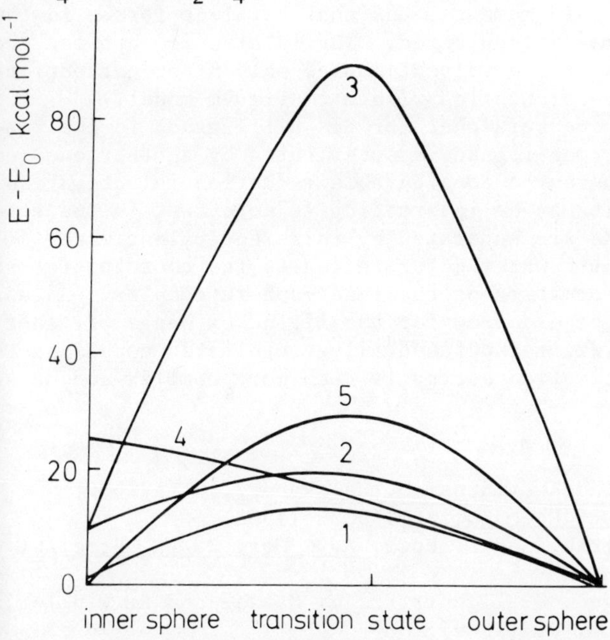

Fig. 4. Energies of activation for substitution reactions on Be^{2+} ions, 1) experimentally obtained $BeSO_4$, 2) calculated, concerted mechanism, $BeSO_4$, 3) calculated, SN 1 mechanism, $BeSO_4$, 4) calculated concerted mechanism, $BeClO_4^+$, 5) calculated, concerted mechanism, water exchange.

such that the free enthalpy of the system is at its minimum. The state of highest energy is shown in Fig. 2 as the transition state. Instead of losing one water molecule the beryllium ion squeezes through the plane of the three water molecules separating it from the anion. The activation energy calculated for this process is 19.3 kcal/mole and 8.3 kcal/mole for the backward reaction. The results are shown in Fig. 3. The calculations demonstrate the existence of two stable species in the model, the inner- and the outer-sphere complex.

The same calculations have been performed for the $BeClO_4^+$ system (Fig. 4, curve 4). Here, the energy increases monotonously when Be^{2+} and ClO_4^- approach more than in the outer-sphere complex. Thus, no inner-sphere complex is predicted by the model in agreement with experiment. Curve 5 in Fig. 4 shows the corresponding calculation for the water exchange,i.e., outer-sphere and inner-sphere complex are identical.

In Fig. 4 the experimentally determined value of E_A = 12 kcal/mole (curve 1) is shown in comparison with $E_A \approx$ 20 kcal/mole for the model discussed here (curve 2) and with $E_A \approx$ 90 kcal/mole for an SN1 mechanism (curve 3). The disagreement between the first two values may be due to the assumptions made: it is assumed that the ions and the water molecules are hard spheres, that there are no electric moments higher than dipole moments, and that covalent forces in the inner-sphere complex may be neglected. The interaction between the complexes and the bulk of the solvent causes only minor corrections as has been shown by a calculation with a continuum model.

Summarizing it can be said that for Be^{2+}all ligands in the inner sphere are involved if one ligand is substituted by another one (concerted mechanism). Thereby a considerable reduction in activation energy is achieved. It may be interesting to note that in the outer sphere complex the ions are separated by three equivalent water molecules and that the fourth water molecule leaves the coordination shell of the cation in the formation of the inner-sphere complex. An analogous behaviour should be expected for the ligand exchange of other weak complexes. However, for octahedrally coordinated complexes the calculation of the activation energy is much more complex and has not yet been performed.

REFERENCES

1. N. Bjerrum; Kgl. Danske Videnskab. Selskab., 7, 9 (1926).
2. R.M. Fuoss; J. Amer. Chem Soc., 80, 5059 (1958).
3. W. Knoche, C.A. Firth, and D. Hess; Adv. Mol. Relax. Proc., 6, 1 (1974).
4. H. Diebler, M. Eigen, G. Ilgenfritz, G. Maass, and R. Winkler; Pure Appl. Chem., 20, 93 (1969).
5. C. Kalidas, W. Knoche, and D. Papadopoulos; Ber. Bunsenges. phys. Chem., 75, 106 (1971).
6. A. Bonsen and W. Knoche; in preparation.
7. W. Knoche; "Pressure-Jump Methods", this volume.
8. W. Knoche; "On Data Processing, Evaluation and Precision of Relaxation Measurements", this volume.

9. C.W. Davies; "Ion Association" - Butterworths, London (1962).
10. G. Kurtze and K. Tamm; Acustica. 3, 34 (1953).
11. H. Strehlow and W. Knoche; Ber. Bungsenges. physik. Chem. 73, 427 (1969).

ULTRASONIC AND LASER T-JUMP STUDIES OF THE NICKEL CARBOXYLATE COMPLEX FORMATION REACTION IN AN AQUEOUS SOLUTION

Tatsuya Yasunaga, Shoji Harada and Kiyoshi Tamura

University of Hiroshima, Japan

INTRODUCTION

The step-by-step mechanism (Eigen mechanism) has been proposed[1] for the complex formation reaction, and the rate-determining step of the reaction has been revealed to be the dehydration process of the metal ion. Most of the kinetic studies of the nickel complex formation reactions were focused only on this rate-determining process of the relatively stable complexes. Then, the systematic studies have been carried out with the help of the ultrasonic absorption measurement and the laser T-jump method on the nickel monocarboxylate complex formation reactions. The results suggested that, in the nickel carboxylate solutions, there are at least two relaxation processes. The authors aimed to clarify the relaxation mechanisms of the nickel carboxylate solutions in connection with the complex formation reactions and to obtain the kinetic values of the reactions.

EXPERIMENTAL

All of the chemicals were of a reagent grade and were used without further purification. Ultrasonic absorption was measured by the following two methods. In the frequency range 10-100kHz, the reverberation method[2] was used. In the frequency range 3.5 MHz to 95 MHz, the pulse method[3] was used. Moreover, the laser T-jump method[4] was applied for the studies of the relaxation phenomena in the low frequency range. Neodymium glass laser flashed 1.06μ laser light as a 3 joule/25n sec pulse. The degree of the temperature-jump of the sample is about 0.1°C. The concentration changes of the species were followed by a change of the conductivity.

Wyn-Jones (ed.), Chemical and Biological Applications of Relaxation Spectrometry, 277–289.

RESULTS AND DISCUSSION

Ultrasonic absorption measurement of the 0.1 M nickel acetate solution was carried out and the result is shown in Fig. 1. The relaxation spectrum is composed of, at least, two relaxation absorptions. (One is in kHz range, the other is in MHz range). The kinetic studies were carried out for each relaxation phenomenon.

1. kHz RANGE RELAXATION BASED ON SLOW REACTIONS

As is seen in Fig. 1, the kHz range relaxation absorption was observed even in the low pH solution. The decrease of the absorption was caused by the decrease of the free ligand concentration. The fact shows that the absorption is based on the nickel acetate complex formation reaction. Ultrasonic relaxation absorption measurements were, however, not applied furthermore because the relaxation frequency is too low to be determined by this method. Then, a laser T-jump method was applied by which the lower frequency range can be studied. A representative relaxation spectrum by a laser T-jump method is shown in Fig. 2. All of the relaxation spectra can be characterized as single relaxations. The experimental conditions and the observed reciprocal relaxation times are shown in Table 1.

In an aqueous solution, nickel acetate complex is considered to be formed by the step-by-step mechanism.

$$\underset{(1)}{Ni^{2+} + Ac^-} \underset{k_{21}}{\overset{k_{12}}{\rightleftharpoons}} \underset{(11)}{Ni^{2+}\, O{<}^{H}_{H}\, Ac^-} \underset{k_{32}}{\overset{k_{23}}{\rightleftharpoons}} \underset{(111)}{NiAc^+} \qquad (1)$$

where (1) is free ion, (11) is outer-sphere complex, (111) is inner-sphere complex. The relationships between the relaxation times and the rate constants for the reaction equilibriae, (1) $\rightleftharpoons$ (11) and (11) $\rightleftharpoons$ (111), are given as follows:

$$2\pi f_{r_1} = 1/\tau_1 = k_{12}\gamma_{Ni}(C_{Ni} + C_{Ac}) + k_{21} \qquad (2)$$

$$2\pi f_{r_2} = 1/\tau_2 = k_{23}\,\{1 + \{K_o\gamma_{Ni}(C_{Ni} + C_{Ac})\}^{-1}\}^{-1} + k_{32} \qquad (3)$$

where K_o is the outer-sphere complex formation constant calculated by Fuoss equation[5]. With reference to the results of other nickel complex formation reactions,[6,7,8] the relaxation phenomena will be reasonably attributed to the rate-determining step of Eq. (1), i.e., (11) $\rightleftharpoons$ (111). Then, the relaxation time is substituted in Eq. (3) and the rate constants, k_{23} and k_{32}, can be obtained by plotting $1/\tau$ against $\{1+\{K_o\gamma_{Ni}(C_{Ni}+C_{Ac})\}^{-1}\}^{-1}$. Unfortunately, however, the ionic strength was not kept constant. K_o, γ_{Ni} and the

TABLE 1. RELAXATION TIMES AND EXPERIMENTAL CONDITIONS FOR THE NICKEL ACETATE SYSTEM AT 25°C

C_o a) (M)	C_{Ni} (10^{-2}M)	C_{Ac} (10^{-2}M)	C_{NiAc} (10^{-2}M)	μ	$1/\tau$ ($10^3 sec^{-1}$)
0.03	2.3	5.3	0.7	0.08	5.8
0.05	3.6	8.6	1.4	0.12	5.5
0.075	5.0	12.5	2.5	0.18	7.8
0.10	6.3	16.3	3.7	0.23	8.2
0.20	10.4	30.4	9.6	0.41	9.6
0.30	13.7	43.7	16.3	0.57	11.4
0.40	16.3	56.3	23.7	0.73	12.6
0.50	18.3	68.3	31.7	0.87	15.1

a) C_o refers to the total stoichiometric concentrations of the nickel acetate

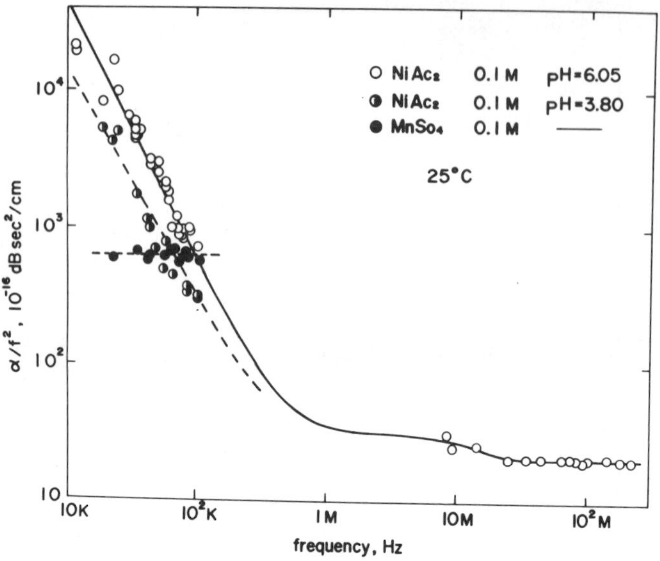

Fig. 1. Ultrasonic absorption in 0.1 M nickel acetate solutions at 25°C : o; pH 6.05, ◑; pH 3.80 ●; 0.1 M $MnSo_4$

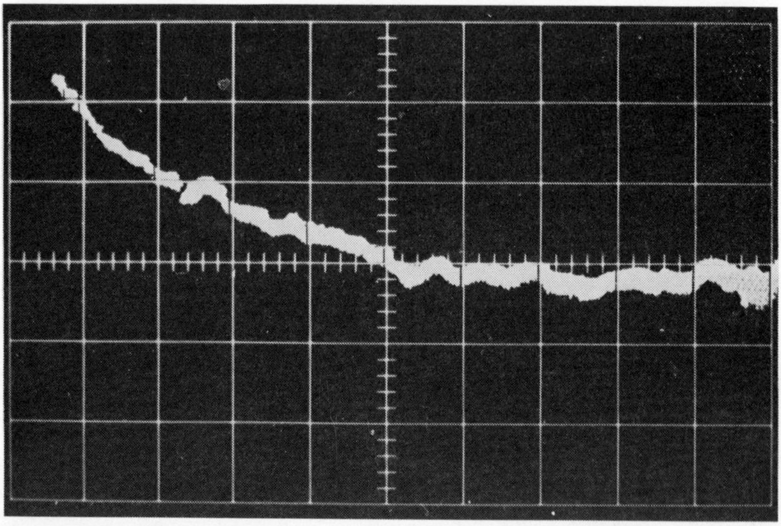

Fig. 2. Representative relaxation spectrum observed by the laser T-jump method in 0.2 M nickel acetate solution at 25°C : sweep; 50 μ sec/div.

stability constant changed with concentration. Therefore, various parameters were determined by the following procedures. At first, the stability constant at ionic strength zero is assumed. $NiAc_2$ concentration and Eqs. (4), (5)[9] give γ_{Ni}, K, μ, and C_i. K_o is calculated by the Fuoss equation[5].

$$K_{\mu=0} = \frac{1}{\gamma_{Ni}} \cdot \frac{C_{NiAc}}{C_{Ni} \cdot C_{Ac}} = \frac{1}{\gamma_{Ni}} . K \qquad (4)$$

$$\log\gamma_i = -\frac{1}{2} \cdot Z_i^2 \left(\frac{\sqrt{\mu}}{1+\sqrt{\mu}} - 0.3\mu \right) \qquad (5)$$

Substituting all of the calculated values in Eq. (3), experimentally obtained $1/\gamma$ values are plotted. If the relaxation phenomenon is attributed to the equilibrium (11)$\rightleftharpoons$ (111) of Eq. (1), the plot will show linear relationship and k_{23} and k_{32} are obtained from the slope and the intercept of the line, respectively. With k_{23}, k_{32} and K_o at ionic strength zero, the value of stability constant at ionic strength zero was calculated by the equation;

$$K_{\mu=0} = K_o\left(1 + \frac{k_{23}}{k_{32}}\right) \qquad (6)$$

and the value should be coincident with that of the first assumed value. Above routine work was repeated until all the data coincide. Final $1/\gamma$ vs. $[1+[K_o\gamma_{Ni}(C_{Ni} + C_{Ac})]^{-1}]^{-1}$ plot is shown in Fig. 3 and the ionic concentrations or other parameters are tabulated in Table 1. From Fig.3, k_{23} and k_{32} were obtained to be $2.7 \cdot 10^4$ and $5 \cdot 10^3 sec^{-1}$, respectively. The value of k_{23} is very close to the rate constants of the rate-determining step of various kinds of nickel complex formation reactions ever studied[6,7,8]. The value is also very close to the rate constant of the water substitution on nickel ion obtained by NMR[10]. These facts show that the rate-determining step of the nickel acetate complex formation is the dehydration process of the nickel ion. From the temperature dependence of the reciprocal relaxation time, activation energy of the reaction was obtained: $E^{\neq}$ = 12 Kcal mol^{-1}. An activation enthalpy of the water substitution, obtained by NMR[10], is reported to be 11.5 Kcal mol^{-1}. This coincidence of the activation parameters also shows that the relaxation phenomenon is based on equilibrium (11) $\rightleftharpoons$ (111) of Eq. (1). The same kinds of studies were carried out on the nickel complex solutions of propionate, butylate and β-chloropropionate. The values of k_{23}, k_{32} obtained are summarized in Table 2. The values of k_{23} are also very close to the rate constant of water substitution of nickel ion obtained by NMR. Since k_{23} values[6,7,8] for many complexes are approximately constant, different values of stability constants will be reflected in the values of k_{32}. It

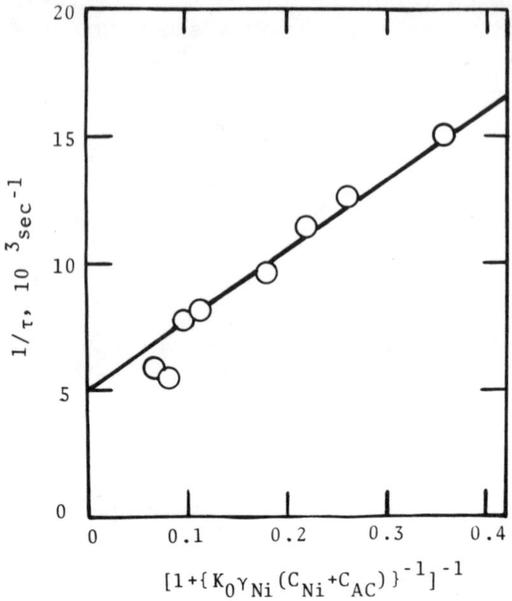

Fig. 3. $1/\tau$ vs. $\{1 + \{K_0\gamma_{Ni}(C_{Ni} + C_{Ac})\}^{-1}\}^{-1}$ plot at 25°C

has been shown that, for a series of related ligands, the stability constants are often inversely proportional to the acid dissociation constants of the protonated ligand. This relation is also found in Table 2.

2. THE RELAXATION PHENOMENA IN MHz FREQUENCY RANGE

The relaxation phenomena in MHz frequency range was studied under various experimental conditions. The results on the solution of the various concentrations of nickel acetate are shown in Fig. 4.

In Figs. 5, 6, the results on many solutions where the ratio of metal and ligand concentrations was changed were shown. All of the absorptions were characterized by a single relaxation equation. In nickel acetate solutions the following complex formation equilibriums exist:

$$Ni^{2+} + Ac^- \rightleftharpoons NiAc^+ \quad (7)$$

$$NiAc^+ + Ac^- \rightleftharpoons NiAc_2 \quad (8)$$

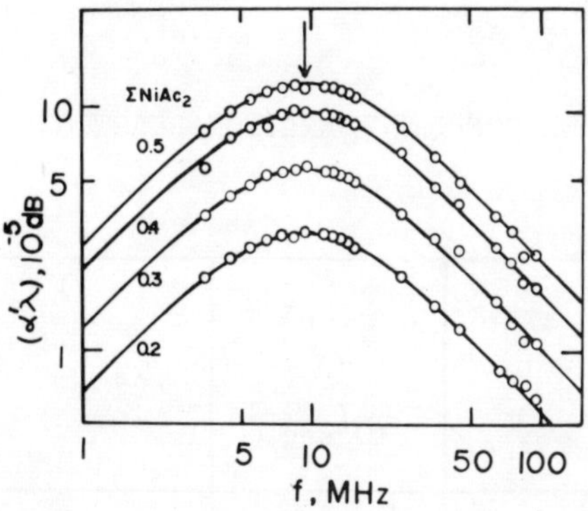

Fig. 4. Excess absorption per wavelength (α'λ) of nickel acetate (C_0) at 25°C.

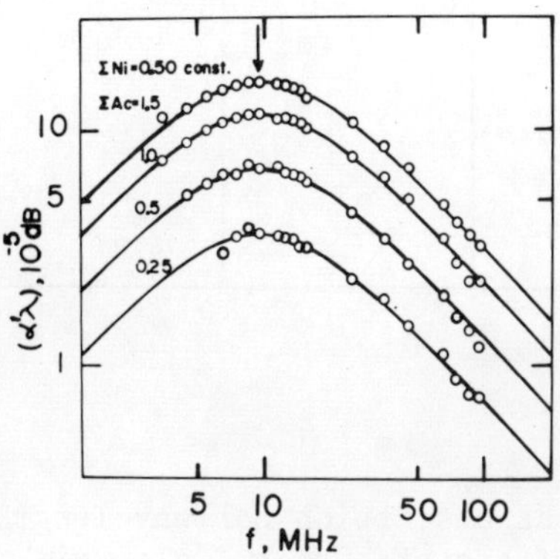

Fig. 5. (α'λ) in excess acetate ion solution at 25°C : ΣNi = 0.50 M.

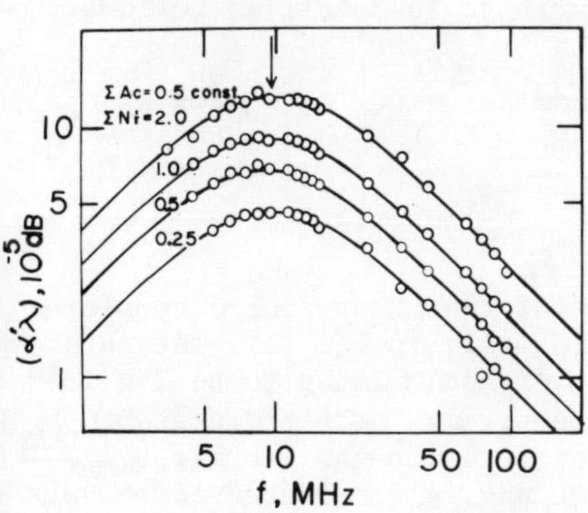

Fig. 6. (α'λ) in excess nickel ion solution at 25°C : ΣAc = 0.50 M

$$NiAc_2 + Ac^- \rightleftharpoons NiAc_3^- \quad (9)$$

TABLE 2. RATE CONSTANTS OF THE NICKEL COMPLEX FORMATION REACTION AT 25°C

ligand	k_{23} a) $10^4 sec^{-1}$	k_{32} a) $10^3 sec^{-1}$	k_{43} b) $10^7 sec^{-1}$	pKa
Acetate	2.7	5.0	6.0	4.76
Propionate	2.4	5.5	4.7	4.88
Butylate	2.4	4.5	4.1	4.82
Acrylate	-	-	6.3	4.26
β-Chloro-propionate	2.6	7.5	8.8	4.08
Chloroacetate	-	-	23	2.81

a) at $\mu = 0$

b) at $\mu = 2.0$

As is seen in Figs. 5, 6, maximum excess absorption per wavelength increases not only with the addition of acetate ion but also with that of nickel ion. This means that the absorption is related to the chemical equilibrium of Eq. (7). Step-by-step mechanism of Eq. (7), i.e., Eq. (1), can be rewritten in detail as follows.

$$Ni^{2+} + Ac^- \underset{k_{21}}{\overset{k_{12}}{\rightleftharpoons}} Ni^{2+}O{<}^{H}_{H}O{<}^{H}_{H}\ Ac^- \underset{k_{32}}{\overset{k_{23}}{\rightleftharpoons}} Ni^{2+} + O{<}^{H}_{H}\ Ac^- \underset{k_{43}}{\overset{k_{34}}{\rightleftharpoons}} NiAc^+ \quad (10)$$

(1) (11) (111) (1V)

where (11) and (111) are two types of the outer-sphere complexes which are represented by the species (11) in Eq. (1). The rate constants k_{12} and k_{21} in Eq. (10) were calculated to be 2×10^{10} $M^{-1}sec^{-1}$ and 3.7×10^9 sec^{-1}, respectively, at $\mu = 0$ and 25°C by the Debye-Eigen equations[11,12]. The reaction equilibrium (11)$\rightleftharpoons$(111) is the anion desolvation step and is considered to be very fast. k_{34} and k_{43} in Eq. (10) correspond to k_{23} and k_{32} in Eq. (1) respectively, and the value is obtained in Part (I) to be k_{34} =

$2.7 \times 10^4\ sec^{-1}$, $k_{43} = 5.0 \times 10^3\ sec^{-1}$. Purdie et al.[13] have attributed the relaxation absorption in nickel acetate (f_r = 65 MHz) to the (11)$\rightleftharpoons$(111) equilibrium of Eq. (10) only because the relaxation frequency is constant irrespective of the nature of the metal ion. Then the authors examined if the MHz region absorption can be attributed to the (11)$\rightleftharpoons$(111) equilibrium of Eq. (10).

The relationships between the rate constants and the relaxation frequencies (or relaxation times) of each equilibrium are given as follows:

$$1/\tau_1 = 2\pi f_{r_1} = k_{12}\gamma_{Ni}(C_{Ni} + C_{Ac}) + k_{21} = k'_{12} + k_{21} \qquad (11)$$

$$1/\tau_2 = 2\pi f_{r_2} = k_{23} \cdot \frac{k'_{12}}{k_{21} + k'_{12}} + k_{32} = k'_{23} + k_{32} \qquad (12)$$

$$1/\tau_3 = 2\pi f_{r_3} = k_{34} \cdot \frac{k'_{23}}{k_{32} + k'_{23}} + k_{43} \qquad (13)$$

If the absorption is attributed to the (11)$\rightleftharpoons$(111) equilibrium in Eq. (10), Eq. (12) is used. As is seen in Figs. 4, 5, 6, the relaxation frequency seems to be constant irrespective of the large ionic concentration change. This constancy of f_r is feasible when one of the following two conditions is fulfilled; (i) $k_{23} \ll k_{32}$ or, (ii) $k'_{12} \gg k_{21}$ or $k'_{12} \ll k_{21}$. In each case, the following investigations were carried out.

(i) Stability constant is related to the rate constants by

$$K = \frac{k_{12}}{k_{21}} \cdot \{1 + \frac{k_{23}}{k_{32}} (1 + \frac{k_{34}}{k_{43}})\} \qquad (14)$$

Here, k_{12}/k_{21} ($=K_o$) is the outer-sphere complex formation constant which is assumed to be about unity at $\mu = 2.0$ by Fouss equation[5]. The term k_{34}/k_{43} is given in Part (I) to be about 6. If we adopt the literature value of K = 5.3, k_{23}/k_{32} can be calculated to be about 6. The result excludes the assumption $k_{23} \ll k_{32}$.

(ii) With use of the calculated values of k_{12}, k_{21}, γ_{Ni} and the ionic concentrations, k'_{12} and k_{21} were compared. The results show that one of them is not always much larger than the others under all experimental conditions. This fact excludes the second possibility. Moreover, the results in Part (I) show that the reciprocal relaxation time ($1/\tau_3$) has the definite dependency on ionic concentration. Referring to Eq. (13), the variables of k'_{23}

fall in the same order of magnitude as k_{32}. This means, from Eq. (12), that f_{r_2} should exhibit some ionic concentration dependence under experimental conditions. Above discussions show that the relaxation absorption in MHz region cannot be ascribed to the (11) $\rightleftharpoons$ (111) equilibrium of Eq. (10).

For the assignment of the relaxation absorption, the author modified the Eq. (1) as follows:

$$Ni^{2+} + Ac^- \underset{k_{21}}{\overset{k_{12}}{\rightleftharpoons}} Ni^{2+}\, O{<}^{H}_{H}\, Ac^- \underset{k_{32}}{\overset{k_{23}}{\rightleftharpoons}} NiAc^+ \underset{k_{43}}{\overset{k_{34}}{\rightleftharpoons}} {*NiAc^+} \qquad (15)$$

(1) (11) (111) (1V)

where the species (111) and (1V) are represented by species (111) in Eq. (1) or species (1V) in Eq. (10). The equilibrium (111)$\rightleftharpoons$(1V) is considered to be very rapid conformation change reaction in the complex. The relaxation frequency and the rate constants of the equilibrium (111)$\rightleftharpoons$(1V) are related by the following equation.

$$2\pi f_r = 1/\tau = k_{34} + k_{43} = k_{43}\,(1 + K_{34}) \qquad (16)$$

where $K_{34}=k_{34}/k_{43}$. The following experimental results are well interpreted by the assignment of the relaxation absorption to the equilibrium (111)$\rightleftharpoons$(1V) in Eq. (15). (a) The fact that f_r is independent of concentration can be easily understood from Eq. (16). (b) Ultrasonic absorption per wavelength is proportional to the $NiAc^+$ concentration. (c) The relaxation time concerned to the rate-determining-step depends on ionic concentration. On the other hand, the MHz relaxation frequency is independent of ionic concentration. (d) The effects of viscosity and dielectric constant are relatively small. This implies that the relaxation is concerned with the intramolecular reaction.

The structure of the species (111) and (1V) are assumed and are shown in Fig. 7 with reference to the crystal structure of $NiAc_2 \cdot 4H_2O$[14] and internal conjugate base mechanism[15]. The model in Fig. 7 is supported by the following experimental facts. (a) The same complexes can also be made in the solvents, which have OH groups, such as alcohols. The same kind of absorptions were observed in methanol and glycerol solution (Fig. 8). In methanol solution, hydrogen bond will be weak. Larger value of f_r in methanol solution than that in aqueous solution implies that k_{43} is much larger than k_{34}. (b) pH dependence of f_r was observed in aqueous solutions and the results are shown in Fig. 9. pH dependence of f_r will be interpreted as a catalytic effect of H^+ on the hydrogen bond formation and breaking[16]. (c) The absorption was observed in the other nickel carboxylate complex solutions; the studies were carried out in the nickel complex solutions of propionate, butylate, acrylate, chloroacetate and β-chloropropionate. In each, f_r was independent of ionic concentration. In Fig. 10, the values of f_r were plotted against the pK_a of the acid. The complex of the

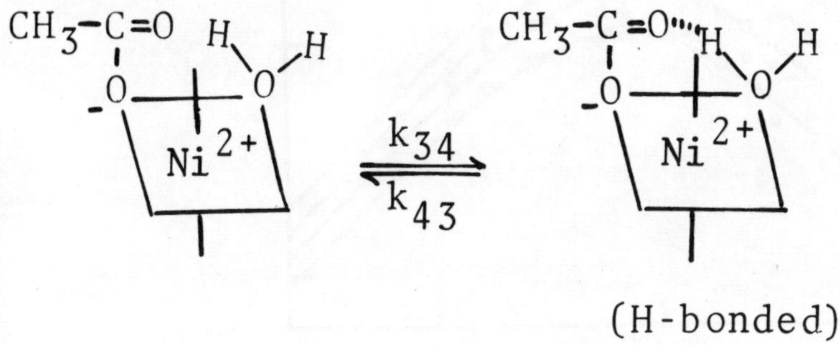

Fig. 7. Proposed model of the two kinds of nickel acetate complexes in an aqueous solution

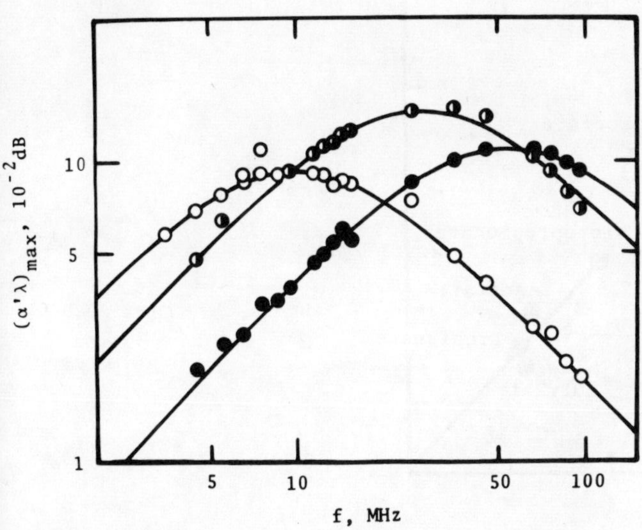

Fig. 8. (α'λ) of 0.4 Nickel acetate in aqueous (o), methanol (◑) and glycerol (●) solutions at 25°C.

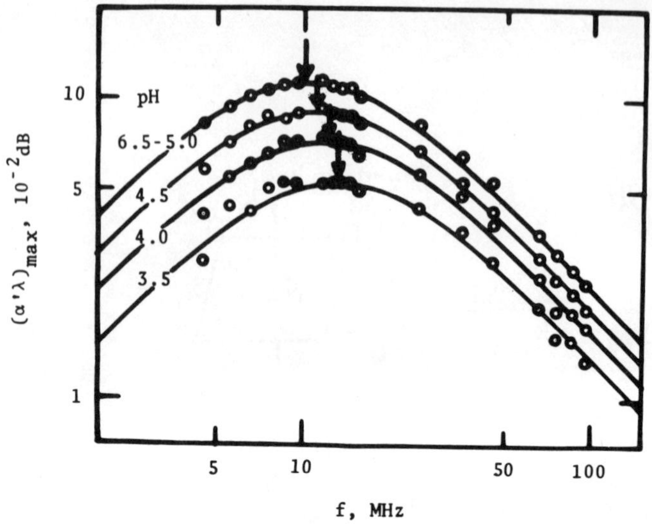

Fig. 9. $(\alpha'\lambda)$ of 0.4 M nickel acetate at various pH

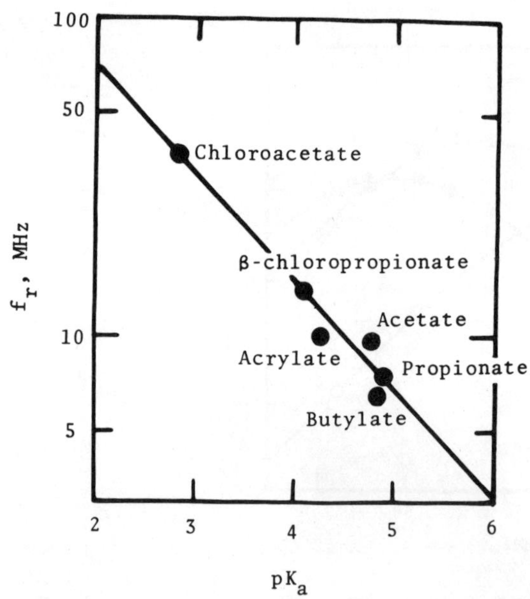

Fig. 10. The plot of f_r vs. pK_a of the acids.

low pK_a carboxylic acid is labile and the hydrogen bond will be weak, too. If we admit the condition $k_{43} \gg k_{34}$, the tendencies of f_r increase and the absorption decrease with decrease of pK_a of the acid will be well understood.(d) From the temperature dependence of f_r', the activation energy was obtained: E = 6 Kcal mol^{-1}. The value is one half of the activation enthalpy of the dehydration process and is approximately equal to the activation energy of the hydrogen bond breaking process.

In consideration of $k_{43} \gg k_{34}$, Eq. (16) is approximated as follows:

$$2\pi f_r = 1/\tau = k_{43} + k_{34} \simeq k_{43} \tag{17}$$

Then, by Eq. (17), the rate constants of the intramolecular hydrogen bond breaking of nickel monocarboxylate complexes can be calculated. The values are listed in Table 2 with the rate constant of the rate-determining process of the complex formation obtained from the data in the kHz range. Summarizing the results, two relaxation phenomena observed by the ultrasonic absorption measurements are attributed to the (11)$\rightleftharpoons$(111) and (111)$\rightleftharpoons$(1V) equilibriums of Eq. (15). The rate constants and activation parameters were obtained.

REFERENCES

1. M. Eigen and L. DeMaeyer, "Technique of organic chemistry", Vol. VII, 2nd Ed., S.L. Friss, Interscience Publishers Inc., New York, N.Y. (1963).
2. T. Yasunaga, N. Tatsumoto and M. Miura, J. Chem. Phys., **43**, 2735 (1965).
3. N. Tatsumoto, J. Chem. Phys., **47**, 4561 (1967).
4. S. Harada, Dissertation (Univ. of Hiroshima).
5. R. Fouss, J. Amer. Chem. Soc., **80**, 5059 (1958).
6. R.G. Wilkins, Accounts Chem. Rev., **3**, 408 (1970).
7. M. Eigen and R.G. Wilkins, Advan. Chem. Ser., **49**, 55 (1964).
8. D.J. Hewkin and R.H. Prince, Coord. Chem. Rev., **5**, 45 (1970).
9. C.W. Davies, "Ion Association," Butterworths, London (1962).
10. T.R. Stengle and C.H. Langford, Coord. Chem. Rev., **2**, 349 (1967).
11. P. Debye, Trans. Electrochem. Soc., **82**, 265 (1942).
12. M. Eigen and L. DeMaeyer, Proc. Roy. Soc., (London) **A247**, 505 (1958).
13. V.L. Garza and N. Purdie, J. Phys. Chem., **74**, 275 (1970).
14. C. Oldham, Prog. Inorg. Chem., **10**, 222 (1968).
15. D.B. Rorabacher, Inorg. Chem., **5**, 1891 (1966).
16. R.D. Corsaro and G. Atkinson, J. Chem. Phys., **55**, 1971, (1971).

THE ASSOCIATION OF OPTICALLY ACTIVE IONS; ULTRASONIC INVESTIGATION OF THE ASSOCIATION OF BIS AND TRIS (ORTHOPHENANTHROLINE) ZINC (11) d-CAMPHORSULFONATE

Paul Hemmes

Department of Chemistry, Rutgers University, Newark, New Jersey 07102

Pfeiffer and Quehl[1,2] discovered that the optical rotation of a solution of zinc d-camphorsulfonate was altered by the addition of three moles of 1,10 phenanthroline. This optical anomaly and similar systems have been designated the Pfeiffer effect. In a further study, Pfeiffer and Nakatsuka[3] showed that when racemic Ni $(phen)_3\ SO_4$ was added to a solution of ammonium d-camphorsulfonate an "instantaneous" change of optical rotation was observed which slowly changed to give an equilibrium rotation significantly different from that of the anion alone. This behavior is seemingly similar to the optical absorbance studies by Posey and Taube[4] on the system Co $(NH_3)_5(H_2O)^{+3} + SO_4^{-2}$. In this system there is an "instantaneous" change in absorbance when the ions are mixed followed by a slow change of absorbance. The mechanism for this latter reaction is the rapid formation of an outer sphere complex, the ion pair $Co(NH_3)_5(H_2O)\ \ SO_4^{+}$, followed by replacement of coordinated water by SO_4^{-2} to form the inner sphere complex. This multistep mechanism of ionic association has been shown by Eigen to have considerable generality[5]. In view of this mechanism the postulated mechanism for Pfeiffer activity (between unlike charged ions) is as follows. The racemic complex which must be more or less labile in terms of racemization reacts with the optically active non-labile anion to form two types of ion pairs. If we designate the racemic complex by D or L and the anion by d the two ion pairs are Dd and Ld. Since these are not mirror images the two ion pairs are diastereomers which may have different formation constants. Since it has been shown that formation of outer sphere complexes may lead to changes in optical rotation[6] we can expect that this step in the reaction can lead to the observed rapid rotation change. Since the diastereomeric ion pairs have different stabilities and the cation is labile, there is an interconversion of less stable diastereomer

Wyn-Jones (ed.), Chemical and Biological Applications of Relaxation Spectrometry, 291–293.

into the more stable one. Due to their nature, the optical rotation of a hypothetical 1.0M solution of pure Dd will not be the same as the corresponding rotation of the species Ld. Hence the interconversion of diastereomers will be accompanied by a further change in optical rotation. A somewhat similar though less detailed postulate has been given by Mayer and Brasted[7]. In order to test these postulates we have begun a study of diastereomeric ion pairs using both classical techniques and relaxation kinetic methods. A preliminary finding of the latter results is presented. We have used ultrasonic absorption to study the formation of the outer sphere complexes between $Zn(phen)_3^{+2}$ and the camphor sulfonate anion CS^-. Since Brasted et al have shown that the species $Zn(phen)_2(H_2O)_2^{+2}$ can also lead to Pfeiffer activity[8], we have studied this species also.

1. EXPERIMENTAL

The ultrasonic instrumentation has been described previously[9]. Solutions were prepared by dissolving weighed quantities of $ZnCO_3$ in solutions containing d-10 camphorsulfonic acid (Aldrich Chemical Co.). To this solution was added the appropriate amount of 1,10 phenanthroline (Fisher). Solutions were made slightly basic by addition of NaOH or $NaHCO_3$. The resulting pH values were in the range 8-9. This was to insure the formation of the complex by eliminating the possible formation of protonated ligand species yet not so basic as to form possible hydroxy complexes of zinc or mixed ligand complexes. The solutions were then diluted to the appropriate volume with distilled water. A solution of $ZnCS_2$ was prepared as above but omitting the ligand. This solution was run in slightly acidic solution (pH 4-5) to avoid hydrolysis of Zn^{+2}

2. RESULTS

A 0.10M solution of $ZnCS_2$ showed no significant excess sound absorption compared to water. A 0.10M solution of $Zn(phen)_3\ CS_2$ on the other hand shows a relaxation in the frequency range 10-150 MHz. The relaxation frequency is 37 MHz and the magnitude of the effect μ_{max} is 120 x 10^{-5}. The system $Zn(phen)_2(H_2O)_2\ CS_2$ also shows a relaxation with a relaxation frequency of 100 MHz and μ_{max} = 110 x 10^{-5}. The data for this system show some deviation at low frequency which may be a second relaxation effect. If so, the valu of the relaxation frequency is ~ 10 MHz. A number of possible causes for a second relaxation in this system exist. The first is the possibility that the formation of two diastereomeric ion pairs will give rise to two distinct relaxations. While this is true in principle we have not observed such behavior in the system d,l $Co(en)_3^{+3}$ + d-tartrate^{-2}. The diastereomeric ion pairs in this system have been reported to differ in stability[10]. Another possible cause of a second relaxation is simply the displacement of one of the co-ordinated water molecules from the cation by the anion to form a contact ion pair. It is interesting that this process leads to a relaxation frequency of ~ 30 MHz for the system $Zn(H_2O)_6^{+2}$.+SO_4^{-2}.It is also possible, though not likely, that $Zn(phen)_3^{+2}$ is formed in the solution and this second relaxation is due to ion pair formation of this ion with CS^-. In order to investigate these possibilities we are setting up an ultrasonic measurement system which is capable of much greater precision than the current instrument. This will enable us to more precisely measure the relaxation effects. A further difficulty is the lack of ion pair formation constants. We are currently trying to evaluate them via a number of techniques.

3. DISCUSSION

While we have not proven the hypothesis involving the cause of Pfeiffer activity it is evident that there exist rapid processes in $Zn(phen)_x$ CS_2 systems which do not occur for the aquo ion. If they are due to ionic association as seems likely then we will have to account for the formation of ion pairs in this case where the aquo ion shows little or no such association. It is evident that such association is not due solely to electrostatic interaction since the smaller ion is less associated. The complete ultrasonic measurements and related thermodynamic studies will be directed in part toward better understanding of the bonding in such systems.

REFERENCES

1. P. Pfeiffer and K. Quehl, Ber. Deut. Chem. Ges., **64**, 2267 (1931).
2. P. Pfeiffer and K. Quehl, ibid, **65**, 560 (1932).
3. P. Pfeiffer and Y. Nakatsuka, ibid, **66**, 410, 415 (1933).
4. F. Posey and H. Taube, J. Am. Chem. Soc., **78**, 15 (1956).
5. M. Eigen and P. Tamm, Z. Elecktrochem., **66**, 93, 107 (1962).
6. S.F. Mason and B.J. Norman, Chem. Commun.,335 (1965).
7. L.A. Mayer and R.C. Brasted, J. Coord. Chem., **3**, 85 (1973).
8. R.C. Brasted, V.J. Landis, E. Kuhajeh, P.E.R. Nordquist and L.A. Mayer, Coordination Chemistry S. Kirschner Ed. Plenum Publishing Co., New York (1969).
9. H-C Wang and P. Hemmes, J. Am. Chem. Soc., **95**, 5115 (1973).
10. K. Ogino and U. Saito, Bull. Chem. Soc., Japan 826 (1967).
11. P. Hemmes, F. Fittipaldi and S. Petrucci, Acustica, **21**, 228 (1969).

MECHANISM OF COMPLEX FORMATION IN NONAQUEOUS SOLVENTS*

H. Hoffmann

Institut für Physikalische Chemie I
Universität Erlangen-Nurnberg,
Erlangen,
Egerlandstr. 3

*This article is a shortened and somewhat modified version of the full paper that will appear in the Journal of Pure and Applied Chemistry.

INTRODUCTION

Most of the kinetic measurements that have been carried out over the years to elucidate the mechanism of complex formation between metal ions and ligands were done in aqueous solutions[1]. These investigations have led to a more or less good understanding of the factors controlling the rate constants of the reactions. It was shown in many cases that the rate determining step for complex formation between an octahedrally co-ordinated metal ion and a ligand was release of a water molecule from the inner co-ordination sphere of the metal ion. The reaction mechanism therefore can be formulated as a two step process:

$$M + L \underset{}{\overset{K_o}{\rightleftharpoons}} M(H_2O)L \underset{k_{-1}}{\overset{k_1}{\rightleftharpoons}} M\text{-}L \qquad (1)$$

where K_o is the outer sphere equilibrium constant and k_1 the rate constant for the controlling step. For many systems this rate constant k_1 was shown to be independent of the entering ligand and was usually close to the rate constant for the solvent exchange process.

$$MS_5S^{*} + S \xrightarrow{k_o} MS_6 + S^{*} \qquad (2)$$

Wyn-Jones (ed.), Chemical and Biological Applications of Relaxation Spectrometry, 295–310.

Even for many reactions with multidentate ligands the same rate determining step was found to control the overall rate of the multistep process[2]. For a number of such complicated reactions it was possible to gain a detailed understanding of the individual processe taking place[3].

With a few multidentate ligands the sequence of the steps could be evaluated and in some cases even the rate constants for the specific steps have been determined[4]. Lately, the research on the kinet of complex formation in aqueous solutions has shifted to more and mo complicated systems in which steric effects and the influence of coordinated groups on the overall reactions are studied[5]. Other research groups became interested in reactions in nonaqueous solvents. Here the emphasis at the moment is still on simple systems that lend themselves to conclusions that are relevant for a general understand of the role the solvent plays in kinetics.

GENERAL CONSIDERATIONS

While most of the kinetic studies in aqueous media could rely on available equilibrium constants[6] for the most favourable concentration range and on pH and redox indicators[7] to follow reaction such information is mostly lacking in nonaqueous systems. Another complicating factor for the measurements is the water content of th solvent. Water being a better donor than most of the solvent molecules is in many systems preferentially co-ordinated and interferes with measuremtns even at very low concentrations. Extreme care mus be taken to dry the solvents or at least it should be assured that water content present in the solvent does not affect the measuremen This can be done by adding small amounts of water. On the other ha there also are conditions that are most favourable for kinetic stud in nonaqueous systems. In aprotic solvents, no hydrolysis and solvolysis reactions need to be considered. Another big advantage of many organic solvents in comparison to water for kinetic studies is their lwo melting point which makes it possible to study reactio over a large temperature range. In a few systems the temperature w varied more than 100° and rate constants could be shifted over 8 or of magnitude[8]. Reactions with some metal ions that are quite fast at room temperature can be made so slow that ordinary mixing technique can be used to follow the rate. An example is shown on the figure 1 which shows the relaxation times for Ni- and Co-Trifluoroa tate change from 10 μsec to thousands of seconds depending on the temperature.

At present the investigations in nonaqueous solvents usually the following objectives:

1. Is mechanism (1) valid for all solvents or only for water?
2. Do the specific rate constants for different metal ions show t same graduation as in water?
3. Can exchange and dissociation rate constants be correlated and understood on the basis of the donor properties of the solvent

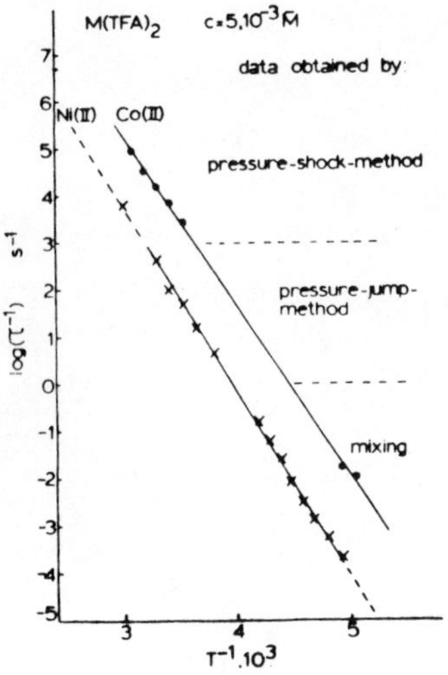

Fig. 1 Temperature dependence of relaxation times in methanol for $Ni(CF_3COO)_2$ and $Co(CF_3COO)_2$. Plot of log 1/τ against 1/T

The validity of mechanism (1) is usually checked by evaluating the rate constant k_1 for the reactions of several ligands with the same metal ion and by comparing k_1 whenever possible with the solvent exchange rate. Fortunately, this rate constant for the solvent exchange process (2) is known from nmr-measurements for many metal ions and solvents*. A recent summary of the available data has been given by Langford[9].

A selection of the available data in the literature is given in table 1. It is noteworthy that the values from different sources that are quoted for the same process vary somewhat, especially in their activation parameters. The discrepancy in the k_o values for 25° is usually small enough (maybe a factor of 2) to make the comparison with the k_1 values relevant. Whenever k_o values are not available the validity of mechanism (1) can be tested by comparing the rate constants k_1 for ligands with varying nucleophilicities. In order to avoid any complications that are due to steric or other effects, only rate constants for reactions with simple monodentate ligands should be used for this purpose.

The rate constants will generally be evaluated from relaxation or from stopped flow measurements. In relaxation experiments, the reciprocal relaxation time τ is given by expression[10] (3)

$$\frac{1}{\tau} = k_{-1} + \frac{K_o \cdot k_1 (c_M + c_L)}{1 + K_o (c_M + c_L)} \tag{3}$$

*See J.J. Delpuech - this publication.

Table 1

Solvent exchange rates at 25° and their activation parameters

solvent	metal-ion	k_0 (sec^{-1})	$\Delta H^{\ddagger}$ (Kcal/Mol)	$\Delta S^{\ddagger}$ e.u.	ref.
methanol	Ni(II)	$1.0 \cdot 10^3$	15.8	8	13
	Co(II)	$1.8 \cdot 10^4$	13.8	7.2	13
	Fe(II)	$5.0 \cdot 10^4$	12	3	14
	Mn(II)	$9.5 \cdot 10^5$	7.4	-5.7	15
		$3.7 \cdot 10^5$	6.2	-12	14
	Mg(II)	$4.7 \cdot 10^3$	16.7	14	16
DMSO	Ni(II)	$5.2 \cdot 10^3$	12.1	-1.3	17
		$3.2 \cdot 10^3$	13.0	1.4	18
		$1.1 \cdot 10^4$	12.3	1.2	19
DMF	Ni(II)	$3.8 \cdot 10^3$	15.0	8.0	20
		$6.9 \cdot 10^3$	14.0	6.0	21
	Co(II)	$4 \cdot 10^5$	13.6	12.6	20
$CH_3 \cdot CN$	Ni(II)	$3.9 \cdot 10^3$	10.9	-8.8	22
		$2.8 \cdot 10^3$	11.7	-3.6	23
		$1.2 \cdot 10^4$	11.8	-0.2	24
		$3 \cdot 10^3$	15.0	8.5	25
		$2.1 \cdot 10^3$	16.4	12.0	26
NH_3	Ni(II)	$1.95 \cdot 10^5$	10.1	-0.5	27

In many cases, it is possible to obtain the rate constant k_1 directly from the kinetic measurements without making calculations or estimations for the size of outer sphere complex*. At high concentrations a levelling off of $1/\tau$ is very often observed which directly indicates a two step mechanism and facilitates the evaluation of k_1 [11]. A typical plot for the levelling off of $1/\tau$ with increasing concentrations is shown in figure 2. In many

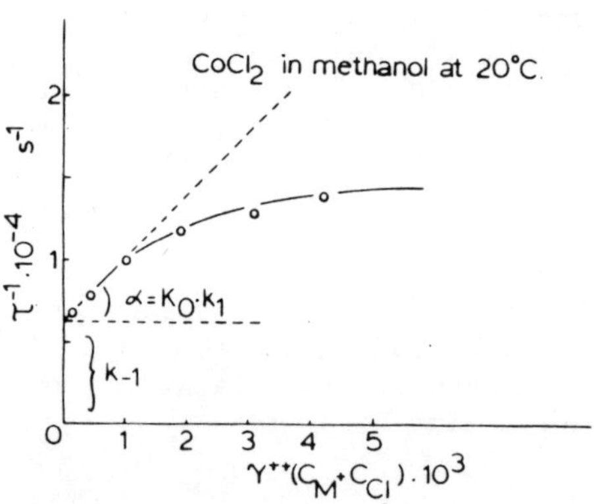

Fig. 2. Evaluation of rate constants from relaxation measurements. Plot of $1/\tau$ against ($c_M + c_{Cl}$) for $CoCl_2$ in methanol.

* See W. Knoche - this publication.

other systems the rate constant k_1 can not be obtained directly for lack of solubility of the compound under investigation or for the reason that the K_o values are not large enough for $K_o(c_M + c_L)$ in the relaxation expression to become comparable to unity. In these circumstances only the product of K_o and k_1 can be obtained from the data and K_o is usually calculated from the Fuoss expression[12]

$$K_o = \frac{4\,\pi}{3}\,a^3.\exp\frac{z_1 z_2\,e^2}{\varepsilon\,k\,T\,a}$$

in order to gain k_1.

RESULTS

A number of systems have been studied during the last years, the results of which can be used to draw general conclusions that are relevant to the raised questions. Whenever possible, very simple chemical systems were used for the investigations in order not to complicate the measurements unnecessarily. Some of the most used solvents in coordination chemistry were chosen. Some results that are presently available are listed in the tables (2 - 8) for the solvents $CH_3.OH$, DMSO, DMF and $CH_3.CN$. For most of the given results the activation parameters for the formation and dissociation rate constants were determined in order to make a comparison between the rate constant for the solvent exchange and the rate determining step possible.

GENERAL CONCLUSIONS

While the rate and the stability constants of the studied systems vary widely, it is nevertheless possible to make some conclusions that seem to be valid for all systems.

1. The rate determining step k_1
   The given results provide substantial and convincing evidence that the rate of complex formation is controlled by the solvent exchange process of a solvated ion. The rate constants k_1 and k_o and their corresponding activation parameter are in good agreement.

   This result seems to be valid quite independently of the solvent and the bulkiness of the molecules that make up the solvent. The rate constant k_1 does not seem to depend on the structure of the solvent as has previously been postulated on the basis of results that were obtained with bipyridine as a ligand[44]. The conclusions that were drawn from these results could not be confirmed by results that were obtained with simple monodentate ligands.

Table 2

Rate constants and activation parameters for the reactions between Ni(II) and different ligands X in methanol at 20°C

X	$K_o \cdot k_1$	k_{-1}	k_1	$\Delta H^{\ddagger}_{K_o \cdot k_1}$	$\Delta S^{\ddagger}_{K_o \cdot k_1}$	$\Delta H^{\ddagger}_{k_{-1}}$	$\Delta S^{\ddagger}_{k_{-1}}$	ref.
	$(M^{-1}s^{-1})$	sec^{-1}	sec^{-1}	Kcal	e.u.	Kcal	e.u.	
bipyr.	51		510	17	7.4			28,31
phen	240		2400					32
terpyr	46		460	16.8	5			32
SCN	$1 \cdot 10^5$	1	750	19.5	30	19	6.6	28,29,30
Cl	$1 \cdot 10^5$	100	800	20	33	15	+2	35
CF_3COO	$0{,}9 \cdot 10^5$	30	600	19	34.5	18.5	6.5	35
CCl_3COO	$0.7 \cdot 10^5$	80	400	18.5	27	18	12	33
$CHCl_2COO$	$1.5 \cdot 10^5$	30	900	19.5	32	16.5	4.8	33
CH_2ClCOO	$1.8 \cdot 10^5$	60	1000	17.5	25	17	7.8	33
$CH_3 \cdot COO$	$2.5 \cdot 10^5$		$2.2 \cdot 10^3$					33

Table 3

Rate constants and activation parameters for the reaction between bivalent metal ions M and chloride in the solvent methanol at 20°

M	$K_o \cdot k_1$	k_{-1}	k_1	$\Delta H^{\ddagger}_{K_o \cdot k_1}$	$\Delta S^{\ddagger}_{K_o \cdot k_1}$	$\Delta H^{\ddagger}_{k_{-1}}$	$\Delta S^{\ddagger}_{k_{-1}}$	ref.
Ni	$1 \cdot 10^5$	100	600	20	33	15	+ 2	34
Co	$3 \cdot 10^6$	$7 \cdot 10^3$	$1.75 \cdot 10^4$	16	26	13.2	4.2	11
Fe	$8.5 \cdot 10^6$	$1 \cdot 10^4$	$5 \cdot 10^4$	13.6	20	12.8	3.7	11
Mn	$6.8 \cdot 10^7$	$2.7 \cdot 10^4$	$4 \cdot 10^5$	10.7	14	11.4	0.8	11
Zn	$4.4 \cdot 10^9$	$7 \cdot 10^5$	$26 \cdot 10^7$	7.5	11	4	-18	35

Table 4

Rate constants and activation parameters for the reactions between bivalent metal ions M and $CF_3 \cdot COO$ in the solvent methanol at 20° (ref. 36)

M	$K_o \cdot k_1$	k_{-1}	k_1	$\Delta H^{\ddagger}_{K_o \cdot k_1}$	$\Delta S^{\ddagger}_{K_o \cdot k_1}$	$\Delta H^{\ddagger}_{k_{-1}}$	$\Delta S^{\ddagger}_{k_{-1}}$
Ni	$0.9 \cdot 10^5$	30	600	19	34.5	18.5	12
Co	$4.5 \cdot 10^6$	$2.7 \cdot 10^3$	$3.1 \cdot 10^4$	16.5	29	16	12
Fe	$1.4 \cdot 10^7$	$8.6 \cdot 10^3$	$7.7 \cdot 10^4$	13.5	18	12.2	1.25
Mn	$1.8 \cdot 10^8$	$5 \cdot 10^5$	$1 \cdot 10^6$	8.5	7	4	-18.6
Mg	$7.6 \cdot 10^5$	$5.4 \cdot 10^2$	$4 \cdot 10^3$	20	36	13.4	0

Table 5

Rate constants and activation parameters for the reaction between Ni(II) with different ligands X in Dimethylsulfoxide at 20°

X	$K_o \cdot k_1$	k_{-1}	k_1	$\Delta H^{\ddagger}_{K_o \cdot k_1}$	ref.
pyridine	$2\cdot10^3$	70	10^4	4	32
bipyr				12.6	32
phen	350			8.3	32
terpyr					
SCN	$8\cdot10^4$	80	$2\cdot10^3$		37,39
Cl	$7\cdot10^4$	150	$18\cdot10^3$		37
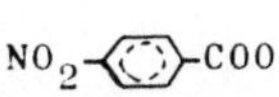	$1.9\cdot10^4$	16			38
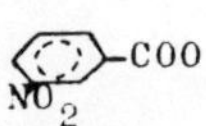	$1.3\cdot10^4$	16			38
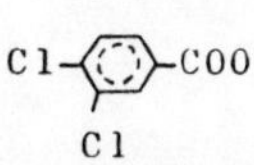	$1.3\cdot10^4$	20			38

Table 6

Rate constants and activation parameters for the reaction between Ni(II) with different ligands X in Dimethylformamide at 20°

X	$K_o \cdot k_1$	k_{-1}	k_1	$\Delta H^{\ddagger}_{K_o \cdot k_1}$	$\Delta S^{\ddagger}_{K_o \cdot k_1}$	$\Delta H^{\ddagger}_{k_{-1}}$	$\Delta S^{\ddagger}_{k_{-1}}$	ref.
bipyr	360		$3\cdot10^3$	12.7	-3.3			31
SCN	$1.4\cdot10^5$	15	$2.2\cdot10^3$	17.2	24	15.7	-1.4	40
Cl	$3.8\cdot10^5$	52	$6.3\cdot10^3$	19.2	33	12.4	-8.4	40
$CF_3\cdot COO$	$1.8\cdot10^5$	280	3.10^3	18.2	29	12.7	-4	41
	$2.4\cdot10^5$	2800	$4\cdot10^3$	18.1	28	16.5	13.7	41

Table 7

Rate constants and activation parameters for the reaction between Ni(II) and different ligands X in acetonitrile at 20° (if not stated otherwise)

X	$K_o \cdot k_1$	k_{-1}	k_1	$\Delta H^{\ddagger}_{K_o \cdot k_1}$	$\Delta S^{\ddagger}_{K_o \cdot k_1}$	$\Delta H^{\ddagger}_{k_{-1}}$	$\Delta S^{\ddagger}_{k_{-1}}$	ref.
	$M^{-1} \cdot s^{-1}$	sec^{-1}	sec^{-1}	Kcal	e.u.	Kcal	e.u.	
pyridine (25°)	$8.3 \cdot 10^2$			14.7				32
bipyr (25°)	$4.1 \cdot 10^3$			6.5				32
phen (25°)	$5 \cdot 10^4$			4.7				32
terpyr (25°)	$2.2 \cdot 10^3$			8.4				32
NO_3	$1.9 \cdot 10^5$	7.5	$2.6 \cdot 10^3$	17.5	24	13.5	-9.5	42
$CF_3.COO$	$1.5 \cdot 10^5$	5	$2.0 \cdot 10^3$					42
CH_3-o-SO_3	$1.3 \cdot 10^5$	26	$1.8 \cdot 10^3$	17	21	15	-3	42
SCN	$1 \cdot 10^5$	0.3		17	22	18	+0.8	43

Table 8

Rate and stability constants for mono- and bis-complexes in different solvents at 20°.

System	K_1	K_2	K_{01}	K_{02}	K_{J1}	K_{J2}	$k_{-1}(s^{-1})$	$k_{-2}(s^{-2})$
$Ni(SCN)_2$ in DMSO	1.10^3	35	45	7	22	5	80	$4 \cdot 10^3$
$Ni(SCN)_2$ in DMF	$8.3 \cdot 10^3$	$2.3 \cdot 10^2$	70	7.5	120	31	15	600
$Ni(CCl_3COO)_2$ in $CH_3.OH$	900	50	170	10	5.3	5	80	$2.2 \cdot 10^3$
$Ni(NO_3)_2$ in $CH_3.CN$	$2.5.10^4$	$1.2 \cdot 10^3$	73	6.3	340	190	7.5	520
$Ni(CH_3-\phi-SO_3)_2$ in $CH_3.CN$	$5.2 \cdot 10^4$	$1.2 \cdot 10^3$	50	6.3	100	190	26	$1.2 \cdot 10^3$
$ZnCl_2$ in $CH_3.OH$	$8.3 \cdot 10^3$	$1.1 \cdot 10^4$	170				8.10^4	$1.6 \cdot 10^5$

The relative unimportance of the structure of the solvent for the rate determining step can also be concluded from kinetic measurements over a wide temperature range in methanol. Although the viscosity changes from 0.5 cp to 4 cp when the temperature is lowered from 30° to -70° mechanism (1) is not altered.

Furthermore it was shown recently by Frankel[45)] that the rate constant k_o is not affected by the composition of the solvent as long as the inner coordination sphere of the metal ion is kept constant.

Both, for DMF and DMSO, it was found that the solvent exchange rates on Ni(II) and Co(II) did not change when the solvents where diluted with noncoordinating solvents like $CH_3.NO_2$ and CH_2Cl_2. For protic solvents, the situation could be different because the rate constant k_o should depend on the strength and the number of hydrogen bonds between the inner and the outer coordination sphere and consequently a change in the composition of the outer sphere would probably affect the exchange rates. But even in this case, the changes are probably quite small. The rate constant k_1 in methanol seems to be fairly independent of the basicity of the entering ligand. On the other hand, the reaction with ethylenediamine for which the conjugate base mechanism was observed in water[46], has not yet been studied in methanol.

Clearly, most of the available evidence is consistent with the solvent exchange process being a SN_1-process.

2. The bis-complex in comparison to the mono-complex.

Another simple result is that wherever it was possible to observe the mono- and the bis-complex it is found that the bis-complex is less stable than the mono-complex. The difference in the stability constants is mainly due to the difference in the outer sphere stability constants while the inner sphere constants do not vary much. Some examples are given in table 8. In this respect the complexes behave the same in non-aqueous solutions as in water. Whenever there are deviations from this normal behavior there is probably a special reason for it like a change in the symmetry or in the electronic state between the mono- and the bis-complex as for instance for $ZnCl_2$ in $CH_3.OH$. While the stability constants for the inner sphere constants are of similar size, the characteristic rate constants are not. The bis-complex was always found to form and dissociate considerably faster than the mono-complex. Correspondingly, the activation energies for the bis-complex are always lower than for the mono-complex.

The change in rate constant is not only caused by a change in the activation enthalpy but also in the entropy.

The consequence of course is that the ratio of the dissociation rate constants of the two complexes is temperature dependent and normally k_2/k_1 becomes smaller with increasing temperatures. This behavior is seen in figure 3. The behavior

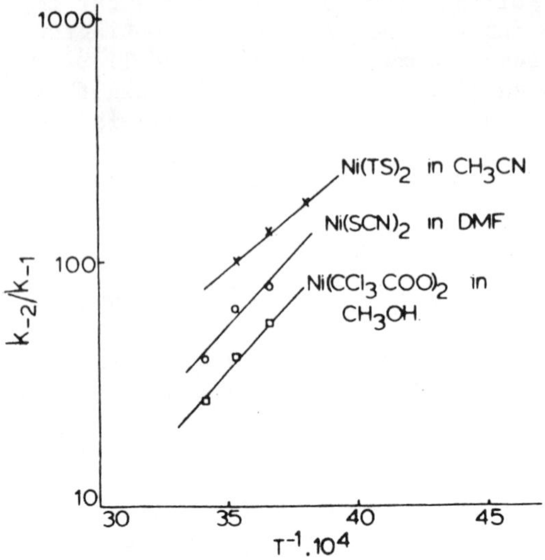

Fig. 3 Dependence of the ratio of the dissociation rate constant k_{-2} and k_{-1} on the temperature. Plot of log k_{-2}/k_{-1} against 1/T.

resembles the pattern observed for the activation parameters of k_o where it was found that a decrease in activation enthalpy is partially compensated by a decrease in activation entropy.

3. The stability constant as a function of the solvent

As regards to the variation of the inner sphere stability constant $K_1^I = k_1/k_{-1}$ for a given complex in different solvents, it was recently pointed out, that the stability constant increases with decreasing donicity of the solvent[42, 43]. Similar conclusions were put forward by Gutmann and Schmid[47]. Qualitatively, such a behavior is to be expected from simple arguments based on the solvating power of the solvent.

Since the equilibrium

$$[MS_5X]\,S \quad \underset{k_1}{\overset{k_{-1}}{\rightleftharpoons}} \quad [MS_6]\,X \tag{9}$$

is the result of the competing reactions with the rate constants k_1 and k_{-1}, a quantitative interpretation of this result has to account for the change of these two rate constants with the variation of the solvent. The two rate constants shall therefore be discussed separately.

4. The dissociation process of a complex in different solvents

The dissociation process

$$MS_5 \cdot X \xrightarrow{k_{-1}} MS_5 + X \qquad (10)$$

can be looked at as a dissociation process of a compound composed from the Lewis acid MS_5 and the Lewis base X. The acidity of the Lewis acid A and the donicity of the Lewis base D determine the activation energy of the process. Both contributions to the activation process can be investigated independently.

a) Influence of the acidity of MS_5 on k_{-1}

The acidity of MS_5 depends on the donicity of the five solvent molecules that are coordinated to M. It is lowered by the increase of the donicity of the coordinated molecules. The influence can qualitatively be studied by measuring the dissociation rate constant for a given complex like Ni-SCN in different solvents. Recently it was assumed that the acidity decreases exponentially with the donicity[43]. With such an assumption it was possible to explain the trend in the activation enthalpies for the solvent exchange. However, judged from the dependence of the dissociation rate constants of a given complex in different solvents, it seems more likely that the acidity decreases linearly with D. For a given leaving group the log k_{-1} values increase about linearly over several orders of magnitude with the donicity values of the solvents[48]. Examples for several complexes are given in fig. 4.

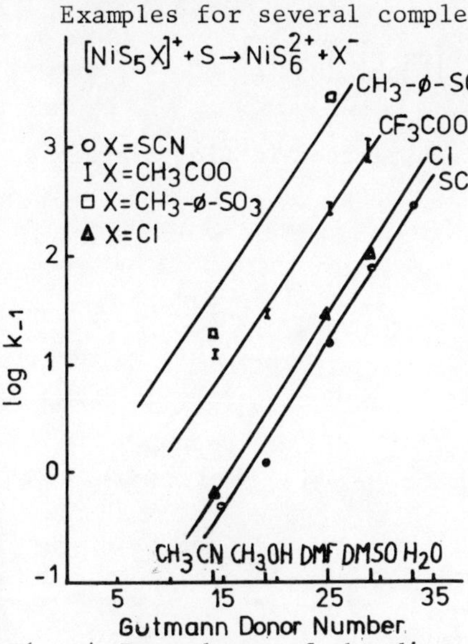

Fig. 4 Dependence of the dissociation rate constant k_{-1} on the donicity of the solvent. Plot of the log k_{-1} against the Gutmann donicity.

Such linear relations can be explained by assuming that the activation energy for the dissociation process can be expressed as the product from the donicity of the leaving group D_x and the activity A of the remaining intermediate complex which in turn decreases linearly with D_s.

$$\Delta G^{\neq} = A \cdot D_x = (A_o - \alpha \cdot D_s)D_x \tag{11}$$

with $k = C \exp -\frac{\Delta G^{\neq}}{RT}$, we obtain

$$\ln k = \ln C - \frac{A_o \cdot D_x}{RT} + \frac{\alpha \cdot D_s \cdot D_x}{RT} \tag{12}$$

As seen from equation (12), the log of the rate constant k_{-1} should linearly increase with the donicity of the solvent with the slope of the plot depending on the donicity of the leaving group.

b) The influence of the leaving group.

The influence of the leaving group on the activation energy of k_{-1} can be studied by keeping the Lewis acid MS_5 constant and varying the donicity of the leaving group. Such measurements have been done for various Lewis acids like $SbCl_5$ and $Ni(H_2O)_5$. Gutmann and Schmid for instance reported the rate constants for the reaction[47].

$$SbCl_5\text{-}S + CPh_3Cl \underset{k_{-1}}{\overset{k_1}{\rightleftharpoons}} [Ph_3C]^+[SbCl_6]^- + S$$

The log of the rate constant k_1 is plotted in fig. 5 against

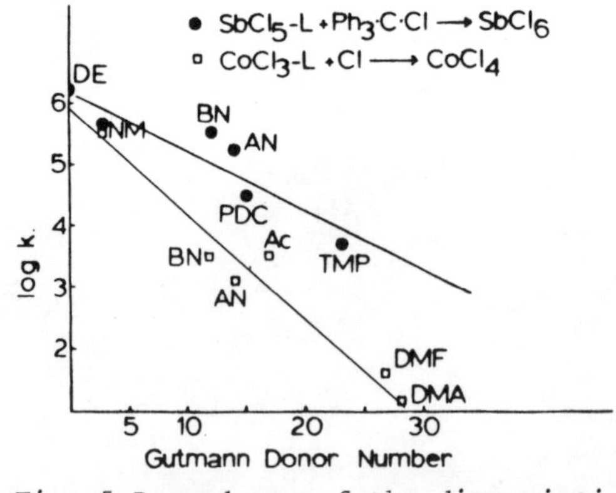

Fig. 5 Dependence of the dissociation rate constant k_{-1} on the donicity of the leaving group.

the donicity of the leaving group S. As expected log k_1 decreases linearly with the donicity of S. The rate determining step for the rate constant k_1 is believed to be the breaking of the $SbCl_5$-S bond and probably independent of the entering ligand CPh_3Cl.

5. The solvent exchange rate constant as a function of the solvent. Rate determining for the solvent exchange process is the dissociation of a solvent molecule from the inner coordination sphere of the metal ion. From the above consideration, it is obvious that for the process;

$$MS_6 \xrightarrow{k_1} MS_5 + S$$

both acidity of MS_5 and donicity of the leaving group are varied when the solvent is varied. As shown, both effects on the rate constant tend to oppose each other.

The increase in the acidity of A is balanced by the decrease in the donicity of the leaving group if the donor properties of the solvent are lowered. Consequently, the rate constants for the solvent exchage are very similar in many solvents. As a result of the formation rate constant of a complex being little dependent on the solvent and the log k_{-1} linearly dependent on the solvent donicity, log K-values are also linearly dependent on the donicity of the solvent. This is shown in the fig. 6. This empirical relation lends itself

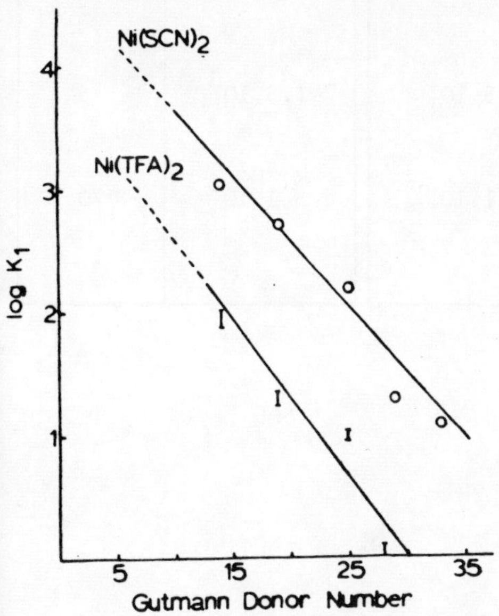

Fig 6 Dependence of the stability constant $K_1 = (k_1/k_{-1})$ on the donicity fo the solvent.

for the extrapolation of equilibrium constants in nonaqueous solvents using the values that have been determined in the solvent water. Therefor it should be possible to make use of the hundreds of stability constants that have been determined for various ligands and metal ions to estimate values in nonaqueous solvents. However, we should be mindful that while equation (12) can certainly be helpful for the estimation of rate constants we should not be too surprised if the agreement with the experiment leaves something to be desired. It should be warned that the mentioned correlations seem to exist only for solvents in which hydrogen bonds play no major role or for ligands that are not well suited for hydrogen bond formation. With simple ligands like Cl no correlation can be observed as shown in table 9.

TABLE 9.

Comparison of stability constants K_1 for $NiCl_2$ and $Ni(SCN)_2$ in protic and aprotic solvents at 20°.

	DMF	DMSO	$CH_3.OH$	H_2O
$NiCl_2$	7.10^3	$4.5.10^2$	$1.5.10^3$	5
$Ni(SCN)_2$	9.10^3	1.10^3	1.10^5	70

REFERENCES

1. M. Eigen, R. G. Wilkins, Mechanism of Inorganic Reactions, R. F. Gould, Ed. Advances Chem. Soc. No. 49, Amer. Soc. Washington, D.C. 1965
2. R. G. Wilkins, Accounts of Chemical Research 3, 408 (1970).
3a D. W. Margerum, D. B. Rorabacher, J. F. G. Clarke, Inorg. Chem. 4, 667 (1963).
b H. Hoffmann, U. Nickel, Ber. Bunsenges. Phys. Chem. 72, 1092 (1968).
4. H. Hoffmann, Ber. Bunsenges. Phys. Chem. 73, 432 (1969).
5a D. W. Margerum, H. M. Rosen, J.A.C.S. 89, 1088 (1967).
b H. Hoffmann, E. Yeager, Ber. Bunsenges. Phys. Chem. 74, 641 (1970).
c A. G. Desai, H. W. Dodgen, J. P. Hund, J.A.C.S. 92, 798 (1970).
6. L. G. Sillen, A. E. Martell "Stability constants of metal ion complexes", The Chemical Society, London (1964).
7. G. G. Hammes, J. J. Steinfeld, J.A.C.S. 84, 4639.
8. H. Hoffmann, unpublished data
9. G. H. Langford, T. R. Stengle, Ann. Rev. Phys. Chem. 19, 193 (1968).
10. G. W. Castellan, Ber. Bunsenges. Phys. Chem. 67, 899 (1963).
11. P. Fischer, H. Hoffmann, G. Platz, Ber. Bunsenges. Phys. Chem. 76, 1060 (1972).
12. R. M. Fuoss, J.A.C.S. 80, 5059 (1958).
13. Z. Luz, S. Meiboom, J. Chem. Phys. 40, 2686 (1964).
14. F. W. Breivogel, J. Chem. Phys. 51, 445 (1969).
15. Z. Luz, H. Levanon, J. Chem. Phys. 49, 2031 (1968).
16. S. Nakamara, S. Meiboom, J.A.C.S. 89, 1765 (1969).
17. N. S. Angermann, G. B. Jordan, Inorganic Chem. 8, 2579 (1969).
18. L. S. Frenkel, Inorg. Chem, 10, 814 (1971).
19. J. C. Boubel, J. J. Delpuech, Mol. Phys. 27, 113 (1974).
20. N. A. Matwiyoff, Inorg. Chem. 5, 788 (1966).
21. L. S. Frenkel, Inorg Chem. 10, 2360 (1971).
22. N. A. Matwiyoff, S. V. Hooker, Inorg. Chem. 6, 1127 (1967).
23. D. R. Ravage, T. R. Stengle, C. H. Langford, Inorg. Chem. 6, 1251 (1967).
24. J. F. O'Brien, W. L. Reynolds, Inorg. Chem. 6, 2110 (1967).
25. I. D. Camplen, J. P. Carver, R. A. Dwek, A. J. Drumelin, R. E. Richards, Mol. Phys. 20, 913 (1971).
26. R. J. West, S. F. Lincoln, Aust. J. Chem. 24, 1169 (1971).
27. T. J. Swift. H. H. Lo, J.A.C.S. 88, 2994 (1966).
28. R. G. Pearson, P. Ellgen, Inorg. Chem. 6, 1379 (1967).
29. F. Dickert, H. Hoffmann, W. Jaenicke, Ber. Bunsenges. Phys. Chem. 74, 500 (1970).
30. J. William, S. Petrucci, J. Phys. Chem. 77, 130 (1973).
31. H. P. Nennetto, E. F. Caldin, J. Chem. Soc. (A) 2191 (1971).

32. P. K. Chattopadhyay, J. F. Coetzee, Inorg. Chem. 12, 113 (1973).
33. F. Dickert, R. Wank, Ber. Bunsenges. Phys. Chem. 76, 1028 (1972).
34. F. Dickert, P. Fischer, H. Hoffmann, G. Platz, Chem. Comm. 106 (1972).
35. M. Franke, H. Hoffmann, G. Platz, Proceedings of the XVI. ICCC, Dublin, 1974.
36. H. Escher, H. Hoffmann, G. Platz (to be published).
37. F. Dickert, H. Hoffmann, Ber. Bunsenges, physik. Chem. 75, 1320 (1971).
38. F. Dickert, H. J. Etzold, A. Wolf, Proceedings of the XV. ICCC, Moscow, 1973.
39. J. Williams, S. Petrucci, B. Sesta, M. Battistini, Inorg. Chem. 13, 74 (1968).
40. F. Dickert, H. Hoffmann, Proceedings of the XIII. ICCC in Poland, Vol. I, p. 201 (1970).
41. R. Sperati, H. Hoffmann, to be published.
42. H. Hoffmann, T. Janjic, R. Sperati, Ber. Bunsenges. phys. Chem. 78, 223 (1974).
43. F. Dickert, H. Hoffmann, T. Janjic, Ber. Bunsenges. phys. Chem. (in press).
44. H. P. Bennetto, E. F. Caldin. J. Chem. Soc. (A), 2198 (1971).
45. L. S. Frenkel, Inorg. Chem. 10, 2360 (1971).
46. D. B. Moss, D. B. Rorabacher, Inorg. Chem. 9, 1314 (1970).
47. V. Gutmann, R. Schmid, Coord. Chem. Rev. 12, 263 (1974).
48. V. Gutmann, Coordination Chemistry in Nonaqueous Solutions, Verlag Springer, Wien-New York 1968.

LIGAND SUBSTITUTION PROCESSES. AN NMR STUDY. I.
SOLVATION SHELLS OF DIAMAGNETIC CATIONS Al^{3+} AND Be^{2+}

J.J. Delpuech*, M.R. Khaddar, A.A. Peguy
and P.R. Rubini

(Laboratoire de Chimie Organique, Equipe de
Recherche Associee au C.N.R.S., Universite de NANCY I,
C.O. 140, 54037 - Nancy Cedex, France)

ABSTRACT

Nuclear magnetic resonance may be used as a relaxation method to study chemical reactions at equilibrium. Continuous wave nmr is a powerful tool to study the first solvation shell of diamagnetic cations for which the exchange of ligands between bound and free solvent molecules fit the nmr time scale. This ligand exchange is studied using not only ^{1}H resonances, but also ^{31}P and ^{27}Al spectra. An associative mechanism is found for the tetrahedral solvate $Al(HMPA)_4^{3+}$ and dissociative processes for octahedral AlA_6^{3+} and tetrahedral Be A_4^{2+} solvates (in nitromethane), where A is an organophosphorus solvent (trialkyl phosphates and phosphonates).

INTRODUCTION

In electrolytic solutions, the presence of ions modifies the nmr spectrum of the solvent in various ways. Under conditions to be defined later, different nmr signals may be observed for bound and bulk solvent molecules[1] (S_b and S_f respectively), yielding solvation numbers from their area. Conversely, solvation may also change the charge distribution of the ion itself: the nmr spectrum of the cation nucleus may therefore be used as well for these studies.

In fact, the possibility of observing separate nmr signals is severely limited by nuclear relaxation and chemical exchange. Conversely, the alteration of nmr lines by these phenomena may be used to estimate the rates of these processes[2]. Let us first recall some results of basic theory of dynamic nmr, a nuclear

Wyn-Jones (ed.), Chemical and Biological Applications of Relaxation Spectrometry, 311–321.

relaxation method to study reactions at equilibrium.

DYNAMIC NMR[2 - 5]

A given type of nucleus (^{1}H, ^{13}C, ^{31}P or ^{27}Al in this paper) may resonate at different frequencies according to its position in the molecule as a result of different chemical shifts and coupling constants[2]. Now, a same nucleus may occupy successively these various locations owing to some chemical exchange processes (intra or intermolecular): these locations are classically named "nuclear sites" or "sites". Let us examine the simplest case of two sites only: A and B.

(A) These two sites are defined by:

- their relative frequency shift: $\Delta\nu = \nu_A - \nu_B$ (Hz) or $\Delta\omega$ (radians sec^{-1}) (Figure 1a).

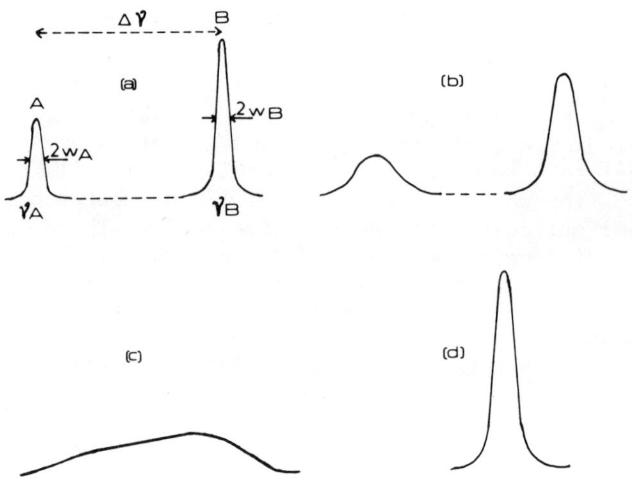

Fig. 1 - Coalescence spectrum for two sites.

- their transversal relaxation times (T_{2A}, T_{2B}), or the corresponding linewidths: $2w_A = \frac{1}{\pi T_{2A}}$; $2w_B = \frac{1}{\pi T_{2B}}$

- their molar fractions (i.e. the relative areas of the corresponding lines): P_A and P_B

(B) The site exchange: $A \underset{kB}{\overset{kA}{\rightleftharpoons}} B$ is characterized by the mean lifetimes of one nucleus in sites A (or B):

τ_A(or B) = $1/k_A$(or B). k_A and k_B are pseudomonomolecular rate constants for magnetization transfer, to be connected to the actual chemical rate constants by the following equation:

$$k_A = \frac{\text{rate of appearance (or of disappearance) of A by all processes}}{\text{stationary concentration [A] of A.}}$$

The exchange reaction being <u>at equilibrium</u>, the molar fractions P_A and P_B are stationary, i.e.

$$k_A P_A = k_B P_B \text{ or } \tau_A/P_A = \tau_B/P_B$$

(C) <u>The effect on lineshapes</u> is described as follows

- <u>for slow exchanges</u> ($\tau_A . \Delta\omega >> 1$), lines are broadened (figure 1b) by a quantity:

$$2\pi\Delta w_A = 1/\tau_A \text{ and } 2\pi\Delta w_B = 1/\tau_B$$

- these lines collapse into one broad signal when $\tau_A . \Delta\omega \cong 1$ ("<u>Coalescence</u>"; figure 1c)

- for fast exchanges ($\tau_A . \Delta\omega << 1$), this unique line sharpens so as to be weighted mean of the initial resonances (figure 1d).

A general theory for exchanges between two sites has been set up by Gutowsky, McCall, Slichter[6], and McConnell[7]. Their treatment has been extended to any number <u>n</u> of sites $j(\nu_j, W_j, P_j)$. The final result is conveniently expressed in a matrix formulation due to Anderson[8], Kudo[9] and Sack[10]: relative intensity of signal = imaginary part of $\|P\| . \|M\|^{-1} . \|1\|$. In this matrix product, $\|P\|$ is a row vector representing the molar fractions of the n sites: $\|P\| = \| P_1, P_2 \text{ .... } P_i \text{ .... } P_n \|$; and $\|1\|$ is the unit column vector; $\|M\|$ is the sum of a diagonal matrix with complex elements equal to: $2\pi\{w_j + \underline{i}(\omega - \omega_j)\}$, and of an exchange matrix built from the exchange probabilities k_{ij} (for a nucleus to jump from site i to site j) in the following manner:

$$\left\| \begin{matrix} -\sum_{j\neq 1}^{n} k_{1j} & k_{12} & \cdots\cdots & k_{1n} \\ k_{21} & -\sum_{j\neq 2}^{n} k_{2j} & \cdots\cdots & k_{2n} \\ \vdots & & & \\ k_{n1} & k_{n2} & \cdots\cdots & -\sum_{j\neq n}^{n} k_{nj} \end{matrix} \right\|$$

Theoretical curves may be computed for a series of trial values of exchange probalities so as to obtain the best fit with experimental spectra.

CHEMICAL EXCHANGE BETWEEN BULK AND BOUND SOLVENT

Solvent molecules are not firmly attached to the solvated ions, and the exchange between free and bound molecules: $S_b \xrightarrow{k} S_f$ is equivalent to a site exchange of solvent nuclei[1]. It is extremely important whether this exchange is fast or not, on the nmr time-scale.

(a) If the exchange is fast, only one line is observed. Its frequency may yield the solvation number if some extrapolation of the resonance frequency of bound molecules is possible. This is the case for all anions, and most cations, except a few ones: Al^{3+}, Ga^{3+}, In^{3+}, Be^{2+}, Zn^{2+}, Mg^{2+} (Ref. 11-16, respectively). For all the other cations, the kinetic information is lost, except for some paramagnetic cations, for which measurements of relaxation times by pulsed nmr may yield the rate constants. Such examples will be given in the second part of this paper.

(b) If the exchange is slow ($k<\Delta\omega$), several lines are observed, from which solvation numbers may be computed. These lines coalesce at higher temperatures, yielding the kinetic parameters of the exchange.

We must remark that the nmr time-scale (i.e. $\Delta\omega$) may be different, for a same cation solvent system, according to the observed nucleus. This explains why we have looked for other nuclei than ^1H for these studies, namely:

- ^{31}P in 1969[17-19], therefore choosing organophosphorus solvents: trialkylphosphates (trimethyl : TMPA, or triethyl : TEPA) ; phosphonates (dimethyl - methyl : DMNP ; diethyl - ethyl : DEEP) ; dialkylhydrogenphosphites (dimethyl : DMHP) ; hexamethylphosphorotriamide (HMPA)

- ^{13}C in 1971[20-21]

- ^{27}Al in 1973[22-23]

The solubility of aluminium or beryllium perchlorates is generally poor in these organo-phosphorus solvents. They were accordingly used as mixtures either with water or nitromethane. In the first case, water competes with the organic solvent to solvate the cation, and results are not easily interpreted. Pure nitromethane is a convenient inert solvent for the Al^{3+} or Be^{2+} solvates prepared in the solid state: AlA_6^{3+}, $3ClO_4^-$, where A =

TMPA, TEPA, DMMP, DEEP or DMHP, and $Al(HMPA)_4^{3+}$, $3ClO_4^-$, and Be A_4^{2+}, $2ClO_4^-$, with A = TMPA, DMMP or HMPA.

These solutions were added, in a further step, with free organic solvent A, or with water.

ALUMINIUM SOLVATES IN PURE NITROMETHANE[24]

^{1}H and ^{13}C nmr yield only one line, and no structural information is obtained from these nuclei. Convenient probes are ^{31}P and ^{27}Al nuclei[23].

(a) For hexacoordinated solvates, AlA_6^{3+}, a sharp heptet is obtained at 25 C by ^{27}Al nmr (figure 2a). The intensities of its components are in the ratio :1:6:15:20:15:6:1, thus revealing a coupling between ^{27}Al and six equivalent ligands around the aluminium cation, with a coupling constant between ^{31}P and ^{27}Al of 19.5 ; 15.0 and 13.4 Hz for TMPA, DMMP and DMHP respectively. The cubic symmetry of these solvates ensures exceptionaly sharp lines in spite of a strong quadrupolar relaxation of the aluminium - 27 nuclei.

These results are confirmed by ^{31}P nmr. Six lines are observed for the solvate (figure 2b), each one corresponding to the six magnetic states of the ^{27}Al nucleus with the same probability.

Spectra are coalescing when temperature is increased (cf. figure 3), revealing an exchange of solvent molecules. Quantitative measurements have been carried out both by Al and ^{1}H nmr, in good mutual agreement: $k(25^{o}C) = 5.0$ and 5.1 sec^{-1}; $\Delta H^{\neq}$ = 20.0 and 18.0 ± 1.5 kcal; $\Delta S^{\neq}$ = 11.4 and 4.4 ± 5 e.u, respectively for $Al\ (DMMP)_6^{3+}$

Two important results are obtained for these ligand substitutions around the aluminium cation:

(i) The rate - law is independent from the amount of added free ligand: a dissociative mechanism[25-26] is accordingly assigned to this exchange:

$$Al\ (TMPA)_5^{3+} \xrightarrow{6k} Al\ (TMPA)_5^{3+} + TMPA$$

$$Al\ (TMPA)_5^{3+} + TMPA * (free) \xrightarrow{fast} Al\ (TMPA)_5\ (TMPA^{*})^{3+}$$

Let us notice that the rate constant computed from ^{27}Al nmr is actually six times as high as the rate constant for one individual solvent molecule, deduced from ^{1}H spectra.

(ii) Activation enthalpies and entropies are high, as expected for such mechanism (table 1).

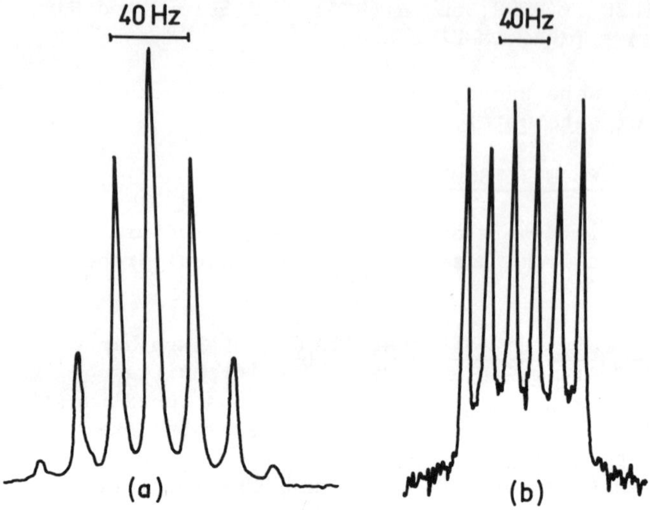

Fig. 2 - ^{27}Al (a) and ^{31}P (b) spectra of $Al(TMPA)_6^{3+}$ in CH_3NO_2

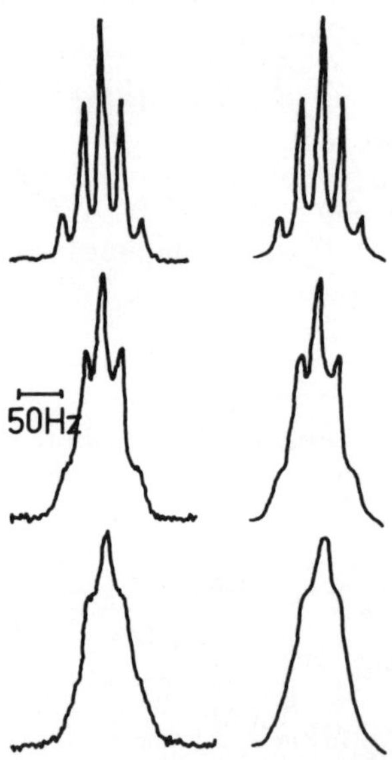

Fig. 3 - ^{27}Al spectrum of $Al(HMPA)_4^{3+}$ at: -32°; -6°; $+5^{\circ}C$ (from top to bottom). Experimental and theoretical curves (left and right respectively).

TABLE 1 - LIGAND SUBSTITUTION REACTIONS OF HEXA AND TETRACOORDINATED Al^{3+} CATION.

SOLVATE	Solvent	$k_{25°C}$ (sec^{-1} or $M^{-1} sec^{-1}$)	$\Delta H^{\neq}$ (kcal $mole^{-1}$)	$\Delta S^{\neq}$ (e.u)	Mechanism
$Al(TMPA)_6^{3+}$	CH_3NO_2	0.38	23.5	18.2	Diss.
$Al(DMMP)_6^{3+}$	CH_3NO_2	5.1	19.0	7.9	Diss.
$Al(DMHP)_6^{3+}$	CH_3NO_2	1.33	19.8	6.9	Diss.
$Al(HMPA)_4^{3+}$	CH_3NO_2	4.8×10^3	7.7	- 10.2	Assoc.
$Al(DMMP)_{0.5}(H_2O)_{5.5}^{3+}$	$DMMP/H_2O$	0.30	20.5	17.8	Diss. ?
$Al(HMPA)_{0.58}(H_2O)_{5.42}^{3+}$	$HMPA/H_2O$	0.24	21	16.1	Diss. ?
From literature :					
$Al(H_2O)_6^{3+}$ (Ref. 28)	H_2O	0.16	27	28	Diss. ?
$Al(DMSO)_6^{3+}$ (Ref. 11d)	DMSO	0.6	20	3.7	Diss. ?
$Al(DMF)_6^{3+}$ (Ref. 29)	DMF	0.15	17.7	4.7	Diss. ?

(b) For the tetrahedral solvate: $Al(HMPA)_4^{3+}$, a sharp quintet is observed by ^{27}Al nmr at $-32\,^\circ C$, with intensities in the ratio 1:4:6:4:1, and again a hexet for ^{31}P. This is clear evidence for a tetrahedral arrangement of four HMPA ligands around Al^{3+}. In sharp contrast with octahedral solvates, the rate - law is found to be proportional to the amount of free HMPA added to the solution, strongly suggesting an associative SN_2 mechanism:

$$Al(MHPA)_4^{3+} + HMPA^* \text{ (free)} \xrightarrow{4k} Al(HMPA)_3(HMPA^*)^{3+} + HMPA$$

with: $k^{27}Al = 4k\,[HMPA^*]$

Much smaller $\Delta H^{\neq}$ and $\Delta S^{\neq}$ values are obtained for such a mechanism (table 1).

BERYLLIUM SOLVATES IN PURE NITROMETHANE

Their ^{31}P spectrum consists of four lines equally spaced and with the same intensity.[24]

This quartet results from a coupling between ^{31}P and ^{9}Be nuclei ($j \simeq 6$ Hz), each line corresponding to the four possible magnetic states of the ^{9}Be nucleus ($I = 3/2$). The sharpness of lines reveals again a cubic symmetry, i.e. a tetrahedral arrangement of ligands around the Be^{2+} cation, cancelling the effects of quadrupolar relaxation.

Kinetic studies have been carried out by proton nmr with the solvates $Be\,A_4^{2+}$, where A = TMPA, DMMP and HMPA. The rate-law was found independent from the concentration of the added free ligand. Dissociative mechanisms have accordingly been assigned to these substitution reactions. The results displayed in table 2 point out another important conclusion: the activation parameters may be high (for HMPA) or small (for TMPA and DMMP), keeping the same dissociative mechanism. Especially, the activation entropy is either positive, or negative, in contrast with the empirical criterion often used to distinguish between SN_1 and SN_2 ligand substitution processes.

AQUEOUS SOLUTIONS

In this case, water is not an inert solvent, so that mixed species:

$S_i = Al\,A_i\,(H_2O)_{6-i}^{3+}$ and $Be\,A_i(H_2O)_{4-i}^{2+}$ ($i = 0$ to 6 or 0 to 4 respectively) are obtained. Proton nmr yields only one line for all the species S_i (the so-called bound signal S_b). However, separate lines are obtained in two cases:

(a) for the system: $Be^{2+}/HMPA/H_2O$ by ^{31}P nmr[19]

TABLE 2 - Ligand substitution reactions on Be^{2+} cation

SOLVATE	SOLVENT	$k_{25^\circ C}$ (sec^{-1})	$\Delta H^{\neq}$ (k cal $mole^{-1}$)	$\Delta S^{\neq}$ (e.u)
$Be(TMPA)_4^{2+}$ (Ref. 27)	CH_3NO_2	3.6	13.6	-10.5
$Be(DMMP)_4^{2+}$	CH_3NO_2	1.3	15.0	-7.4
$Be(HMPA)_4^{2+}$	propylene carbonate	2.7×10^{-4}	27.9	18.6
$Be(DMMP)_{0.15}(H_2O)_{3.85}^{2+}$	$DMMP/H_2O$	85	19	-10
$Be(HMPA)_{0.94}(H_2O)_{3.06}^{2+}$	$HMPA/H_2O$	0.072	14.5	-10

(b) for the systems: $Al^{3+}/A/H_2O$, where A = TMPA, TEPA, DMMP or DMHP[22-24]

For these systems, the following results were obtained:

(a) The mean solvation number A of the organic solvent, and, for some solutions, the solvation number H_2O of water, checking that the sum ($n_A + n_{H_2O}$) is 6 and 4 for aluminium and beryllium perchlorates respectively.

(b) The distribution of these solvates, i.e. the percentages of species s_i, within certain limits (i = 0 to 3).

(c) The equilibrium constants for substitution reactions: $S_i + H_2O \rightleftharpoons S_{i-1} + A$.

(d) Some kinetic data (tables 1 and 2) by proton nmr.

The rate - law can not be derived in those mixtures. The similarity of activation parameters only suggests a dissociative mechanism for the aluminium cation. No conclusion can be drawn for Be^{2+} solutions.

CONCLUSIONS

Nuclear relaxation methods may yield a wealth of information about moderately fast ligand exchange processes. Ionic solvation in the first coordination sphere of Al^{3+} and Be^{2+} cations can be described quantitatively using simultaneously the resonances of nuclei either of the solvent molecules or of the cation.

ACKNOWLEDGEMENT

Financial support from the Centre National de la Recherche Scientifique and from the Direction des Recherches et Moyens d'Essais is gratefully acknowledged.

REFERENCES

1. For a review, sec: (a) J.F. Hinton and S. Amis, Chem. Rev. 67, 367 (1967); (b) C. Deverell, "Progress in Nuclear Magnetic Resonance Spectroscopy", (J.W. Emsley, J. Feeney, and L.H. Sutcliffe, Eds.), Vol.4,p.235, Pergamon Press, New York, 1969; (c) K. Kustin and J. Swinehart, "Inorganic Reaction Mechanism" (J.O. Edwards, Ed.) Part 1, p.107, Wiley-Interscience, New York 1970, (d) J.J. Delpuech, A. Peguy and M.R. Khaddar, J. Electroanal. Chem. 29, 31 (1971); (e) J.F. Hinton and S. Amis, Chem. Rev. 71, 627 (1971); (f) M. Szwarc, "Ions and Ion Pairs in Organic Reactions", p.311, Wiley-Interscience, New York, 1972; (g) A. Fratiello, "Inorganic Reaction Mechanisms" (J.O. Edwards, Ed.) Part 2, p.57, Wiley-Interscience, New York, 1972; (h) J.W. Akitt, "Annual Reports on nmr Spectroscopy" (E.F. Mooney, Ed.), Vol. 5A, p.465, Academic Press London, 1972.
2. J.A. Pople, W.G. Schneider and H.J. Bernstein, "High-resolution nuclear magnetic resonance", 1959, McGraw-Hill, New York.
3. A. Loewenstein and T.M. Connor, Ber. der Bunsen Gesell. fur Phys. Chem., 67, 280 (1963).
4. J.J. Delpuech, Bull. Soc. Chim. France, 1964, p.2697.
5. C.S. Johnson, "Advances in Magnetic Resonances", Vol.1, Academic Press, New York, 1963, p.33-102.
6. H.S. Gutowsky, D.W. McCall and C.P. Slichter, J. Chem. Phys., 21, 279 (1953); H.S. Gutowsky and A. Saika, Ibid., 21 1688 (1953); H.S. Gutowsky and C.H. Holm, Ibid, 25, 1228 (1956).
7. H.M. McConnell, J. Chem. Phys., 28, 430 (1958).
8. P.W. Anderson, J. Phys. Soc. Japan, 9, 316 (1954).
9. R. Kubo and K. Tomita, Ibid., 9, 88 (1954); R. Kubo, Ibid., 9, 935 (1954).
10. R.A. Sack, Mol. Phys. 1, 163 (1958).
11. (a) R.E. Connick and D. Fiat, J. Chem. Phys., 39, 1349 (1963); (b) M. Alei, Jr. and J.A. Jackson, J. Chem. Phys., 41, 3402 (1964); (c) R. Schuster and A. Fratiello, J. Chem. Phys., 47, 1554 (1967); (d) S. Thomas and W. L. Reynolds, J. Chem. Phys., 44, 3148 (1966); W. G. Movius and N.A. Matwiyoff, Inorg. Chem., 6, 847 (1967); (f) L. Supran and N. Sheppard, Chem. Commun., 832 (1967); (g) J.F. Hon, Mol. Phys., 15, 57 (1968); (h) J.F. O'Brien and M. Alei Jr., J. Phys. Chem., 74, 743 (1970).
12. (a) D. Fiat and R.E. Connick, J. Amer. Chem. Soc., 88, 4754 (1966); (b) T.J. Swift, O.G. Fritz and T.A. Stephenson, J. Chem. Phys., 46, 406 (1967); (c) M. Alei Jr. and J.A. Jackson, J.Chem. Phys., 41, 41, 3402 (1964).

13. (a) T.H. Cannon and R.E. Richards, Trans. Faraday Soc., 62, 1378 (1966); (b) R.E. Schuster and A. Fratiello, J. Chem. Phys. 47, 1554 (1967); A. Fratiello, R.E. Lee, V.M. Nishida, and R.E. Schuster, Ibid., 48, 3705 (1968); (c) J. Crea and S.F. Lincoln, Inorgan. Chem. 11, 1131 (1972), (d) H. Haraguchi; K. Fuwa, and S. Fujiwara, J. Phys. Chem., 77, 1497 (1973).
14. Ref. 11a; 11b and 12d.
15. S.A. Al-Baldawi and T.E. Gough, Canad. J. Chem., 47, 1417 (1969); S.A. Al-Baldawi, M.H. Brooker, T.E. Gough and D.E. Irish, Ibid., 48, 1202 (1970).
16. (a) J.H. Swinehart and H. Taube, J. Chem. Phys., 37, 1579 (1962); (b) S. Nakamura and S. Meiboom, J. Amer. Chem. Soc., 89, 1765 (1967); (c) N.A. Matwiyoff and H. Taube, Ibid., 90, 2796 (1968); (d) T. Alger, Ibid., 91, 2220 (1970).
17. C. Beguin, J.J. Delpuech, and A. Peguy, Mol. Phys., 17, 317 (1969).
18. J.J. Delpuech, A. Peguy, and M.R. Khaddar, J. Electroanal. Chem., 29, 31 (1971).
19. J.J. Delpuech, A. Peguy, and M.R. Khaddar, J. Magn. Resonance 6, 325 (1972).
20. J.C. Boubel, J.J. Delpuech, M.R. Khaddar, and A. Peguy, Chem. Commun., 1263 (1971).
21. D. Canet, J.J. Delpuech, M.R. Khaddar, and P. Rubini, J. Magn. Resonance, 15, 325 (1974).
22. D. Canet, J.J. Delpuech, M.R. Khaddar, and P. Rubini, J. Magn. Resonance, 9, 329 (1973).
23. J.J. Delpuech, M.R. Khadar, A. Peguy, and P. Rubini, Chem. Commun., 154 (1974).
24. J.J. Delpuech, M.R. Khaddar, A. Peguy, and P. Rubini, in the press.
25. C.H. Langford and H.B. Gray, "Ligand substitution Processes", Benjamin, New York, 1966.
26. F. Basolo and R.G. Pearson, "Mechanisms of Inorganic Reactions" Wiley, New York, 1958.
27. J. Crea and S.F. Lincoln, J.C.S. Dalton, 2075 (1973).
28. D. Fiat and R.E. Connick, J. Amer. Chem. Soc., 90, 608 (1968).
29. W.G. Movius and N.A. Matwiyoff, Inorg. Chem., 8, 925 (1969).

LIGAND SUBSTITUTION PROCESSES II - SOLVATION SHELLS OF PARAMAGNETIC CATIONS Ni^{2+} AND Co^{2+}

J. C. Boubel, Jean-J. Delpuech* and Alain A. Peguy

(Laboratoire de Chimie physique Organique, Equipe de Recherche Associée au CNRS, Université de NANCY I, C.O. 140, 54037 - Nancy Cédex, France)

ABSTRACT

Nuclear relaxation rates are measured on the following systems: nickel (II) and cobalt (II) perchlorates in dimethylsulfoxide (DMSO); nickel (II) perchlorate in hexamethylphosphorotriamide (HMPA). Application of pulsed nmr to chemical exchange in paramagnetic systems is discussed on these examples. Kinetic parameters are obtained for the solvent substitution reaction around the Ni^{2+} or Co^{2+} cations.

INTRODUCTION

NMR is widely used to determine relaxation times of organic solvents containing paramagnetic cations[1-4], and therefore substitution rates between bylk and bound solvent molecules (S_f and S_b). The table 1 shows a summary of the main results in this field[5]. From these data, we observe that cations may be classified according to the mobility of their solvation shells, whatever the solvent. This fact may probably be assigned to a predominant crystal field energy contribution to the activation energy[18]. This paper deals with two octahedral solvates of dimethylsulfoxide around Ni^{2+} and Co^{2+} cations; and with tetrahedral solvate $Ni(HMPA)_4^{2+}$.

GENERAL CONDITIONS[6]

The conditions for nmr work are very special for these solutions. A strong interaction takes place between nuclear and electronic spins[19], thus bringing forth very short relaxation times T_1 and T_2, and therefore a considerable line broadening.

Wyn-Jones (ed.), Chemical and Biological Applications of Relaxation Spectrometry, 323–331.

TABLE 1 - Solvation of paramagnetic cations. Ligand exchange rate constants : k_{25^oC} (sec^{-1}; 1st line); $\Delta H^{\neq}$ (kcal. $mole^{-1}$; 2nd line); $\Delta S^{\neq}$ (e.u.; 3rd line).

Solvent / Ion	H_2O	DMSO	DMF	CH_3CN	MeOH	EtOH
Ni^{2+}	$2.7x10^4$ 11.6 0.6 (4)	$1.14x10^4$ 13.4 5.2 (6)	$3.8x10^3$ 15 8 (7)	$3.9x10^3$ 10.9 - 8.8 (8)	$1.0x10^3$ 15.8 8 (9)	$1.1x10^4$ 10.8 - 4 (10)
Co^{2+}	$1.1x10^6$ 8.0 - 4.1 (4)	$3.4x10^5$ 9.1 - 2.4 (6)	$3.9x10^5$ 13.6 12.6 (7)	$1.4x10^5$ 8.1 - 7.5(8)	$1.8x10^4$ 13.8 7.2 (9)	
Mn^{2+}	$3.1x10^7$ 8.1 2.9 (4)				$3.7x10^5$ 6.2 - 12.0 (11)	
Fe^{2+}	$3.2 x 10^6$ 7.7 - 3 (4)				$5.0 x 10^4$ 12.0 3 (11)	
Fe^{3+}	$1.5 x 10^2$ - - (12)	$5.0x10^1$ - - (13)	$3.3x10^1$ 12.5 - 10 (10)	<40 (10)	$2.4x10^3$ 10.7 - 7 (10)	$2.0x10^4$ - - (10)
Cu^{2+}	10^4 and $2x10^8$ 11 and 5 - 4 and -4 (4)					
V^{2+}	9 and 10^1 16.4 5.5 (14)					
VO^{2+}	$7.4x10^2$ 13.3 - 1.5 (15)	$>1.5x10^3$ (15)	$5.7x10^2$ 7.2 - 16 (16)	$2.85x10^3$ 7.0 - 20 (15)	$3.3x10^2$ 12 - 1.3(15)	
Ti^{3+}	$1.0x10^5$ 6.2 - 15 (17)				$1.9x10^5$ 3.3 -24 (17)		

Accordingly, it is necessary to use very small quantities of paramagnetic cation. The solvation shell always contains a very small fraction of the total number of solvent molecules. In this manner, only one line is observed for the solution being studied. This line is usually broad and shifted from the signal of the pure solvent.

The free solvent S_f is a diamagnetic site D and the bound solvent a paramagnetic site M. Let us recall that two nuclear sites (D and M) are characterized by three quantities (cf. Part I):

(a) The frequency shift: $\Delta\nu = \nu_D - \nu_M$(Hz) or $\Delta\omega_M$ (radians sec^{-1}) between the two sites. $\Delta\omega_M$ is called the knight shift and is proportional to the reciprocal of temperature, according to:

$$\frac{\Delta\omega_M}{\omega} = - \frac{Ahg_e\beta_e}{g_n\beta_n} \qquad \frac{S(S+1)}{3\,k\,T} \qquad (1)$$

where : A = hyperfine coupling constant; g_e, g_n: electronic and nuclear Landé factors: β_e, β_n: electronic and nuclear magnetons; h and k: Planck and Boltzmann constants; S: electronic spin number of the cation. In case of fast exchange, only one line is observed, therefore shifted by a quantity:

$$\Delta\nu = \nu - \nu_D = P_M\Delta\nu_M \quad \text{(contact shift)} \quad (2)$$

This law allows computing $\Delta\omega_M$ from $\Delta\omega$ at high temperatures (fast exchange), and then to extrapolate $\Delta\omega_M$ at any temperature T and nmr frequency ω by means of equation (1).

(b) The relaxation times T_1 and T_2 for these two sites. Values T_{1D} and T_{2D} are those of the pure solvent. Relaxation times T_{1M} and T_{2M} are very short on account of electron-nucleus interactions. The linewidth for the bound solvent is therefore very large (10^2 to 5 x 10^2 Hz). In fact, T_{1M} and T_{2M} are unknown and difficult to extrapolate.

(c) The molar fractions P_D and P_M:

$P_D \simeq 1$ and $P_M = \frac{n_M C_M}{[S]}$, where n_M is an assumed solvation number, C_M and [S] the concentrations of the salt and the pure solvent.

It is difficult to operate with those solutions by the same methods as those described for diamagnetic cations solvation. CW-nmr is not the most adequate technique, except for measuring $\Delta\omega$, because lines are often too broad to allow accurate measurements. The most reliable method consists in using simultaneously both CW - nmr to measure $\Delta\omega$, and pulsed nmr to measure T_1 and T_2.

Relaxation times of the ^{1}H (DMSO) or ^{31}P (HMPA) nuclei use 180° - 90° pulsed sequences for T_1, or Carr - Purcell - Gill - Meiboom sequences for T_2. These measurements were carried out on a Bruker B - KR 303s spectrometer, with a pulse-gated integrator to improve the signal/noise ratio, at 8, 14 and 60 MHz. Contact shifts were measured at 60 MHz on a jeol C - 60 apparatus and at 90 MHz on a Bruker HX - 90 spectrometer.

To go further, we must now examine how to relate these values to quantities of interest, i.e. to chemical rate - constants for the ligand exchange between free and bound solvent molecules : $S_b(M) \xrightarrow{k=1/\tau_M} S_F(D)$.

THE CHEMICAL EXCHANGE:

Let us define three quantities :

$$\frac{1}{T_{1r}} = \frac{1}{P_M}\left(\frac{1}{T_1} - \frac{1}{T_{1D}}\right) \qquad (3)$$

$$\frac{1}{T_{2r}} = \frac{1}{P_M}\left(\frac{1}{T_2} - \frac{1}{T_{2D}}\right) \qquad (4)$$

$$\Delta\omega_r = \Delta\omega/P_M \qquad (5)$$

These three quantities are built using the measured values : T_1, T_2, T_{1D}, T_{2D}, and P_M. They are connected to the four unknowns[1]: T_{1M}, T_{2M}, τ_M (or $1/k$), $\Delta\omega_M$, by a set of equations due[4] to Swift and Connick:

$$\frac{1}{T_{1r}} = \frac{1}{T_{1M}+\tau_M} \qquad (6)$$

$$\frac{1}{T_{2r}} = \frac{1}{\tau_M}\left[\frac{1/T_{2M}^2 + 1/T_{2M}\tau_M + \Delta\omega_M^2}{(1/T_{2M} + 1/\tau_M)^2 + \Delta\omega_M^2}\right] \qquad (7)$$

$$\Delta\omega_r = \frac{\Delta\omega_M}{(\tau_M/T_{2M} + 1)^2 + \tau_M^2\,\Delta\omega_M^2} \qquad (8)$$

Before discussing the solution of these equations, let us remark that their validity implies that $1/T_{1r}$ and $1/T_{2r}$ are constant for various values of the molar fraction P_M. Therefore these quantities are actually obtained by a plot of $1/T_1$ and $1/T_2$ as a function of P_M, and computed from the slope of the resulting straight lines[6]. The accuracy is: 1% for $1/T_{1r}$ and 3% for $1/T_{2r}$.

Coming back now to the solution of the above equations, there are more unknowns than relationships : 4 against 3. However, T_{1M} and T_{2M} actually depend upon common correlation times τ_o [1,21-23]. In the so called conditions of motional narrowing : $T_{1M} = T_{2M}$. This assumption allows a numerical computation of T_{1M} (or T_{2M}), τ_M and $\Delta\omega_M$. The accuracy on these values is discussed in a full paper[6] : 1%, 1% and 5%, respectively.

THE Co^{2+}/DMSO SYSTEM

A solvation number of 6 was assumed on the following grounds:[24]

(a) Complexes Co $(ClO_4)_2$, 6 DMSO existing in the solid state

(b) A solvation number of 6 in a mixture : DMSO/CH_2Cl_2 measured by cw - nmr[25].

(c) The electronic spectrum characteristic of the octahedral Co^{2+} cation.

The results are summarised in Table 2. T_{1M} is effectively independent from the nmr frequency. An Arrhenius plot leads to the following parameters:
$k_{25°C} = (3.40 \pm 0.4) \times 10^5\ sec^{-1}$; $\Delta H^{\neq} = 9.1 \pm 1.0$ kcal.mole^{-1}; $\Delta S^{\neq} = -2.4 \pm 3.0$ e.u.

These measurements are only possible in the pure solvent, and do not yield the rate-law, in contrast with the Al^{3+} and Be^{2+} solvates examined in Part I. Nothing sure can therefore be said about the mechanism of substitution of DMSO molecules around the Co^{2+} cation.

THE Ni^{2+}/DMSO SYSTEM.[6]

The results are summarised in Table 3. $1/T_{1M}$ increases with frequency. We must then revise our assumption : $T_{1M} = T_{2M}$, and therefore, examine the relaxation processes in the paramagnetic site.

The main contribution to nuclear relaxation arises from a dipole interaction between nuclear and electronic spins (The scalar contribution may be shown to be negligible). Its value is given by equations due to Solomon and Bloembergen[1,21,22,26-28]:

$$\left(1/T_{1M}\right)_{DD} = K_{DD}\left(3\,T_{1e} + \frac{7\,T_{2e}}{1+\omega_e^2\,T_{2e}^2}\right)$$

$$\left(1/T_{2M}\right)_{DD} = \frac{K_{DD}}{2}\left(7\,T_{1e} + \frac{13\,T_{2e}}{1+\omega_e^2\,T_{2e}^2}\right)$$

where:
$K_{DD} = 1.1 \times 10^{13}\ sec^{-2}$ for an assumed geometry of the complex T_{1e} and T_{2e} : Longitudinal and transversal relaxation times of the electronic spin.
ω_e : resonance frequency of the electron.

The assumption : $T_{1M} = T_{2M}$ demands that : $\omega_e^2\,T_{2e}^2 \ll 1$ and $T_{1e} = T_{2e}$. The observed inequality between these two relaxation times implies that : $\omega_e^2 T_{2e}^2 \simeq 1$. However such a hypothesis could only account for decreasing values of $1/T_{1M}$ with frequency, in sharp contrast with experimental data. We must therefore admit that T_{1e} and T_{2e} are a function of the frequency, according to a theory due to McLachlan[29]:

TABLE 2 - Measured values for the Co $(ClO_4)_2$/DMSO system (From Ref. 6)

Operational frequency	T/°C	$\left(\frac{1}{T_{1r}}\right)/s^{-1}$	$\left(\frac{1}{T_{2r}}\right)/s^{-1}$	$\Delta\omega_M$ (Hz)	$\frac{1}{T_{1M}} = \frac{1}{T_{2M}}/s^{-1}$	$\frac{1}{\tau_M}/10^{-6}s^{-1}$
60 MHz	15	107.2	171.5	632	107.2±1.0	2.45±0.20
	20	102.0	151.9	620	102.0±1.0	3.0±0.3
	25	100.2	143.6	612	100.2±1.0	3.4±0.4
	35	96.9	118.9	593	96.9±1.0	6.3±0.8
	45	90.6	101.8	573	90.6±1.0	11.6±3.0
	70	79.7	79.8	532	79.7±1.0	——
14 MHz	15	112.9	113.8	147	112.9±1.0	
	35	98.5	97.9	138	98.5±1.0	
	70	81.5	82.7	124	81.5±1.0	
8MHz	15	116.1	116.9	84.3	116.1±1.0	
	35	98.3	——	79.1	98.3±1.0	

TABLE 3 - Measured values for the Ni $(ClO_4)_2$/DMSO system (From Ref. 6)

Operational Frequency	T/°C	$\left(\frac{1}{T_{1r}}\right)/s^{-1}$	$\left(\frac{1}{T_{2r}}\right)/s^{-1}$	$\Delta\omega_M$ (Hz)	$\frac{1}{T_{1M}} = \frac{1}{T_{2M}}/s^{-1}$	$\frac{1}{\tau_M}/10^{-4}s^{-1}$
60 MHz	15	422	1558	533	452±5	0.65±0.10
	25	413	1200	515	429±4	1.10±0.10
	36	382	690	496	388±4	3.0±0.3
	41	384	607	489	387±4	4.1±0.5
	45	357	512	482	359±4	5.8±0.6
	61	359	427	459	360±4	12.2±3.5
	80	326	333	435	——	——
14 MHz	15	255	346	124	265±5	0.58±0.07
	25	258	290	120	264±5	1.30±0.40
	45	257	288	112	258±5	——
8 MHz	15	241	292	70	250±3	
	25	244	258	69	249±3	
	45	249	285	64	250±3	

$$1/T_{1e} = \frac{\Delta^2\tau}{10}\left[\frac{2}{1+\omega_e^2\tau^2} + \frac{8}{1+4\omega_e^2\tau^2}\right]$$

$$1/T_{2e} = \frac{\Delta^2\tau}{10}\left[3 + \frac{5}{1+\omega_e^2\tau^2} + \frac{2}{1+4\omega_e^2\tau^2}\right]$$

where:
Δ^2 is the trace of the square of the zero-field splitting tensor; τ is a correlation time that characterizes the frequency of impacts of free DMSO molecules against the solvated cation[22]:

From the variation of $1/T_{1M}$ with frequency, we may compute:

(a) $\Delta = 1.90 \pm 0.15\ cm^{-1}$, a value close to that obtained by e.s.r. on the solid salt $NiK_2(SO_4)_2 \cdot 6\ H_2O$.[30]

(b) $\tau = 3.8 \times 10^{-12}$ sec at 25°C, a value close to that obtained by e.s.r. with aqueous solutions of Fe^{3+}, Mn^{2+}, Cr^{3+} cations[22].

(c) In turn, a revised value of $1/T_{2M}$ and therefore of $1/\tau_M$: $1/T_{2M}$(25°C; 60 MHz) = 461 sec^{-1} (against 429 sec^{-1}); $k_{25^oC} = 1.14 \times 10^4\ sec^{-1}$ (vs. 1.10×10^4); $\Delta H^{\neq} = 13.4 \pm 0.6$ kcal. $mole^{-1}$ and $\Delta S^{\neq} = 5.2 \pm 1.5$ e.u. (vs. 12.3 and 1.5 respectively).

Rates for Co^{2+} cation are thirty times as large as those for Ni^{2+} cation. This result is also observed for other solvents (Table 1), and for other complexation reactions in aqueous solutions. The difference arises from a decrease of the activation enthalpy, generally assigned to a decreased crystal field activation energy in SN_2- type reactions on going from a d^7 to a d^8 electronic configuration[39].

THE Ni^{2+}/HMPA SYSTEM.[31]

This system offers the same peculiarities as the previous one. The points of interest are:

(a) The study uses ^{31}P nmr at 8 and 14 MHz.

(b) This a first example of a substitution rate on a paramagnetic tetrahedral[32] nickel (II) complex: $Ni\ (HMPA)_4^{2+}$

This point is of fundamental importance, since it may be inferred from crystal field theory that the crystal field activation energy is very low for such an arrangement[18,33]. Effectively, the kinetic parameters are the following ones:

$k_{25^oC} = 9.68 \times 10^5\ sec^{-1}$
$\Delta H^{\neq} = 0.87 \pm 0.5\ kcal.mole^{-1}$
$\Delta S^{\neq} = -28.2 \pm 1.5$ e.u.

Other parameters are:
$\Delta\omega_M = 41.81$ kHz, at 25°C and 14 MHz.
$\tau = 5.6 \times 10^{-12}$ at 25°C
$\Delta = 3.3\ cm^{-1}$

CONCLUSIONS

Nuclear relaxation measurements allow obtaining kinetic data for paramagnetic systems. Unlike the case for diamagnetic cations, measurements of relaxation rates can be more easily interpreted in terms of molecular motion. This explains the interest for these paramagnetic cations, which are used as probes to obtain kinetic and structural information, chiefly when applied to biochemical systems[34].

ACKNOWLEDGEMENTS

Financial support from the Centre National de la Recherche Scientifique and from the Direction des Recherches et Moyens d'Esssais is gratefully acknowledged. We thank Pr. Rivail for the pulsed nmr spectrometer.

REFERENCES

1. N. Bloembergen, E.M. Purcell, and R.V. Pound, Phys. Rev., 73, 679 (1948).
2. I. Solomon, Phys. Rev., 99, 559 (1955).
3. N. Bloembergen, J. Chem. Phys., 27, 572 (1957).
4. T.J. Swift and R.E. Connick, J. Chem. Phys., 37, 307 (1962).
5. J.J. Delpuech, A. Peguy and M.R. Khaddar, J. Electroanal. Chem. 29, 31 (1971).
6. J.C. Boubel and J.J. Delpuech, Molec. Phys., 27, 113 (1974).
7. N.A. Matwiyoff, Inorg. Chem., 5, 788 (1966).
8. N.A. Matwiyoff and S.V. Hooker, Inorg. Chem., 6, 1127 (1967).
9. Z. Luz and S. Meiboom, J. Chem. Phys., 40, 2686 (1964).
10. F.W. Breivogel Jr., J. Phys. Chem., 73, 4203 (1969).
11. F.W. Breivogel Jr., J. Chem. Phys., 51, 445 (1969).
12. M.R. Judkins, Ph.D. Thesis, Lawrence Radiation Lab. Rept. UCRL - 17561, 1961, quoted in Ref. -1.
13 C.H. Langford and F.M. Chung, J. Amer. Chem. Soc., 90, 4485 (1968).
14. M.V. Olsen, Y. Kanazawa and H. Taube, J. Chem. Phys., 51, 289 (1969).
15. N.S. Angerman and R.B. Jordan, Inorg. Chem., 8, 65 (1969).
16. N.S. Angerman and R.B. Jordan, J. Chem. Phys., 48, 3983 (1968).
17. A.M. Chmelnick and D. Fiat, J. Chem. Phys., 51, 4238 (1969).
18. F. Basolo and R.G. Pearson, "Mechanisms of Inorganic Reactions" Wiley, New York, 1958.
19. A. Abragam, "The principles of nuclear magnetism", Clarendon Press, Oxford, 1961.
20. A. Carrington, and A.D. McLachlan, "Introduction to Magnetic Resonance", Chap. 8 and 10, Harper International Edition, 1967.

21. I. Solomon and N. Bloembergen, J. Chem. Phys., 25, 261 (1956).
22. M. Rubinstein, A. Baram and Z. Luz, Molec. Phys., 20, 67 (1971).
23. H.G. Hertz, "Progress in NMR Spectroscopy", Vol. 3, (W. Emsley), J. Feeney and L.H. Sutcliffe, Eds.), Pergamon Press, 1960.
24. F.A. Cotton and R. Francis, J. Amer. Chem. Soc., 82, 2986 (1960).
25. L.S. Frankel, Inorg. Chem., 10, 814 (1971).
26. L.O. Morgan and A.W. Nolle, J. Chem. Phys., 31, 365 (1959).
27. A.W. Nolle and L.O. Morgan, J. Chem. Phys., 26, 642 (1957).
28. N. Bloembergen and L.O. Morgan, J. Chem. Phys., 34, 842 (1961).
29. A.D. McLachlan, Proc. R. Soc., A, 280, 271 (1964).
30. P.L.Z. Mukherjee, Kristallographie, 91, 504 (1935).
J.H.E. Griffiths and J. Owen, Pric. R. Soc. A, 213, 459 (1952; M. Date, Sci. Rep. Res. Inst., Tohottu Univ., A, 6, 390 - (1954); S. A. Altschuler and B. M. Kosyrew, "Election Paramagnetic Resonance", Academic Press, New York, 1964.
31. J.C. Boubel, J.J. Delpuech and A. Peguy, in the press.
32. J.T. Donaghue and R.S. Drago, Inorg. Chem., 1, 866 (1962), Ibid., 2, 572 (1963); Ibid, 2, 1158 (1963).
33. L.E. Orgel, "An Introduction to Transition - Metal Chemistry; Ligand Field Theory", Methuen and Co. Ltd., London, 1960, W. Manch and W.C. Fernelius, J. Chem. Educ., 38, 192 (1961); C. J. Ballhausen and C. Klixbull Jørgensen, Kgl. Danske Videnskab. Selskab. Mat. fys. Medd., 29, 14 (1955).
34. R.A. Dwek, "Nuclear Magnetic Resonance in Biochemistry : Applications to Enzyme Systems", Clarendon Press, Oxford, 1973.

ULTRASONIC STUDIES OF POLYION - COUNTERION INTERACTIONS IN POLYELECTROLYTES SOLUTIONS

R. ZANA and C. TONDRE

C.R.M., C.N.R.S., 6, rue Boussingault, STRASBOURG, FRANCE

I - INTRODUCTION

Polyelectrolytes are polymers whose monomer units include one or several ionizable groups. Two essential differences exist between polyelectrolytes and simple electrolytes in solution. First, interionic interactions are always reduced upon dilution in the latter but not in the former because the distance between like charges on the polymer chain is determined by bond distances and angles and not by the overall polyelectrolyte conformation. Second, the electrostatic potential describing the interactions between ions of opposite charge is of spherical symmetry for simple electrolytes and of cylindrical symmetry with polyelectrolytes.[1]

At high polyion charge density, the cylindrical potential gives rise to the so-called condensation phenomenon: part of the counterions condense on the polyion in order to reduce its charge parameter to a value just below 1.[1]

The fraction ρ of condensed counterions can be calculated from Manning's theory.[1] On the other hand, the interaction between uncondensed counterions and polyions can be accounted for in terms of Debye-Huckel potential.[1]

There are however a number of questions that Manning's phenomenological theory cannot answer. The first one concerns the fraction of bound counterions and of polyion charged sites which undergo a change of hydration upon binding. There is now a large body of results[2-4] which do show that counterion binding is accompanied by a volume change which is usually larger than that associated with ion-pair formation in solution of simple electrolytes, owing to the larger electrostriction of polyions as compared with simple ions. The second question concerns the rate of exchange between the different types of counterions (condensed

Wyn-Jones (ed.), Chemical and Biological Applications of Relaxation Spectrometry, 333–341.

hydrated, condensed unhydrated, uncondensed).

We thought that ultrasonic absorption methods would prove useful in providing us some answers to the above questions for the following reasons:

(1) Counterion binding by polyion gives rise to a sizeable volume change, therefore the necessary condition is fulfilled for this process to give rise to a relaxational ultrasonic absorption.[5]

(2) Counterion binding by polyion must be very similar to ion-pair formation in solution of simple electrolytes.

(3) Ultrasonic absorption methods have proved very successful in solving problems such as those raised above, in the case of ion-pair formation in solutions of simple electrolytes.[5]

II - REVIEW OF THE ULTRASONIC ABSORPTION STUDIES DEALING WITH COUNTERION-POLYION INTERACTION IN POLYELECTROLYTE SOLUTIONS

The first evidence of a countribution of counterion binding to the absorption of a polyelectrolyte solution was reported in 1970, in a study dealing with polyethylenesufonic (PES) acid.[6a] It was observed that the absorption of the polyacid solution does not change upon neutralization by tetramethylammonium (TMA)-hydroxyde and shows a large increase upon neutralization by NaOH. Such results can be easily understood in terms of counterion binding. Indeed, the binding of the protons initially present into the solution by the PES polyanions gives rise to a negligible volume change and, in turn, to a negligible excess absorption. The same is true for TMA^+ because the resulting effect of this ion on the volume of the surrounding water molecules is negligible.[7,8] Therefore, no change of absorption can occur upon neutralixation of PES-acid by TMA hydroxyde because in this process protons are replaced by TMA^+. On the contrary, the binding of Na^+ by PES^- gives rise to a sizeable volume change[2c,2d] and in turn to a significant excess absorption. Similar results have been obtained with a large number of polyacids.[6b,6f] It has been found that upon neutralization by TMA-hydroxyde the excess absorption of strong polyelectrolytes does not change and that of weak polyelectrolytes changes only slightly.[6] Likewise, it has been reported that no excess absorption occurs upon chelation of TMA^+ by strong chelating agents in contrast to what is found upon chelation of alkali metal ions by the same compounds.[8] This led us to use TMA^+ as a reference ion to separate the contribution due to counterion binding from the contributions due to other processes. The absorption due to the binding of a counterion C by a polyion P has been taken equal to the difference between the absorptions of equimolar solutions of the polysalts C-P and TMA-P. This definition implies that the contributions of processes other than site binding[9] are not modified when replacing C^+ by TMA^+. This assumption appears to be valid to a very good approximation, at least at low frequency where the contribution of counterion binding is much larger than that of the other processes, essentially those involving segmental motion of the polyion backbone and/or of the side group of the polymer molecule.[10] Indeed the absorption of a polysalt C-P solution was changed only very slightly upon addition of C^+ halides.[6b,6c] Also, such additions were found to result in negligible volume changes[2a] although they are known to bring about drastic changes of the polysalt intrinsic viscosity and thus of the chain stiffness.[10]

Systematic studies of the excess absorption due to counterion binding yielded the following results:

(1) The absorption is proportional to the polysalt concentration c and increases with the polyion charge density.[6]

(2) The ionic sequences obtained when writing the alkali metal ions in the order of increasing absorption[6b,6c] are very close to those obtained when measuring volume changes upon binding.[2c]

(3) The results obtained with carboxymethylcellulose of various substitution degree[6c] and a copolymer of maleic acid and methylvinylether[6b] clearly show that the absorption due to counterion binding only occurs when the charge parameter of these polyions becomes larger than one, as to be expected from the condensation theory.[1]

(4) Counterion binding by polyion is a multistep process. In the frequency range between 1 and 155 MHz, the curves $\Delta\alpha/f^2$ vs f ($\Delta\alpha$ = difference between the absorption coefficient α of equimolar solutions of polysalts C-P and TMA-P; f = ultrasonic frequency) for a series of polysalts could be fitted by the relaxation equation (1). The values of the relaxation frequencies f_1 and f_2, relaxation amplitudes A_1 and A_2 and of the constant B are listed in Table I for polyethylenesulfonates and polyphosphates (PP).

$$\frac{\Delta\alpha}{f^2} = \frac{A_1}{1 + f^2/f_1^2} + \frac{A_2}{1 + f^2/f_2^2} + B \qquad (1)$$

For all the polysalts except Co-PP, the B values are greater than the experimental error on $\Delta\alpha/f^2$ at high frequency ($\pm 2.10^{-17}$ cm^{-1} sec^2). This may indicate either the existence of an additional relaxation process at f > 100 MHz, or more likely, the failure of our method of evaluating the contribution of counterion binding at high frequency. Indeed, this contribution is then rather small and thus, comparable to the absorption due to other processes.[9,10] These processes may then make different contributions to the absorption of equimolar solutions of C-P and TMA-P.

From the data listed in Table I, it appears that within the experimental error A_1 and A_2 are proportional to c and f_1 and f_2 are indenpedent of c. These results are consistent with that reported under point (1) above. On the other hand, the low frequency relaxation process appears to be mostly dependent on the nature of the counterion and practically independent of the polyion while the opposite is true for the high frequency relaxation process.

Table I : Values of the relaxation parameters[(a)]

Polysalt	concentration equiv./liter	10^{17} A_2 cm^{-1} s^2	f_2 MHz	10^{17} A_1 cm^{-1} s^2	f_1 MHz	10^{17} B cm^{-1} s^2
Li - PP	0.115	10	5	30	18	8
Na - PP	0.033	53	4	32	14	3
"	0.062	87	3.7	60	19	5.5
"	0.107	120	4.5	105	22	10
K - PP	0.117	10	6.5	30	16	4
Rb - PP	0.110	-	-	17	17	3
Mn - PP	0.071	620	1.4	30	12	6
Co - PP	0.068	500	1.05	44	11	1
Co - PP	0.125	700	1.4	70	10	2
	"	950	1.05	90	10	2
	"	1500	0.72	100	10	2
	"	1000	1.05	75	12	2
Na - PESA	0.095	22	5.5	12	45	2.5
"	0.19	40	6	18	40	6
"	0.36	75	6	42	50	9

(a) The above values were determined by a trial and error procedure, in order to best fit the experimental results. An indication on the accuracy on the values of A_1, A_2, f_1 and f_2 is given by the four sets of values of the relaxation parameters for Co-PP 0.125 N, which all fit the results within experimental error, with the second set giving the best fit.

All the above results are strikingly similar to those reported for ion-pair formation in solutions of divalent sulfates[5,12-15].

Note that in other studies,[16] no clear conclusion could be reached concerning the characteristics of the excess absorption due to polyion-counterion interaction. However, the authors attributed to counterion binding the total excess absorption of the polysalts investigated with respect to solvent, i.e., water, although it is known[6c,6d] that for polycarboxylates this quantity also includes contributions due to other equilibria. This provides a reasonable explanation for the difference between their results and ours.

III - A KINETIC MODEL OF COUNTERION BINDING BY POLYIONS

In view of the above results the following scheme has been proposed for counterion binding by polyions.[6e,6f]

$$C + P \rightarrow CP_1 \underset{k_2}{\overset{k_1}{\rightleftharpoons}} CP_2 \underset{k_4}{\overset{k_3}{\rightleftharpoons}} CP_3 \qquad (2)$$

This mechanism is very similar to that originally proposed for ion-pair formation in solution of divalent sulfates.[12] However, the reaction $C + P \rightarrow CP_1$ has been written as an irreversible reaction in order to include in the above model the counterion condensation,[1] which is specific to polyelectrolyte solutions. The condensed counterions are those in states CP_1, CP_2, and CP_3. They probably exchange very fast with free counterions but, at any moment, the fraction of condensed counterions is constant and independent of the polyion concentration if this concentration is not too large.

In reaction (2), the CP_i's represent different states of hydration of the complex formed by one counterion C and that part of the polyion where it is fixed. As in ion-pair formation, CP_2 and CP_3 are the outer-sphere and inner-sphere complexes, respectively, between C and the "ligand" P. The high and low frequency relaxation processes are assigned to the formation of CP_2 and CP_3, respectively. In state CP_1, the hydration shells of C and P are in contact but still intact. Only three states of hydration appear in reaction (2) because our data in the range 1 - 155 MHz indicate two relaxation processes.

The expressions of the relaxation amplitudes A_1 and A_2 for the system of coupled reactions (2) have been obtained[6e] using the matrix method,[17,18] and modified as to include only three unknown quantities; the rate constant k_1 (see eq. (2)) the sum $k_1 + k_2$ and the volume change ΔV_{21} upon outer sphere complex formation. Note that these expressions include also the fraction ρ of condensed counterions which can be calculated from Manning's treatment,[1] the total volume change δV_{CP} upon binding which can be determined from density measurements or dilatometry[2d,2h] and the relaxation frequencies f_1 and f_2 which can be determined experimentally.

The results relative to Co-PP have been used in an attempt to solve the two equations relating A_1 and A_2 to k_1, $k_1 + k_2$ and ΔV_{21}, because for this system the experimental values of A_1, A_2 and δV_{CP} are large and, thus, fairly accurately determined. Also NMR studies on either $CoSO_4$[19] or Co-PP[2h] provided us with the additional result which was required to permit the resolution of the above equations (see reference 6e). Indeed, NMR studies[19,20] on Co^{2+} complexation by small ligands have yielded the value of k_3, rate constant for the exchange of a water molecule between the inner hydration shell of the ion and the bulk of the solution. This result can be used in the calculations relative to the Co-PP solutions on the assumption that k_3 has the same value whether the substituting ligand is small or is part of a polyion. On the other hand, the fraction of Co^{2+} bound by polyphosphate with complete dehydration has been recently obtained[2h] and can be taken as equal to the fraction of Co^{2+} ions in state CP_3. Using either one of these two data made it possible to obtain for the Co-PP system the volume changes upon outer sphere and inner sphere complex formation (ΔV_{21} and ΔV_{32}), the four rate constants k_1, k_2, k_3 and k_4 and the fraction of counterions in states CP_1, CP_2 and CP_3. The values are listed in Table II for the four sets of relaxation parameters that were found to fit the ultrasonic absorption data for the Co-PP 0.125 M solution.

These results show the following salient features:

(1) The two methods of calculation yield very similar results when using the third set of relaxation parameters. Thus, within the experimental error, NMR, density and ultrasonic absorption data appear consistent.

(2) The fraction of Co^{2+} ions in state CP_1 is very small thereby indicating a very tight binding of this ion by the polyphosphate.

(3) The volume changes upon outer-sphere and inner-sphere complex formation are quite insensitive to the method of calculation and are found to be in a ratio of 4 to 5. This result is strikingly similar to that for ion-pair formation in solution of divalent sulfates, when a three state model is used[14,15]. Note however that the volume changes are much larger for Co^{2+} binding by polyphosphate than by SO_4^{2-} because of the much larger electrostriction of the polyion.

(4) Finally, the comparison of ΔV_{21} and ΔV_{32} to the values of the contributions of the Co^{2+} and polyphosphate ions to the total volume change δV_{CP} upon binding[2h] has provided us with the first experimental evidence that the two ionic species contribute to ΔV_{21} and ΔV_{32}, as this was recently theoretically predicted.[21]

CONCLUSIONS

Ultrasonic absorption methods appear as a tool capable of providing us with a number of informations on counterion binding by polyions, when used in conjunction with other methods such as densitometry or dilatometry and NMR. In view of these results a systematic ultrasonic absorption study of the binding of divalent

Table II : Results of the calculations for Co-PP 0.125 N

	$10^{17}A_1$ (cm^{-1}	$10^{17}A_2$ s^2)	f_1 (MHz)	f_2	10^7k_1	10^7k_2 (s^{-1})	10^7k_3	10^7k_4	ΔV_{21} (cm^3/mole)	ΔV_{32}	$[CP_1]/\rho c$	$[CP_2]/\rho c$	$[CP_3]/\rho c$
(a)	70	700	10	1.4	6.2	0.05_5	0.24	0.64	23	4.6	0.006	0.72	0.27
	90	950	10	1.05	6.2	0.10	0.24	0.42	22.9	4.4	0.01	0.63	0.36
	100	1500	10	0.72	6.1	0.14_5	0.24	0.21	22.3	4.2	0.01	0.47	0.52
	75	1000	12	1.05	7.4	0.11	0.24	0.42	22.8	4.3	0.01	0.63	0.36
(b)	70	700	10	1.4	6.2	0.1	0.52	0.37	22.3	3.4	0.006	0.41	0.58
	90	950	10	1.05	6.1	0.17	0.39	0.28	21.9	4.3	0.01	0.41	0.58
	100	1500	10	0.72	6.1	0.18	0.27	0.19	21.8	4.4	0.01	0.41	0.58
	75	1000	12	1.05	7.4	0.15	0.39	0.27	22.1	3.6	0.01	0.41	0.58

(a) The calculations have been performed using the value $k_3 = 0.24.10^7\ sec^{-1}$ (19)

(b) The calculations have been performed using the value $r = [CP_3]/\rho c = 0.58$ (2h)

metal ions by polyphosphate appears worthwhile.

REFERENCES

1. G. Manning, J. Chem. Phys., 51, 924, 933 and 3249 (1969); Biopolymers, 9, 1543, (1970).
2. (a) U. P. Strauss and Y. P. Leung, J. Amer. Chem. Soc., 87, 1476, (1965);
(b) A. Begala and U. P. Strauss, J. Phys. Chem., 76, 254, (1972);
(c) J. Hen and U. P. Strauss, ibid., 78, 1013, (1974);
(d) C. Tondre and R. Zana, ibid., 76, 3451, (1972);
(e) V. Crescenzi, F. Delben, S. Paoletti and J. Skerjanc, ibid., 78, 607 (1974);
(f) F. Delben and S. Paoletti, ibid., 78, 1487, (1974);
(g) F. Delben, S. Paoletti, V. Crescenzi and F. Quadrifoglio, Macromolecules, 7, 538, (1974);
(h) P. Spegt, C. Tondre, G. Weill and R. Zana, Biophysical Chem., 1, 55 (1973).
3. A. Ikegami, J. Polymer Sci., A - 2, 907, (1964) and Biopolymers, 6, 431, (1968).
4. H. Asai, J. Phys. Soc. Japan., 16, 761, (1961).
5. J. Stuehr and E. Yeager, "Physical Acoustics", Vol. IIA, W. P. Mason Ed., Academic Press, New York, N.Y., 1965, p. 351, and references therein.
6. (a) C. Tondre and R. Zana, Proc. IUPAC Symp., Leiden 1970, Vol. I, p. 387;
(b) C. Tondre and R. Zana, J. Phys. Chem., 75, 3367, (1971);
(c) R. Zana, C. Tondre, M. Rinaudo and M. Milas, J. Chim. Phys. Physicochim. Biol., 68, 1258 (1971);
(d) R. Zana and B. Michels, Proc. 7th Int. Cong. Acoustics, Budapest 1971, Ed. Akademiei Kiado, Vol. 2, p 41;
(e) R. Zana and C. Tondre, Biophysical Chemistry, 1, 367, (1974)
(f) C. Tondre and R. Zana, in "Polyelectrolytes", E. Selegny, Ed., D. Reidel Publ. Co., Holland., (1974), p. 323.
7. B. Conway, J. Desnoyers and R. Verrall, J. Phys. Chem., 75, 3031 (1971) and references therein.
8. M. Eigen, Pure Appl. Chem., 6, 97 (1963).
9. R. Zana, J. Macromol. Sci. - Rev. Macromol. Chem., in the press.
10. R. A. Pethrick, ibid., C9, 91 (1973).
11. R. W. Armstrong and U. P. Strauss, "Polyelectrolytes" in "Encyclopedia of Polymer Science and Technology", Interscience, Wiley and Sons, New York, 1969, Vol. 10, p. 781.
12. M. Eigen and K. Tamm, Z. Elektrochem., 66, 107, (1962).
13. A. Bechtler, K. Breitschwerdt and K. Tamm, J. Chem. Phys., 52, 2975 (1970).
14. L. G. Jackopin and E. B. Yeager, J. Phys. Chem., 74, 3766 (1970
15. W. Knoche, G. Firth and D. Hess, Adv. Molecular Relaxation Processes, 6, 1 (1974).
16. G. Atkinson, E. Baurngartner and R. Fernandez-Prini, J. Amer. Chem. Soc., 93, 6436 (1971) and E. Baumgartner, G. Atkinson and M. Emara, ibid., 95, 5881 (1973).

17. M. Eigen and L. de Maeyer, Techniques of Organic Chemistry, Vol. VIII, Part. 2, Interscience, New York, 1963.
18. G. Hammes and A. Park, J. Amer. Chem. Soc., 90, 4151 (1968).
19. A. Chmelnik and D. Fiat, J. Chem. Phys., 47, 3986, (1967).
20. T. J. Swift and R. E. Connick, J. Chem. Phys., 41, 2553 (1963).
21. P. Hemmes, J. Phys. Chem., 76, 895, (1972).

THE USE OF THE TEMPERATURE-JUMP METHOD IN THE STUDY OF ORGANIC REACTION MECHANISMS

Claude F. Bernasconi

Thimann Laboratories, University of California,
Santa Cruz, California 95064

Although the temperature-jump method has been around for over 15 years[1] and commercial apparatuses have been available for almost 10 years it is still not a popular tool with organic chemists. My own experience in applying this method to mechanistic and reactivity problems of organic reactions has more and more convinced me that the problem is not with the organic reactions or the temperature-jump method but with the organic chemists.

This paper is an attempt to show how well suited the method in fact is, particularly when it comes to the detailed understanding of a reaction mechanism. I shall illustrate this by choosing nucleophilic aromatic substitution reactions as an example.

The basic mechanism of nucleophilic aromatic substitutions[2] is shown in eq. 1 where S represents one or several electron

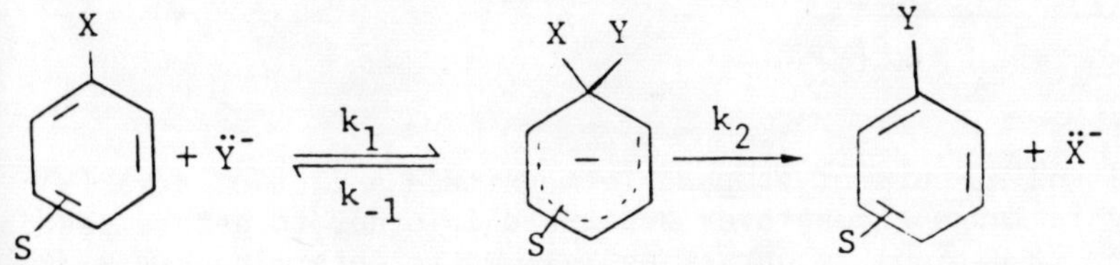

1 2 3 (1)

Wyn-Jones (ed.), Chemical and Biological Applications of Relaxation Spectrometry, 343–351.

withdrawing substituents, Y^- is an anionic nucleophile (e.g. RO^-, HO^-, RS^-, etc.) and X is the nucleofugic leaving group (e.g. Halogen, PhO, RO, etc). Since in most instances the intermediate (anionic σ-complex, Meisenheimer complex) does not accumulate (k_{-1}, $k_2 \gg k_1$) these systems are commonly treated by the steady state approximation. Thus a rate study only provides an overall rate constant, $k = k_1k_2/(k_{-1} + k_2)$. When $k_2 \gg k_{-1}$ we have $k = k_1$, when $k_{-1} \gg k_2$, $k = k_1k_2/k_{-1}$; since the rate law provides no information about the relative magnitude of k_2 and k_{-1}, these limiting conditions have to be inferred indirectly. Anyhow, k provides little information about the rates of the individual steps in the mechanism.

The situation is somewhat better when the nucleophile is a primary or secondary amine because of the possibility of base catalysis in the conversion of the intermediate to products ($k_3^B\{B\}$ step in eq. 2).

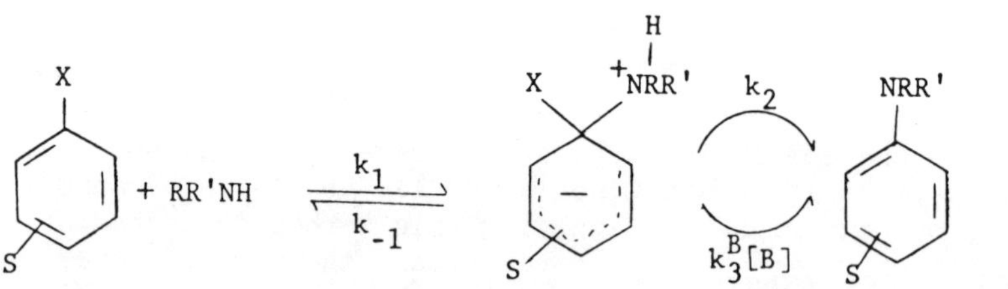

(2)

4 5 6

Here the steady state constant is given by

$$k = \frac{k_1k_2 + k_1k_3^B\{B\}}{k_{-1} + k_2 + k_3^B\{B\}} \tag{3}$$

When $k_3^B\{B\}$ and k_{-1} are of comparable magnitude a plot of k vs {B} (B usually in large excess over substrate in order to assure pseudo-first order condition) is curved and allows to determine k_1, k_2/k_{-1} and k_3^B/k_{-1}. Absence of base catalysis permits the inference $k_3^B\{B\}+ k_2 \gg k_{-1}$ ($k = k_1$) while a linear dependence of k on {B} means $k_{-1} \gg k_2 + k_3^B\{B\}$ ($k = k_1k_2/k_{-1} + k_1k_3^B\{B\}/k_{-1}$).

Some of the problems which have intrigued numerous workers in the field concern (a) the absolute magnitudes of the rate constant k_{-1}, k_2, k_3^B (diffusion controlled), (b) the dependence on structure (S, X, Y^-, RR'NH, solvent) of the various elementary rate processes, (c) the question why some reactions with amines are based catalyzed ($k_{-1} > k_2 + k_3^B\{B\}$) but others not ($k_2 + k_3^B\{B\} >> k_{-1}$), (d) the details of the mechanism of the base catalyzed step k_3^B.

Although none of these questions can be answered from steady state kinetic data, this has not prevented a lot of speculation about them. It is evident that in order to find firm answers the individual rate processes have to be measured separately. I believe our work of the last 6 years has provided some of these answers.

Apart from finding a kinetic method able to measure the processes in question -- the temperature-jump method turned out to cover the time scale of many of them -- systems had to be found where the intermediate(s) accumulate to detectable levels. Strongly electron withdrawing substituents on the aromatic substrate, strong nucleophiles and dipolar aprotic solvents are all known to increase the stability of complexes like **2** and **5** relative to their respective starting materials. However, unless the leaving group is very slow in departing from the complex, the complex may only appear as a short lived transient before being rapidly and practically irreversibly converted to the frequently thermodynamically much more stable final product, thus precluding an investigation by the standard temperature-jump technique (in contrast to the combined stopped-flow-temperature-jump technique, vide infra).

To circumvent some of these difficulties we have studied the kinetics of equilibria like **1** ⇌ **2** or **4** ⇌ **5** by using model compounds. An early model compound has been 1,3,5-trinitrobenzene (TNB) which lacks a leaving group, thus providing nicely reversible systems as shown in eq. 3 and 4.

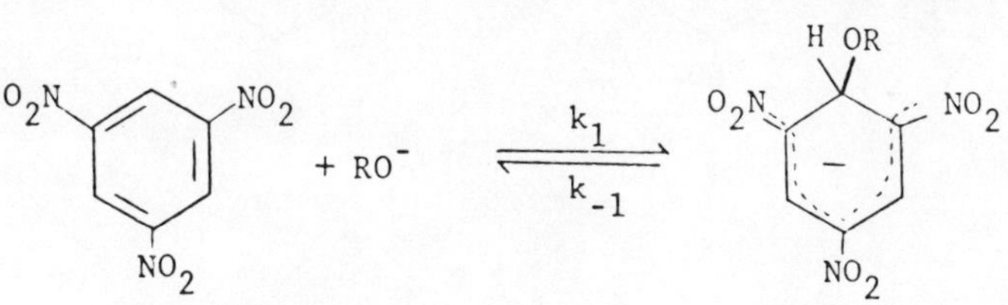

(3)

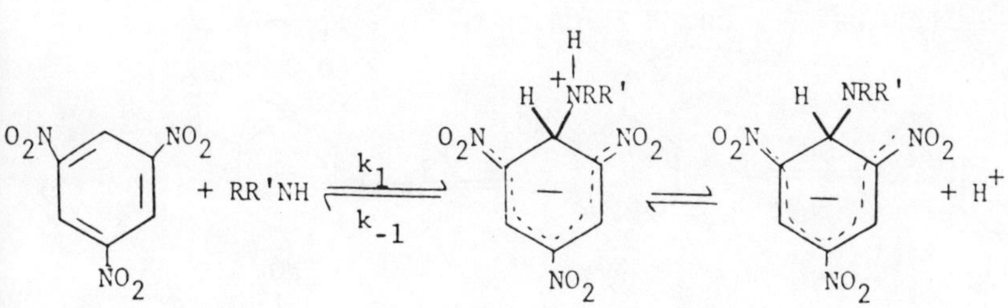

(4)

These studies have provided valuable insights into relative leaving group reactivities of alkoxide ions[3] and amines[4] from aromatic carbon. On the other hand it is relatively difficult to draw reliable conclusions from these model reactions about the absolute magnitude of say k_{-1} to be expected in "real" substitution reactions. This is because in TNB the leaving group X is replaced by a hydrogen atom, a change which has a significant electronic and probably steric effect on the reactivity towards nucleophiles.

Examples of model reactions which mimic more closely the real substitution processes and nevertheless are reversible are shown in eq. 5, 6 and 7.

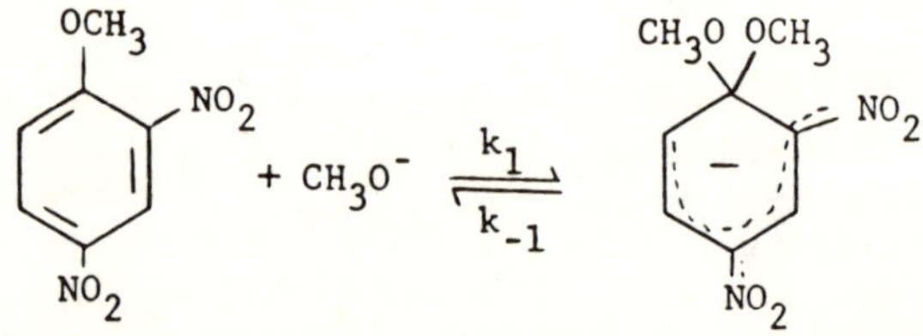

(5)

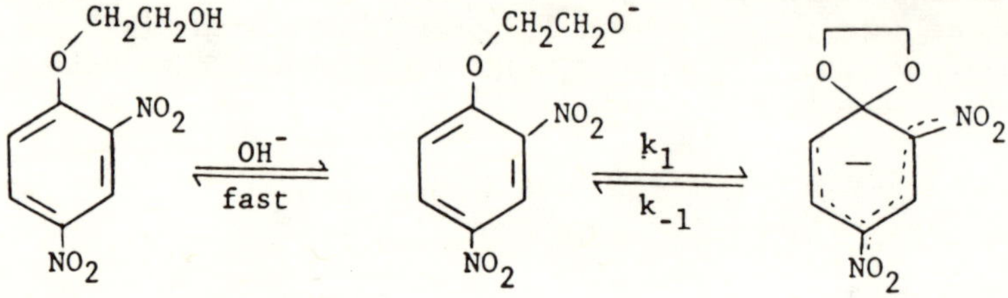

(6)

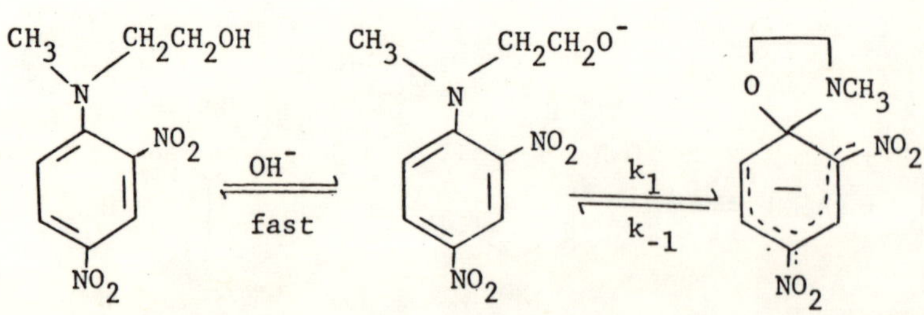

(7)

Because methoxide is a very poor leaving group, the study of reaction 5 provided an estimate of the lower limit for the rate of leaving group expulsion from σ-complexes in the popular 2,4-dinitrophenyl series[2a] (k_{-1} = 42 sec^{-1} in methanol solution[5]). From a comparison of reaction 6 with 5 we learned how intramolecularly conducted reactions affect k_1 and k_{-1}[6]; from a comparison of reaction 7 with 6, conducted in various DMSO-water mixtures[6,7], conclusions about the effect of groundstate resonance stabilization (6a ↔ 6b ↔ 6c) on k_1 and k-1 could be evaluated.[6]

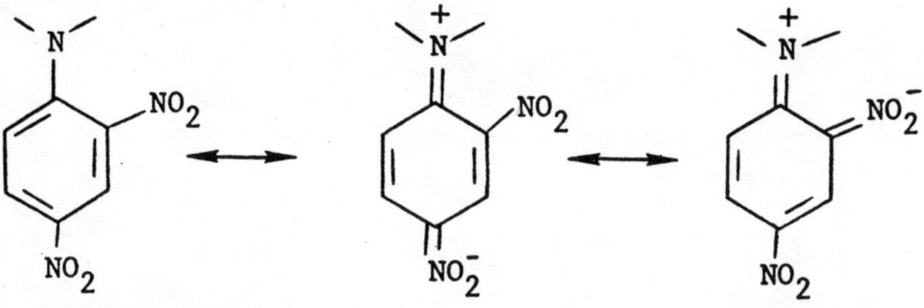

6a 6b 6c

A question of considerable interest and debate has been the mechanism of the base catalyzed conversion of 5 to 6 (eq. 2)[2a]. The mechanism which has received widest support is the one proposed by Bunnett[8] and is shown in eq. 8 where

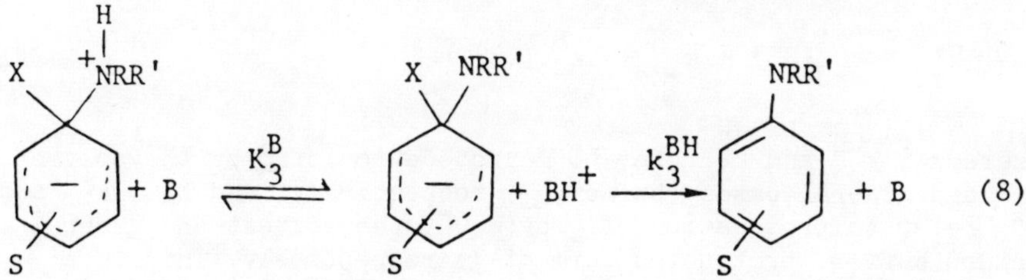

5 5a 6

the deprotonation of 5 is visualized as being a rapid equilibrium reaction followed by rate limiting, general acid catalyzed leaving group expulsion- k_3^B of eq. 2 thus becomes $K_3^B k_3^{BH}$. An earlier mechanism, also suggested by Bunnett[9], is shown in eq. 9. Here deprotonation of 5 is the rate limiting step, followed by rapid

$$5 + B \overset{k_3^B}{\rightleftharpoons} 5a \xrightarrow{\text{fast}} 6 \qquad (9)$$

leaving group expulsion. This mechanism lost favor when it became generally known that proton transfer rates are usually very fast.[10] Another mechanism, involving leaving group departure concerted with proton removal by the base has also been advocated by some authors[11] but has not been as widely accepted as scheme 8.[2a]

The study of the model reaction 10 in water[12] has shed

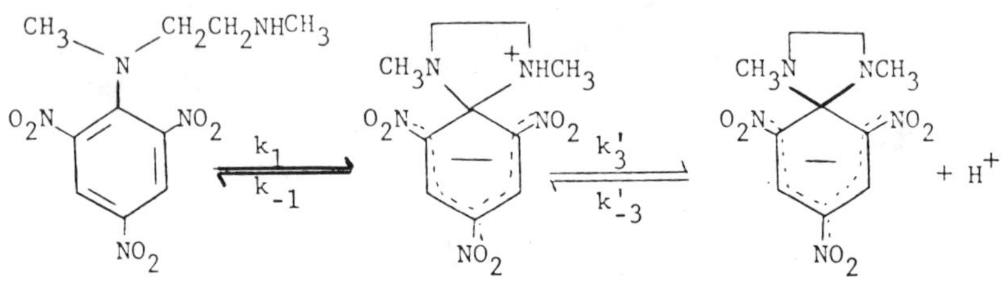

(10)

7 8 9

some light on the question of the mechanism of base catalysis. We found that the deprotonation-protonation equilibrium, $8 \rightleftharpoons 9 + H^+$, is not rapidly established on the time scale of the reaction $7 \rightleftharpoons 8$; it is kinetically significant with k_3' and k_{-3}' defined as

$$k_3' = k_3 + k_3^{OH}\{OH^-\} + \Sigma k_3^B\{B\} \qquad (11)$$

$$k_{-3}' = k_{-3}\{H^+\} + k_{-3}^{OH} + \Sigma k_{-3}^B\{BH\} \qquad (12)$$

where k_3, k_3^{OH} and k_3^B refer to deprotonation of 8 by the solvent, OH^- and general bases (buffer, 7) respectively, and k_{-3}, k_{-3}^{OH} and k_{-3}^B refer to protonation of 9 by H_3O^+, the solvent and general acids (buffer, protonated form of 7) respectively. This is a consequence of a high value of k_{-1} (2×10^5 sec^{-1}) and a low pK of 8 (6.64) which makes $k_{-1} > k_3' + k_{-3}'$ under typical reaction conditions.

By suitable extrapolation to "real" substitution reactions we come to the conclusion that deprotonation of 5 to form 5a is probably rate limiting at least in some cases and that the reverse reaction, 5a → 5 is slower than the reaction 5a → 6, at least with weakly activated systems (e.g. 4-nitro-benzene derivatives) and relatively good leaving groups. This means that the reaction 5 → 5a becomes over-all rate limiting, in disagreement with the widely accepted mechanism of eq. 8, but in agreement with the old

Bunnett mechanism of eq. 9.

Our conclusions have been reenforced by studying reactions analogous to 10 in aqueous DMSO solution but with less activated aromatic systems (2,6-dinitro-4-trifluoromethylphenyl and 2,4-dinitrophenyl derivative)[13], and also by studying the reactions of scheme 13 (in aqueous solution).[14]

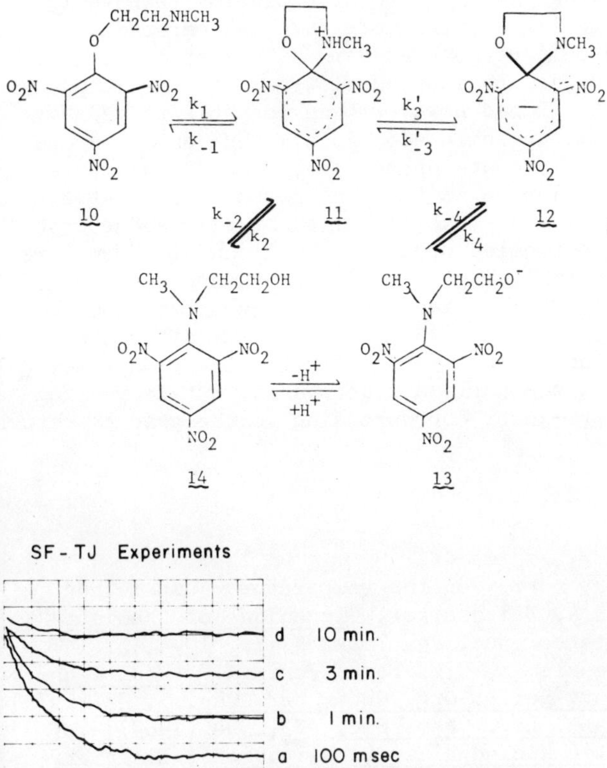

(13)

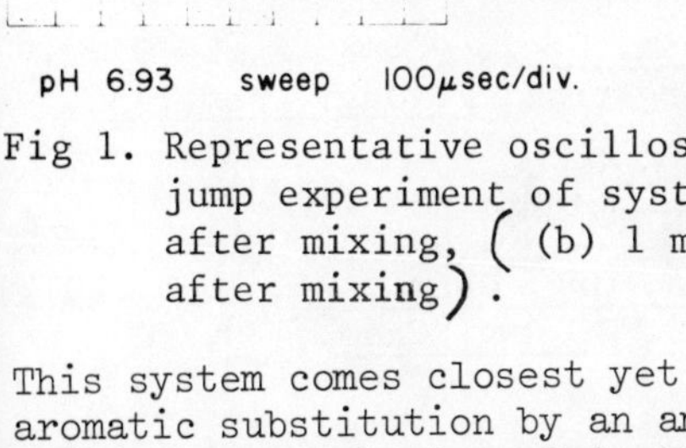

Fig 1. Representative oscilloscope traces of a stopped-flow-temperature-jump experiment of system 13; (a) temperature-jump 100 msec after mixing, ((b) 1 minute, (c) 3 minutes, (d) 10 minutes after mixing).

This system comes closest yet to mimicing a real nucleophilic aromatic substitution by an amine. We were able to evaluate all rate constants indicated in the scheme (twelve constants; k_3' and k_{-3}' account for three each) which to the best of our knowledge makes it one of the most (the most) thoroughly characterized multi-step reaction in organic chemistry.

This reaction is also interesting from a purely experimental point of view because it poses some problems not encountered with reactions 3 through 7 or 10. First there are two significant relaxation times instead of one. The first is associated with the equilibration 10 ⇆ 11 ⇆ 12 and is the exact analogue to the one for the equilibration 7 ⇆ 8 ⇆ 9 (eq. 10). The second (slower) relaxation time is associated with the equilibration 12 ⇆ 13 ⇆ 14 and is analogous to the one of system 7 except that it is coupled to the fast processes 10 ⇌ 11 ⇌ 12. The greatest experimental difficulty was in studying the first relaxation time because 13 and 14 are thermodynamically so much more favored compared to 10 that 10 (and 11) could only be generated in situ with a life time of seconds or a few minutes. By stopped-flow-temperature-jump combination experiments[1] it has however been possible to measure the first relaxation time as follows.[14] An alkaline solution made from 14 which above pH 11.5 contains mainly 12 was mixed with a slightly acidic buffer solution so that the mixture would have a pH ~ 7 to 9. At these pH-values the system eventually returns to 14 (thermodynamically controlled product) but 11 and 10 form more rapidly (kinetically controlled products)[15] and hence the equilibrium 10 ⇆ 11 ⇆ 12 established itself for a short time. Thus a temperature-jump applied within a short time after the mixing process allowed us to measure the first relaxation time; as is shown in Figure 1, the amplitude decreases with increasing delays of the temperature-jump, corresponding to the gradual return of the system to 14.

REFERENCES

1. For the most recent review on the temperature-jump method see G.G. Hammes in A. Weissberger, "Technique of Chemistry", Vol. VI, Wiley-Interscience, New York, N. Y., 1974, p. 147.
2. For recent reviews see (a) C.F. Bernasconi, MTP (Med. Tech. Publ. Co.) Int. Rev. Sci. : Org. Chem. Ser. One, 3, 33 (1973); (b) F. Pietra, Quart. Rev. Chem. Soc., 23, 504 (1969); (c) J. Miller, "Aromatic Nucleophilic Substitution", Elsevier, New York, N. Y., 1968.
3. (a) C.F. Bernasconi, J. Amer. Chem. Soc., 92, 4682 (1970); (b) C.F. Bernasconi and R. G. Bergstrom, J. Org. Chem., 36, 1325 (1971).
4. C.F. Bernasconi, J. Amer. Chem. Soc., 92, 129 (1970).
5. C.F. Bernasconi, J. Amer. Chem. Soc., 90, 4982 (1968).
6. C.F. Bernasconi and H.S. Cross, J. Org. Chem., 39, 1054 (1974).
7. C.F. Bernasconi and R.H. de Rossi, J. Org. Chem., 38, 500 (1973).
8. J.F. Bunnett and R.H. Garst, J. Amer. Chem. Soc., 87, 3879 (1965).
9. J.F. Bunnett and J.J. Randall, J. Amer. Chem. Soc., 80, 6020 (1958).
10. M. Eigen, Angew. Chem. Int. Engl. Ed., 3, 1 (1964).
11. (a) See e.g. E. Buncel, A.R. Norris and K.E. Russell, Quart. Rev., Chem. Soc., 22, 123 (1968); (b) C.R. Hart and A.N. Bourns, Tetrahedron Letters, 2995 (1966).

12. C.F. Bernasconi and C.L. Gehriger, J. Amer. Chem. Soc., **96**, 1092 (1974).
13. C.F. Bernasconi and F. Terrier, to be published.
14. C.F. Bernasconi, C.L. Gehriger and R. H. de Rossi, to be published.
15. C.F. Bernasconi, R.H. de Rossi and C.L. Gehriger, J. Org. Chem., **38**, 2838 (1973).

CONFORMATIONAL ANALYSIS USING ULTRASONIC RELAXATION SPECTROMETRY

E. Wyn-Jones

Department of Chemistry, University of Salford,
Salford, U.K.

As a result of the pioneering work of Hasall[1] and Barton[2] the significance in the relationship between the conformation of a molecule and its physical properties, stability and reactivity was realized and the term conformational analysis was coined. This subject now ranks as one of the most important topics in organic stereochemistry and the widespread interest in this field is reflected in the number of general review articles and text books on the subject[3-9]. In addition to physical organic chemistry conformational analysis has important implications in polymer chemistry and biochemistry. This article gives a brief review of the way in which the ultrasonic relaxation method is used in studies on the conformational analysis of small molecules[10-14].

In this field several different experimental techniques have been employed[6-8] and the object in the majority of the investigations is to determine, by direct or indirect means, the energies associated with the conformational changes in question. The role of the ultrasonic method can be illustrated if we consider the conformational equilibrium (1) in 2-methyl-1,3-dioxan. As a result of ring inversion there is a dynamic equilibrium between the equatorial and axial chair conformers of this molecule.

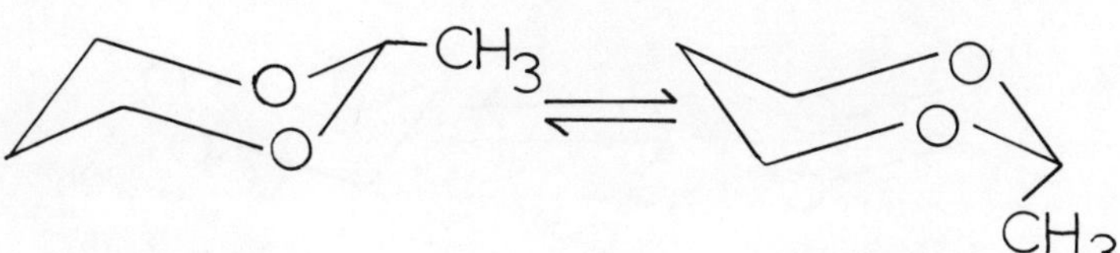

(1)

equatorial axial

Wyn-Jones (ed.), *Chemical and Biological Applications of Relaxation Spectrometry*, 353–357.

Equilibrium (1) is essentially a two state unimolecular process which is characterized by a single relaxation time, . In addition there is an enthalpy difference ΔH^o (and possibly a volume difference ΔV^o) between the two conformers which means that, in principle, an ultrasonic wave can perturb the equilibrium. Provided the correct frequency range is chosen ($\simeq 1/2\pi\tau$) a single relaxation will be observed experimentally. As shown in the introductory lecture[15] the relaxation time and amplitude μ_m of a two state equilibrium are related, respectively, to the rate constants and equilibrium parameters (ΔH^o, ΔV^o and K, the equilibrium constant). From the Eyring rate equation the temperature dependence of the rate constants yield the activation energy between the tansition state and the stable forms. In many cases it is possible (sometimes with the aid of semi-empirical calculations) to predict the conformation of the transition state[16]. Thus we have seen that, in principle, the ultrasonic method provides a direct way of determining the energies associated with a conformational change.

Most of the studies have been carried out with the pulse technique in the frequency range 10-200 MHz[10-14]. These measurements are normally made on pure liquid samples because (1) μ_m for intramolecular conformational equilibria attains its maximum value which in turn facilitates the analysis of the single relaxation equation $\alpha/f^2 = (A/(1 + (f/f_c)^2)) + B$. In practice 5 - 8 values of α/f^2 are available at different frequencies and the accuracy in the three disposable parameters A, B and f_c is highest when the dispersion in α/f^2 with frequency is large, and (2) it is more convenient to carry out measurements over a wide temperature range (173-373K). Before proceeding with the actual measurements it is necessary to establish the intramolecular nature of the relaxation process. This is done in a preliminary study by checking the concentration dependence of the relaxation parameters. For equilibrium (1) μ_m is directly proportional to the total concentration of both conformers and τ is independent of concentration. In order to obtain an exact description of the equilibrium being perturbed by the sound wave measurements are carried out on a related series of molecules. For example, in 1,3-dioxans a relaxation was observed in the 2- and 4-methyl derivatives but not in 1,3-dioxan itself. In this latter molecule ring inversion leads to two identical chair conformers (equilibrium (2)) which means

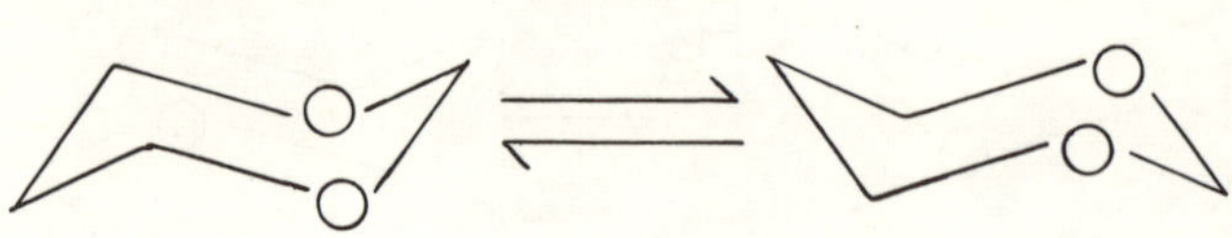

(2.

that $\Delta H^o = \Delta V^o = 0$ and no coupling with a sound wave was observed as expected. In addition no relaxation was observed in cis-2,4,-dimethyl-1,3-dioxan. Here ring inversion is essentially "blocked" because the energy of the diaxial conformer is much greater than the diequatorial form (8-11 kcal mol^{-1}) as a result of syn-axial methyl interactions[17].

Once the molecular origin of the relaxation has been established the absorption measurements are carried out on pure liquids over the temperature range spanning the relaxation. A general survey of the types of conformational systems that have been studied is given in several recent review articles[10-14]. The conformational energies derived from these measurements are mainly those for two state equilibria.

In molecules such as 2-bromo butane the conformational equilibrium is a three state coupled process which is characterized by two relaxation times. Experimentally only a single relaxation was observed and the consequence of this observation has been discussed[17a]. Multiple relaxations have, however, been observed[18]. In 2-chloromethyl-1,3-dioxan two relaxations arising from ring inversion and internal rotation about the exocyclic single bond were observed. In the cis 4-methyl-2-chloromethyl-1,3-dioxan it was possible to block ring inversion and to leave the internal rotation unaffected. By measuring the temperature dependence of the ultrasonic parameters and using infra-red data a complete analysis of this system was possible[18]. Recently we have attempted to distinguish between ring inversion and nitrogen inversion in some N-methyl piperidines and related compounds[19]. By synthesising several compounds in which systematically ring inversion is blocked and also the relaxation associated with nitrogen inversion removed from the ultrasonic time scale it has been possible to study the dynamics associated with both these processes.

Provided the anlysis of the single relaxation equation is satisfactory the energy barrier opposing a two state conformational change can be obtained with good accuracy from the temperature dependence of the relaxation times. In some cases these values compare well with dynamic n.m.r. data[8,12,16]. On the other hand the methods of deriving enthalpy differences, ΔH^o, from the ultrasonic data involve several approximations and assumptions and in most cases the resulting values differ markedly from those found using spectroscopic methods[11,12]. As a result most ultrasonic conformational studies have been confined to determining energy barriers. Presumably a combination of the approximation and assumptions that are required in relating the temperature dependence of μ_m to ΔH^o as well as uncertainties in the μ_m values from the analysis of the single relaxation equation account for these discrepancies. In order to avoid these difficulties we reconsidered the problem of evaluating the energetic data ΔH^o and activation energy E from ultrasonic experiments. For a two state process equation (3) was constructed which relates the temperature

$$\frac{\alpha}{f^2} = \frac{P_1(P_2)^{T_o/T}}{uT^2\left[P_4(P_3)^{T_o/T} + 4\pi^2 f^2\right]} + B \qquad (3)$$

and frequency dependence of α/f^2 with the conformational energies in question[20]. In expression (3) u is the velocity of sound at temperature T, P_1 and P_4 are constants, $P_2 = \exp(-\Delta H^o + E)/RT_o$, $P_3 = \exp(-E/RT_o)$ and T_o is a fixed temperature chosen as 298K. In practise some 30-50 values of α/f^2 are available at different temperatures and frequencies and the constants P_1 to P_4 are derived from this data using equation (3) and a computer programme using a least mean square minimization procedure. This method of analysis has been used for some 40 compounds and, so far, all the data are consistent with other methods. A typical example where this method was recently used was in an attempt to settle a debate concerning the magnitude of the conformational energy of the N-methyl group in piperidine[21]. Since 1973 independent estimates of 0.7, 1.8 and 3.8 kcal $mole^{-1}$ have been quoted for this quantity. In several N-methyl derivatives of piperidines, pyrrolidones, morpholines and piperazines where the segment of the molecule in which nitrogen inversion takes place is very similar, values of ΔH^o in the range 0.7 - 1.1 kcal mol^{-1} were found using equation (3). These values agree well with Katritzky's earlier dipole moment data.

By using the pulse technique in the frequency range 10 - 200 MHz and temperature range 173 - 373K it is possible to study conformational equilibria with energy barriers in the range 3 - 10 kcal $mole^{-1}$. Provided the enthalpy difference is of the order of a few hundred calories to 5 kcal $mole^{-1}$ a relaxation process with a reasonable amplitude parameter can be observed. In favourable cases it is possible to derive both kinetic and equilibrium energy parameters.

REFERENCES

1. O. Hassel, Tidsskr. Kjemi Bergvesen Met 3, 32 (1943). (English Translation Topics in Stereochemistry, 6, 11 (1971))
2. D.H.R. Barton, Experienta 6, 316 (1950).
3. E.L. Eliel, N.L. Allinger, S.J. Angyal and G.A. Morrison "Conformational Analysis" London, Interscience 1965.
4. M. Hanack, "Conformational Theory" London Academic Press 1965
5. See, for example, Topics in Stereochemistry, editors E.L. Eliel and N.L. Allinger.
6. S. Mizushima "Structure of Molecules and Internal Rotation" New York, Academic Press 1954.
7. "Internal Rotation in Molecules" ed. W.J. Orville-Thomas John Wiley and Sons 1974.

8. R.A. Pethrick and E. Wyn-Jones, Quart Rev. 23, 301 (1969).
9. M.V. Volkenstein "Conformational Statistics of Polymeric Chains" New York Interscience 1963.
10. J. Lamb in "Physical Acoustics" Chapter 4, ed. W.P. Mason Vol. 11A (1965).
11. W.J. Orville-Thomas and E. Wyn-Jones in "Transfer and storage of energy by molecules" eds. A.M. North and G.M. Burnett, Vol. 2, p.265 (1969).
12. E. Wyn-Jones and R.A. Pethrick, Topics in Stereochemistry 5, 205 (1970).
13. A.M. North and R.A. Pethrick in "Molecular Structure and Properties" ed. G. Allen, Vol. 3 of M.T.P. Intern. Rev. of Science, Butterworth 1972.
14. S.M. Walker, p.285 reference (7).
15. J. Rassing, this publication.
16. V.M. Gittins, E. Wyn-Jones and R.F.M. White p.425 reference (2)
17. G. Eccleston and E. Wyn-Jones, J. Chem. Soc. 6, 3417 (1971).
17a. E. Wyn-Jones and J. Rassing Adv. in Molecular Relaxation processes. 2, 227 (1972).
18. G. Eccleston, B. Walsh, E. Wyn-Jones and H. Morris, Trans. Faraday Soc., 67, 3223 (1971).
19. V.M. Gittins, P.J. Heywood and E. Wyn-Jones, unpublished data.
20. P.J. Heywood, J.E. Rassing and E. Wyn-Jones, Adv. in Molecular Relaxation Processes, 6, 307 (1975).
21. P.J. Crowley, M.J.T. Robinson and M.G. Ward, J.C.S. Chem. Comm. 825, 1974 and references quoted therein.

ULTRASONIC ABSORPTION DETERMINATION OF THE THERMODYNAMICS OF FAST ISOMERIZATION PROCESSES

P. Hemmes, L. Oppenheimer and F. Jordan

Department of Chemistry, Rutgers University, Newark, New Jersey, 07102

Ultrasonic absorption measurements have been used in the study of a number of chemical phenomena. These include studies on ion pair formation in solution,[1] cation[2] and anion[3] desolvation, rotational energy barriers,[4] vibrational relaxation processes,[5] viscoelastic properties of liquids[6] and molecular association processes.[7]

The ultrasonic technique is applicable to a range of time (10^{-6} - 10^{-9} sec) in which a multitude of chemical and physical processes occur. The physical processes include rotational and conformational changes. Since nmr techniques do not measure such fast processes, ultrasonics promises to be a valuable supplementary technique.

We wish to report an extension of the ultrasonic technique which enables measurements of the thermodynamics of fast isomerization processes. Such measurements are ordinarily very difficult to obtain for the following reason. Assume some measurable property of the system P which is dependent upon the concentrations of the isomeric species A and B with constants of proportionality p_A and p_B. Then the observable

$$P = p_A A_{eq} + p_B B_{eq}$$

Since the total (initial) concentration $C_T = A_{eq} + B_{eq}$

and $$K = \frac{B_{eq}}{A_{eq}}$$

$$P = (p_A/(1 + K) + p_B K/(1 + K))\, C_T$$

Wyn-Jones (ed.), *Chemical and Biological Applications of Relaxation Spectrometry*, 359–363.

This equation has three unknowns unless p_A and p_B can be determined independently. For a process A + X $\rightleftharpoons$ B, the quantities can be determined by using judicious amounts of X leading to varying extents of conversion to B. For an isomerization process this conversion cannot be brought about isothermally, hence the problem is without solution. Suppose, however, that the isomerization process has a non-zero ΔV and/or has a moderate value of ΔH and the process kinetically is in the ultrasonic time domain. A solution of the isomerizing solute will show an excess sound absorption compared to the solvent. The experimental quantity μ is given by

$$\mu = (\alpha - \alpha_o)\lambda$$

where α and α_o are the sound absorption coefficient of solution and solvent respectively, λ is the wavelength of the sound. The frequency dependence of μ is given by

$$\mu = 2\mu_{max}\ \omega\tau/\ (1 + \omega^2\tau^2)$$

where μ_{max} is the maximum value of μ, ω the angular frequency and τ the relaxation time of the system. The quantity μ_{max} is determined by the thermodynamics of the system and the solvent properties.

$$\mu_{max}^{8} = \frac{\pi}{\beta RT} (\Delta V_s)^2\ \Gamma^{-1} \tag{1}$$

$$\text{and}\quad \Delta V_s^{8} = \Delta V - \frac{\alpha}{\rho C_p}\ \Delta H \tag{2}$$

and for an isomerization process

$$\Gamma^{-1} = KC_T/(1 + K)^2$$

where β is the solvent compressibility; Γ^{-1} a concentration function such that

$$\Gamma = \Sigma_i \frac{\nu_i^2}{c_i}$$

with ν the stoichiometric coefficient of species i and c_i the concentration of species i; ΔV_s the adiabatic volume change, ΔV the isothermal volume change and ΔH the enthalpy for the reaction; α the thermal expansivity of the solution, ρ its density and C_p its heat capacity; C_T the total solute concentration. Eq. 1 has three unknowns ΔH, ΔV and K. There are two distinct solutions depending on whether or not $\Delta V \equiv 0$.

1. If $\Delta V=0$ one has two remaining unknowns. Assuming that all thermodynamic function but K are temperature independent within a narrow range of temperatures, one can investigate the effect of temperature on μ_{max}.

Define a new quantity G such that

$$G = \beta RT\mu_{max}/\pi C_T;$$

and with $\Delta V = 0$

$$G = \frac{\alpha^2}{\rho^2 C_p^2} (\Delta H)^2 \frac{K}{(1 + K)^2}$$

the temperature dependence of G then is

$$\frac{d \ln G}{d\ 1/T} = 2 \frac{d \ln Q}{d\ 1/T} - \frac{\Delta H}{R} \frac{1 - K}{1 + K} \qquad (3)$$

and $Q = \dfrac{\alpha}{\rho C_p}$

a quantity tabulated as a function of temperature for most common solvents. Measurement of

$$\frac{d \ln G}{d\ 1/T}$$

at several temperatures then leads to a second independent equation in ΔH and K. Hence from Eqs (1) and (3) one can determine these quantities. As a check on the method one can employ the assumption that ΔH should be reasonably constant with temperature. Furthermore, calculation of the quantity

$$\frac{d \ln K_{calc}}{d\ 1/T}$$

by plotting lnK vs 1/T should be equal

$$\frac{-\Delta H}{R}$$

Such a ΔH value should be close to ΔH, the average of the ΔH values at each temperature. If the ΔH values found by the two approaches are significantly different the assumption $\Delta V \equiv 0$ is probably invalid. For aqueous systems one tests the assumption by making measurements at 4^oC. At this temperature G is zero for $\Delta V = 0$ (since α vanishes at 4^oC), and any excess absorption must be due to the ΔV term.

Since the absorption of water is large at $4^{\circ}C$ a small ΔV may, unfortunately, be difficult to detect. In these cases comparison of $\overline{\Delta H}$ and ΔH from the log K_{calc} vs 1/T plot is a better criterion for indication of $\Delta V \neq 0$. Even a small ΔV causes inconsistencies in the two ΔH values especially at low temperatures in aqueous medium.

2. If $\Delta V \neq 0$ we suggest the following procedure. Under these conditions equation (3) becomes:

$$\frac{d \ln G}{d1/T} = \frac{2|\Delta H|}{\left|\Delta V - \frac{\alpha}{\rho C_p}\Delta H\right|}\frac{\delta Q}{\delta 1/T} - \frac{\Delta H}{R}\frac{(1 - K)}{(1 + K)} \qquad (4)$$

then define

$$W = \frac{d \ln G}{d1/T} - \frac{2|\Delta H|}{\left|\Delta V - \frac{\alpha}{\rho C_p}\Delta H\right|}\frac{\delta Q}{\delta 1/T}$$

and $W = -\dfrac{\Delta H}{R}\dfrac{1 - K}{1 + K}$

or, $$K = \frac{\Delta H - WR}{\Delta H + WR} \qquad (5)$$

Substitution of (5) into (1) leads to a fourth degree equation in ΔH. Small (0.1 cc/mole) incremental ΔV values are now tried with the total range of ΔV estimated from the assumptions $\Delta H = 0$ and Γ^{-1} (equal to $C_T/4$). With these values ΔV is calculated, the range being $\pm \Delta V$. To avoid possible errors the range of ΔV can be expanded by taking $\Gamma^{-1} = C_T/10$, which is about as small a concentration factor as can be measured experimentally. Assuming $\Delta V = 0$, a limit can also be placed on ΔH. For each ΔV one calculates four ΔH values from Eq 1 (employing a digital computer) at each temperature. Of the four ΔH roots, two will be erroneous usually (imaginary or clearly outside the range of ΔH values allowed). The other two roots are $r_1 \cong -r_2$ (of course $r_1 = -r_2$ when $\Delta V = 0$). Assuming, as previously, that ΔH and ΔV are nearly temperature independent in the short range which can be studied, the solution lies in finding a ΔV which produces nearly temperature independent ΔH (parenthetically, usually only a narrow range of ΔV values will produce such a ΔH). From such a set of ΔV and ΔH values one can calculate K from eq 5 at each temperature and check for internal consistency of the data by showing that ΔH from the log K vs (1/T) plot agrees with the $\overline{\Delta H}$ calculated above.

Acknowledgement. Thanks are due to the Rutgers University Research Council and the Petroleum Research Fund (to P. H.) for financial support of this research.

REFERENCES

1. S. Petrucci, Ionic Interactions Vol. 11, S. Petrucci, ed., Academic Press, New York 1971.
2. P. Hemmes, F. Fittipaldi and S. Petrucci, Acustica, 21, 228 (1969).
3. H.C. Wang and P. Hemmes, J. Amer. Chem. Soc., 95, 5115 (1973).
4. M.S. deGroot and J. Lamb, Trans. Faraday Soc., 51, 1676 (1955).
5. J.H. Andreae, E.L. Heasell and J. Lamb, Proc. Phys. Soc., 59B, 625 (1956).
6. G.J. Gruber and T.A. Litovitz, J. Chem. Phys., 47, 2185 (1967)
7. G.G. Hammes and H.O. Spivey, J. Am. Chem. Soc., 89, 1621 (1967)

A COMPARISON OF PROTON TRANSFER RATES IN AQUEOUS AND NON-AQUEOUS SOLUTIONS

B. Bianchin, J. Chrisment, J.J. Delpuech*, M.N. Deschamps, D. Nicole and G. Serratrice

(Equipe de Recherche Associeé au C.N.R.S., Laboratoire de Chimie Physique Organique, Université de Nancy I, base officielle no 140,54037 NANCY CEDEX, France)

ABSTRACT

Proton transfer mechanisms are described in either water or anhydrous dimethylsulphoxide as solvents, using the same acidic substrates in each case : ammonium salts, acetylenic compounds, thiols or alcohols. Rate measurements were performed as a function of the pH by means of dynamic NMR of proton or carbon -13 nuclei at 25°C. From these experiments, it seems that the presence of water is not necessary to obtain fast proton transfers.

INTRODUCTION

To date, no reliable comparison has been made of the rate of a simple proton transfer in aqueous and non-aqueous solutions. This is probably due to the fact that the same substrates and mechanisms are needed in both media for this purpose. Such a comparison however is fundamental, since water is generally thought to be exceptional in this respect on account of the abnormal proton conductivity. We report first examples of such comparative experiments using pure water and anhydrous dimethylsulphoxide as solvents. A variety of proton transfers are being studied in our laboratory. They may be classified according to the nature of the atom yielding or receiving the exchanging proton:

- nitrogen in ammonium salts
- carbon in acetylenics
- sulphur in thiols, or oxygen in alcohols

Wyn-Jones (ed.), Chemical and Biological Applications of Relaxation Spectrometry, 365–373.

Each of these transfers is a research theme by itself. We wish to report here only a selection of our results illustrating solvent effects, and also of our methods chiefly using proton nmr, with or without ^{14}N - decoupling, and carbon -13 nmr.

TRANSFERS FROM NITROGEN IN AMMONIUM SALTS

The mechanism of such transfers has been studied in slightly acidic aqueous solutions by Grunwald, Loewenstein and Meiboom[1]. The exchange rate is slower in more acidic solutions, so that the pH must be lowered to 3-4 to fit the nmr time-scale. In these conditions, the amount of free amine is very small (10^{-7} - 10^{-6} M) and is not detected by nmr. However, the nmr spectrum of the protonated ammonium salt is modified when the acidic proton is exchanged, and transfer rates may be derived from the observed lineshapes.

The various mechanisms observed in water are summarized in Table 1, together with the values of rate constants in Table 2.

Table 1. Mechanisms for deprotonation of methylammonium ions BH^+ in water (pH = 3 to 5).

$$BH^+ + H_2O \xrightarrow{k_4} B + H_3O^+ \qquad (I)$$

$$BH^+ + OH^- \xrightarrow{k_5} B + H_2O \qquad (II)$$

$$BH^+ + B \xrightarrow{k_6} B + HB^+ \qquad (III)$$

$$BH^+ + \underset{\substack{|\\ H}}{O}\text{-}H + B \xrightarrow{k_7} B + H - \underset{\substack{|\\ H}}{O} + HB^+ \qquad (IV)$$

$$\left[\frac{1}{[BH^+]}\right]\left[-\frac{d[BH^+]}{dt}\right] = k_4 + k_5\,[OH^-] + (k_6 + k_7)\,[B]$$

(from Ref. 1)

Table 2. Rate constants for deprotonation of methylammonium ions in water at 25°C (From Ref. 1)

	k_4 (sec^{-1})	k_5 ($10^{10}M^{-1}\ sec^{-1}$)	k_6 ($10^8M^{-1}\ sec^{-1}$)	k_7 ($10^8M^{-1}\ sec^{-}$
NH_4^+	24.4	3.0	11.7	1
$CH_3NH_3^+$	0.90	3.7	4.0	5.3
$(CH_3)_2NH_2^+$	0.52	3.1	0.5	5.6
$(CH_3)_3NH^+$	4.0	2.1	0.3	3.1

In slightly acidic solutions, only the last two are important. Their relative contribution depends on steric hindrance to the approach of the base into the vicinity of the acidic N-H protons of the conjugate ammonium ion. For NH_4^+, the third reaction (III) is predominant with a rate constant[14]:

$$k_{H_2O} = 1.17 \times 10^9 \ M^{-1} \ s^{-1} \text{ at } 25^oC$$

The same experiment was carried out in anhydrous DMSO[2]. Ammonium chloride is dissolved in DMSO, containing very small amounts (10^{-5} - 10^{-6}M) of dry hydrochloric acid.

The ammonium protons are represented by a triplet, one component for each magnetic state of nitrogen - 14 nucleus. These lines coalesce into a singlet when the pH is raised from 4.5 to 5.2 (Figure 1). From the lineshape, we may derive the mean lifetime τ of one proton in a given ammonium cation. The exchange rate τ^{-1} is then found directly proportional to the ratio : $[NH_4^+] / [H^+]$ (figure 2).

This result is accounted for by reaction (III):

$$NH_4^+ + NH_3 \longrightarrow NH_3 + NH_4^+;$$

$$\tau^{-1} = k_{DMSO} [NH_3] = k_{DMSO} \ K_A \frac{[NH_4^+]}{[H^+]}$$

where K_A is the ionization constant of NH_4^+ in DMSO:pK_A = 10.48. The value obtained for k_{DMSO} : $1.21 \times 10^9 \ M^{-1} \ sec^{-1}$ is surprisingly close to that obtained in water.

Therefore, water has no special role in this transfer. For sterically hindered amines, however, we may expect a decreased proton transfer rate in DMSO, since the reaction (IV) involving a small bridging water molecule is impossible in anhydrous DMSO. We have used for this purpose a piperidinium salt, the 1, (2,6 cis) - trimethylpiperidinium ion, which may exist under two isomeric forms AH and BH[3] :

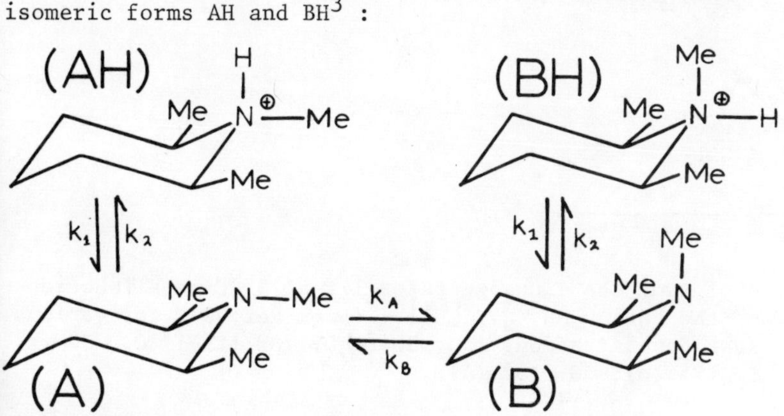

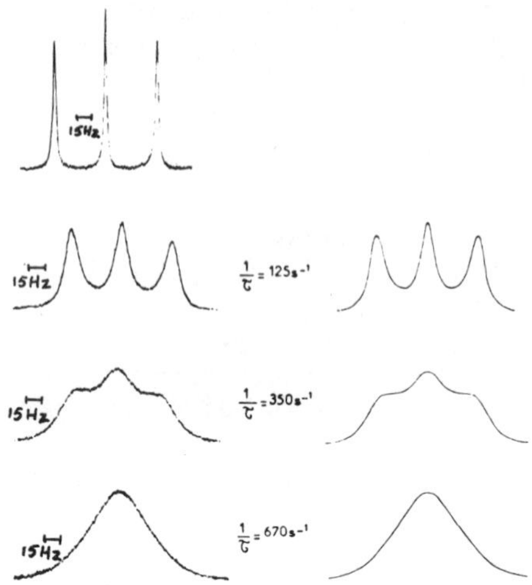

Fig. 1. ^{1}H nmr spectra of ammonium chloride in DMSO at 60 MHz for several exchange rates $1/\tau$. Experimental and theoretical curves, left and right respectively (from Ref. 2).

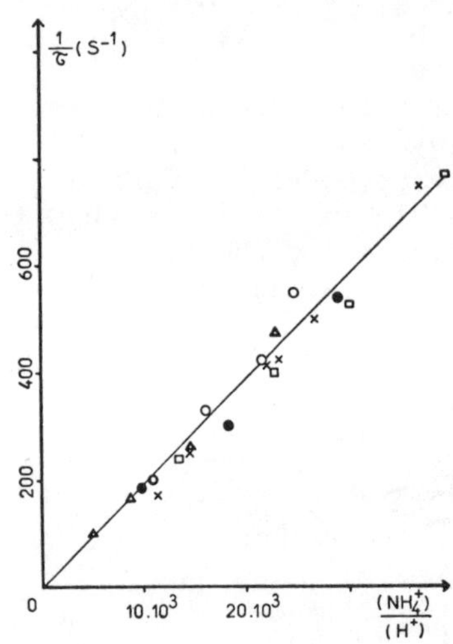

Fig. 2. A plot of the exchange rates $1/\tau$ at $25^{\circ}C$ as a function of the ratio $[NH_4^+] / [H^+]$. (From Ref. 2), for various salt concentrations : 0. 304 (●); 0. 283 (□); 0. 237 (x); 0. 175 (o); 0. 152 (Δ)M.

This salt is also a good probe to observe simultaneously nitrogen inversion[4] on the very small amount of free-amine (A or B), bringing forth another type of exchange in the spectrum of the cations. The kinetic scheme is complex[5] and different laws are obtained according as the reprotonation rate k_2 is fast or slow compared to nitrogen inversion rate k_A or k_B.

In water, both isomers are observed by their N and C - methylic spectrum[3,4]. When the solution is progressively made less acidic, a first coalescence is obtained for each N-methylic doublet independently (Figure 3)[3,6]. This is an evidence for a deprotonation of each isomeric cation without any nitrogen inversion on the nmr time-scale: this is demonstrated by the presence of two C- methyl doublets at pH = 7.2 (Figure 4). For more basic solutions, a simultaneous coalescence of C - and N - methyl lines is observed, revealing nitrogen inversion. From these coalescences, we may derive deprotonation rates, through mechanisms (III) and (IV):

$$k_{H_2O} = 1.13 \times 10^8 \ M^{-1} \ sec^{-1}$$

and nitrogen inversion rate : $k_A = 10^3 \ sec^{-1}$ at 25^oC.

In DMSO, we cannot observe the N-methyl doublets obscured by the solvent lines. We use then the exchanging proton N-H itself[7], only sensitive to deprotonation, and the C- methyl lines, only sensitive to nitrogen inversion.

For pH = 6 to 8, both exchange rates are found equal[8]:

$$\tau^{-1} = k'_{DMSO} \, [salt] \, / \, [H^+], \text{ with } k'_{DMSO} = 6.21 \times 10^{-6} \ sec^{-1}$$

Two hypotheses may account for this result: either, (a) a proton transfer brings forth a nitrogen inversion simultaneously, or (b) the reprotonation is so slow that the free amine obtained after deprotonation suffers a great number of nitrogen inversions before its reprotonation.

The distinction between these two assumptions results from the study of the coalescence of the two N- methyl doublets in deuterated DMSO : this time, all four lines simultaneously coalesce into a broad band. Moreover, a theoretical simulation of such a curve shows that the observed lineshapes are only consistent with the second hypothesis.

Therefore, we actually found a deprotonation rate constant:

$$k_{DMSO} = k'_{DMSO} \, / \, K_{A\ (DMSO)} = 2.58 \times 10^4 \ M^{-1} \ sec^{-1} \text{ at } 25^oC.$$

The corresponding reprotonation rate k_2 may be deduced from k_{DMSO}: $k_2 = 1.03 \times 10^4 \ sec^{-1}$. We must then admit that the nitrogen inversion rate constant k_A is far larger than $10^4 \ sec^{-1}$ in DMSO. From a more complete analysis, we may presume that $k_A = 10^6 \ sec^{-1}$, a value relatively close to that obtained from ultrasonic studies of the pure amine[9] (i.e. in a quite different solvent).

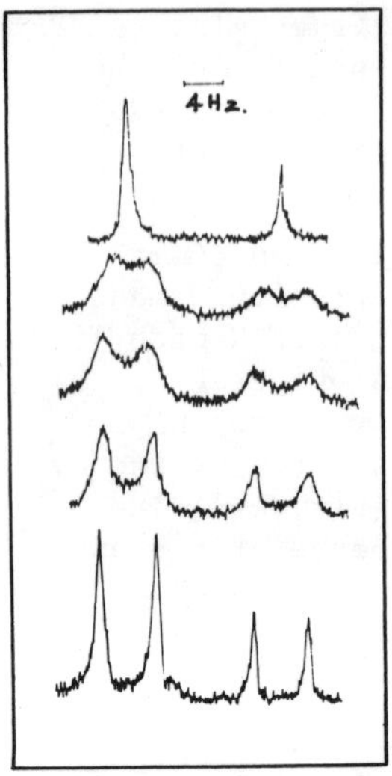

Fig. 3. nmr spectrum of the N-methylic protons of a 0.3 molar aqueous solution of 1, (2, 6-cis) - trimethylpiperidinium chloride, at 33°C, for pH = 0; 3.6; 3.94; 4.10 and 7.30 respectively (from the bottom to the top) (From Ref. 6)

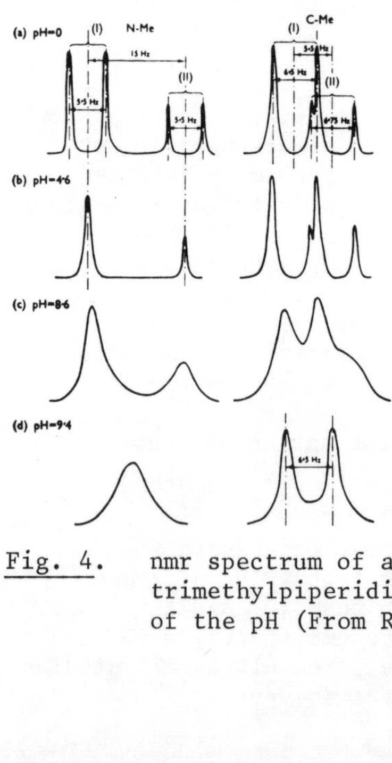

Fig. 4. nmr spectrum of aqueous solutions of 1, 2, 6-trimethylpiperidinium chloride at 33°C, as a function of the pH (From Ref. 4).

These results illustrate clear-cut differences between aqueous and DMSO solutions; very slow deprotonation and reprotonation rates are obtained in DMSO, since a small water molecule cannot play the role of an intermediate bridge in a hydrate $BH^+ \ldots OH_2$. In water, this hydrate persists after deprotonation : $B \ldots OH_2$, explaining an abnormally slow nitrogen inversion : 10^3 in water against $10^6\ s^{-1}$ in DMSO.

TRANSFERS FROM ACETYLENIC CARBON

N, N'-dipropargylamine : $Me_2\ N\text{-}\underline{\underline{CH_2}} - C \equiv C - \underline{\underline{H}}$ or $R - C \equiv C - H$, was used for this purpose because of its solubility both in water and in DMSO.[10]

In basic water[11] (pH $\simeq$ 12), the acetylenic proton is mainly abstracted through the reaction:

$$R - C \equiv C - \underline{H} + OH^{\ominus} \xrightarrow{k} RC \equiv C^{\ominus} + H_2O$$

with : $k^{OH^-}_{(H_2O)} \simeq 10^3\ M^{-1}\ s^{-1}$ at 25^oC

In DMSO containing ca. 10^{-2} M dimsyl anion, the predominant transfer reaction as deduced from the nmr spectrum of the underlined protons, is carried out by the conjugate acetylide ion:

$$RC \equiv C - H + RC \equiv C^{\ominus} \xrightarrow{k'} R - C \equiv C^{\ominus} + RC \equiv C\text{-}H$$

where : $k' = 4.08 \times 10^3\ M^{-1}\ s^{-1}$ at 25^oC

However, the reaction with hydroxide ion may also be studied by adding small quantities of water (0.1 to 1 M) to DMSO, and observing the width of the water line. Preliminary results show that:

$$k^{OH-}_{(DMSO)} = 1.35 \times 10^8\ M^{-1}\ s^{-1}$$

Thus, hydroxide ion is much more reactive in DMSO than in water with respect to proton abstraction. This is a familiar result for SN_2 reactions, but shows that the unfavourable solvation of hydroxide ion is not counterbalanced by the possibility of proton transfer in the hydrogen - bonded network of water.

Let us notice the necessity of using ^{13}C nmr to study proton transfers to and from the carbon atom of compounds such as phenylacetylene : $C_6H_5 - C \equiv C - \underline{H}$ where the exchanging proton is not coupled to any other proton. We gave the first example of such kinetics when studying the deprotonation rate of basic DMSO itself [12].

TRANSFERS TO AND FROM ALCOHOLIC OXYGEN

With these amphoteric compounds, we may observe either their protonation or their deprotonation, in acidic or basic solutions respectively.

In aqueous solutions[13] of alcohols ROH, results of literature show that the predominant reactions are:
(a) in acidic conditions:

$$ROH + ROH_2^+ \longrightarrow ROH_2^+ + ROH, \text{ with}$$

$$k^{ROH_2^+}_{(H_2O)} = 10^8\ M^{-1} \text{ at } 25^oC, \text{ for EtOH.}$$

and : $ROH + H_3O^+ \longrightarrow ROH_2^+ + H_2O$

with $k^{H^+}_{(H_2O)} = 10^6\ M^{-1}sec^{-1}$

(b) in basic conditions:

$$ROH + RO^- \longrightarrow RO^- + ROH, \text{ with:}$$

$$k^{RO^-}_{(H_2O)} = 10^6\ M^{-1}\ s^{-1}$$

Preliminary results[14] using benzylic alcohol show that the predominant reactions are:
(a) in acidic DMSO (pH $\simeq$ 5) :

$$ROH + H^+ \ldots\ldots DMSO \longrightarrow ROH_2^+ + DMSO, \text{ with:}$$

$$k^{H^+}_{(DMSO)} \simeq 10^5 - 10^6\ M^{-1}\ sec^{-1}$$

(b) in basic DMSO (pH $\simeq$ 13)

$$ROH + RO^- \longrightarrow RO^- + ROH$$

where : $k^{RO^-}_{(DMSO)} \simeq 10^8\ M^{-1}\ sec^{-1}$

Similar results have been obtained with the corresponding thiol[15]. From all these results, we may tentatively conclude that the presence of water, or more generally, of a hydrogen-bonded network, is not compulsory to promote fast proton transfers.

ACKNOWLEDGEMENTS. Financial support from the Direction des Recherches et Moyens d'Essais and from the Centre National de la Recherche Scientifique is gratefully acknowledged.

REFERENCES

1. E. Grunwald, A. Loewenstein and S. Meiboom, J. Chem. Phys., 27, 630 (1957); A. Loewenstein and S. Meiboom, J. Chem Phys., 67, 2208 (1963).
2. B. Bianchin and J. J. Delpuech, Bull. Soc. Chim. France, 34 (1973).
3. J. J. Delpuech and M. N. Deschamps, Tetrahedron, 26, 2723, (1973).
4. J. J. Delpuech and M. N. Deschamps, Chem. Commun., 1188 (1967).
5. J. J. Delpuech, Org. Magn. Resonance, 2, 91 (1970).
6. J.J. Delpuech, F. Siriex, and M. N. Deschamps, Org. Magn. Resonance, 4, 651 (1972).
7. B. Bianchin and J. J. Delpuech, Tetrahedron, 30, 2859 (1974).
8. B. Bianchin and J. J. Delpuech in the press.
9. E. Wyn-Jones, this publication.
10. J. Chrisment and J. J. Delpuech, Tetrahedron Letters, 6, 521 (1972).
11. J. Chrisment and J. J. Delpuech, in the press.
12. J. Chrisment, J. J. Delpuech and P. Rubini, Molec. Phys., 27, 1663 (1974).
13. E. Grunwald, C. F. Jumper, and S. Meiboom, J. Amer. Chem. Soc., 84, 4664 (1962).
14. J. J. Delpuech and D. Nicole, in the press.
15. J. J. Delpuech and D. Nicole, J. Chem. Soc., Perk. 2, 1025 (1974).

ELECTRIC FIELD AND TEMPERATURE-JUMP STUDY OF THE COLOR CHANGE KINETICS OF ETHANOLIC AND AQUEOUS PHENOLPHTHALEIN.[1]

M.W. Massey and Z.A. Schelly*

Department of Chemistry, University of Georgia,
Athens, Georgia 30602, U.S.A.

Electric field and temperature-jump kinetic results, using optical detection[2] (Fig. 1), indicates that the hydrolysis mechanism (see also Fig. 2)

$$PhH^- + OH^- \underset{k_{21}}{\overset{k_{12}}{\rightleftharpoons}} PhH^- \, . \, . \, . \, OH^- \underset{k_{32}}{\overset{k_{23}}{\rightleftharpoons}} Ph^= + H_2O \qquad (1)$$

proposed previously in combination with the steady state approximation for the pseudo acid phenolphthalein in water[3], is not valid in pure ethanol. In this solvent

$$PhH^- + OH^- \underset{k_{21}}{\overset{k_{12}}{\rightleftharpoons}} \underline{Ph}^= \, . \, aq \underset{k_{32}}{\overset{k_{23}}{\rightleftharpoons}} Ph^= \, . \, aq \qquad (2)$$

describes the reaction, and the steady state approximation cannot be applied. The first step represents the diffusion controlled deprotonation of PhH^- resulting in the still colorless $\underline{Ph}^=$. aq. The much slower intramolecular electronic rearrangement and the opening of the lactone ring take place in the second step, where the pink $Ph^=$. aq is formed. Since the colored form was monitored, only one of the two reciprocal relaxation times,

$$\tau^{-1}_{slow} \approx k_{32} = 1.25 \times 10^6 \ sec^{-1}$$

Wyn-Jones (ed.), Chemical and Biological Applications of Relaxation Spectrometry, 375–377.

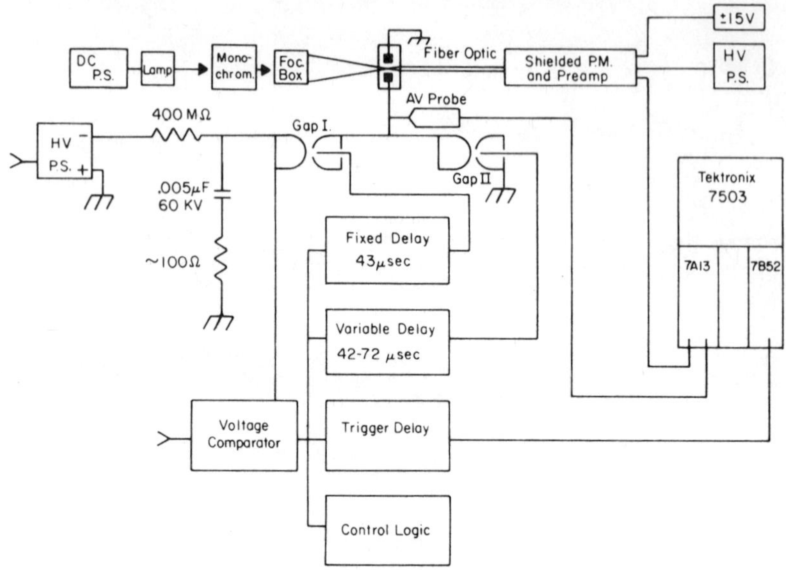

Fig. 1. Block diagram of apparatus.

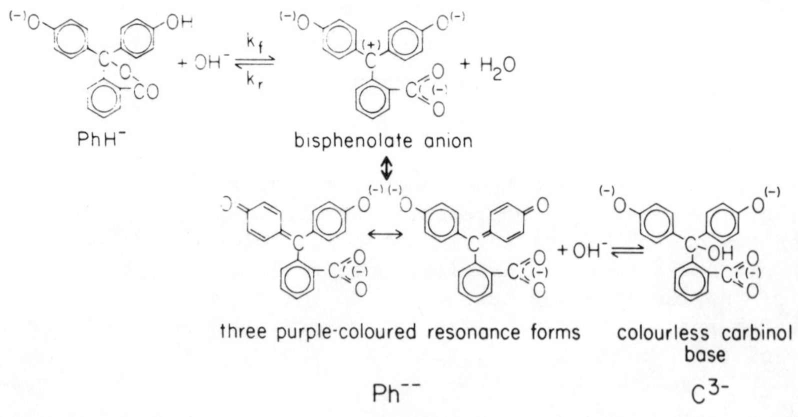

Fig. 2. Hydrolysis mechanism.

could be determined, directly yielding the rate of polarization of the resonance system and the closing of the lactone ring.

REFERENCES

1. M.W. Massey, Z.A. Schelly, J. Phys. Chem., 78, 2450 (1974).
2. G. Ilgenfirtz, Dissertation, Gottingen, 1966.
   S.L. Olsen, R.L. Silver, L.P. Holmes, J.J. Auborn, P. Warrick and E.M. Eyring, Rev. Sci. Instr., 42, 1247 (1971).
3. M.C. Rose, J. Stuehr, J. Am. Chem. Soc., 90, 7205 (1968).

RATES OF PROTON TRANSFERS IN 70% DMSO - 30% WATER. MECHANISM OF DEPROTONATION OF INTRAMOLECULARLY HYDROGEN BONDED SPECIES

Claude F. Bernasconi and F. Terrier

Thimann Laboratories, University of California,
Santa Cruz, California

In connection with work described elsewhere in this volume[1] we carried out temperature-jump studies on systems such as eq. 1;

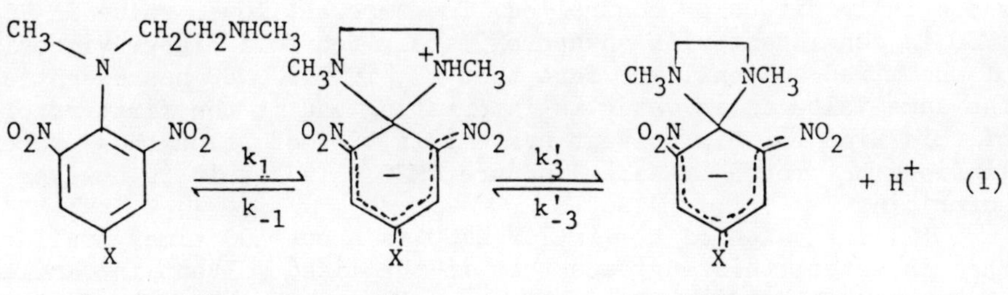

1a X = NO_2 2a 3a

1b X = CF_3 2b 3b

$$k'_3 = k_3 + k_3^{OH}\{OH^-\} + \Sigma k_3^B\{B\} \quad (2)$$

$$k'_{-3} = k_{-3}\{H^+\} + k_{-3}^{OH} + \Sigma k_{-3}^B\{BH\} \quad (3)$$

k'_3 and k'_{-3} are defined by eq. 2 and 3 with k_3, k_3^{OH} and k_3^B being

Wyn-Jones (ed.), Chemical and Biological Applications of Relaxation Spectrometry, 379–384.

the rate constants of deprotonation of 2 by the solvent, hydroxide ion, and buffer bases respectively, and k_{-3}, k_{-3}^{OH} and k_{-3}^{B} being the rate constants for protonation of 3 by the lyonium ion, the solvent, and buffer acids respectively.

Compound 1a ($X = NO_2$) was studied in aqueous solution[2], compound 1b ($X = CF_3$) in 70% DMSO-30% water (v/v)[3]. In both cases the reactions are characterized by one measurable relaxation time (2 can be treated as a steady state intermediate because $k_{-1} + k'_3 >> k_1 + k'_{-3}$ under all experimental conditions) is given by

$$\frac{1}{\tau} = \frac{k_1 k'_3}{k_{-1} + k'_3} \frac{K_A}{K_A + \{H^+\}} + \frac{k_{-1} k'_{-3}}{k_{-1} + k'_3} \qquad (4)$$

where K_A is the acid dissociation constant of the protonated (at the other nitrogen) form of 1. By measuring $1/\tau$ as a function of pH and of {B}, {BH^+}, and determining the pK of 2 by a spectrophotometric method[2] we were able to evaluate all rate constants of the reactions (1).

In this paper we are only concerned with the proton transfer processes. Our results are summarized in Table I and II. The following points are noteworthy. (1) The high value of k_{-3} in both solvents indicates that the reaction $3 + H^+ \rightarrow 2$ is essentially diffusion controlled. The somewhat lower value in 70% DMSO is consistent with an approximately fourfold higher viscosity of the mixed solvent. The fact that k_{-3} in 70% DMSO has essentially the same value as in water is interesting and is the first report of this kind in this solvent system. In view of Delpuech's[4] recent findings on proton transfer in pure DMSO this result is however not surprising.

(2) The value of k_3^{OH} in 70% DMSO is about 100 times smaller than in water; this suggests that in the mixed solvent intramolecular hydrogen bonding to one of the ortho nitro groups becomes

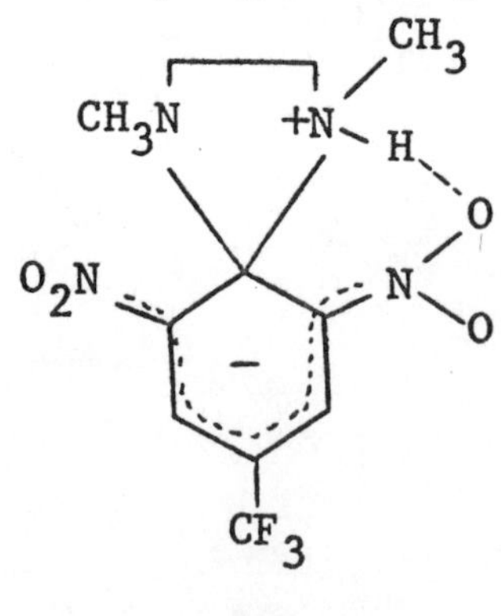

2b

TABLE I. SOLVENT, HYDROXIDE AND LYONIUM ION PROMOTED PROTON TRANSFERS

	1a in H_2O[a]	1b in 70% DMSO[b]
k_3^{OH}, M^{-1} sec^{-1}	5.2×10^{9}	4.2×10^{7}
k_{-3}^{OH}, sec^{-1}	4.45×10^{2}	1.0×10^{-3}
k_3, sec^{-1}	1.35×10^{4}	9.3×10^{2}
k_{-3}, M^{-1} sec^{-1}	5.9×10^{10}	2.9×10^{10}
pK(2)	6.64	7.49[d]
K_S^c	1.94×10^{-14}	7.85×10^{-19}
K_A	2.23×10^{-9}	6.30×10^{-9} [d]

[a] At 25°, μ = 0.5 M maintained by NaCl. [b] At 20°, μ = 0.5 M maintained by $(CH_3)_4NCl$. [c] K_S = selfionization constant of solvent. [d] Determined in 70% DMSO.

TABLE II. PROTON TRANSFER WITH BUFFERS IN 70% DMSO-30% WATER (V/V), COMPOUND 1b.[a]

BH	pK(BH)[b]	pK(2b)[b]	ΔpK = pK(BH)-pK(2b)	$10^{-6} \times k_3^B$ $M^{-1}\,sec^{-1}$	k_{-3}^B $M^{-1}\,sec^{-1}$
p-Cyanophenol	9.24	7.49	1.75	5.7	1.0×10^{5}
o-Bromophenol	10.58	7.49	3.09	10.2	8.3×10^{3}
p-Chlorophenol	11.54	7.49	4.05	6.9	6.1×10^{2}
Phenol	12.48	7.49	4.99	7.3	7.4×10^{1}
Benzimidazole	13.01	7.49	5.52	7.2	2.2×10^{1}
Indazole	14.52	7.49	7.03	7.5	7.0×10^{-1}

[a] At 20°, μ = 0.5 M maintained by $(CH_3)_4NCl$. [b] Determined in 70% DMSO

significant,[5] but is relatively unimportant in water.

(3) The values for k_3^B are relatively low which is also consistent with intramolecular hydrogen bonding in 2b. There are two mechanistic possibilities for the deprotonation of a hydrogen bonded species[6]. The first involves a direct attack of the base on the bonded hydrogen, eq. 5. In the second an equilibrium

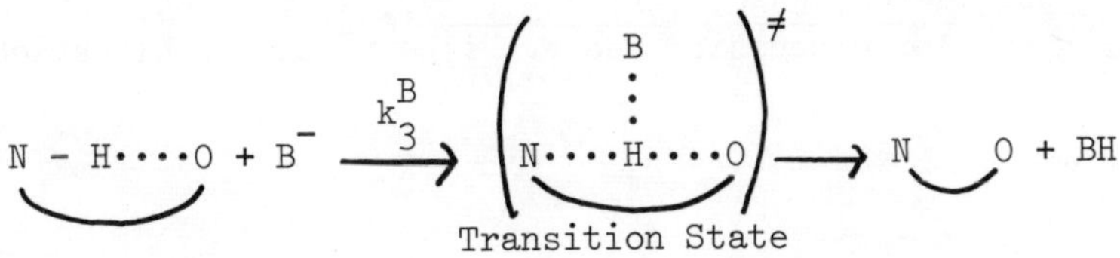

(5)

(independent of base) between the hydrogen bonded and an open form precedes the actual deprotonation step (eq. 6); for this mechanism

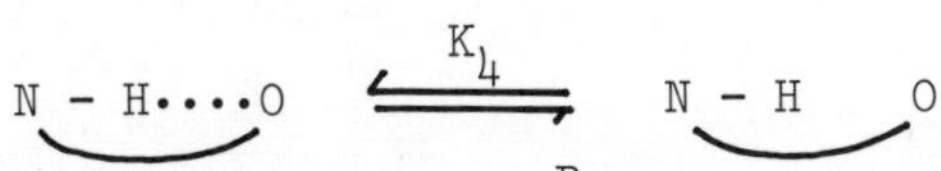

(6a)

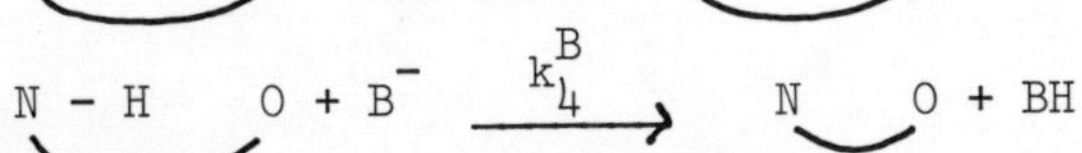

(6b)

which is the one favored by Eigen[5] we have $k_3^B = k_4^B K_4$.

Our results tend to support the second mechanism. If the first mechanism prevailed one would expect that the energy of the transition state depends on the base strength of B, leading to higher rates as pK(BH) and with it ΔpK increases, at least up to a certain point. Our k_3^B-values are essentially pK-independent which is contrary to expectation.

The constancy of k_3^B can be rationalized in terms of the second mechanism as follows. Let us assume the reaction 6b with $B^- = OH^-$ is diffusion controlled and has a rate constant, k_4^{OH}, at least as large as $5 \times 10^9\ M^{-1}sec^{-1}$ (5×10^9 is the value of k_3^{OH} for the deprotonation of 2a in water, presumably diffusion controlled or nearly so). In 70% DMSO we have then $k_3^{OH} = 4.2 \times 10^7 = k_4^{OH}K_4 \approx 5 \times 10^9\ K_4$ or $K_4 \approx 4.2\ 10^7/5 \times 10^9 \approx 0.01$. The acid dissociation constant of 2b as determined in this study is then an apparent dissociation constant, defined as the acid dissociation constant of the open form multiplied by K_4; the deprotonations then refer to the reaction of B^- with an acid (open form) whose pK is at least two units lower than pK(2), making all ΔpK-values of Table II at least two units larger. Thus even for p-cyanophenol we now have ΔpK ≈ 4. Since Eigen plots[5] (log k vs. ΔpK) usually level off at ΔpK ≈ 4 it is not surprising that our k_3^B-values are all ΔpK-independent.

REFERENCES

1. C.F. Bernasconi, this publication.
2. C.F. Bernasconi and C.L. Gehriger, J. Amer. Chem. Soc., 96, 1092 (1974).
3. C.F. Bernasconi and F. Terrier, to be published.
4. J.J. Delpuech, this publication.
5. M. Eigen, Angew. Chem. Int. Engl. Ed., 3, 1 (1964).
6. For a similar discussion see F. Hibbert, this publication.

SOLVENT-JUMP RELAXATION KINETICS OF THE ASSOCIATION OF RHODAMINE TYPE LASER DYES [1]

M.M. Wong and Z.A. Schelly*

Department of Chemistry, University of Georgia,
Athens, Georgia 30602, U.S.A.

Equilibrium spectral photometric and solvent-jump relaxation kinetic results of the monomer (M) - dimer (D) equilibrium

$$2MH^{+} \underset{{}_{o}k_{-1}}{\overset{{}_{o}k_{1}}{\rightleftharpoons}} D^{++} \qquad (1)$$

of aqueous and ethanolic rhodamine B, rhodamine 3B, rhodamine 6G, and rhodamine 110 (Fig. 1) have been obtained, which indicate that, contrary to previous suggestions, association occurs in both solvents with the exception of rhodamine 3B and 6G in ethanol.

Although, in the case of ethanolic rhodamine B and 110 equilibrium (1) is coupled with the acid-base reaction

$$MH^{+} + EtOH \underset{k_{-2}}{\overset{k_{2}}{\rightleftharpoons}} M^{++} + EtOH_{2}^{+} \qquad (2)$$

only one relaxation time has been observed in all cases, given by

$$\tau^{-2} = {}_{o}k_{-1} \left[\frac{f_D}{f_{\neq}}\right]^2 + 8\,{}_{o}k_{1}\,{}_{o}k_{-1}\; f_D \left[\frac{f_M}{f_{\neq}}\right]^2 C_T \qquad (3)$$

as a function of the total concentration C_T and activity coefficients f_i.

Wyn-Jones (ed.), Chemical and Biological Applications of Relaxation Spectrometry, 385–386.

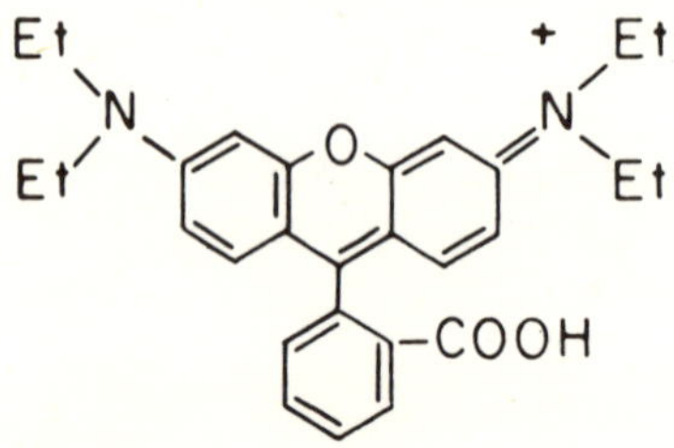

Rhodamine B

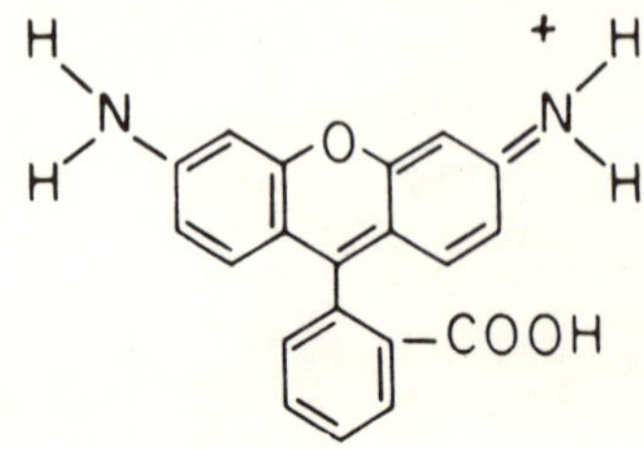

Rhodamine 110

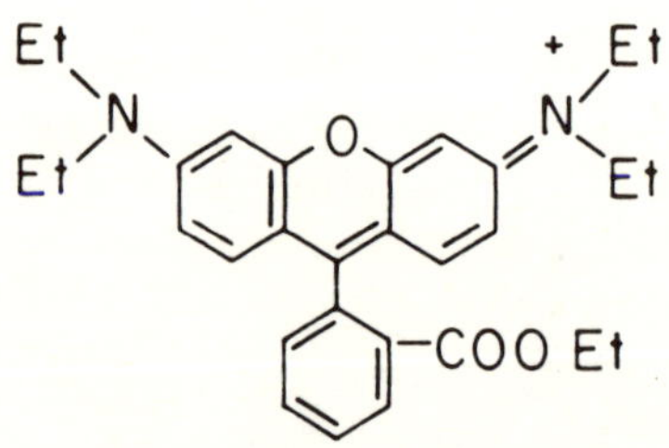

Rhodamine 3B

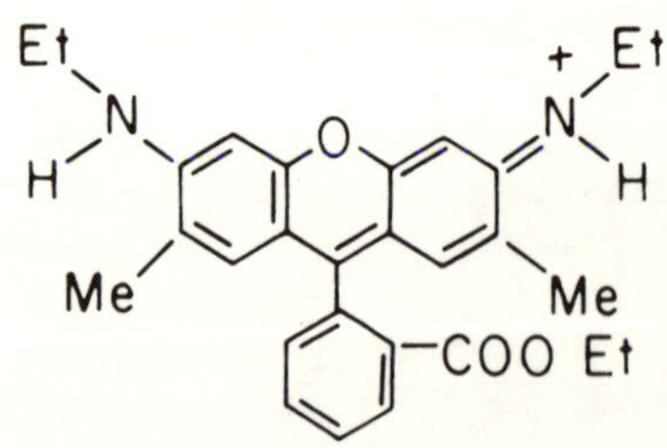

Rhodamine 6G

Fig. 1. Rhodamine Type Laser Dyes.

The equilibrium constants of 22° are between 10^3 and 10^4 M^{-1} the $_{o}k_{1}$'s are around 10^6 M^{-1} sec^{-1} and the $_{o}k_{-1}$'s around 10^2 sec^{-1} in both solvents.

Association in water is caused by hydrophobic interaction, resulting in plane-plane dimers. Head-to-tail aggregation if found in ethanol, caused by partial overlap of the xanthene planes and a contribution of hydrogen bonds.

REFERENCE

1. M.M. Wong, Z.A. Schelly, J. Phys. Chem., 78, 1891 (1974).

TEMPERATURE-JUMP STUDY OF THE RATES OF PROTON TRANSFER AND THE STRENGTH OF INTRAMOLECULAR HYDROGEN BONDS IN PERI-SUBSTITUTED NAPHTHALENES

Frank Hibbert

Department of Chemistry, Birkbeck College, University of London, Malet Street, London WC1E 7HX

ABSTRACT. Proton transfers from protonated 1,8-bis(dialkylamino) naphthalenes to hydroxide ion are slow, largely because of strong intramolecular hydrogen bonds in the protonated amines. The results are discussed with reference to the general conclusion that the extent to which the rate of thermodynamically favourable proton transfer from an intramolecularly hydrogen bonded acid is lowered below the diffusion limited value gives a measure of the strength of the hydrogen bond. The several assumptions leading to this conclusion are described. The N....H....N hydrogen bond in 1,8-bis (dimethylamino)naphthalene which results in a lowering of rate of ca.10^5 is particularly strong and is probably an important factor in determining the unusual basicity of 1,8-bis(dimethylamino) naphthalene.

INTRODUCTION

The report[1] that 1,8-bis(dimethylamino)naphthalene is a stronger base than expected for an aromatic amine by more than six orders of magnitude, and stronger than many aliphatic amines has prompted us to begin a kinetic and thermodynamic acid-base study of this and other peri-substituted naphthalenes.[2] A summary of our studies with compounds (I), (II) and (III) is presented in this paper. Details of the experiments and results are omitted since a full description will be given elsewhere;[3] the general conclusions will be developed here.

The usual basicity of 1,8-bis(dimethylamino)naphthalene (pK_4 ca.12.3)[1] compared, for example, with 1-dimethylaminonaphthalene (pK ca.4.4) could arise from several factors. Crowding of groups around the peri-positions and the nitrogen lone pair interaction

Wyn-Jones (ed.), Chemical and Biological Applications of Relaxation Spectrometry, 387–393.

could result in considerable strain in the amine. Relief of this strain on protonation may contribute towards the high pK.

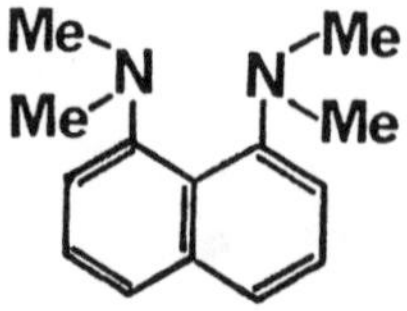

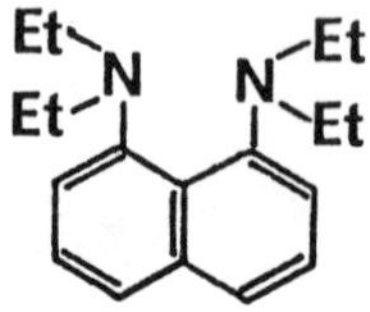

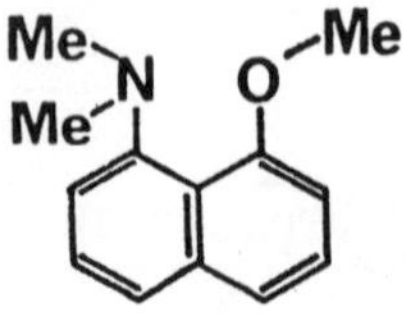

(I)1,8-bis(dimethyl amino)-naphthalene (II)1,8-bis(diethyl amino)-naphthalene (III)1-methoxy-8-dimethyl aminonaphthalene

An alternative explanation notes that conjugation between the amino groups and aromatic rings, which is often used to explain the low basicity of aromatic compared with aliphatic amines may be weakened in 1,8-bis(dimethylamino)naphthalene by steric effects. In addition there is evidence for an intramolecular hydrogen bond in the protonated amine[1,5] which cannot be present in the free amine, and this would also lead to a higher pK than normal.[6] Each of these factors is compatible with the observation[1] that introduction of up to three methyl groups into 1,8-diaminonaphthalene has a fairly small effect on the pK but introduction of the fourth methyl group increases the pK by about six units.

DISSOCIATION CONSTANTS

The pK values shown in Table 1 were determined spectrophotometrically for (I) and (II) in sodium hydroxide solutions and for (III) in phosphate buffers. The value for (I) in aqueous solution is within experimental error of the previous determination[1,7]. The larger pK for (II) compared with (I) is similar to the effect observed for other amines on replacing methyl by ethyl groups.[8] This means that if strain is an important factor in determining the pK values, there is no great increase in strain on substituting ethyl for methyl groups in 1,8-bis(dimethylamino)naphthalene.

RATES OF PROTON TRANSFER

Chemical relaxation times for equilibrium (1) with B as (I) and (II) have been determined using the temperature-jump method. Under our conditions the reciprocal relaxation time is given by $1/\tau = k_f[OH^-] + k_r[H_2O]$. The measured values of the forward and reverse rate coefficients are shown in Table 2.

$$BH^{+} + OH^{-} \underset{k_r}{\overset{k_f}{\rightleftharpoons}} B + H_2O \qquad (1)$$

TABLE 1

$$BH^{+} + H_2O \overset{K_a}{\rightleftharpoons} B + H_3O^{+}$$

B	pK_a at 25.0° and ionic strength 0.1M	
1,8-bis(dimethylamino) naphthalene (I)	12.1 ± 0.1	{aqueous}
1,8-bis(dimethylamino) naphthalene (I)	11.5 ± 0.2	{20% dioxan/water(v/v)}
1,8-bis(diethylamino) naphthalene (II)	12.7 ± 0.2	{20% dioxan/water(v/v)}
1-methoxy-8-dimethyl aminonaphthalene (III)	7.75± 0.03	{aqueous}

TABLE 2

B	$10^{-5}k_f/\text{lmol}^{-1}\text{s}^{-1}$	$10^{-2}k_r/\text{lmol}^{-1}\text{s}^{-1}$	
1,8-bis(dimethylamino) naphthalene	1.9 ± 0.4	0.70±10	{aqueous}
1,8-bis(dimethylamino)-naphthalene	4.6 ± 0.8	*7.3 ±0.6	{20% dioxan/ water (v/v)}
1,8-bis(diethylamino)-naphthalene	4.6 ± 1.3	*1.14±0.08	{20% dioxan/ water (v/v)}

*The results given in 20% dioxan/water (v/v) are values of $k_r\{H_2O\}$ (s^{-1}).

The extremely low values for the rates of proton transfer from the protonated amines are noteable. The reactions are all thermodynamically favourable in this direction but occur at rates which are well below the diffusion limited values (ca.3 x 10^{10} $\text{lmol}^{-1}\text{s}^{-1}$) expected for proton transfer from a normal nitrogen acid[9]. The factor most responsible for slow proton transfer is a strong intramolecular hydrogen bond in the protonated amines. The effect an internal hydrogen bond has on rates of proton transfer is well established,[9,10] but the magnitude of the effect observed here is unusually large.

SCHEME

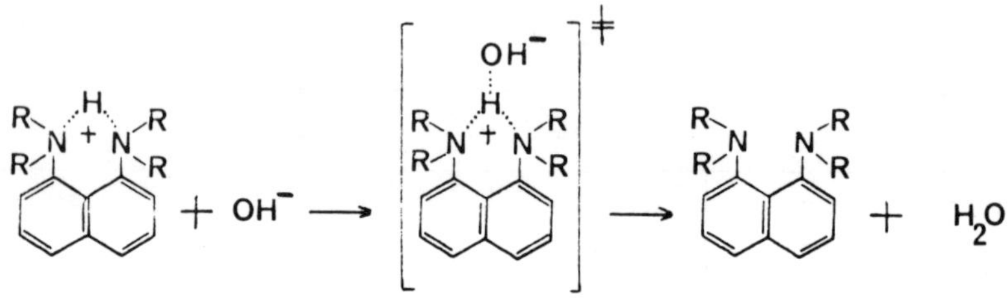

(2)

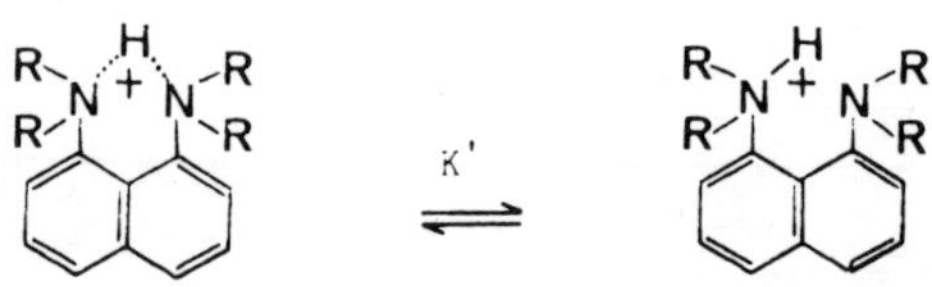

(3)

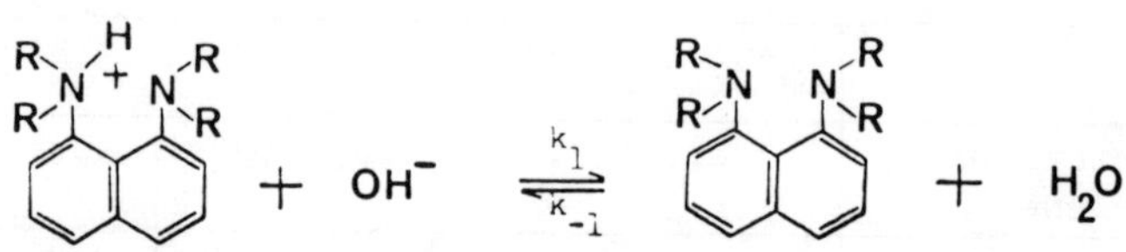

(4)

Two possible mechanisms can be written for proton removal from an acid with an intramolecular hydrogen bond. The first[11] which is shown in equation (2) in the Scheme involves direct attack on the hydrogen bonded proton. In the transition state, the hydrogen bond will be partly broken and the bond to hydroxide ion partly formed. An alternative mechanism, as originally suggested by Eigen,[9] involves a non-chelated form of the acid as an intermediate and the proton is removed from this species as shown in equations (3) and (4) of the Scheme. If we adopt this second mechanism, assuming K' is much less than unity and that the unimolecular breaking of the hydrogen bond in step (3) occurs more rapidly than step (4), the measured rate coefficients are $k_f = k_1 \times K'$ and $k_r = k_{-1}$. If we make the further assumption that attack by hydroxide ion on the nonchelated form of the acid is diffusion controlled (k_1 ca. 3×10^{10} $lmol^{-1}s^{-1}$) then a value for the equilibrium constant K' can be calculated. For 1,8-bis(dimethyl amino)naphthalene the value K' ca. 6×10^{-6} (ΔG^o = 30 $kJmol^{-1}$) is obtained. This analysis shows that the extent to which thermodynamically favourable proton transfer from a chelated acid is reduced below the diffusion limited value gives a measure of the strength of the intramolecular hydrogen bond. It is worthwhile to consider the usefulness of this observation as a general method of determining intramolecular hydrogen bond strengths.

The method is useful only when a proton transfer from the chelated acid can be studied which is thermodynamically favourable. Where proton transfer from the chelated acid is strongly unfavourable thermodynamically, rates of proton transfer similar to those for a normal acid will be observed. This conclusion will not be discussed at length here but is illustrated by the following results[10a] for the enol form of acetylacetone (EH). For (5) where proton transfer from the internally hydrogen bonded acid is thermodynamically favourable, a rate ca. 10^3 below the diffusion limit is observed characteristic of an internal hydrogen bond. However for

$$EH + OH^- \rightleftharpoons E^- + H_2O \qquad (5)$$

$$EH + H_2O \rightleftharpoons E^- + H_3O^+ \qquad (6)$$

(6) where the forward reaction is now unfavourable, the reverse rate is diffusion limited and the forward rate has the value expected for a normal acid with the pK of EH. If reaction (6) alone were studied, the kinetic results would give no indication of the presence of the intramolecular hydrogen bond in EH.

The reliability of this method of determining hydrogen bond strengths depends upon the validity of the assumptions made in the analysis. One of these assumptions, namely that k_1 has the diffusion limited value can be checked in the present case by comparing the relative acidities and rates of proton transfer for (I) and (II). The rates of proton transfer from the protonated amines to hydroxide ion differ by a factor of 10^2. Making the earlier assumptions this would mean that the hydrogen bond strengths as measured by the values of K' for (I) and (II) differ by a similar factor. The measured equilibrium constants (K) of equilibrium (1) differ for (I) and (II) by a factor of sixteen (Table 1). Providing K' is much less than unity, the value of K is given by the product K' x K", where K" is the equilibrium constant for equation (4) in the Scheme. Hence we obtain the unlikely result that the value of K" for 1,8-bis(dimethylamino)naphthalene is about six times lower than for 1,8-bis(diethylamino)naphthalene. This result is not compatible with the effect observed for other amines on replacing methyl by ethyl groups.[8] It is also incompatible with the reasonable assumption that formation of the free bases from the non-chelated acids (4) results in an increase (if anything) in strain which will be the same or greater for (II) compared with (I). We therefore conclude that one of the original assumptions is incorrect; it is probable that the rate of proton removal from the non-chelated form of protonated 1,8-bis(diethylamino)naphthalene is lower than the diffusion limit. This could arise from transition state strain caused by the diethylamino groups being forced into an unfavourable conformation in order to permit attack by hydroxide ion. If however the value of k_1 for (I) is diffusion limited, the analysis will give a reliable estimate of the strength of the intramolecular hydrogen bond in this case. Peri-substituted naphthalenes provide a wide range of molecules suitable for measuring relative hydrogen bond strengths using this method and further studies are in progress.

The presence of a strong hydrogen bond in protonated 1,8-bis (dimethylamino)naphthalene is an important factor in determining the unusually high thermodynamic basicity of this amine. Our result K' ca. 6×10^{-6} indicates that about five units of the enhanced pK may be attributed to the intramolecular hydrogen bond. It is not to be concluded however that strain is unimportant. For example strain in the non-chelated form of the acid may promote formation of the strong hydrogen bond. For 1-methoxy-8-dimethylamino naphthalene, the pK is only slightly higher than normal for an aromatic amine. In the absence of a kinetic study we cannot be certain that this is because the N—H....O hydrogen bond in this molecule is weak, but it seems likely.

REFERENCES

1. R.W. Alder, P.S. Bowman, W.R.S. Steele and D.R. Winterman, Chem. Comm., 723 (1968).
2. A preliminary report has been published; F. Hibbert, J. Chem. Soc. Chem. Comm., 463 (1973).
3. F. Hibbert, J. Chem. Soc. Perk II (1975) in press.
4. A. Fischer, G.J. Sutherland, R.D. Topson and J. Vaughan, J. Chem. Soc., 5949 (1965).
5. M.R. Truter and B.L. Vickery, J. Chem. Soc. Dalton, 395 (1972); E. Haselbach, A. Henriksson, F. Jachimowicz and J. Wirz, Helv. Chim Acta, 55, 1757 (1972).
6. E.S. Gould, 'Mechanism and Structure in Organic Chemistry' Holt, New York, pp. 30 and 209 (1960).
7. R.W. Alder, Bristol University, personal communication.
8. D.D. Perrin, 'Dissociation Constants of Organic Bases in Aqueous Solution', Butterworths, London (1965) supplement (1972).
9. M. Eigen, Ang. Chem. Internat. Edn., 3, 1 (1964).

10. See for example: (a) M. Eigen, W. Kruse, G. Maass and L. de Maeyer, Progr. Reaction Kinetics, 2, 285 (1964); (b) M. Eyring and D.L. Cole, 'Nobel Symposium 5', Interscience, 255 (1967); J.P. Birk, P.B. Check and J. Halpern, J. Amer. Chem. Soc., 90, 6959 (1968); M.C. Rose and J.E. Stuehr, J. Amer. Chem. Soc., 94, 5532 (1972); T. Fueno, O. Kajimoto, Y. Nishigaki and T. Yoshioka, J. Chem. Soc. Perk II, 738 (1973).
11. M.H. Miles, E.M. Eyring, W.W. Epstein and R.E. Ostlund, J. Phys. Chem., 69, 467 (1965).

FAST PROTON-TRANSFER REACTIONS IN APROTIC SOLVENTS

B.H. Robinson,
Chemical Laboratory, University of Kent at Canterbury

The content of the lecture, which dealt with some aspects of the kinetics of proton-transfer reactions between oxygen and nitrogen centres, has been reported and so no duplication is attempted. The relevant references are:

DIFFUSION-CONTROLLED PROTON-TRANSFER

1. R.M. Noyes, Prog. Reaction Kinetics, 1, 129 (1961).
2. K.S. Smitz and J.M. Schurr, J. Phys. Chem., 76, 534 (1972).
3. K. Sole and W.H. Stockmayer, Int. J. Chem. Kin. V, 733 (1973).
4. E.F. Caldin, J.E. Crooks and D. O'Donnell, J.C.S. Farad. 1, 69, 993, 1000 (1973).
5. G.D. Burfoot, E.F. Caldin and H. Goodman, J.C.S. Farad. 1, 70, 105 (1974).

'SLOW' PROTON-TRANSFER (Role of the Solvent and Isotope-Effects).

1. J.E. Crooks and B.H. Robinson, Trans. Farad. Soc., 67, 1707 (1971).
2. G. Gammons, B.H. Robinson and M.J. Stern, J.C.S. Chem. Comm. 1157 (1972)
3. E.F. Caldin and S. Mateo, J.C.S. Chem. Comm, 854 (1973).

Wyn-Jones (ed.), Chemical and Biological Applications of Relaxation Spectrometry, 395.

ULTRASONIC ABSORPTION STUDY OF AQUEOUS SOLUTIONS OF NUCLEOTIDES AND NUCLEOSIDES

J. LANG

C.N.R.S., CENTRE DE RECHERCHES SUR LES MACROMOLECULES,
6, rue Boussingault, 67083 STRASBOURG CEDEX - FRANCE

The ability of ultrasonic waves to perturb proton transfer equilibria has been widely used for the kinetic study of such equilibria.[1-9] In addition to rate constants, ultrasonic absorption measurements can also be used for the determination of the volume changes associated to proton transfer equilibria. Here we describe the results of an ultrasonic absorption study of proton transfer reactions occurring in aqueous solutions of nucleotides and nucleosides. It will be shown that proton transfer equilibria also explain some excess of ultrasonic absorption found for aqueous solutions of DNA and RNA. Finally the results of an ultrasonic study of aqueous solutions of adenosine 5'-monophosphate will be given. In this study, dimerization of the nucleotide coupled with proton transfer reactions have been used to interpret the data.

A brief description of the two types of proton transfers which have appeared in these studies will be given first. These two types of proton transfers are:

A. Proton transfer reactions with the solvent.

These reactions are the following:

$$XH \rightleftharpoons X^- + H^+ \qquad \text{Protolysis (1)}$$

$$XH + OH^- \rightleftharpoons X^- + H_2O \qquad \text{Hydrolysis (2)}$$

where XH and X^- stand, respectively, for the acid and the basic forms of the conjugate acid-base pair XH,X^- which will be called in the following the acid-base group X. One can show that for an aqueous solution of X the variation of α/f^2 (where α is the

Wyn-Jones (ed.), Chemical and Biological Applications of Relaxation Spectrometry, 397–405.

ultrasonic absorption coefficient of the solution and f the ultrasonic frequency) vs pH presents two maxima (see figure 1).

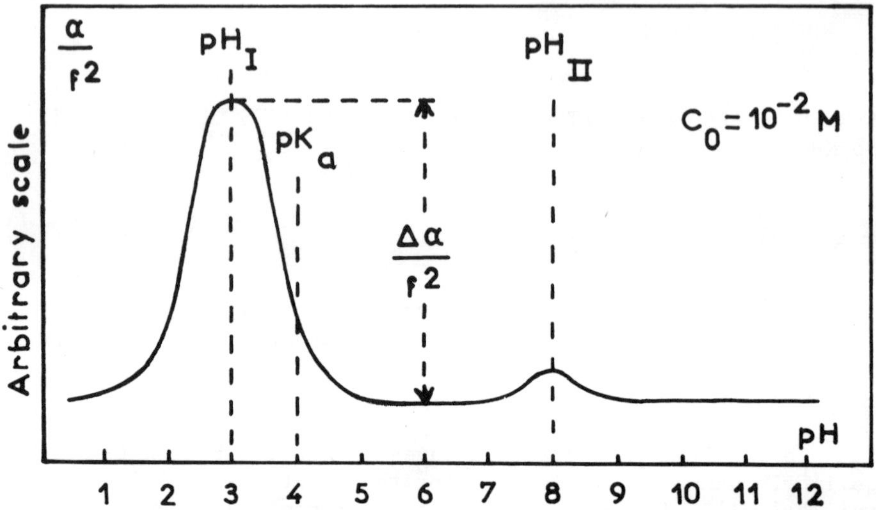

Fig. 1. Variation of α/f^2 with pH for reactions (1) and (2).

If the dissociation constant K_a of XH is much lower than the concentration C of X in the solution then these maxima appear at $pH_I = (pK_a - \log C)/2$ and $pH_{II} = (14 + pK_a + \log C)/2$. pH_I and pH_{II} are associated to the protolysis and hydrolysis reaction respectively. In most cases studied to date both pH_I or pH_{II} appeared around neutral pH. In the first situation the concentration of species X^- and H^+ are not large enough to give rise to a detectable ultrasonic absorption maximum within the accuracy of the experiments. In the second situation the concentrations of species XH and OH^- are too low for a maximum to be detected. For this reason the experimental variation α/f^2 vs pH for an acid-base group in aqueous solution usually gives rise to only one maximum. If the same molecule carries two acid-base groups then theoretically four maxima should appear but only two of these will be detected experimentally in most of the cases.

B. Proton transfer reactions between two acid-base groups

If two acid-base groups X and Y are present in the solution, beside the protolysis and the hydrolysis of X and Y, the following direct proton transfer reaction will occur:

$$XH + Y^- \rightleftharpoons YH + X^- \qquad \text{Proton Exchange} \qquad (3)$$

It can be shown for this proton exchange reaction that α/f^2 vs pH goes through a maximum for a value of the pH given by pH_{III} = $(pK_X + pK_Y)/2$ if the concentrations of X and Y are equal. K_X and K_Y stand for the dissociation constants of XH and YH, respectively.

There are two important differences between these two types of proton transfer reactions with respect to the ultrasonic absorption:

(i) The values of pH_I and pH_{II} depend on the concentration of the acid-base group X, whereas pH_{III} is independent of the concentrations of the acid-base groups X and Y as long as these two groups are kept at equal concentration. This difference may be used as a test to dintinguish which of these two types of proton transfer reactions cause the observed ultrasonic absorption maximum.

(ii) If one takes into account the concentrations only, reactions (3) may give rise to a detectable maximum at neutral pH since it involves the concentrations of the X and Y species only and not the H^+ or OH^- concentrations as it is the case for reactions (1) and (2).

1. ULTRASONIC ABSORPTION AS A FUNCTION OF THE pH

Aqueous solutions of eleven nucleotides, seven nucleosides, thymine and D-ribose 5-phosphate have been studied at 2.82 MHz and 25° as a function of pH between 1 and 13 and in some cases as a function of the concentration and the ionic strength.[10]

Table I summarizes the results obtained with 0.02M solutions and without added salt. The eleven nucleotides given in this table are the following: desoxyadenosine 5'-monophosphate (5'dAMP), adenosine 5'-monophosphate (5'AMP), adenosine 5'-phosphoramide (5'AMP NH_2), xanthosine 5'-monophosphate (5'XMP), deoxycytidine 5'-monophosphate (5'dCMP), cytidine 5'-monophosphate (5'CMP), inosine 5'-monophosphate (5'IMP), thymidine 5'-monophosphate (5'TMP) thymidine 5'-monophosphate diammonium salt (5'$TMPNH_4^+$) uridine 5'-monophosphate (5'UMP) and guanosine 5'-monophosphate (5'GMP). Table I gives the pH values at which ultrasonic absorption maxima appear in the acid (pH_A Expt) and in the alkaline (pH_B Expt) range. The amplitude of the corresponding maxima are given by $(\Delta\alpha/f^2)_A$ and $(\Delta\alpha/f^2)_B$. $(\Delta\alpha/f^2)_{A,B}$ is equal to $(\alpha/f^2)_{A,B} - (\alpha/f^2)_N$ where $(\alpha/f^2)_A$ and $(\alpha/f^2)_B$ are the values of α/f^2 at pH_A and pH_B, respectively, and $(\alpha/f^2)_N$ is the value of α/f^2 where no maximum occurs (see figures 1 and 2).

These different maxima have been attributed to the A or B types of proton transfer equilibria by comparing the variation of α/f^2 vs pH to the potentiometric titration curves which give the value of pK_N, pK_{P1} and pK_{P2}. pK_N, pK_{P1} and pK_{P2} are the pK'_as of the ionizable nitrogen of the base, of the first phosphoric acid function and of the secondary phosphoric acid

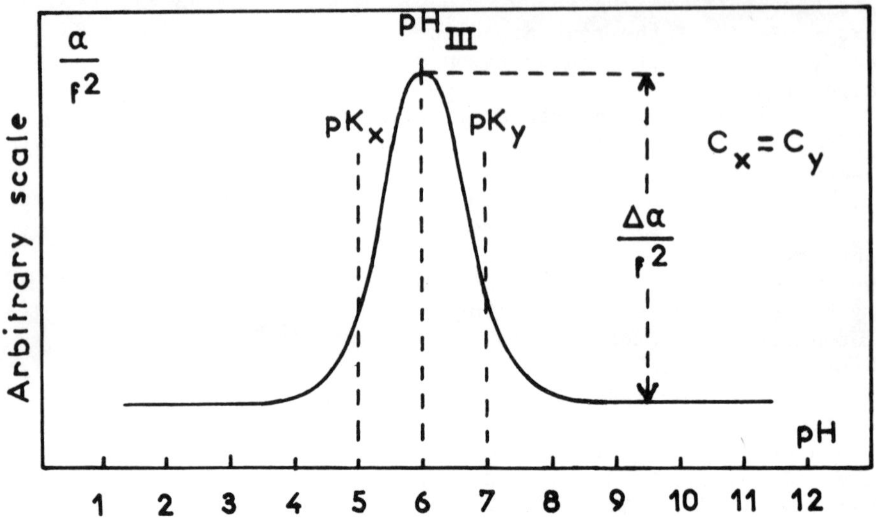

Fig. 2. Variation of α/f^2 with pH for reaction (3).

function, respectively. These pK'$_a$s have been used to calculate the value of pH_A and pH_B with the relations (a), (b), (c), (d) and (e) given in Table I, where C is the concentration of the compound studied. These relations are similar to pH_I, pH_{II} or pH_{III} depending on whether a protolysis (c) or an hydrolysis (e) or a proton transfer reaction between two acid-base groups (a, b, d) is at the origin of the maximum. The agreement between the experimental and calculated values of pH_A and pH_B is satisfactory.

The comparison of the curves α/f^2 vs pH with the potentiometric titration curves is not the only test used in order to assign the different maxima. The amplitude of the maxima can be calculated according to:

$$\frac{\Delta\alpha}{f^2} = 1.18 \times 10^{-7}(\Delta V_o)^2 \frac{\Gamma\tau}{1+4\pi^2 f^2 \tau^2}$$

where ΔV_o and τ are the volume change and the relaxation time associated to the reaction and Γ a term depending on the equilibrium concentrations and on the stoichiometric coefficients of the different species involved in the reaction.

From different data available in the literature[10] it has been possible, in some cases, to evaluate ΔV_o and τ and therefore to predict the value of $\Delta\alpha/f^2$ for a given reaction. For instance the maximum which appears for thymidine at $pH_B = 11.1$

TABLE I: Summary of the Results Obtained with 0.02 *M* Solutions at 2.82 MHz and 25°

	Acid range			Alkaline range		
	pH_A		$(\Delta\alpha/f^2)_A \times 10^{17}$ sec²/cm	pH_B		$(\Delta\alpha/f^2)_B \times 10^{17}$ sec²/cm
	Expt	Calcd[a]		Expt	Calcd[a]	
5′dAMP	5.05 ± 0.1	5.12 (b)	76 ± 4			0
5′AMP	5.05 ± 0.1	5.12 (b)	76 ± 4			0
5′AMPNH$_2$	3.15 ± 0.1	3.1 (a)	13.5 ± 2			
5′XMP	5.9 ± 0.1	5.88 (b)	102 ± 5			0
5′dCMP	5.5 ± 0.1	5.42 (b)	141 ± 7			0
5′CMP	5.35 ± 0.1	5.37 (b)	103 ± 5			0
5′IMP	4.1 ± 0.2	3.97 (c)	9 ± 2	10.55 ± 0.3	10.8 (e)	11 ± 2
5′TMP	4.2 ± 0.2	4.07 (c)	9 ± 2	11.1 ± 0.15	11.22 (e)	24 ± 3
5′TMPNH$_4^+$	4.1 ± 0.2	4.09 (c)	8.5 ± 2	9.7 ± 0.1	9.69 (d)	315 ± 10
5′UMP	4.3 ± 0.3	4.04 (c)	7.5 ± 2.5	10.8 ± 0.3	11.02 (e)	12 ± 3
5′GMP	4.2 ± 0.2	4.1 (c)	12.5 ± 2.5	10.9 ± 0.2	11.02 (e)	19 ± 3
		4.3 (b)				
Deoxyadenosine			0			0
Xanthosine	Precipitation occurs at pH <4.9					0
Deoxycytidine			0			0
Inosine			0	10.6 ± 0.2	10.6 (e)	11 ± 3
Thymidine			0	11.1 ± 0.2	11.05 (e)	20 ± 2
Uridine			0	10.45 ± 0.3	10.7 (e)	12.5 ± 2.5
Deoxyguanosine			0	10.7 ± 0.15	10.8 (e)	23 ± 2
Thymine			0	11 ± 0.1	11.1 (e)	30.5 ± 3
				11.9 ± 0.1		31.5 ± 3
D-Ribose 5-phosphate	4.15 ± 0.2	3.98 (c)	6 ± 2			

[a] The calculated values of $pH_{A,B}$ have been obtained using (a) $pH_A = \frac{1}{2}(pK_{P1} + pK_N)$; (b) $pH_A = \frac{1}{2}(pK_{P2} + pK_N)$; (c) $pH_A = \frac{1}{2}(pK_{P2} - \log c)$; (d) $pH_B = \frac{1}{2}(pK_N + pK_{NH_4^+})$ with $pK_{NH_4^+} = 9.26$; and (e) $pH_B = \frac{1}{2}(14 + pK_N + \log c)$.

has been attributed to the hydrolysis reaction of the lactam group of the base according to reaction (4);

$$- NH - CO + OH^- \rightleftharpoons - N - CO^- - + H_2O \qquad (4)$$

The calculated value of $\Delta\alpha/f^2$ has been found to be (20±7) x 10^{-17} cm^{-1} sec^2 in good agreement with the experimental one (20±2) x 10^{-17} cm^{-1} sec^2. When such a prediction was possible the agreement between the experimental and the calculated value of $\Delta\alpha/f^2$ was satisfactory.

From the results given in Table I some general features appear:

(1) The nucleosides do not show any maximum in the acid range. Inosine, thymidine and uridine do not have an ionizable group in the acid range and therefore no maximum is expected to appear in this pH range.

Deoxyadenosine, deoxycitidine and deoxyguanosine comprise a nitrogen atom protonable in the acid range and could therefore show an absorption maximum in this pH range. An evaluation of the amplitude of the maximum associated with the protolysis reaction for these three nucleosides gives a value which is less than 3 x 10^{-17} cm^{-1} sec^2 and thus will barely show on the plot α/f^2 vs pH. The nucleosides which present a lactam group ionizable in the alkaline range, i.e. iosine, thymidine and uridine, present in this pH range a maximum due to the hydrolysis given by reaction (4).

(2) For the nucleotides the maxima of relatively small amplitude which appear in the acid and in the alkaline range are due to protolysis and hydrolysis reactions. Maxima of larger amplitudes are due to proton transfer reactions between two acid-base groups. If one excepts 5'TMP NH_4^+ these maxima appear in the acid range for nucleotides which have a nitrogen atom protonable in this pH range. A more detailed study of this type of proton transfer reaction has been undertaken in the case of 5'AMP and will be described in section II.

Aqueous solutions of DNA also present a maximum in the variation of α/f^2 vs pH[11,12]. This maximum occurs at around pH 11.7. It has been suggested that this maximum might be associated with the perturbation by the sound wave of some equilibria between the native and the denatured forms of the DNA molecule. Experiments undertaken on denatured DNA suggested that this maximum was more likely due to proton transfer equilibria. The results obtained on nucleotides and nucleoside clearly indicate that, indeed, proton transfer reactions are responsible for this maximum. In DNA base pairing through H bonds involves the lactam groups of thymidine and guanine. Proton transfers on these groups remain, however, possible as shown by potentiometric studies[13], but the pK'_as are higher than those of the corresponding nucleotides. For example, the pK'_as of the lactam group of 5'TMP and 5'GMP are 10.2 and 9.8, respectively.[13] As a result the absorption maximum is found at around pH 11 for nucleotides and 11.7 for DNA, in excellent agreement with the values calculated using the relation giving pH_{II}.

In aqueous solutions of RNA a maximum also appears in the alkaline range which is due to the hydrolysis of the lactam group of uracil and guanine.

II - ULTRANONIC ABSORPTION OF AQUEOUS SOLUTIONS OF 5'AMP

The maximum which appear for 5'AMP at around pH 5.05[13] has been attributed essentially to a proton exchange reaction between different ionized forms of the nucleotide molecule[14] according to reaction (5).

$$A + D \rightleftharpoons 2C \qquad (5)$$

where A, D and C represent the three ionized forms of 5'AMP which are in a much higher concentration at around pH 5 than all other ionized forms of the nucleotide. These three ionized forms are:

$$A = N_1 - R - PO_4^{2-}$$

$$C = N_1 - R - PO_4H^-$$

$$D = {}^{+}HN_1 - R - PO_4H^-$$

where N_1 is the protonable nitrogen atom of the base and R the non-ionized part of the molecule. Reaction(5) represents the proton transfer reaction between the N_1 nitrogen and the secondary phosphoric acid function. A is a proton acceptor and D a donor. It must be pointed out that although the acid-base groups X and Y belong here to the same molecule, the proton exchange reaction occurs according to reaction III and not to an intramolecular reaction.

The variation of $(\Delta\alpha/f^2)_A$ has been studied as a function of the ultrasonic frequency for four concentrations of 5'AMP between 0.016 and 0.148M. For each concentration this variation is characterized by a single relaxation frequency f_R. However reaction (5) alone cannot account for the experimental variations of f_R and of the relaxation amplitude A_R vs. C. From several models tried it appears that model (6) can account satisfactory for all the experimental results found with 5'AMP if one assumes that H^+

$$\begin{array}{ccc} & A + H^{+} + C & \\ (1) \rightleftharpoons \; k_{13} & & k_{13'} \rightleftharpoons \; (2) \\ A + D & \underset{(4)}{\overset{k_{33'}}{\rightleftharpoons}} & 2\,C \underset{(3)}{\rightleftharpoons} C_2 \end{array} \qquad (6)$$

is in a steady state and that the dimerization is much faster than the proton transfer reactions. In this model C_2 represents a dimer.

Model (6) is characterized by two relaxations processes. The fast one, due to the dimerization, cannot fit the experimental results. The slowest which is principally due to the exchange reaction (4) of model (6) can account for the experimental results. Using Debye equation for diffusion controlled reaction it has been shown[14] that k_{13}'/k_{13} can be taken equal to 3. With this value the slowest process permits one to obtain the different parameters given in Table 2. $\Delta V_1 = \bar{V}_{C_2} - 2\bar{V}_C$ and $\Delta V_2 = \bar{V}_A + \bar{V}_D - 2\bar{V}_C$ are the volume changes associated to reaction (3) and (4) of model (6), respectively. K_D is the dimerization constant.

Table 2

Kinetic and Equilibrium data of reaction (6)

k_{13}' $\times 10^{-10}, M^{-1}$	k_{33}' $\times 10^{-9}, M^{-1}$	K_D M^{-1}	ΔV_1 cm^3/mol	ΔV_2 cm^3/mol
6.5±1	4.8±0.5	7±1	5±2	-26±2

The results on Table 2 demand the following remarks:

1 - The value of K_D $(7\pm1)M^{-1}$ is in a good agreement with the association constant of 5'AMP found to be $(8\pm2)M^{-1}$ from the results of Rossetti and Van Holde[15] using the method that these authors reported elsewhere.[16]

2 - The value found for k_{13}', is almost the 3/4 of the value 10^{11} M^{-1} sec^{-1} calculated for a diffusion controlled recombination between H^+ and $-O-PO_3^{2-}$. This value of 3/4 can account for the fact that the proton can bind on any of the three oxygen atoms of the phosphate group, the fourth being involved in the phosphate sugar linkage.

3 - The value found for k_{33}', is smaller by a factor of only 1.5 than the calculated one, thus showing that the proton exchange is almost diffusion controlled. This result is in agreement with the prediction of Eigen's theory[17] for a proton exchange between a donor and an acceptor characterized by pK_a's differing by more than 2.5 units.

4 - The values of ΔV_1 and ΔV_2 are given with their sign since the value of ΔV_2 is known from other techniques. ΔV_1 is small as expected for a nonionic reaction, and positive thus indicating that dimerization is accompanied by an increase of volume as for reactions involving the formation of hydrophobic bonds, such as

micelle formation in soap solutions[18] or association between potassium 3,5 dinitrobenzoate and 1- naphtol.[19] However these results are at variance with those reported by Porschke and Eggers[20] who found a negative volume change upon stacking of N_6 - N_9 dimethyladenine. More results will be necessary before this difference can be explained.

REFERENCES

1. M. Eigen and L. De Maeyer, in "Techniques of Organic Chemistry", Vol. III, Part II, S. L. Friess, E. S. Lewis and A. Weiss Berger, Ed., Interscience, New York, N.Y., (1963), p. 895.
2. J. Stuehr, in "Techniques of Chemistry", Vol. VI. Part II, Investigation of Rates and Mechanisms of Reactions, Gordon G. Hammes, Ed., Wiley-Interscience, New York, N.Y., (1974), p. 237.
3. H. Inoue, J. Sci. Hiroshima Univ., Ser. A-II, Vol. 34, 17, (1970).
4. M. Eigen, G. Maass and G. Schwarz, Z. Phys. Chem. NF 74, 319, (1971).
5. G. Maass and F. Peters, Ang. Chem. Int. Ed., 11, 428, (1972).
6. R. D. White and L. J. Stutsky, J. Phys. Chem., 76, 1327 (1972).
7. S. Nishikawa, T. Yasunaga and K. Takahashi, Bull. Chem. Soc. Jap., 46, 2992 (1973).
8. T. Sano and T. Yasunaga, J. Phys. Chem., 77, 2031 (1973).
9. S. Brun, J. E. Rassing and E. Wyn-Jones, Adv. Mol. Relax. Proc., 5, 313 (1973).
10. J. Lang, J. Sturm and R. Zana, J. Phys. Chem., 77, 2329, (1973).
11. J. Lang and R. Cerf, J. Chim. Phys., 66, 81, (1969).
12. J. Sturm, J. Lang and R. Zana, Biopolymers, 10, 2639 (1971).
13. A. Peacocke, Chem. Soc., Spec. Publ., No8, 139 (1957); L. Cavalieri and A. Stone, J. Amer. Chem. Soc., 77, 6499 (1955).
14. J. Lang, J. Sturm and R. Zana, J. Phys. Chem., 78, 80, (1974).
15. G. P. Rossetti and K. E. Van Holde, Biochem. Biophys. Res. Commun., 26, 717 (1967).
16. K. E. Van Holde and G. P. Rossetti, Biochemistry, 6, 2189 (1967).
17. M. Eigen, Angew. Chem., Int. Ed. Engl., 3, 1 (1964).
18. L. Benjamin, J. Phys. Chem., 70, 3790 (1966); K. Shinoda and T. Soda, ibid., 67, 2072 (1963); J. Corkill, J. Goodman and T. Walker, Trans. Faraday Soc., 63, 768 (1967).
19. A. K. Colter and E. Grunwald, J. Phys. Chem., 74, 3637, (1970).
20. D. Porschke and F. Eggers, Eur. J. Biochem., 26, 490, (1972).

ULTRASONIC ABSORPTION IN PROTEIN SOLUTION

L.J. Slutsky and R.D. White

Department of Chemistry, University of Washington,
Seattle, Washington 98195

INTRODUCTION

Since Carstensen and Schwan's work on aqueous solutions of hemoglobin[1] there have been a number of studies of excess acoustic absorption in solutions of biologically important macromolecules[2-7] and their synthetic analogs[8-10]. Relaxation times or relaxation spectra adequate to characterize the observed frequency dependence of the ultrasonic velocity and attenuation have been deduced[1-3,8,9], but the variety of chemical and structural equilibria possible in a solution of flexible polyelectrolytes is great enough to render the unambiguous identification of the microscopic process responsible for the excess absorption a matter of some difficulty. In individual cases arguments have been advanced nominating solvation[1-3,8] ionization[4,6], and conformational change[5,9] as the principal contributors to the observed relaxation spectrum in aqueous protein solution.

There exists a fair body of experimental information on the kinetics[11-14] and thermodynamics[15] of the acidic and basic ionization reactions of amino acids and simple polypeptides. In these systems the Debye-Smoluchowsky[16-17] theory of diffusion-controlled reaction rates has given a good account of the kinetic parameters. The data on simple systems constitutes a basic for the estimation of the contribution of perturbation of acid-base and internal charge -transfer equilibria to the acoustic absorption in protein solutions. We wish to explore here the extent to which it is possible to account for the frequency and pH dependence of the excess acoustic absorption in aqueous bovine hemoglobin on this basis, and to briefly discuss some convenient mathematical procedures for computing relaxation spectra in complex systems.

Wyn-Jones (ed.), Chemical and Biological Applications of Relaxation Spectrometry, 407–421.

IONIZATION REACTIONS OF HEMOGLOBIN

In a system of c components with r chemical or structural equilibria it will be possible to write r independent equations the form

$$\sum_i \mu_{ij} B_i = 0 \qquad (1)$$

where μ_{ij} is the coefficient of the ith chemical species in the jth reaction (μ_{ij} is here taken to be positive for products negative for reactants) and B_i designates the ith component. The change in the concentration of the ith component when the jth reaction is displaced from equilibrium serves to define the "degree of advancement", dx_j, of the jth reaction

$$\mu_{ij}\, dx_j = dc_i \qquad (2)$$

and the total change in the concentration of the ith component when all reactions are displaced from equilibrium is

$$dc_i = \sum_j \mu_{ij} dx_j \qquad (3)$$

The degree of advancement of the autoionization of water and the reactions

$$-AH + H_2O \underset{k_b}{\overset{k_f}{\rightleftharpoons}} A^- + H_3O^+ \qquad (4a)$$

$$-BH^+ + OH^- \underset{k_{b'}}{\overset{k_{f'}}{\rightleftharpoons}} B + H_2O \qquad (4b)$$

where -AH represents the protonated form of an acid residue (aspartyl, glutamyl, terminal carboxyl) and BH^+ the protonated form of a basic residue (histidyl, lysyl, arginyl, etc.) constitute a basis set sufficient to describe both the internal and overall state of ionization of the hemoglobin molecule. The volume changes for the proton-transfer reactions of amino acids and simple polypeptides are known. Orttung[18] has made a detailed calculation based on the Kirkwood-Tanford[19] theory from which the degree of ionization of individual residues or groups of residues in hemoglobin can be estimated.

The volume changes for the basis reactions are given in Table 1 as are the degrees of protonation of each residue as a function of pH.

Measured values[11] of k_f in Eq. 4a for the carboxylic acids range from $3.5 \times 10^{10} M^{-1} sec^{-1}$ for benzoic acid to $5 \times 10^{10} M^{-1} sec^{-1}$ for formic acid. We have measured k_f in Eq. 4b for a number of

TABLE 1. VOLUME AND ENTHALPY CHANGES FOR THE IONIZATION OF ACIDIC RESIDUES. FRACTION (f) OF RESIDUES PROTONATED AS A FUNCTION OF pH.

RESIDUE	BASE	ΔV IN CC/MOLE	ΔH[15] IN KCAL	FRACTION OF RESIDUES PROTONATED[18] pH 6	7.5	9	12	NUMBER OF RESIDUE/MOLECULE
ARGINYL$^+$	OH^-	25[9]	0	1.0	1.0	1.0	0.959	14
LYSYL$^+$	OH^-	25[9]	- 3	1.0	1.0	0.996	0.863	44
CYSTINYL	OH^-	6[20]	- 6	1.0	0.997	0.979	0.720	4
TYROSYL	OH^-	4[21]	- 6	1.0	.99	0.840	0.314	12
n-TERMINAL AMINO$^+$	OH^-	25[9]	- 3	0.930	0.87	0.711	0.058	4
HISTIDYL$^+$	OH^-	25[9]	- 6	0.784	0.67	.529	0.022	38
COOH	H_2O	- 10[15]	1.5	0.112	0.013	$5x10^{-4}$	0	72

amino acids and simple peptides[13]. Typical results in units of $10^{10}M^{-1}sec^{-1}$ are 1.9, 2.0, 2.1, 2.2, and 2.2 for glycine, threonine, triglycine, serine, and piperidine respectively. If the effective radius for reaction, r_d, is taken to be equal to an N-H--O hydrogen bond distance (2.7Å and a geometrically reasonable steric factor is chosen (0.6), Debye's result

$$k_{12} = \frac{\sigma 4\pi N z_A z_B e_0^2 (D_A + D_B)}{\epsilon kT\{\exp(z_A z_B e_0^2/\epsilon r_d kT)-1\}} \tag{5}$$

where N is Avogadro's number, σ a steric factor, e_0 is the electronic charge, z_A and z_B are the algebraic charges of the ions, ϵ is the dielectric constant of the solvent, D_A and D_B are the diffusion coefficients of the reacting ions, and r_d is an effective radius for reaction predicts values between $2.0 \times 10^{10}M^{-1}sec^{-1}$ (triglycine) and 2.2×10^{10} (glycine).

A plausible approach to the computation of the contribution of ionization and internal charge transfer reactions to the relaxation spectrum of hemoglobin is then to estimate the forward rate constants of Eqs. 4 from the conventional theory of diffusion controlled reaction rates, to obtain the reverse rate constants and equilibrium concentrations at any pH from Orttung's computed degree of ionization, and to solve the system of coupled first-order equations obtained by linearizing the kinetic equations (one for each acidic or basic residue) implied by Eqs. 4.

For a system of two reactions, if δ and ϵ represent respectively the degrees of advancement of 4a and 4b, then the linearized kinetic equations corresponding to Eqs. 4a and 4b are:

$$\begin{aligned} -\frac{d\delta}{dt} &= \left[k_b([H_3O^+] + [A^-]) + k_f\right]\delta + k_b[A^-]\epsilon \\ -\frac{d\epsilon}{dt} &= \left[(k_{b'}([H_3O^+] + [B]) + k_{f'}\right]\epsilon + k_{b'}[B]\delta \end{aligned} \tag{4c}$$

and the secular equation that determines the (circular) relaxation frequencies (ω) is

$$\begin{vmatrix} k_b([H_3O^+] + [A^-]) + k_f - \omega & k_b[A^-] \\ k_{b'}[B] & k_{b'}([H_3O^+] + [B]) + k_{f'} - \omega \end{vmatrix} = 0 \tag{4d}$$

where [X] represents the equilibrium molar concentration of species X.

In the simplest case, when the pK's of AH and BH^+ are equal, the relaxation frequencies are given by $\omega = k_b([H_3O^+] + [A^-] + [B]) + k_f$ and $\omega = k_b[H_3O^+] + k_f$. The corresponding normal reactions are $\frac{1}{2}(AH + BH^+) + H_2O = \frac{1}{2}(A^- + B) + H_3O^+$ and $AH + B \rightarrow A^- + BH^+$. In general the normal reactions will be mixtures of dissociation and internal transfer.

For example if BH^+ is a protonated imidiazole residue with a pK_a of 6.1 and AH a relatively acid phenolic residue (i.e. di-iodotyrosyl) with $pK_a = 6.5$, taking k_b to be $2 \times 10^{10} M^{-1} sec^{-1}$, the relaxation frequencies at pH 6.1 in 0.01 M solution properly computed from Eq. 4d are 26 MHz (principally ionization) and 4.4 kHz (principally internal proton transfer). If the off-diagonal elements of the kinetic matrix are arbitrarily set equal to zero as in the simplified treatments of Hussey and Emonds[3d] and O'Brien and Dunn[5c], the normal reactions are of necessity the independent acid dissociations of AH and BH^+ with relaxation frequencies of 9.2 and 16.0 MHz. It is we believe clear that within the framework of the Debye-Smoluchowski and Kirkwood-Tanford theories the normal proton-transfer reactions in protein solution are not simply individual group dissociations and the off-diagonal terms can not be justifiably eliminated from the calculation.

The eigenvalues of the kinetic matrix directly give the relaxation times. In order to calculate the acoustic absorption (α/f^2) it is necessary to compute the standard volume change, (ΔV), standard enthalpy change (ΔH), and chemical factor $\Gamma_r = dx_r/d\ln K_r$ (where K_r is the equilibrium constant for the rth normal reaction) for each of the normal reactions from the eigenvectors of the kinetic matrix and ΔV, ΔH, and K for the basis reactions. The absorption is then calculated by summing over the independent normal reactions.

$$\frac{\alpha}{f^2} = A + \sum_j \frac{C_j \tau_j}{1 + (2\pi\tau_j f)^2} \tag{6}$$

$$C = 2\pi^2 \rho C_o \bar{V}^2 RT \Gamma_M \left[(\beta\Delta H/C_p RT) - (\Delta V/\bar{V}RT)\right]^2$$

where the ρ is the density, C_0 the velocity of sound, $\bar{V}$ the volume per mole of solution, β the coefficient of thermal expansion, C_p the molar heat capacity at constant pressure.

The results of such a computation for a 0.0023 M solution of hemoglobin at pH 12 and 0°C is compared with experiment in Figure 1. The computed relaxation frequencies and the corresponding values of C are given in Table 2. The important contributors to the excess absorption over most of the range displayed in Figure 1 are normal reactions 6 (qualitatively the reaction between protonated crystal residues and lysyl or OH^-) and 7 (principally proton transfer between lysyl and tyrosyl residues). Normal reaction 1 (ionization of the carboxyl group) contributes nothing

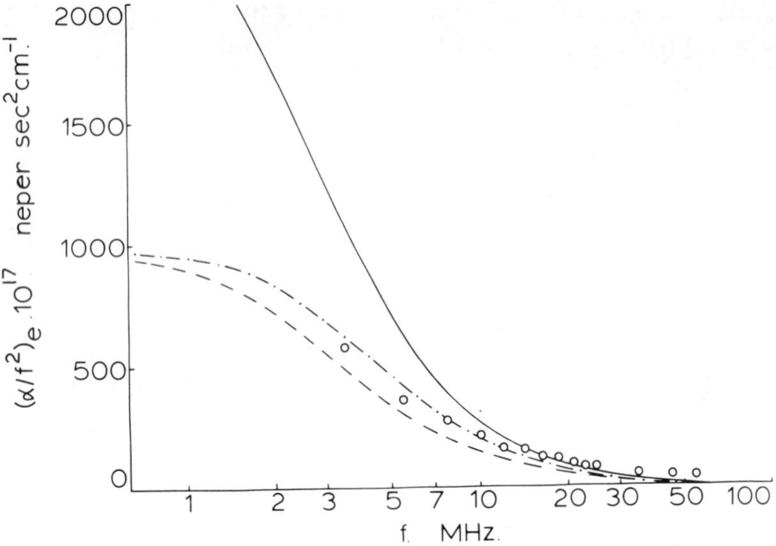

Fig. 1. The frequency dependence of the excess acoustic absorption of .0023 M oxyhemoglobin at 0°C and pH 12. The open circles represent our data while the broken curve (----) shows the calculated absorption due to dissociation only. The curve (-.-.-.-) represents the effect of including direct transfers between tyrosyl-lysyl and tyrosyl-cystinyl pairs assuming independent probabilities for simultaneous proton occupation of a pair of neighbouring residues, and the solid curve shows the calculated result when direct transfers are included and the joint probabilities are computed in such a way as to maximize the chance that a pair is only single ionized.

to the relaxation spectrum at high pH. Since it is presumed here that ΔV for the protonation of all $-NH_2$ groups is the same, normal reaction 3 (principally proton exchange between lysyl and arginyl residues) makes only a small contribution; ΔV for this transfer being zero. At the higher frequencies ($\gtrsim$100 MHz) normal reaction 2 (hydrolysis of lysyl and arginyl residues) becomes the principal contributor to the excess absorption. The remaining eigenvectors do not have simple character approximate chemical equations being: $-(4CYS^- + 3HIS + NTA + 4LYSH^+ + ARCH^+ \rightarrow CYSH + 3HISH^+ + NTAH + 4LYS + ARG)$; 5. $3CYS^- + LYSH^+ + NTAH^+ + 2HIS + 3H2O \rightarrow 3CYSH + 3OH^- + LYS + NTA + 2HISH^+$; 8. $3COOH + OH^- + 2LYS \rightarrow 3COO^- + H_2O + 2LYSH^+$).

At pH 12 this calculation gives quite a good account of the experimentally determined excess absorption; however, the procedure outlined above cannot account for the absorption near neutral pH.

O'Brien and Dunn[5c] have suggested that dissociative proton

TABLE 2. RELAXATION SPECTRUM OF HEMOGLOBIN, pH 12 AND 0°C

	f_R in MHz	C sec cm.$^{-1}$	$C\tau$ $sec^2 cm^{-1}$
1	1800	1.2×10^{-17}	1.1×10^{-27}
2	179	1.66×10^{-7}	1.5×10^{-17}
3	15	3.4×10^{-10}	0.38×10^{-17}
4	5.115	3.3×10^{-10}	1.0×10^{-17}
5	5.110	0.27×10^{-10}	0.08×10^{-17}
6	4.6	1.3×10^{-7}	464×10^{-17}
7	2.9	$.88 \times 10^{-7}$	483×10^{-17}
8	.004	0.012×10^{-10}	4.6×10^{-17}

transfers of the histidyl residues contribute to a broad peak in an acoustic absorption at megacycle frequencies in the neighbourhood of neutral pH. However, at neutral pH, 10^{-7} is a strict upper limit to Γ_r, $2\pi^2\rho c_o(\Delta V/RT)^2$ is approximately 4×10^{-5}, C in equation 6 is then $\approx 4 \times 10^{-12}$, and even for a 1 megacycle relaxation frequency the excess value of α/f^2 will not exceed 10^{-18} neper sec^2/cm and thus it seems unlikely that simple ionization of histidyl residues can contribute significantly at neutrality.

Near neutrality the important eigenvectors of the kinetic matrix will correspond to reactions of the form,

$$-AH + B = BH^+ + A^- \qquad (7)$$

that is to intramolecular proton transfers with -AH a carboxyl or tyrosyl and B a histidyl group. However, intramolecular proton transfers based on Eq. 4 proceed of necessity by the mechanism

$$-AH + H_2O = A^- + H_3O^+$$

$$H_3O^+ + B = BH^+ + H_2O$$

or the equivalent with hydroxyl ion acting as the base. At neutral pH, where the concentration of hydronium and hydroxide ions are in the neighbourhood of 10^{-7} M even the most optimistic assumptions about the reaction radii and steric factors in the Debye theory of diffusion-controlled dissociation and recombination reactions predict relaxation frequencies of the order of 10 KHz and, while internal transfers by a dissociative mechanism may contribute to the very low frequency spectrum observed by Carstensen[1], they do not appear to be important at frequencies higher than 1 MHz.

There remains the possibility of a more direct proton-transfer mechanism. The histidyl residues G19β, E13β, H5α, D6α, EF1α, all have carboxyl neighbours within a 5Å radius, tyrosine H23α has a cystine neighbour at 5.5Å. There is some reason to expect that proton exchange between near neighbours is significantly faster than dissociation and recombination in dilute solution, but there is not a large body of information on the rates of direct proton exchange. The isomeric aminobenzoic acids constitute a possible set of model compounds. Ultrasonic attenuation in 0.5 M o-amino benzoic acid in methanol is given in Fig. 2. The relaxation frequency is 20 MHz, presumably corresponding to direct transfer in a cyclic hydrogen bonded system. The relaxation frequencies in methanolic solutions of the meta and para isomers are, as would be expected, too low to be conveniently determined by ultrasonic techniques.

Inoue[11b] estimates the relaxation frequency for the unimolecular transfer of a proton between the phenol and amino groups of tyrosine to be 18.3 MHz, and finds a value of 0.35 MHz for the $-SH + NH_2 \rightarrow S^- + NH_3^+$ transfer in cysteine .

Although there is justification for the assignment of megacycle relaxation frequencies to proton exchange between proximal amino and carboxyl groups the exact dependence of the rate on distance is still a matter of conjecture. Moreover, the set of basis reactions given in Eq. 4 is complete and introduction of the equations for direct proton-transfer reactions results in a system of linearly dependent kinetic equations. Before considering the possible role of direct transfer mechanisms in the relaxation spectrum of proteins we wish to briefly discuss the development of a convenient algorithm for the computation of α due to chemical relaxation in a system of kinetically distinct but thermodynamically redundant reactions.

ABSORPTION DUE TO CHEMICAL RELAXATION IN COMPLEX SYSTEMS

Starting with all species in Eq. 1 at their equilibrium concentration ^oC_i and introducing the definition $R_j = k_{fj}\prod_i {}^oC_i^{\mu_{ji}}$ the mass-action expression for the rate of the jth reaction becomes for small displacements from equilibrium,

$$\frac{dx_j}{dt}(t) = -R_j \sum_{ij'} \frac{\mu_{ij}\mu_{ij'}}{C_i} x_{j'}(t) \qquad (8)$$

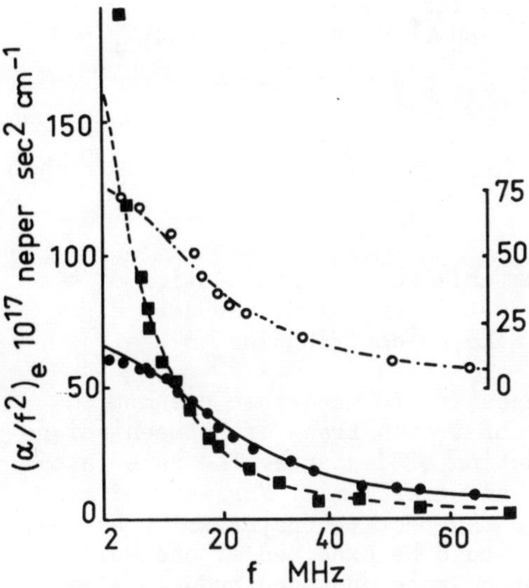

Fig. 2. Frequency dependence of the excess acoustic absorption in 0.5 M solutions of o-aminobenzoic acid in methanol at 25°C (solid circles) and 0°C (squares). The open circles and right-hand scale give the experimental results for o-aminobenzoic acid 1 M in acetone at 25°C. The continuous curves are generated from Eqs. 28-3 using the parameters in Table 2.

or, in matrix notation,

$$\frac{d\tilde{x}}{dt}(t) = -\tilde{R}\tilde{A}\tilde{X}(t) \tag{9}$$

where $\tilde{R}$ is a diagonal matrix with elements R_j and $A_{jj'} = \sum_i \mu_{ij}\mu_{ij'}/{}^oC_i$. With the usual substitution $\tilde{X}(t) = \tilde{X}e^{-\lambda t}$, Eq. 9 becomes

$$\lambda\tilde{X} = \tilde{R}\tilde{A}\tilde{X} \tag{10}$$

For computational purposes it is convenient to symmetrize via the substitution[23] $\tilde{Y} = \tilde{R}^{-1/2}\tilde{X}$ and obtain the eigenvalues λ_i and diagonalizing matrix $\tilde{S}$ of

$$\lambda\tilde{Y} = (\tilde{R}^{1/2}\ \tilde{A}\tilde{R}^{1/2})\tilde{Y} \tag{11}$$

The ultrasonic absorption is given by[25]

$$\frac{\alpha}{f^2} = \frac{2\pi^2 fC_o}{RT(1000)}\ \bar{V}^2 \sum_k \left(\frac{\Delta V_k^*}{\bar{V}} - \frac{\beta}{C_p}\ \Delta H_k^*\right)^2 \frac{\lambda_k^{-2}}{1 + \left[\left(\frac{2\pi f}{\lambda_k}\right)\right]^2} \tag{12}$$

If the volume and enthalpy changes for the basis reactions are represented as column vectors $\Delta\tilde{V}$ and $\Delta\tilde{H}$ with components ΔV_j and ΔH_j then[24]

$$\Delta\tilde{V}^* = (\tilde{S}^{-1}\tilde{R}^{1/2})\Delta\tilde{V}$$

$$\Delta\tilde{H}^* = (\tilde{S}^{-1}\tilde{R}^{1/2})\Delta\tilde{H} \qquad (13)$$

Equations 12 and 13 are suitable for the computation of α when a linearly-dependent basis-set is employed; the zero eigen values obtained in such a case make no contribution to the absorption.

To proceed further it is necessary to make some assumptions about the variation in the rate of proton transfer between neighbouring groups ℓ and m as a function of distance. The rate in o-aminobenzoic acid is presumably an upper limit. For want of more detailed information we have assumed that the rate varies as the inverse cube of the distance as would be expected if one were dealing with a diffusion-controlled rate enhanced by a locally high concentration of hydrogen ion due to the presence of a neighbouring acid group.

Potentially the acoustic absorption associated with a reaction of the form of Eq. 7 is quite high. One transfer per α-β dimer with an equilibrium constant of one and a relaxation frequency of 3 MHz could account for about 1/4 of the observed excess absorption in neutral solution. Most of the 38 histidyl residues have at least one carboxyl neighbour within 8A.

However, in any analysis based on the ionization of a single group in the average electrostatic potential of its neighbours, the fraction, $f_{\ell m}$, of protonated residues ℓ with unprotonated neighbours m will be given by $f_m = f_\ell(1 - f_m)$. Groups of similar pKa will be either simultaneously protonated or simultaneously unpopulated and the concentration of neighbouring pairs with an approximately even distribution of protonated and unprotonated forms will be high only in the narrow range of pH in which $f_\ell \approx f_m \approx 1/2$.

The effect of including direct transfers subject to the approximations discussed in the foregoing on the calculated acoustic absorption is shown in Figs. 3-5. At high pH the important transfers are between the tyrosine-lysine $H_{22\alpha}$-$H_{23\alpha}$ pair and the tyrosine cystine $H_{23\beta}$-$F_{9\beta}$ pairs. At lower pH, histidine-tyrosine pairs (B1α-B5α, $F_{8\alpha}$-$H_{23\alpha}$ and $E_{20\beta}$-$H_{8\beta}$) become important contributors. Near neutrality a large number of carboxylhistidine pairs must be considered.

These calculations are expected to underestimate the contribution of internal transfers. Protonating the near neighbour of a given residue obviously reduced the probability that the residue itself is protonated and taking the joint probability to be the product of the individual probabilities does not properly count

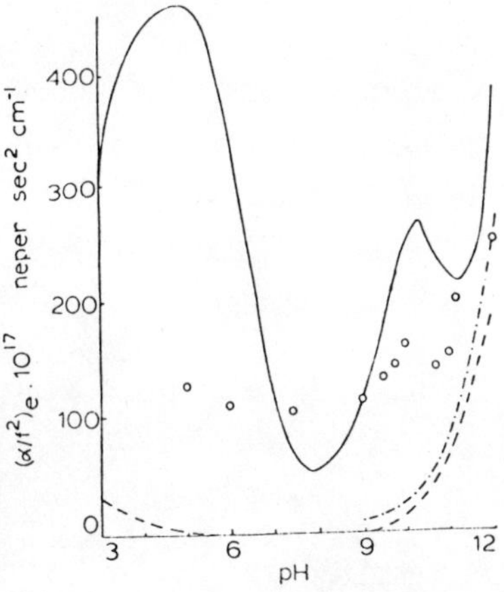

Fig. 3. Experimental and Theoretical pH-Dependence of the Acoustic Absorption in 0.0023 M Bovine Hemoglobin at 0°C and 7.9 MHz. The broken curve represents the effect of dissociative transfer only, -·-·-· includes direct transfers between tyrosyls - lysyl, and histidyl-cystyl pairs calculated assuming independent probabilities of protonation of adjoining residues. The solid curve includes all neighbouring pairs assuming that no protonated residue has a protonated neighbour.

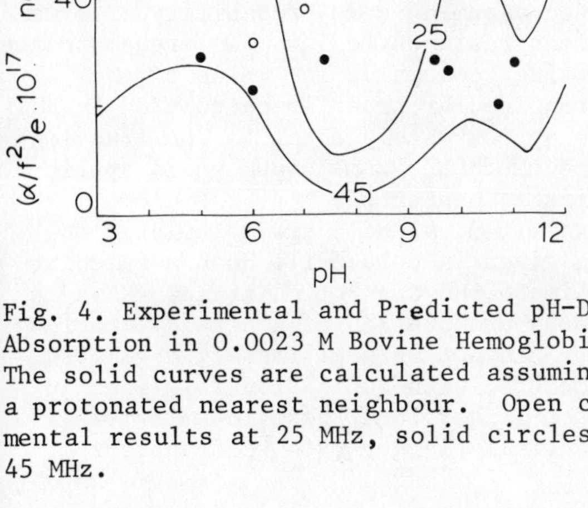

Fig. 4. Experimental and Predicted pH-Dependence of the Acoustic Absorption in 0.0023 M Bovine Hemoglobin at 0°C at 25 and 45 MHz. The solid curves are calculated assuming no protonated residue has a protonated nearest neighbour. Open circles represent the experimental results at 25 MHz, solid circles the experimental results at 45 MHz.

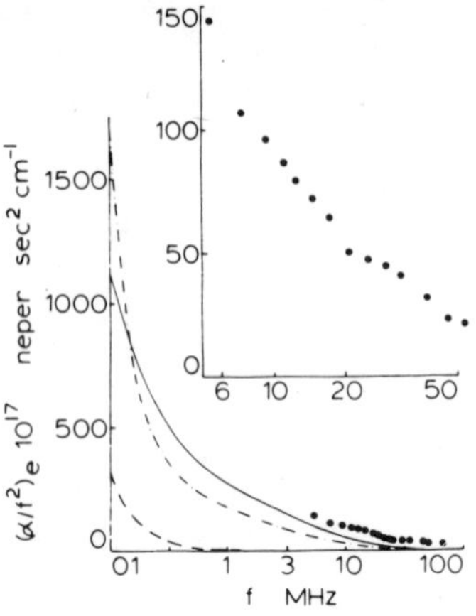

Fig. 5. Excess Ultrasonic Absorption in 0.0023 M Aqueous Bovine Hemoglobin at 0°C and Neutral pH. The lowest dashed curve represent the effect of dissociation, the broken curve includes the effect of direct intramolecular transfers calculated assuming independent probability of protonation of adjacent groups. The upper solid curve is calculated on the basis of the assumption that no carboxyl group immediately adjacent to a protonated histidyl is protonated at neutral pH. The points represent experimental results.

the number of pairs of unoccupied sites near protonated sites. Near neutral pH most carboxyl sites are not protonated. At separations of 5-10 Å the electrostatic potential due to a proton charge is sufficiently great so that it is more reasonable to assume that none of the few protonated carboxyl sites is adjacent to a positively charged histidine than to assume random mixing of occupied and unoccupied sites for proton binding. In Table 3, the probabilities of the four possible states of ionization of a carboxyl-histidyl pair calculated assuming zero probability of simultaneous protonation of adjacent residues (column II) are contrasted with the result of random mixing (columnI). The value of the chemical factor for the intramolecular transfer calculated in this way, $f_C f_H/(f_C + f_H)$, will be approximately equal to the fraction of protonated carboxyl groups. Much the same argument is applicable to tyrosyl-histidyl pairs at higher pH.

The solid curves in Figs. 1, 3, 4 and 5 are calculated on the basis of the populations given in column II. Such a calculation will necessarily overestimate the absorption at low pH where the probability of simultaneous protonation of a histidyl residue and a carboxyl near neighbour can not be legitimately neglected.

The volume changes associated with ionization reactions in nonaqueous media are very much larger than those in water[15,22]. If one were to assign substantially larger ΔV's to transfers

TABLE 3. PROBABILITY OF STATES OF PROTONATION OF PROXIMAL CARBOXYL-HISTIDYL PAIRS

State of Pair		Probability of State	
		I	II
-COOH	$\overset{+}{\text{H-N}}$	$f_C f_H$	0
$-COO^-$	$\overset{+}{\text{H-N}}$	$(1 - f_C)f_H$	f_H
-COOH	N	$f_C(1 - f_H)$	f_C
$-COO^-$	N	$(1 - f_C)(1 - f_H)$	$1 - (f_C + f_H)$

between pairs of interior residues which are in an essentially nonaqueous environment, the calculated absorption would be considerably larger. Moreover no volume change is assumed here for transfers between pairs of histidyl residues. However, since the volume change for ionization reactions is primarily electrostrictive in origin, one would in fact expect a substantial volume change for proton-transfer between residues in different electrostatic environments. The contribution of such transfers to the acoustic absorption is potentially quite large, both in protein solution and in solutions of uniform polypeptides in the region of the helix-coil transition where it may legitimately be supposed that the change in pK between helical and random-coil residues is small, yet significant differences in local dielectric properties exist.

The calculations given here account for the broad distribution of relaxation times necessary to explain the observed acoustic absorption, but underestimate the observed amplitudes. It has been argued in the foregoing that these calculations are likely to underestimate the amplitude and thus that perturbation of internal charge-transfer equilibria constitutes a plausible explanation of the ultrasonic absorption in protein solutions at megacycle frequencies near physiological pH.

CONCLUSIONS

Dissociative ionization reactions are not likely to make an important contribution to the relaxation spectrum in protein solutions at megacycle frequencies in the neighbourhood of physiological pH, but they are likely to be important contributors to the relaxation spectrum at high and low pH. If one accepts the premise that charge transfer between proximal acidic and basic residues occurs at megacycle frequencies, then internal ionization reactions involving histidyl-carboxyl pairs and histidyl-tyrosyl pairs are likely to be significant sources of absorption at megacycle frequencies even near physiological pH. There is not at present a sufficient body of experimental results on simple compounds to allow one to arrive at a kinetic model with any degree of confidence, but such results as are available support the proposition that relaxation of internal charge-transfer equilibria is a major contributor to the ultrasonic relaxation spectrum in protein solution at neutral pH.

REFERENCES

1. E.L. Carstensen and H.P. Schwan, J. Acoust. Soc. Amer., 31, 305 (1959). F.W. Kremkau, E.L. Carstensen and W.G. Aldridge, ibid., 53, 1448 (1973).
2. (a) F. Schneider, F. Muller-Landau and A. Mayer, Biopolymers, 8, 537 (1969).
   (b) P.D. Edmonds, Biochim. Biophys. Acta, 63, 216 (1962).
   (c) M. Hussey and P.D. Edmonds, J. Phys. Chem., 75, 4012 (1971).
3. H. Inoue, J. Sci., Hiroshima Univ., A34, 37 (1970).
4. R. Zana and J. Lang, J. Phys. Chem., 74, 2735 (1970).
5. (a) F. Dunn and L.W. Kessler, J. Phys. Chem., 74, 2736 (1970).
   (b) L.W. Kessler and F. Dunn, J. Phys. Chem., 73, 4256 (1969)
   (c) W.D. O'Brien, Jr. and F. Dunn, J. Phys. Chem., 76, 528 (1972).
6. R.D. White and L.J. Slutsky, J. Col. Inter. Science, 37, 727 (1971). Biopolymers, 11, 1973 (1972).
7. S.A. Hawley and F. Dunn, J. Chem. Phys., 50, 3523 (1969).
8. J. Burke, G. Hammes, and T. Lewis, J. Chem. Phys., 42, 3520 (1965). G.G. Hammes and P.N. Roberts, J. Am. Chem. Soc., 91, 1812 (1969).
9. (a) R. Parker, L.J. Slutsky and J. Applegate, J. Phys. Chem. 72, 3177 (1968); 70, 3018 (1966).
   (b) A.D. Barksdale and J.E. Stuehr, J. Am. Chem. Soc., 94, 3334 (1972).
10. R. Zana and C. Tondre, Biopolymers, 10, 2635 (1971).
11. (a) M. Eigen and L. de Maeyer, "Technique of Organic Chemistry", Vol. VIII, Part II, A. Weissberger, Jr., Ed., John Wiley and Sons, Inc., N.Y., 1961.
   (b) H. Inoue, J. Sci. Hiroshima Univ., A34, 17 (1970).
   (c) D. Grimshaw, P.J. Heywood and E. Wyn-Jones, J.C.S., Faraday Soc. II, 69, 168 (1973), ibid., 69, 756 (1973).
   (d) S. Brun, J.E. Rassing and E. Wyn-Jones, Advances in Molecular Relaxation Processes., 5, 313 (1973).
12. M. Eigen and E. Eyring, J. Am. Chem. Soc., 84, 3254 (1962).
13. (a) K. Applegate, L.J. Slutsky and R.C. Parker, ibid., 90, 6909 (1968).
   (b) R.D. White, L.J. Slutsky and S. Pattison, J. Phys. Chem. 75, 161 (1971).

14. R.D. White and L.J. Slutsky, J. Phys. Chem., 76, 1327 (1974).
15. E.J. Cohn and J.T. Edsall, Proteins, Amino Acids and Peptides as Ions and Dipolar Ions, Reinhold Publishing Corp., New York, 1943.
16. P. Debye, Trans. Electrochem. Soc., 82, 265 (1942).
17. M.V. Smoluchowski, Z. Physik. Chem., 92, 129 (1917).
18. W.H. Orttung, J. Am. Chem. Soc., 91, 162 (1969). Biochem., 9, 2394 (1970).
19. C. Tanford and J.G. Kirkwood, J. Am. Chem. Soc., 79, 5333 (1957).
20. A. Elis, J. Chem. Soc., 1961, 4687.
21. S. Hamman and S. Lim, Austral. J. Chem., 1, 329 (1934).
22. G. Castellan, Ber. Bunsenges. Physik. Chem., 67, 898 (1963).
23. P.R. Schimmel, J. Chem. Phys., 54, 4136 (1971).
24. R.D. White (to be published).

FAST RELAXATIONS IN HUMAN CARBONIC ANHYDRASE COMPLEXES WITH SULFONAMIDES

I. Giannini and G. Sodini

Laboratori Ricerche di Base, Snamprogetti
Monterontondo, Rome, Italy

ABSTRACT. The relaxation spectra of complexes of human carbonic anhydrase with aromatic sulfonamides was investigated by standard T-Jump and laser-photolysis techniques. A quite fast conformational effect occurs in the 200 μs region. Two more faster relaxation effects are seen in the 0.1 to 10 μs region in photolysis experiments. The effect of substitution of the active site Zn with various metals (Co, Mn, Cu, Ni) on the slowest effect was systematically investigated. The time constant for this effect and enzyme activity in the various substituted enzymes are correlated. This provides direct experimental support for the hypothesis that greater catalytic activity of the Zn and Co enzymes is connected with their capacity of undergoing faster changes in the coordination pattern. The spectral changes related with the slowest relaxation in the Co enzyme suggest that the observed event is a transition between distorted and undistorted tetrahedral structures. The faster relaxation effects are buffer dependent and can be tentatively ascribed to protonation of the essential group at the active site with pH $\sim$ 7.

INTRODUCTION

Specificity is a quite well known general feature of enzyme catalyzed reactions. Specific probes are therefore particularly useful in the study of their kinetic properties. Ultrasonic methods, which are frequently used in the analysis of very fast reactions, are not satisfactory in this connection.

In this work we study some kinetic properties of a very efficient enzyme, namely Human Carbonic Anhydrase (HCA), using pulsed laser perturbation[1] and other standard techniques which

Wyn-Jones (ed.), *Chemical and Biological Applications of Relaxation Spectrometry*, 423–432.

allow spectrophotometric detection of events occuring at the active site. HCA is a well known metal enzyme[2,3,4] from red cells, which catalyses the hydration of CO_2, presumably by some kind of activation of water molecules "frozen" at the active site[2,5]; some important aspects of its activity remain unsolved, namely: (i) the function and high specificity of the metal atom (Zn or Co) in the active site[3,6]; (ii) the mechanism of proton transfer at the active site[2,7,8,9].

We present here some preliminary results of kinetic studies on complexes of HCA with aromatic sulfonamides; the latter are very good inhibitors of this enzyme[2,10,11] and were chosen because, as was stated a long time ago by Pauling[12], good inhibitors can be tentatively considered as transition state analogues. Kinetic data on the interaction between sulfonamides and HCA were obtained in the past[10,11] by stopped flow measurements; only one slow binding step was observed with the time constant in the range of some sec^{-1}; however the dependence of the relaxation times upon pH, enzyme subforms and on the structure of the sulfonamides suggested that the slow process observed was coupled to faster relaxations. The binding constants of the sulfonamides used in this work were $\sim 10^6\ M^{-1}$; it was therefore possible to perform experiments at a concentration range where the coloured inhibitor is all bound to the enzyme. The preliminary results reported below could partially answer to the above questions.

RESULTS

(a) Relaxation spectrum of complexes between HCA with Neoprontosil Laser photolysis experiments.

Neoprontosil is a azosulfonamide presenting an intense absorption band in the 500 nm region[9]; it has the following formula

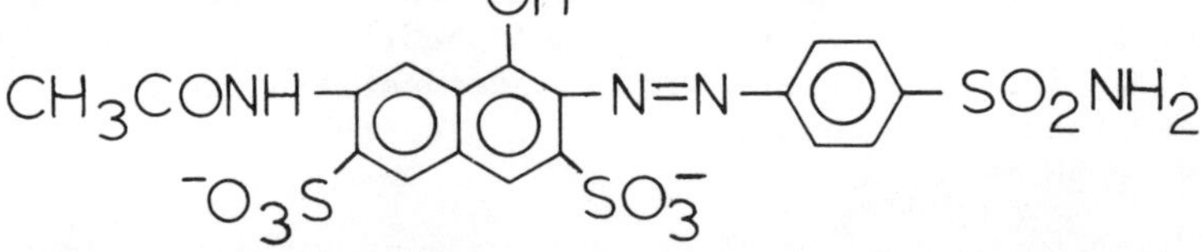

X-ray studies[13] suggest that the sulfonamide group binds directly to the metal atom at the active site; hydrophobic interactions between the substrate benzene ring and the protein also contribute to binding. The centre of substrate chromophore is presumably positioned inside the active site cavity, while some parts of the naphthol ring is probably in contact with the solvent. Hydrogen bonds could be formed between hystidyl side chains of the enzyme and OH or SO_3^- groups of the substrate. The dye absorption spectrum is affected by binding to carbonic anhydrase and the spectral change differs for the different forms of the enzyme[10]. It is difficult to interpret such spectral observations in terms of structural changes. The spectral changes, however, could be used to follow the kinetics of events occurring at the active site.

Enzyme-bound substrate was excited in the pulsed laser apparatus using a light of 532 nm wavelength; this wavelength corresponds to the lower energy edge of the first singlet $\pi\pi^*$ transition band.

The irradiation pulse induces an instantaneous variation of the enzyme-bound but not of the free chromophore spectrum in the 405-435 nm region, which corresponds to the second singlet transition band. This instantaneous variation cannot be explained by a photophysical effect (e.g. triplet transition) or by photodissociation because the sign of the effect is different in the B and C subforms and differs from that expected in dissociation. The observed effect can therefore be tentatively ascribed to an "isomerization". The quantum yield of the effect is roughly estimated to be a few part per thousand, assuming that the absorption variations are of the same order of magnitude as those due to the binding of the chromophore to the enzyme. A very rapid relaxation effect follows the instantaneous spectral variation described above; in the case of Zn-HCA-B the time constant for this effect is in 0.2 - 2.0 μs range (τ_1, fig. 1a). A second slower effect (fig. 1b) is observed with relaxation time ranging between 2 and 10 μs (τ_2). The system then reaches equilibrium with a longer time constant (τ_3) of approximately 200 μs. The effect of pH, buffer composition, etc. on τ_1 and τ_2 are reported in Table 1 and discussed below.

Table I

	k_- sec^{-1}	k'_- sec^{-1}	pK_1	pK_2	k_{1f} M^{-1} sec^{-1}	k_{2f} M^{-1} sec^{-1}
Zn-HCA-B	2000.	4500.	7.1	7.4	1.6×10^7	0.8×10^6
Co-HCA-B	2000.	4200.	7.1	7.2	2.0×10^7	1.0×10^6
Mn-HCA-B	1500.	600.	~7.	-	-	-
Cn,Ni-HCA-B	< 100.	< 100.	-	-	-	-
Zn-HCA-C	800.	1700.	6.8	7.25	2×10^6	3×10^5

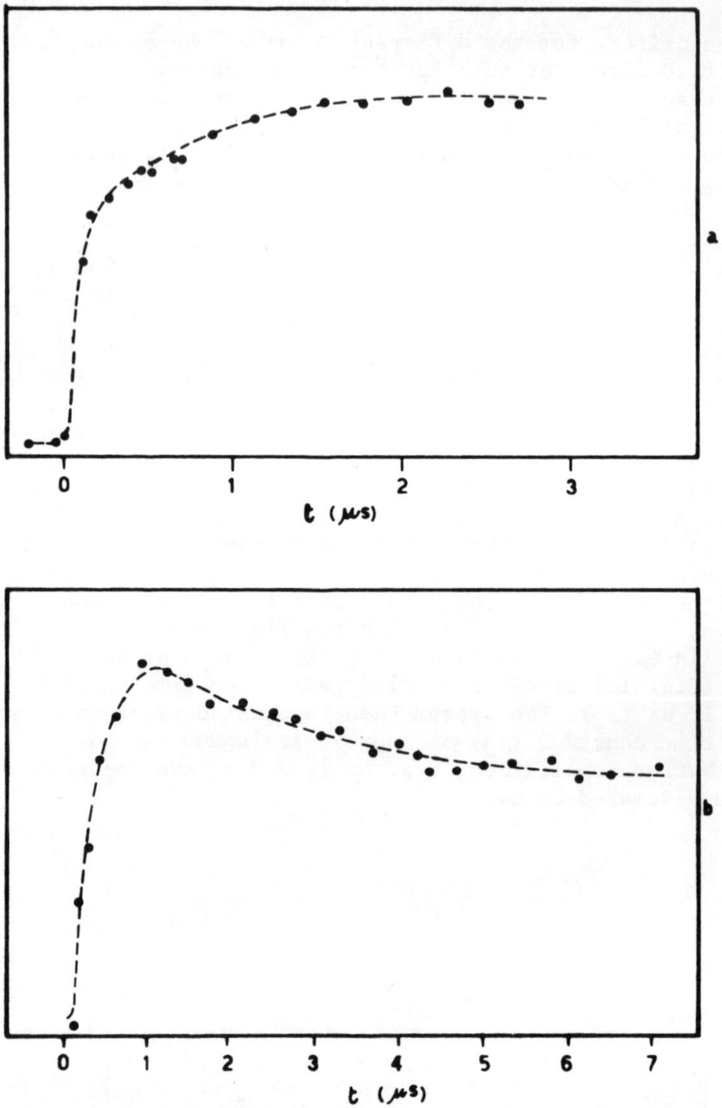

Fig. 1. Fast relaxations of HCA-B-Neoprontosil complexes. Enzyme was prepared from human blood red cells following[14,15,16]. The laser apparatus is illustrated elsewhere in these proceedings[1]. Excitation wavelength was 532 nm, observation at λ = 405 nm in this picture. The laser photochemical perturbation is followed first by two fast relaxation effects and after by a slower one ($\tau_3 \sim$ 200 μs) until the equilibrium is reached back. Conditions are: enzyme concentration 1.5×10^{-4}M; dye concentration 1.2×10^{-4}M; pH = 7.1: K phosphate 0.086M in (1a) and 0.47M in (1b).

(b) T-Jump experiments

The slowest effect (τ_3) was studied also by classical T-Jump experiments (cf. fig. 2). The dependence of τ_3 on substrate and enzyme concentrations, pH, ionic strength, active site metal and the enzyme subform was investigated. The effect can be attributed to a conformational transition because: (i) the dye-protein complex is practically the only species present in the solution, (ii) the relaxation time is independent from enzyme and dye concentrations, (iii) the same relaxation times were measured in T-Jump and Laser experiments and in the latter case the effect cannot be due to substrate binding to the enzyme. The pH dependence of the reciprocal relaxation times τ_3^{-1} is shown in Fig. 3 for HCA-B having different metal at the active site. With all the metals τ_3^{-1} varies as a function of pH with an apparent pK $\sim$ 7. Furthermore τ_3 is strongly dependent upon metal. The observed pH dependence is in agreement with models which postulate a conformational transition coupled to the fast protonation of a group[20] and is in keeping with the current interpretation of the pH dependence found in stopped flow experiments[11]. The effect of metal substitution upon τ_3 can be better appreciated in fig. 4 which reports also some data with Ni and Cu enzymes; these data are lower limit values for the time constant. Because of the lower affinity of these metal substituted enzymes for the dye an appreciable amount of free dye is present in solution; consequently the relaxation effect observed in the 20 ms region in T-jump experiments cannot unequivocally ascribe to the conformational effect.

A conformational effect can be detected even in Ni and Cu enzymes in photolysis experiments; however the time constants are too long to be properly measured by this technique[1]. The values for the specific activity for Cu and Ni enzymes should also be considered as upper limits because the measured value includes the residual activity of traces of Zn enzyme. The plot of isomerization rates against esterase activity[16] shows good correlation between the two kinetic parameters. These results provide direct experimental support for the hypothesis that greater catalytic activity of Zn and Co enzymes is connected with their capacity of undergoing faster changes in coordination pattern.

(c) The problem of which are the structures involved in the above postulated conformational transition should now be considered. The relaxation spectrum of complex between sulfanilamide[21] and Co-HCA-B was studied, monitoring the effect at wavelength corresponding to the absorption bands of the metal. In this system the isomerization process has time constants of $\sim$ 1.0 ms, and is coupled to faster processes. The amplitudes of all effects was studied as a function of the monitoring light wavelength (fig. 5). A linear combination of the amplitudes of respectively the faster and slow effect fit the difference spectrum between Co-HCA-B in

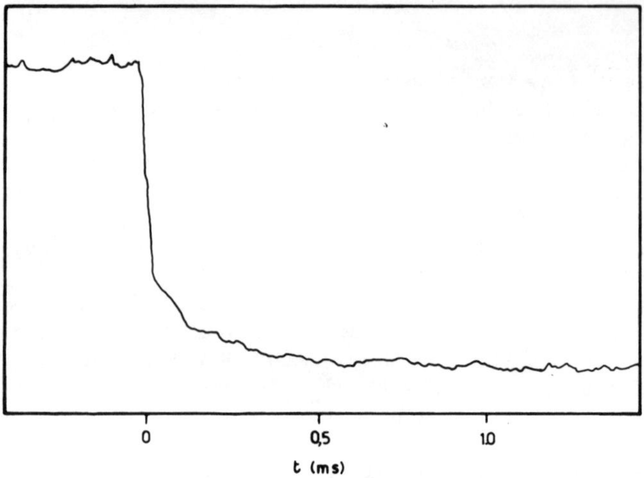

Fig. 2. Isomerization effects observed in HCA-B-Neoprontosil complex by standard T-Jump technique. The observation wavelength is λ= 546 nm. Messanlagen Transient Spectrophotometer is here used, connected to a Biomation mod. 810 transient recorder and punched tape unit. The T-Jump was $4^{o}C$ and the remaining conditions as in fig. 1b.

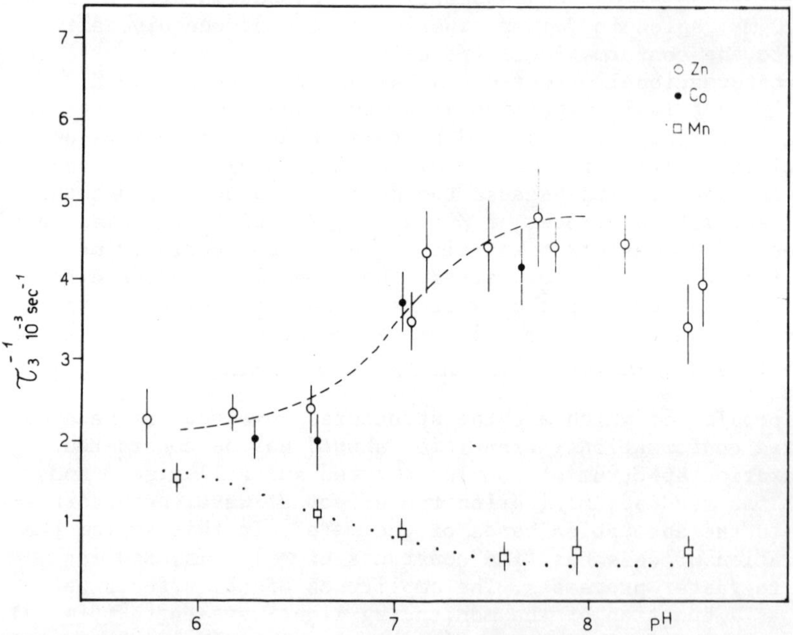

Fig. 3. Isomerization rates of HCA-B-Neoprontosil complex plotted vs. pH for different metal substitutions. The sample was prepared following[16,17,18,19].

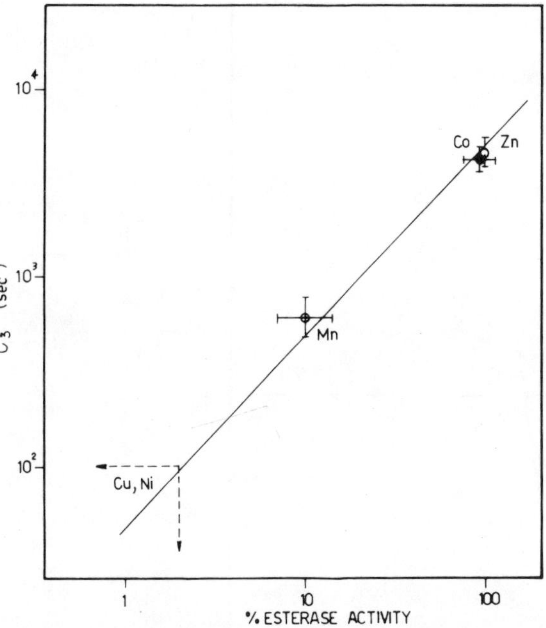

Fig. 4. HCA-B-Neoprontosil isomerization rates plotted against esterase activity for different metal substitution. The data are taken at pH = 7.8.

H_2O at high pH and the Co-HCA-B-sulfanilamide complex. In the two latter forms of the enzyme the metal is generally considered to be coordinated respectively in a "distorted tetrahedral" and "pure tetrahedral" form. Our data suggest that the isomerization effect which we have studied corresponds to a transition between these two forms.

(d) Isomerization model

The simplest model consistent with the above experimental data on τ_3 is shown in the following scheme where SH is sulfonamide, ESH_o, ES_o are "pure tetrahedral" structures, ESH and ES "distorted tetrahedral" structured complexes as defined in the previous paragraph; k_+, k_-, k'_+, k'_- are the kinetic constants associated with the isomerization, K_1, K_2 are the dissociation equilibrium constants of ESH and ESH_o. As shown in the following discussion the proton transfer steps coupled with the isomerization can be tentatively identified with the two fast effects observed. The equilibrium constants favour the formation of ESH_o and ES_o complexes, therefore k_-, $k'_- \gg k_+$, k'_+, and the observed pK in the titration curves of τ_3^{-1} is coincident with pK_1. The kinetic and

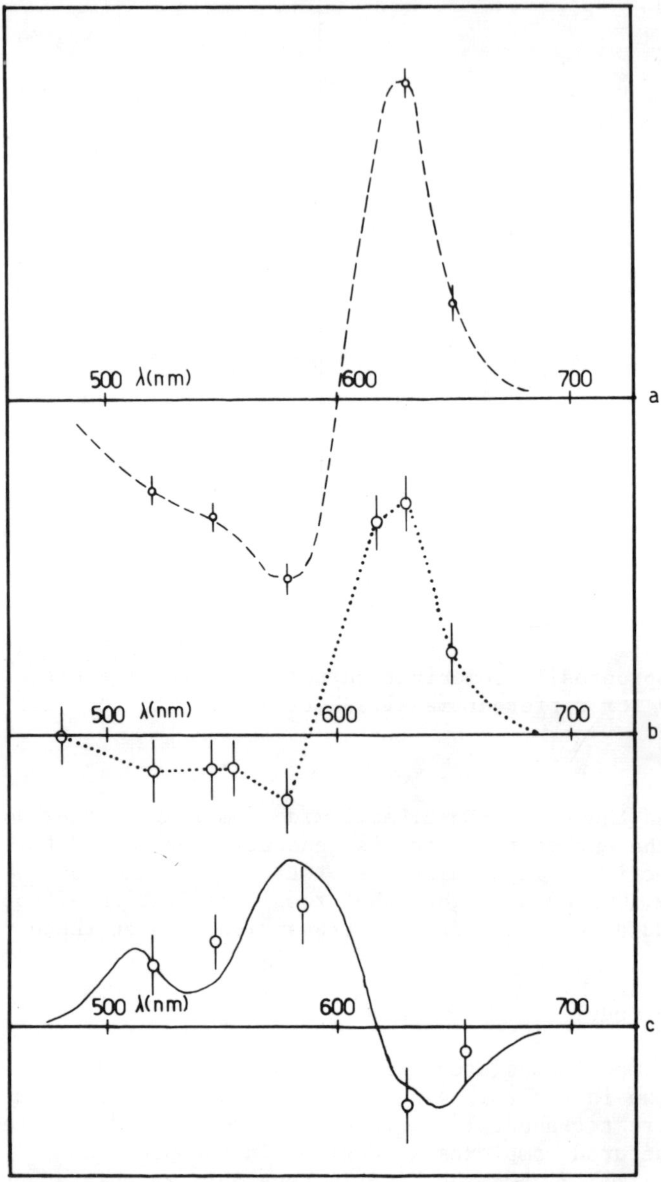

Fig. 5. Co-HCA-B-Sulfanilamide relaxation amplitudes vs. wavelength. The conditions were: pH = 8.0; enzyme concentration 1.4 x 10^{-3}M with sulfanilamide in excess, Tris HCl 0.1M. a) fast step b) isomerization slow step; c) a linear combinaiton of two above experimental curves is fitting the difference spectrum between the "pure tetrahedral" and "distorted tetrahedral" structures.

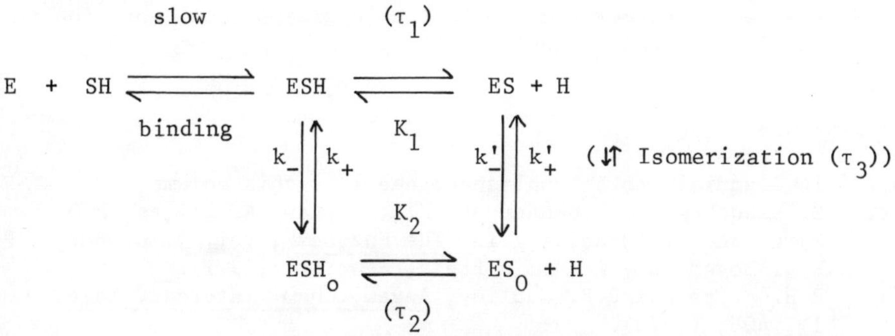

equilibrium constants calculated for different metals and enzyme assuming this model are reported in Table 1. It must be pointed out that the proton in ESH_o cannot be bound to a group very close to the metal because the experimental absorption spectra of the metal bands does not show a pH dependence[11]. The reverse is perhaps true in the ESH complex, and some fast internal equilibrium of protons at the active site can be reasonably postulated.

(e) Some very preliminary data are available now on the behaviour of τ_1 and τ_2 under different conditions of pH, ionic strength, concentrations, etc. A direct linear dependence of both τ_1 and τ_2 upon ionic strength (at constant pH) is observed between 0.02 and 0.2 M K phosphate. Experiments at lower or no ionic strength are now in progress (with desalting procedures and under controlled atmosphere). The best interpretation of the data presently available is consistent with a direct proton transfer to or from buffer molecules. A different mechanism is probably operative at low ionic strength since the observed straight lines do not extrapolate to near zero value expected for proton diffusion in the neutral pH region. It is perhaps worthwhile to emphasize here the importance of clarifying the role of this kind of fast process, which relates specific groups at the active site to the general proton exchange processes with the surrounding ionic atmosphere. In table 1 the data are preliminary interpreted according to the following simplified scheme:

$$AH + E \underset{k_r}{\overset{k_f}{\rightleftharpoons}} EH + A$$

from which K and k_f equilibrium and kinetic constants, were calculated for both τ_1 and τ_2; K is defined $K = \{E\}\{H\}/\{EH\} = K_b . k_f/k_r$ and K_b is the dissociation constant of the buffer.

ACKNOWLEDGEMENTS

This work was performed with the invaluable collaboration of R. Colilli, A. Bernardi and A. Bittoni.

REFERENCES

1. I. Giannini, to be published; see also this volume.
2. S. Lindskog, L.E. Henderson, K.K. Kannan, A. Liljas, P.O. Nyman and B. Strandberg in "The Enzymes", vol. 5, p. 587, P.D. Boyer ed., Academic Press, New York (1971).
3. R.H. Price and P.R. Woolley, Angew. Chem. Internat. Edit. 11, 408, (1972).
4. B.G. Malmstrom, Proc. F.E.B.S., 8 Meeting Amsterdam, p. 119 (1972).
5. A. Liljas et al., Nature New Biology, 235, 131 (1972).
6. P.H. Hapfner and J.E. Coleman, J. Biol. Chem., 248, 6630 (1973)
7. R.J. Khalifah, Proc. Nat. Acad. Sci. U.S.A., 70, 1986 (1973).
8. S. Lindskog and J.E. Coleman, Proc. Nat. Acad. Sci., U.S.A. 70, 2505 (1973).
9. S.H. Koenig, R.D. Brown, T.E. Needham and W.A. Matwiyoff Biochim. Biophys. Res. Comm., 53, 624 (1973) and (1974) in press.
10. J.E. Coleman, J. Biol. Chem., 243, 4574 (1968).
11. P.W. Taylor, R.W. King and A.S. Burger, Biochemistry, 9, 2638, (1970) and ibid. 9, 3894 (1970).
12. G.E. Lienhard et al. in "Cold Spring Harbor Symposia" vol. 36, p.45, Cold S.H. Laboratories, N.Y., (1972); and R. Wolfenden, Nature, 223, 704 (1969).
13. A. Liljas et al, NASA Spec. Publ. NASA - 188, 89 (1969); L. Waara et al., Advances in Exp. Medicine and Biol., 28, 169 (1972).
14. L.E. Henderson and D. Heriksson, Anal. Biochem., 51, 288 (197
15. J. Mc D. Armstrong et al., J. Biol. Chem., 241, 5137 (1966).
16. R.F. Chen and J.J. Kennohan, J. Biol. Chem., 242, 5813 (1967).
17. A. Lanir and G. Navon, Biochemistry, 11, 3536 (1971).
18. S. Lindskog and B.G. Malstrom, Biol. Chem., 237, 1129 (1962).
19. S. Lindskog and P.O. Nyman, B.B.A., 85, 462 (1964).
20. T.C. French and G.G. Hammes, J. Am. Chem. Soc., 87, 4669 (196
21. S. Lindskog and A. Thorslund, Eur. J. Biochem., 3, 453 (1962)

KINETICS OF GLUTAMATE DEHYDROGENASE SELF-ASSEMBLY

Darwin Thusius

Unite de Physico-Chimie des Macromolecules
Biologiques Institut Pasteur, Paris, France

INTRODUCTION

For many years it has been known that beef liver glutamate dehydrogenase (GDH) undergoes a rapid, reversible self-association to form high molecular weight polymers. Because of its bearing on the nature of protein-protein bonds and its possible relevance to metabolic control, the association-dissociation reaction has been studied intensively with a wide variety of physical chemical methods including electron microscopy, light scattering and ultracentrifugation (for a review, see Eisenberg[1]). It is now generally accepted that polymerization proceeds in a linear fashion, with the formation of rigid rods.

The mode of self-assembly has been represented as a sequential addition of monomer units to a growing polymer chain with identical equilibrium constants for each step[1-4].

$$P_1 + P_i \rightleftarrows P_{1+i} \qquad K = \bar{P}_{1+i} / \bar{P}_1 \bar{P}_i \qquad \text{I}$$

$$i = 1, 2 \ldots\ldots \infty$$

Based as it is on equilibrium and structural results, scheme I may or may not be the actual mechansim of aggregation. It is therefore of interest to further probe this system with kinetic techniques. The present communication summarizes results of a detailed temperature jump study of GDH self-assembly[5], together with the numerical simulation of the relaxation spectrum of the sequential addition model[6].

Wyn-Jones (ed.), Chemical and Biological Applications of Relaxation Spectrometry, 433–436.

COMPUTER SIMULATION

In order to rigorously test scheme I with relaxation methods we must predict the concentration dependence of the relaxation times and amplitudes in the expression

$$\delta S(t) = \sum_{i=1}^{N} \delta S_i^o e^{-t/\tau_i} \tag{1}$$

where S is the observed signal and N is the number of elementary steps. This was done with a program based on a series of subroutines written by Dr. G. Ilgenfritz. The calculations assumed that the forward and reverse rate constants of all association-dissociation reactions are independent of chain length. Although an infinite number of elementary steps is implied in the molecular mechanism, a large but reasonable N value (20-30) was sufficient for convergence of the times and amplitudes.

Relaxation amplitudes are defined in terms of the detection system used to monitor concentration changes[7,8]. In scattered light measurements the "specific signals" of the amplitude expressions are[9]

$$\phi_i = \gamma M_i^2 \tag{2}$$

where γ is an instrument constant and M_i is the molecular weight of polymer P_i. Equation 2 assumes that angular dependence and thermodynamic non-ideality are negligible. This proves to be the case for GDH self assembly up to about 3 mg/ml[10,11].

The individual relaxation times and amplitudes computed at different total protein concentrations were used to simulate scattered light decay curves according to equation (1). The computations revealed two criteria for the experimental verification of scheme I. Firstly, above 1 mg/ml first-order plots of the experimental decay curves should be non-linear. Secondly, the overall half-life should increase with increasing protein concentration.

EXPERIMENTAL RESULTS

The above predictions are not, however, in agreement with the observed kinetic behaviour. Both stopped-flow and temperature-jump results are consistent with strictly linear first-order plots over a wide temperature and concentration range[5,12,13]. Furthermore, the experimental relaxation time decreases with increasing protein concentration. These observations argue strongl against scheme I and suggest that monomer plays no special role in GDH self-assembly. Although a sequential addition of monomers may be used to represent the assocation-dissociation stoichiometry, the microscopic mechanism appears to be more complex.

A RANDOM ASSOCIATION DISSOCIATION

A reasonable alternative model is a random association-dissociation in which reactions occur between any two polymer chains without discrimination.

$$P_i + P_j \underset{Kd}{\overset{Ka}{\rightleftarrows}} P_{i+j} \qquad \text{II}$$

$$i = 1,2.....\infty$$

Again we assume an infinite number of steps and identical forward and reverse rate constants. Scheme II is thermodynamically equivalent to scheme I, and can therefore rationalize the known molecular weight distribution of GDH solutions. Although at the molecular level a random association-dissociation is more complex than a sequential addition of monomers, scheme II formally reduces to a simple two state model

$$2A \rightleftarrows B$$

where A denotes protein contact sites and B represents bonds between nomomer units. Since all pathways implied in the microscopic mechanism are accounted for in a single formal reaction, the kinetics are defined by a single relaxation time. It can be shown that the square of the reciprocal relaxation time is given by

$$({}^{1}/\tau)^2 = 4K_aK_dM_1^{-1}C_T + K_d^2 \qquad (3)$$

where C_T is protein concentration in mg/ml and M_1 is monomer molecular weight. Our temperature-jump results[5] as well as those of Kempfle and Winkler[12] are in excellent agreement with equation (3). In addition the random association-dissociation quantitatively accounts for the amplitude concentration dependence[5].

CONCLUSION

Scattered light kinetic measurements independently corroborate earlier thermodynamic data indicating that GDH monomers undergo an open, linear aggregation with identical equilibrium constants for each step.[1-4] The microscopic mechanism apparently involves random associations between all polymer forms - a result inaccessible to non-kinetic methods. Finally, the present study illustrates the potential of scattered light detection in chemical relaxation studies.

ACKNOWLEDGEMENT

This work was supported by the Centre National de la Recherche Scientifique and the Max-Planck-Institut fur Biophysikalische Chemie (Gottingen). The author is grateful to the European Molecular Biology Organization for a short term fellowship.

REFERENCES

1. H. Eisenber, Account Chem. Res., 4, 379 (1971).
2. H. Sund, K. Markau, M. Minssen, and J. Schneider, in "Structure and Function of Oxidation-Reduction Enzymes," A. Akeson and A. Ehrenberg, eds. p. 681, New York, Pergamon Press (1972).
3. D. A. Malencik, and B. R. Anderson, Biochem. 11, 3022 (1972).
4. P. N. Chun, and S. J. Kim, Biochem. 8, 1633 (1969).
5. D. Thusius, P. Dessen, and J. M. Jallon, J. Mol. Biol., in press (1975).
6. D. Thusius, (1974) submitted for publication.
7. D. Thusius, J. Am. Chem. Soc. 94, 356 (1972).
8. D. Thusius, G. Foucault, and F. Guillain, "Dynamic Aspects of Conformation Changes in Biological Macromolecules", Sadron, ed., Reidel Publishing Co., Boston, pp 271 (1972).
9. D. Thusius, this volume.
10. K. Markau, J. Schneider, and H. Sund, Eur. J. Biochem. 24, 393 (1971).
11. E. Reisler, and H. Eisenber, Biochem. 10, 2659 (1971).
12. M. Kempfle, and H. Winkler, Hoppe Seylers Zoit. Physiol. Chem. 354, 816 (1973).
13. H. F. Fisher, and J. R. Berd, Biochim. Biophys. Acta, 188, 168 (1969).

INTERACTION BETWEEN CATIONIC DYES AND POLYELECTROLYTES

Vincenzo Vitagliano

Istituto Chimico, Universita di Napoli,
Naples, Italy

The interaction between cationic dyes and biological polymers has been widely studied in recent years [1-52] because of the great interest of these dyes in various fields of biology. A number of these dyes are used for staining living tissues, some of them have pharmacological properties and their molecules bear similarities with antibiotic molecules such as daunomicine and actinomicine. Many of them have a remarkable mutagenicity being able to interfere in the replication of DNA by causing deletion or insertion of base pairs[52,53]. However, as Bradley and Lifson already pointed out some years ago, the accumulation of even this much experimental knowledge has not yet led to a unified picture of the binding process[54].

On the other hand, the binding of dyes to synthetic polyelectrolytes has been neglected and limited mainly to synthetic polypeptides[14,16,55-80]. Synthetic polyelectrolyte molecules are simpler systems and, in our opinion, the study of their interaction with dyes should give interesting information of relevance to problems connected with biological polymers.

We present here a set of experimental spectrophotometric results on the behaviour of some metachromatic dyes in aqueous solution and on their interaction with vinyl chain polyelectrolytes. Metachromatic effects are commonly shown by all dyes having an electric charge partially delocalized into the chromophore group (Fig. 1)[81].

The spectrum of these dyes in the visible region exhibits, generally, three maxima corresponding to different aggregation forms of the molecules[73,81-97]. The change from one spectrum to the other may be induced by changes in concentration, by addition of micelles, polyelectrolytes, salts etc.

Wyn-Jones (ed.), Chemical and Biological Applications of Relaxation Spectrometry, 437–466.

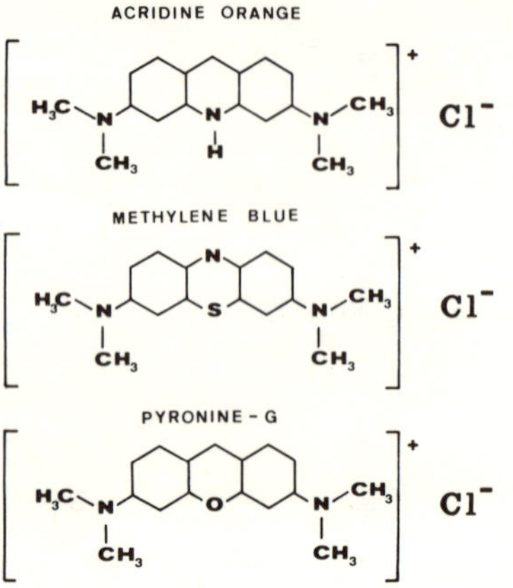

Fig. 1. Metachromic dyes

DYE BEHAVIOUR IN AQUEOUS SOLUTIONS

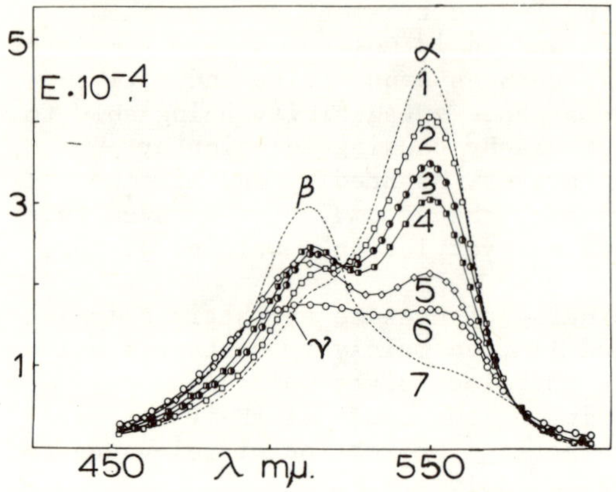

Fig. 2. Extincion Coefficients of Pyronine - G in aqueous HCl solutions (10^{-3}m) at various concentrations: 1, limit curve of monomer dye, 7, limit curve of dimer dye; 2 to 6: 1.486 x 10^{-4}, 4.406 x 10^{-4}, 9.18 x 10^{-4}, 18.36 x 10^{-4} mol/l (from ref. 73)

Figure 2 shows the extinction coefficients of pyronine-G (PG) aqueous solutions as a function of wave length at various concentrations. By increasing concentration the gradual disappearance of the absorption maximum at $\lambda \simeq 550$ nm (α band) and its substitution with a maximum at $\lambda \simeq 512$ nm (β band) can be seen. The isosbestic point at $\lambda \simeq 520$ nm is evidence of an equilibrium between two forms which have been already recognized as monomer and dimer[83,86,92]. At dye concentration higher than 2 - 3 x 10^{-4} mol/l the β band is substituted by a new diffuse band (γ) showing a further blue shift of the maximum. This fact has been attributed to formation of molecular aggregates. Similar behaviour is shown by a number of other dyes, such as methylene blue (MB), acridine organge (AO), thionine, proflavine, crystal violet and triphenylmethane dyes etc.

In dilute solution, where only the α and β bands are present, a dimerisation constant of the type:

$$K = \frac{1 - \alpha}{2\alpha^2 c} \qquad (1)$$

gives a good agreement between computed and experimental degree of dissociation, α, and molar extinction coefficients, as it can be seen in Fig. 3. At concentration higher than 2 - 3 x 10^{-4} mol/l the computed values of the extinction coefficient diverge rapidly from the experimental ones, because of the onset of higher order aggregates (a possible aggregation mechanism is discussed elsewhere in these Proceedings[98].)

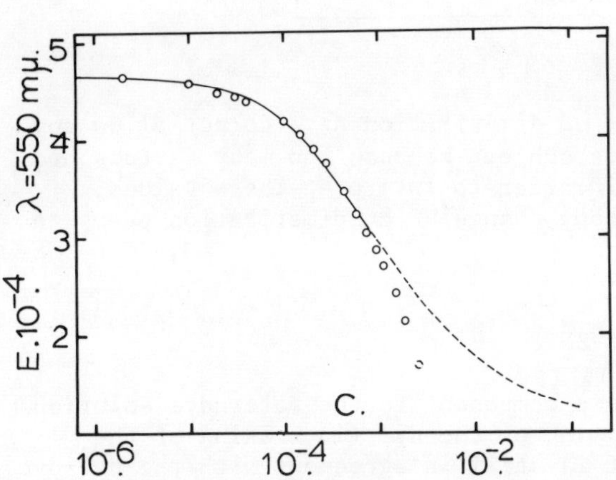

Fig. 3. Comparison of experimental Extinction Coefficients of Pyronine - G in aqueous HCl (10^{-3}m) solutions with the values computed through eq. 1 using K given in Table 1 (from ref. 73).

The aggregates probably contain a limited number of molecules because counterion activity measurements did not show any association phenomena (polyelectrolyte effect) as one should expect in the presence of large aggretates[99]. The interactions between π -

electrons of the aromatic rings must be mainly responsible for the dye association in aqueous solution, however some authors suggest that hydrophobic interactions play a dominant role on this effect[88,100 - 103]. Two points that we may add in favour of this suggestion are: (1) The fact that dyes, such as proflavine or thionine respectively similar to acridine orange and methylene blue but lacking the $-CH_3$ hydrophobic groups, have much lower dimerization constants (see Table 1); (2) The low value of dimerization constant of pyronine-G to be attributed to the presence of the oxygen atom (Fig. 2) which may promote specific interactions with water molecules weakening hydrophobic bonds.

TABLE I

DIMERIZATION THERMODYNAMIC DATA OF SOME DYES IN AQUEOUS SOLUTION AT 20 °C

Dye	$K \times 10^{-3}$ (eq. 1)	ΔH° kcal/mole	ΔS° e.u.
Acridine Orange	10.5(71) 22(86) 9.3(62)	-8.3(86) -6.9(71)	-5.1(71) -8.5(86)
Proflavine	0.48(72)	-6.0	-8.2
Pyronine-G	0.83(73)	-3.9(*)	0.0(*)
Methylene Blue	6.5(96) 4.5(105) 2.14(102)	-5.9(102)	-4.9(102)
Thionine	1.1(83,96)	—	—
Crystal Violet	0.60(106)	—	—

(*) V.Vitagliano and L.Costantino, unpublished results

The available entropy of dimerization data do not allow any reliable discussion on this subject because too many factors had to be taken into account in order to interpret these values, however the "unitary" entropy change[103] on dimerization seems to be, in general, positive[102].

DYE BEHAVIOUR IN MIXED SOLVENTS

Addition of a non-ionic component to the water-dye solution always promotes a "salting in" of the dye (increasing of the monomer α band). This fact is still in agreement with the dye-dye association in terms of hydrophobic bonding, the non-ionic solvent crystallizes in a modified water structure preventing those interactions which promote hydrophobic association[104]. This behaviour is typical both of structure maker compounds, such as hexamethylenetetramine[107,108] and of compounds, such as urea[61], whose interaction with water can be interpreted in a different way[109-112].

The dye "salting - in" may be correlated to a hyperchromism of the α band; this effect was shown by solvents having methyl groups in the molecule, although it may not be a rule. Small blue or red shifts of the α band maximum can be shown.

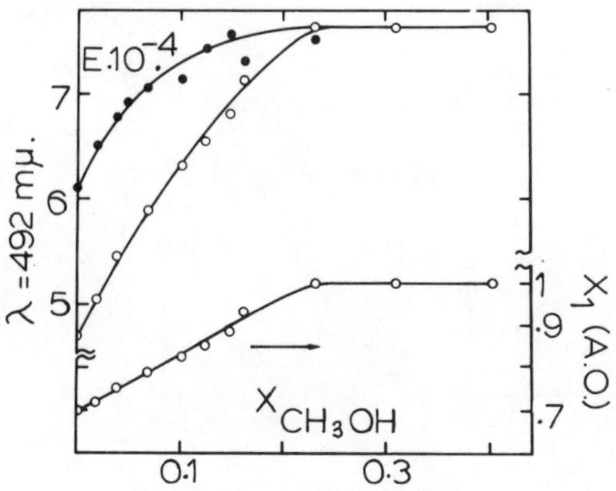

Fig. 4. Extinction Coefficients of Aqueous Acridine Orange solutions and Fraction of Monomer Dye X_1 as a function of Mole Fraction of added Methyl Alcohol:

—●— AO 1.43 x 10^{-6} M; —○— AO 3.67 x 10^{-5} M

X_1 was computed by assuming that the extinction coefficient of monomer dye was that given by the graph at 1.43 x 10^{-6} M, and the extinction coefficient of dimer was 14500 as in water.

Fig. 4 shows the extinction coefficients of 1.43 x 10^{-6} and 3.67 x 10^{-5} molar AO at 492 nm (α band maximum) as a function of added methyl alcohol, the fraction of monomer dye is also given. It can be seen that the dye salting-in is parallel to its hyperchromism and that the dimer disappears completely at a methyl alcohol mole fraction of about 0.2. It may be noted here that in ethyl alcohol - water mixtures the original water structure is completely destroyed at the same alcoholic mole fraction value[135].

An "efficiency scale" of different substances in promoting the AO and MB "salting-in" is: Acridine Orange, CH_3OH < C_2H_5OH ≃ Urea(*) < Hexamethylenetetramine (*) < Methylurea < Dimethylsulfoxide < Acetone < Tetramethylurea; Methylene Blue, Dimethylsulfoxide ≃ CH_3OH < Acetone ≃ Hexamethylenetetramine (*) ≃ C_2H_5OH < Urea(*) < Tetramethylurea. The compounds marked with (*) do not promote appreciable hyperchromism of the α band.

A completely opposite effect is shown by addition of salts to the dye - water solution. In this case we have a "salting-out" of the dye[71,72,88,105]. This effect may be due to the presence of positive ions which build up solvent regions from which the dye molecules are excluded because of electrostatic repulsion; as a consequence dye concentrates in the remaining free volume.

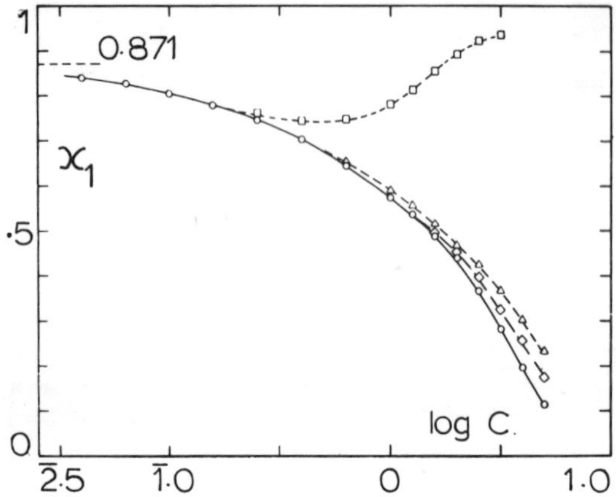

Fig. 5. Effect of added Salts on the Aggregation Equilibrium of Acridine Orange (10^{-5}m) in aqueous solution. Fraction of Monomer AO as a function of added salt concentrations: -Δ- LiCl, -O- NaCl, -◇- CsCl, -□- $N(CH_3)_4Cl$ and $N(CH_3)_4Br$) (from ref. 105).

The behaviour of tetralkylammonium salts is typical[105]; in dilute solutions they act as electrolytes promoting the dye aggregation, in concentrated solutions the hydrophobic component of the molecule prevails and the monomerization of the dye molecules is observed (Fig. 5).

INTERACTION BETWEEN DYES AND POLYELECTROLYTES

If we add a polyelectrolyte solution to a dilute dye solution we can see the gradual substitution of the α absorption band with the γ band characteristic of the aggregate dye. The spectrophotometric graphs are similar to those obtained by increasing the dye concentration in water. When the concentration of absorption sites on the polyelectrolyte molecules becomes greater than the dye concentration the α absorption band may appear again; generally it exhibits a 8 - 10 nm red shift indicating the presence of monomer bound dye. The appearance of the α band parallels the substitution of the γ with the β band.

This effect is very well shown by native and denaturated DNA and it was utilized to titrate DNA solutions[9,113]. The effect was interpreted by Bradley and Wolf[8] as a dilution of the dye along the polyelectrolyte molecule with the increasing of the available binding sites (see Fig. 6); a statistical treatment allowed defining and computing a "stacking coefficient" which accounts for the tendency of the dye molecules to aggregate when bound along

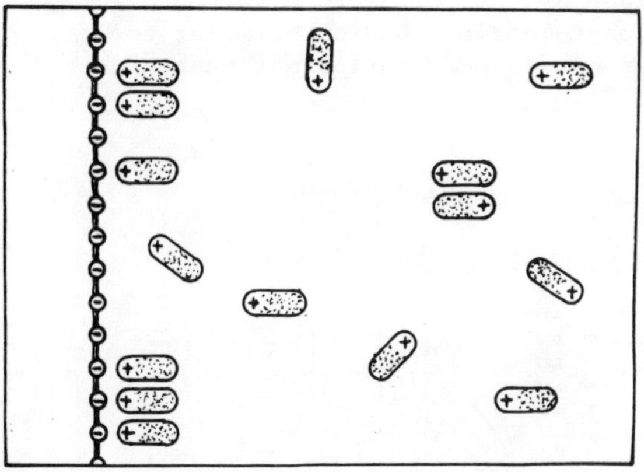

Fig. 6. Bradley and Wolf's model of dye binding on polyelectrolyte.

the polymer chain. With synthetic polyelectrolytes such as polyacrylic acid, PA, polymethacrylic acid, PMA, polyphosphoric acid, at neutral pH the dye dilution along the polymer chain is not so evident as in the DNA case and it is necessary to reach very high values of the ratio P/D of polymer equivalent concentration to dye concentration in order to see the reappearance of the α absorption band (Fig. 7).

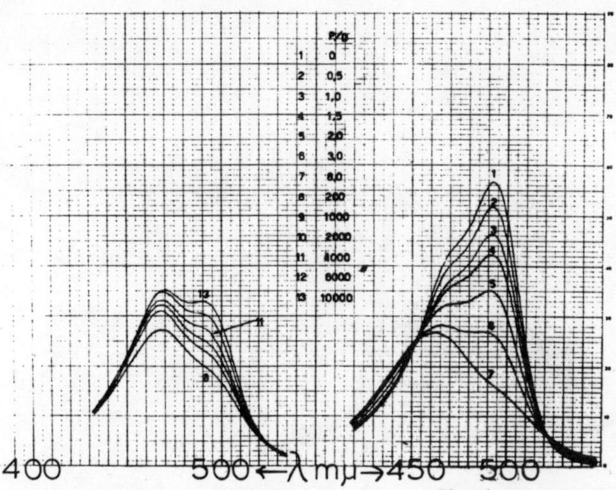

Fig. 7. Absorption Spectra of Acridine Orange in the Presence of Various Amounts of Polymethacrylic Acid at neutral pH.

A peculiar case is given by the polystyrene sulfonic acid, PSS, which shows a spectrophotometric behaviour similar to DNA, exhibiting the α absorption band at P/D ratios not much higher than unity (see Fig. 8 and 9).

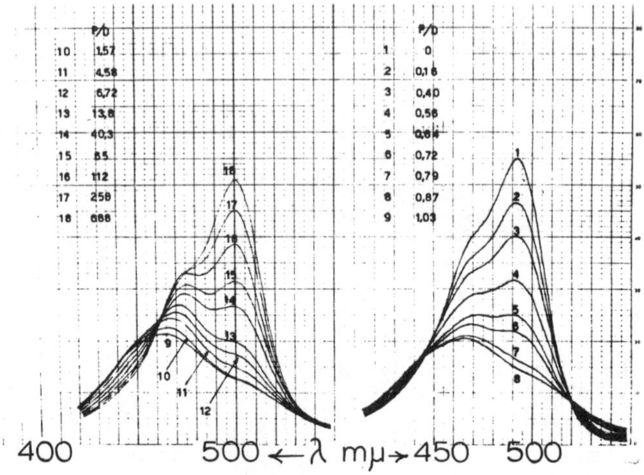

Fig. 8. Absorption Spectra of Acridine Orange (10^{-5}m) in the Presence of Various Amounts of Polystyrenesulfonic acid).

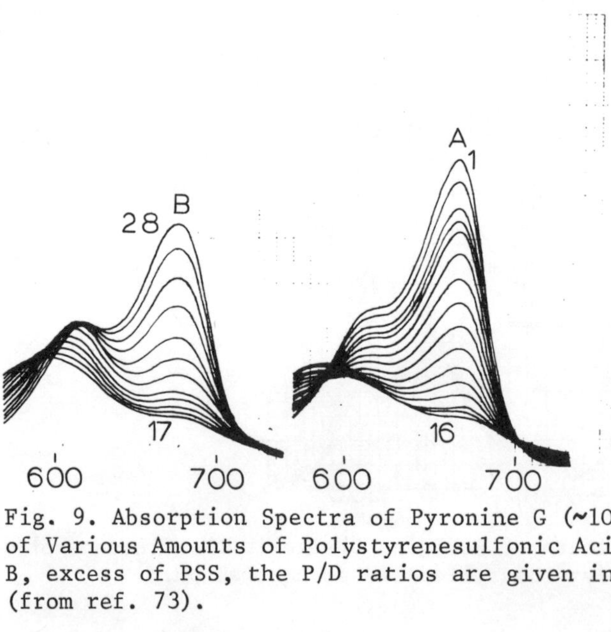

Fig. 9. Absorption Spectra of Pyronine G ($\sim 10^{-5}$m) in the presence of Various Amounts of Polystyrenesulfonic Acid (A, excess of dye, B, excess of PSS, the P/D ratios are given in ref. 73) (from ref. 73).

This fact may be attributed to the benzene rings on the PSS molecule. Sulfonic groups can be easily titrated, by using a dye, to an accuracy of 1 - 2%[70,73,79] (see fig. 10 and 11).

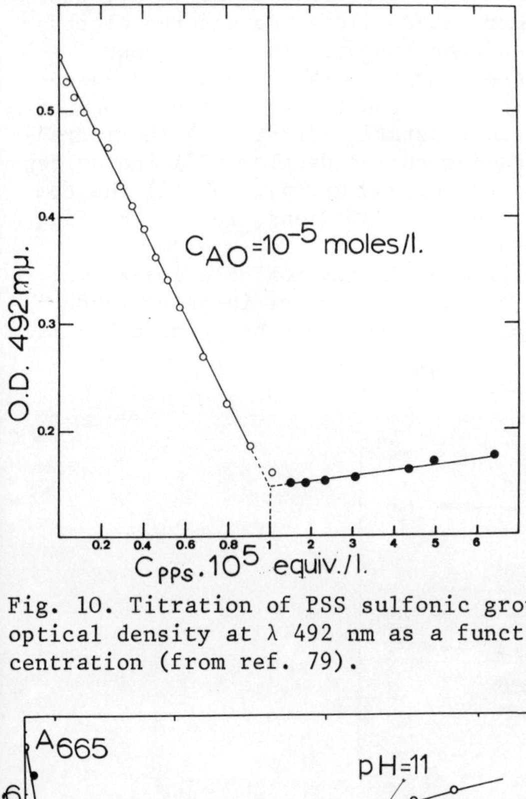

Fig. 10. Titration of PSS sulfonic groups with Acridine Orange, optical density at λ 492 nm as a function of PSS equivalent concentration (from ref. 79).

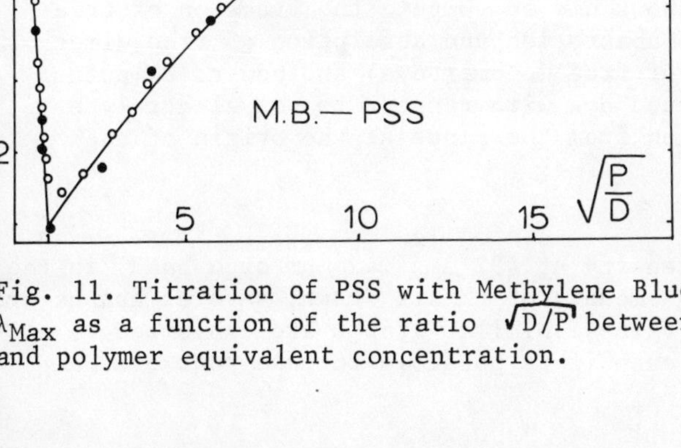

Fig. 11. Titration of PSS with Methylene Blue, Dye Absorption at λ_{Max} as a function of the ratio $\sqrt{D/P}$ between dye concentration and polymer equivalent concentration.

Two different contributions can be recognized for the dye binding to polyelectrolytes: electrostatic interactions and specific interactions which may also include hydrophobic effects. The fact that large flat molecules are bound to the polyanion molecules much more strongly than small ions such as Li^+, Na^+, etc., is a strong evidence that other interactions besides the electrostatic ones play a relevant role on the dye binding process. However, the electrostatic effect must be the main one because unionized polyelectrolyte molecules bind a very little amount of dye[64]. We can assign to the electrostatic effect: (1) The increasing binding strength by increasing charge density. (2) The evidence that binding sites correspond to ionized groups, and (3) the competition of metal ions, including alkaline ions, for the binding sites.

(1) By increasing the degree of ionization of a polyanion, i.e. its charge density, the amount of free dye in solution decreases, at constant P/D ratio. This effect may be seen in Fig. 12,

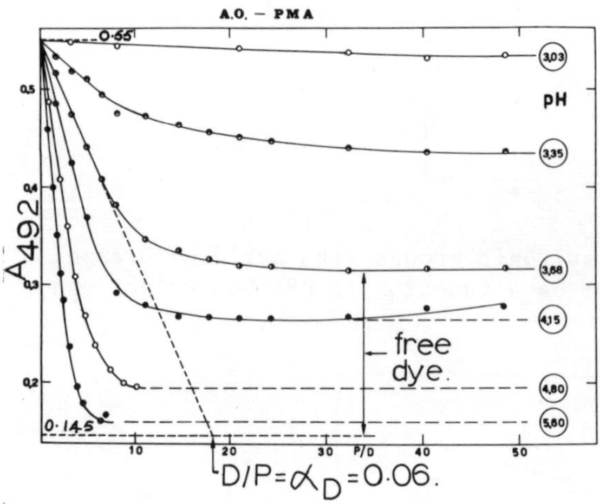

Fig. 12. Acridine Orange (10^{-5}m) Absorption at λ = 492 nm in the presence of increasing amount of polymethacrylic acid at different pH's. The figure shows how to compute the fraction of free dye (0.145 = bound dye absorption and absorption of free dimer dye; 0.610 absorption of free monomer dye) and how to compute the fraction α_D, of bound dye with respect to polyelectrolyte equivalent concentration from the slope at the origin of each graph.

where the AO optical density at 492 nm (maximum of α band) in the presence of increasing amounts of PMA is drawn; several graphs are shown at different pH values. Similar graphs are exhibited by other dyes (in the MB case it is possible to take runs even at higher pH values because the MB molecule is ionized also in basic medium).

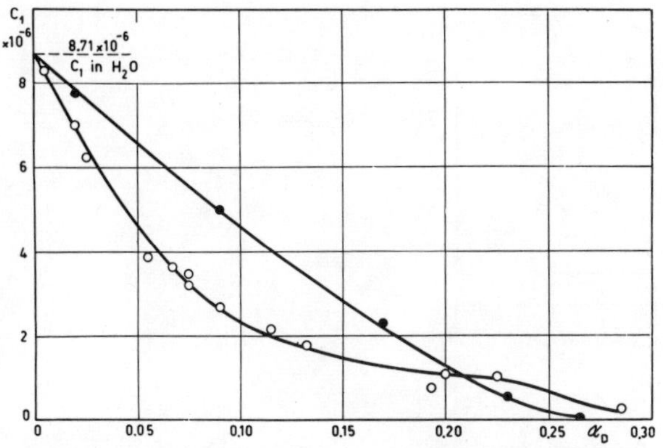

Fig. 13. Concentration of free monomer AO as a function of the degree of ionization of the polyelectrolyte (assumed equal to α_D): ● polyacrylic acid; o polymethacrylic acid. (from ref. 64).

Figure 13 shows the concentration of free AO as a function of the degree of ionization for PA and PMA as obtained from graphs such as those of Fig. 12.

(2) The end point titration corresponds to a very good approximation to the number of ionized groups on the polyelectrolyte molecules. This was accurately established for DNA phosphoric groups[9] and for PSS sulfonic groups[73,79]. In the case of weak polyelectrolytes, such as PA and PMA, the number of available sites on the polymer chain can be obtained by drawing the origin slope of optical density graphs, drawn as a function of P/D, as shown in Fig. 12. Also in these cases the fraction of available binding sites corresponds to a fairly good approximation to the fraction of ionized groups as measured from potentiometric titrations[64]. Figure 14 shows the comparison between potentiometric titrations and dye adsorption titration for PA and PMA. Identification of binding sites with ionized groups was also made for the polyglutamic acid - AO system by other authors[75].

(3) There is competition with metal cations, even with those such as alkali metals which are thought to bind to polyelectrolyte molecules mainly or only by electrostatic interactions.[114-117] Metal cations cause, in fact, a release of dye from the polyelectrolyte to solution. The efficiency in releasing the dye is strongly dependent on the cation charge, although some slight differences can be observed among similar ions with the same charge[60,62].

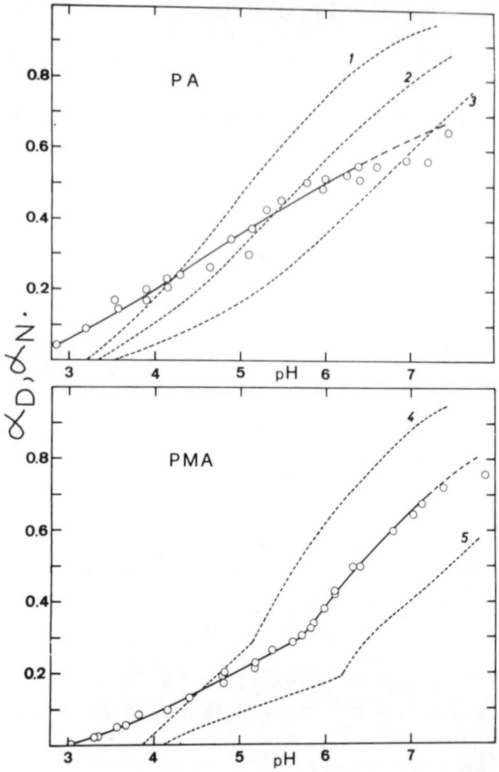

Fig. 14. Comparison between fraction of available binding sites, α_D, and degree of neutralization, α_N, as measured through potentiometric titrations (curves 1 - 5). -O- α_D, Acridine Orange 10^{-5}M, buffer ionic strength 0.001,-----potentiometric titrations, PA 0.01 eq/lit, PMA $\simeq$ 0.0012 eq/lit., added NaCl:(1) 0.3 N, (2) 0.01 N, (4) $\simeq$ 0.3 N, (3 and 5) no added salt. Note that even in acid medium ($\alpha_N \leqslant 0$) the carboxylic groups are partially ionized and they bind some dye molecules.

Figure 15 summarizes the effect of electrostatic interactions on the dye binding to polyelectrolytes: it shows the titration of MB OH 10^{-5} molar with PA and with PSS. In the PSS case two different runs were interrupted at P/D = 1 and P/D = 4 and a NaCl solution was added stepwise. The figure shows the titration of sulfonic groups and the MB^+ ion release from the polyanion by effect of the Na^+ ion competition. In the PA case, the MBOH base is not strong enough to titrate all carboxylic groups and the charge density of the partially ionized polyacid molecules is not high enough to bind all the MB^+ ions; consequently the titration curve does not exhibit end points and a relevant amount of free MB^+ ions is still present at a P/D = 4 ratio. By adding NaOH, at

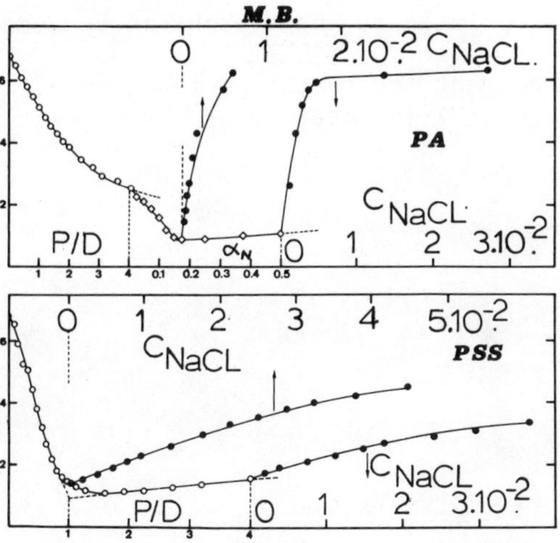

Fig. 15. Titration of Polyacrylic acid and Polystyrenesulfonic acid with the Methylene Blue base (obtained by passing the MB chloride on anion exchange resin) and effect of added sodium chloride.

P/D = 4, there is a charge density increase and a consequent adsorption of the free MB^+ ions which allows the titration end point to be observed. Addition of NaCl releases the bound dye both in the absence and in the presence of added NaOH. A comparison of PSS and PA cases shows that PSS binds the dye more strongly than PA.

STATISTICAL MODEL OF THE POLYELECTROLYTE - DYE SYSTEM

Bradley and Wolf's interpretation[8] of the dye polyelectrolyte interaction (fig. 6) is suitable for a statistical treatment according to an Ising model[118,119]; the polymer chain is represented by a long sequence of available binding sites either empty (0) or occupied by a dye molecule (A):

...00A000AAA000AA00000A00A000AA00AAAA000000AA000AAAA00AAAAA... (2)

In view of the stacking tendency it is necessary to account for first neighbour interactions ..AA.. and eventually second neighbour ..AAA.. or higher order ..AA$_{n-1}$A.. interactions.

There are several mathematical approaches to the statistical treatment of such a system[120-124]. A very convenient sophisticated one was suggested by Lifson[54,125]; the limitation of the method is that it is only applicable to infinite chain problems because end effects are neglected.

According to Lifson's treatment the partition function of a system such as that described by (2) is obtained as the highest root, z_1, of the equation:

$$f(z) = U(z)V(z) - 1 = 0 \tag{3}$$

where

$$U(z) = \sum_{i=1}^{\infty} q_o^i z^{-i} = q_o/(z - q_o) \tag{4}$$

and

$$V(z) = q_a\lambda_a z^{-1} + q_1 q_a^2 \lambda_a^2 z^{-2} + \ldots + q_{m-1} \cdots q_1^{m-1}(q_a\lambda_a)^m z^{-m} \times$$

$$\sum_{j=1}^{\infty} (q_m q_{m-1} \cdots q_1 q_a \lambda_a z^{-1})^{j-m} \tag{5}$$

where

$$v = q_a\lambda_a z^{-1} \tag{6}$$

and q_o is the partition function of a desorbed state site, q_a is that of a state in an adsorbed state, λ_a is the absolute activity of the dye in solution. As pointed out before, it is assumed that neighbouring adsorbed molecules interact with each other such that a pair of first neighbours has a partition function of pairwise interaction (stacking coefficient) q_1, a pair of second neighbours has a partition function q_2, and so on up to the m-th order neighbourhood. The q_1 parameter is related to the free energy of dimerization of the dye along the polymer chain: $\Delta F_D = -RT \ln q_1$. Interactions of bound dye molecules through one or more empty sites, ..AOA.., ..AOOA.., are excluded in equations (4) and (5), although it would not be difficult to include such interactions if there was a reason for doing so[125].

Equation (5) becomes[79]

$$V(z) = v/(1 - v) \quad \text{(random distribution)} \tag{7}$$

$$V(z) = v/(1 - q_1 v) \quad \text{(first neighbour interaction)} \tag{8}$$

$$V(z) = v\,[1 + [q_1 v/(1 - q_1 q_2 v)]] \quad \text{(second neighbour interaction)} \tag{9}$$

$$V(z) = v\,[1 + q_1 v + [q_2 q_1^2 v^2/(1 - q_1 q_2 q_3 v)]] \quad \text{(third neighbour interaction)} \tag{10}$$

The versatility of Lifson's method is connected with the properties of the free energy function of any system:

$$F = kT \ln Z_1 = \sum_i X_i \bar{F}_i \tag{11}$$

where the X_i are the mole fractions of the various species which constitute the system and the $\bar{F}_i$ are their partial molar free energies. Remembering that $f(z) = 0$ is an implicit function of Z_1 and of the various $\bar{F}_i$, from (3) and (11) it is easy to obtain the following expression:

$$X_i = \left(\frac{\partial F}{\partial \bar{F}_i}\right)_{\bar{F}_j} \frac{\dfrac{\partial f\ (z)}{\partial \ln p_i}}{\dfrac{\partial f\ (z)}{\partial \ln z^{-1}}} \tag{12}$$

where P_i is the partition function of species i.

By applying eq. (12) it is possible to obtain information about the dye distribution along the polymer chain without actually solving eq. (3) for the partition function Z_1. Thus, the partition function of the dye in bound state is $P_i = q_a\lambda_a$, and the ratio between bound dye concentration and available binding site concentration is given by:

$$\frac{D}{P} = (\partial f(z)/\partial \ln q_a\lambda_a)/(\partial f(z)/\partial \ln z^{-1}) \tag{13}$$

Furthermore, the fraction of monomer bound dye with respect to the total dye is:

$$X_1 = (\partial f(z)/\partial \ln q_{mon}\lambda_a)/(\partial f(z)/\partial \ln q_a\lambda_a) \tag{14}$$

with q_{mon} being the partition function of sites occupied by the dye in the monomeric state, that is the q parameter of <u>only</u> the first term of eq. (5):

$$\partial f(z)/\partial \ln q_{mon}\lambda_a = U(z)q_a\lambda_a z^{-1} \tag{15}$$

By writing:

$$\begin{aligned} A(v) = 1 &+ 2q_1 v + (3q_1^2 q_2 v^2/(1 - q_1q_2q_3v)) + \\ &(q_1^3 q_2^2 q_3 v^3/(1 - q_1q_2q_3v)^2) \end{aligned} \tag{16}$$

and

$$B(v) = 1 + 2v + 3q_1v^2 + (4q_1{}^2q_2v^3/(1 - q_1q_2q_3v)) + (q_1{}^3q_2{}^2q_3v^4/(1 - q_1q_2q_3v)^2) \qquad (17)$$

one obtains

$$D/P = vA(v)/B(v) \qquad (18)$$

The fraction of monomer bound dye is:

$$X_1 = 1/A(v) \qquad (19)$$

and the fraction of molecules in the m-th aggregate state is:

$$X_m = mX_1\, q_3^{m-3}\, q_2^{m-2}\, (q_1v)^{m-1} \qquad (20)$$

where, actually,

$$\sum_1^{\infty} X_m = 1$$

The fraction of dye molecules bound as a dimer is:

$$X_2 = 2X_1\, q_1\, v \qquad (21)$$

A limiting expression for X_1 at $D/P \to 0$ is

$$\lim_{D/P \to o} X_1 = 1 - 2\, q_1\, (D/P) + \ldots \qquad (22)$$

Figure 16 shows the theoretical values of X_1 and X_2 as a function of $\sqrt{D/P}$ for different values of q_1 and for $q_m = 1$ $(m > 1)$.

Lifson's treatment was successfully applied to some polystyrene sulfonic acid-dye systems by using a first and a second neighbour interaction parameter, and a dye distribution among monomer, dimer and aggregate species was obtained in very good agreement with experimental data (see Fig. 17 and 18)[70,73,79]. While the use of two parameters, q_1 and q_2, was found necessary to describe the dye distribution, the use of a third parameter, q_3, allowed only a slightly better fitting of the experimental data. As shown in Table II, $q_1 = 30$ for the AO - PSS system at 20°C; the q_1 value increases up to about 100 by adding NaCl up to 0.1 molar.

The q_1 values for polyacrylic and polymethacrylic acids, on the other hand, are much higher; they were roughly estimated by extrapolating the experimental extinction coefficients to D/P = 0.

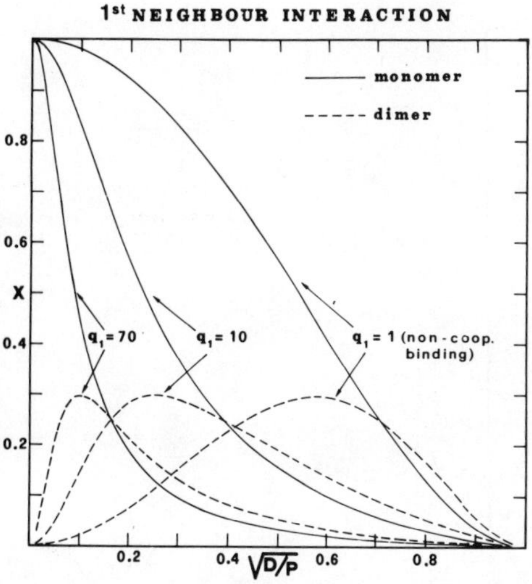

Fig. 16. Computed values of monomer and dimer fractions for different values of the q_1 stacking coefficient (eq. 19 and 21).

Poly(styrenesulfonic acid)

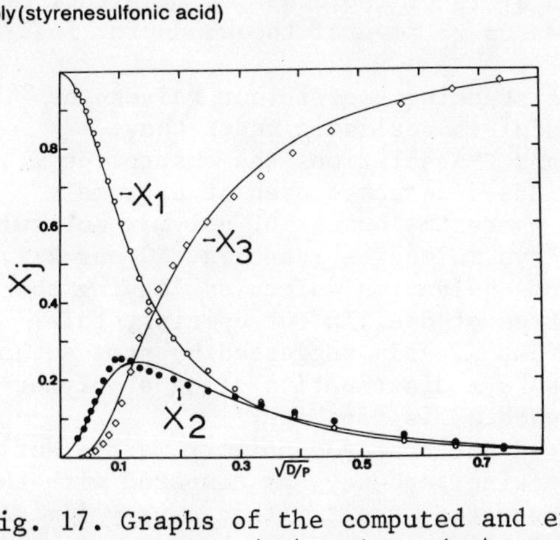

Fig. 17. Graphs of the computed and experimental fraction of AO bound as monomer (X_1), dimer (X_2), and aggregate (X_3) at 20° (from ref. 79)

q_1	q_2	q_3	$10^2\overline{\Delta X}_1$	$10^2\overline{\Delta X}_2$	$10^2\overline{\Delta X}_3$	Temp. °C
30.0	1.64	0.825	0.95	1.44	2.23	20

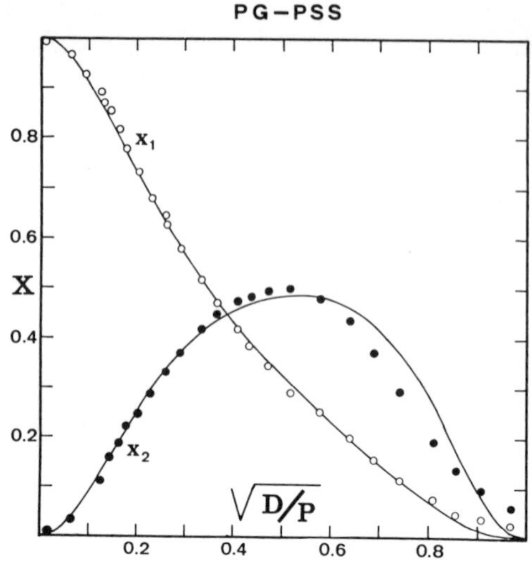

Fig. 18. Comparison of experimental monomer and dimer fractions with the values computed through eq. 18, 19, 21, $q_1 = 5.5$ and $q_2 = 0.26$ (from ref. 73).

The results are shown in Fig. 19 where it can be seen that even at a low degree of ionization q_1 is of the order of magnitude of some hundreds and it increases up to several thousands for fully ionized polyelectrolyte.

Such a high value of the stacking coefficient raises the question whether the Ising model is realistic under these conditions; with ionized PA and PMA solutions the absorption spectra of aggregate dye is indeed detected even at extremely high values of the P/D ratio where the number of polymer molecules is much greater than that of dye molecules (see Fig. 20 and 21), the dye molecules stack on some polyanion molecules leaving the great majority of macroions free of dye. In our opinion, this experimental evidence, which has already suggested by some authors does not allow one to discuss "dye distribution along a polymer chain" in any realistic way when q_1 is too high.

Finally it must be pointed out that any polymer matrix acts on the dye decreasing its stacking tendency, as compared with that in water. The bound dye concentrates itself within a very limited volume, i.e. the volume[116] actually occupied by the polymeric phase in solution. The real dye concentration inside the polymer phase is almost a linear function of the ratio D/P and a rough computation indicates that it is of the order of magnitude of 10^3 with respect to the stoichiometric dye concentration at D/P ≃ 0.00 under these conditions if the dye would keep the same stacking tendency as in water, a q_1 parameter of the order of magnitude of

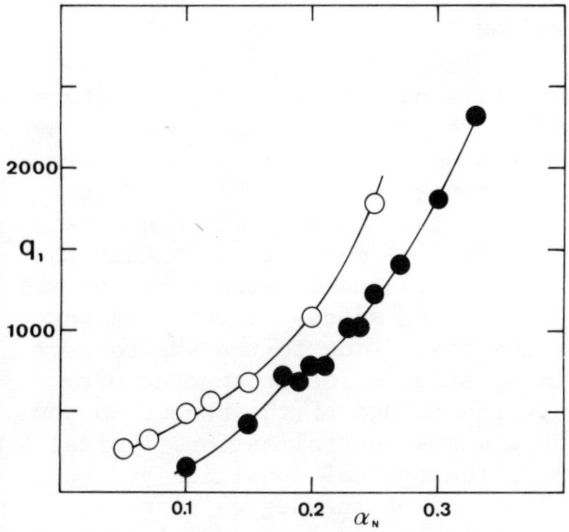

Fig. 19. First neighbour interaction parameter (stacking coefficient) for PA —o—, and for the PMA —●— as a function of the degree of neutralization. The q_1 values were computed by applying eq. (22) to the experimental extinction coefficients, E, at a given wave length:

$$E = E_1 - 2q_1(E_1 - E_2)(D/P) + \dots \qquad (23)$$

E_1 and E_2 being the extinction coefficients of monomer and dimer dye at the same wave length. The wave length of the absorption maximum of monomer bound dye (500 nm) was chosen and it was assumed that $(E_2)_{500} \simeq (E_{aggr})_{500}$ as for PSS.

$10^5 - 10^6$ should be detected.

TABLE II

COMPARISON BETWEEN DIMERIZATION CONSTANT OF SOME DYES IN WATER AND THEIR STACKING COEFFICIENT ON PSS at 20 °C

Dye	1/K (eq. 1) $l \cdot mole^{-1}$	q_1 (PSS)
Acridine Orange	9.3×10^3	~30
Methylene Blue	4.5×10^3	~18
Pyronine-G	0.83×10^3	6.5

EFFECTS DUE TO POLYMER CONFORMATION

The polymer dye interaction is sensitive to polymer conformation. As shown in Fig. 13 and 14, the PMA conformational transition in the neutralization range $\alpha_N = 0.1 - 0.3$[127,128] affects both the binding strength and the number of available binding sites. As may be expected the PMA in the compact coil conformation with higher charge density binds dyes more strongly than the open coil PMA; on the other hand, the last one being more ionized has a higher number of available binding sites. Recently it was also found that dye may induce the transition to the PMA compact coil conformation[80]. Another interesting effect connected with the conformational transition is the change of the monomer bound dye extinction coefficient[64] on the PMA neutralization. Typical curves are shown in Fig. 20 where the optical densities at the monomer λ_{max} are plotted as a function of the degree of neutralization, both for a PMA and a PA sample at constant P/D. A different behaviour is clearly shown by the two polymers. At low α_N the dye is mainly in its monomeric state. By increasing α_N the dye optical density gradually decreases, in the presence of excess PA, because of a gradual aggregation connected with the stacking coefficient increase (Fig. 19). In the presence of excess PMA an increase of the optical density can be observed up to $\alpha_N \simeq 0.1$, followed by a very sharp decrease, through the conformational transition region, connected with the dye aggregation.

Fig. 21 is a similar graph for a low molecular weight PMA sample, Mw $\simeq$ 8500), the absorption spectra of bound AO can be seen at various degrees of neutralization. It is also interesting to note that, with the degree of polymerization about 90 - 100, at P/D = 1000, there are more than 10 polymer molecules per dye molecule and even under these conditions at $\alpha_N > 0.4$ the AO is in its aggregation form.

The limiting extinction coefficients of monomer bound dye were computed by proper extrapolation to D/P = 0 for PA and PMA, as a function of α_N and the results are shown in Fig. 22. It is of considerable interest to note that monomeric dye bound to PMA at low α_N values exhibits a hyperchromic effect with respect to monomer dye in water. This effect is absent for PMA at $\alpha_N > 0.2$ and for PA. The same hyperchromic effect is shown by dyes dissolved in methyl alcohol (see Fig. 4) and other non-aqueous solvents. Present knowledge of PA and PMA aqueous solutions suggests a possible interpretation of the different behaviour of dyes bound to PA and PMA. Since the polyelectrolyte ionized groups preferably occupy regions of the macromolecules exposed to water, the aggregations of dye bound in these regions is favoured even at high values of the P/D ratio. For PMA, however, much experimental evidence indicates that the polyelectrolyte chains are tightly coiled in water at low α_N because of hydrophobic interactions.[68,127-129]

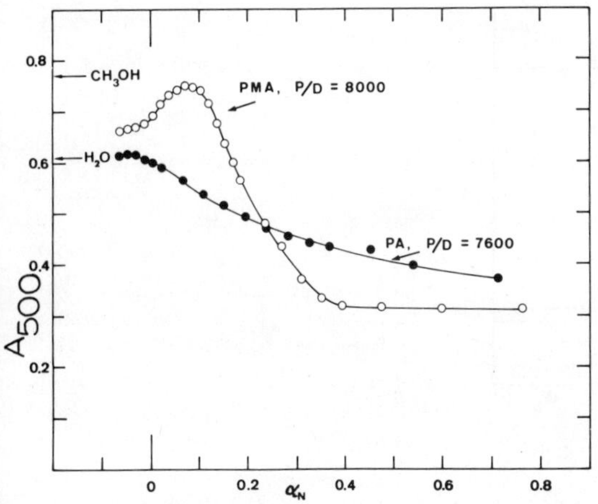

Fig. 20. Optical densities of AO (10^{-5}m) in the presence of excess polyacrylic acid and polymethacrylic acid as a function of the polyacid degree of neutralization (α_N < 0 corresponds to solutions with excess HCl) PMA degree of polymerization ≃ 5000, PMA molecules/dye molecules ≃ 1.6, PA degree of polymerization ≃ 2000, PA molecules/dye molecule ≃ 4.0.

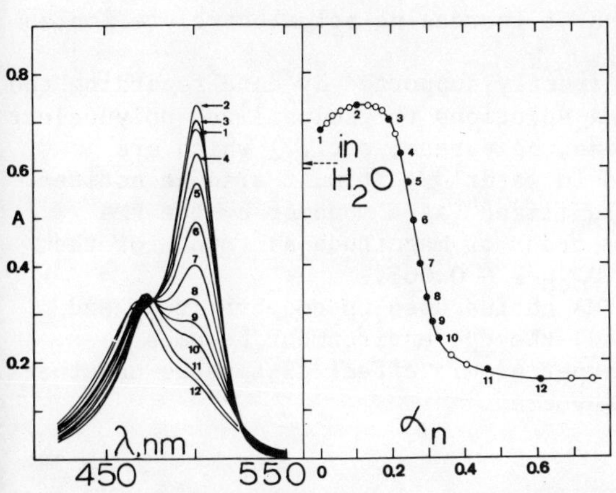

Fig. 21. Absorption Spectra of Acridine Orange (10^{-5}m) in the presence of excess syndiotactic polymethacrylic acid as a function of the polyacid degree of neutralization. (P/D = 1000, degree of polymerization ≃ 85, polymer molecules/dye molecules ≃ 12).

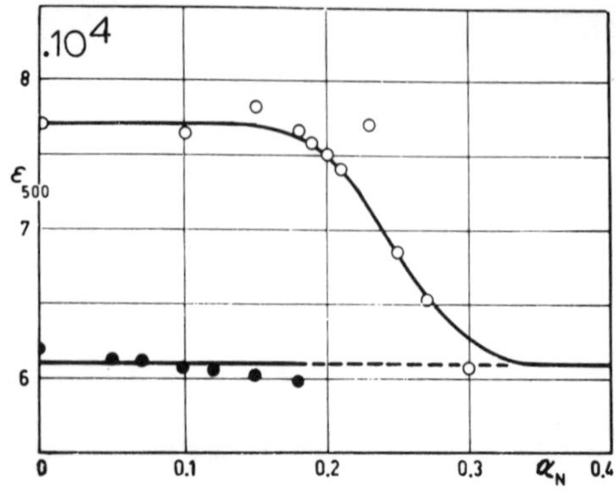

Fig. 22. Molar extinction coefficient at 500 mμ of monomer AO bound to the poly-electrolyte as a function of the degree of neutralization: ● polyacryclic acid; ○ polymethacrylic acid. (from ref. 64).

Some dye molecules may be "solubilized" inside the compact coil in a hydrophobic region not much different from an alcoholic solution. Such an environment might be responsible for the observed hyperchromic effect and for the higher adsorbing power of PMA with respect to PA (Fig. 13). The optical density maxima exhibited by the plots of Fig. 20 and 21 could be simply due to a decrease of free dye in solution because of increasing polyelectrolyte ionization.

This hypothesis is indirectly supported by data regarding the ability of unneutralized PMA solutions to "solubilize" polynuclear aromatic hydrocarbons (pyrene, bezpyrene, etc...) which are extremely sparingly soluble in water[130]. It must also be noticed that the amount of dye "solubilized" as a monomer by the PMA compact coil is of the same order of magnitude as found for the aromatic compounds, namely $D_{mon}/P \simeq 0.001$.

At $\alpha_N > 0.1 - 0.2$ the PMA chains open up cooperatively and become increasingly solvated, the dye environment becomes increasingly aqueous, the hyperchromic effect disappears and the aggregation of the dye is favoured.

EFFECTS DUE TO POLYMER TACTICITY

The polyelectrolyte chain tacticity affects the dye binding process[60,67]; isotactic polymers with vinyl chain bind dyes more strongly than atactic polymers (which are known to be mainly syndiotactic) and the bound dye exhibits a lower stacking tendency.

This last effect can be seen in Fig. 23 where some absorption

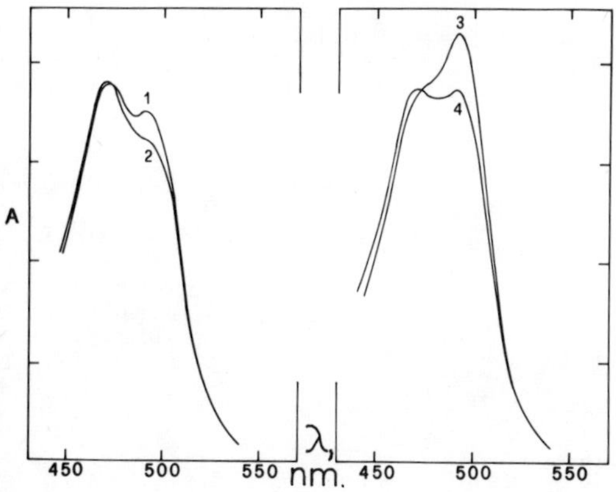

Fig. 23. Comparison between atactic and isotactic polymethacrylic acid at neutral pH. Absorption spectra of 10^{-6} molar AO bound to PMA. 1 atactic PMA (Mw ~ 5 x 10^5), P/D = 8000; 2 atactic PMA (Mw ~ 5 x 10^5), P/D = 4800; 3 isotactic PMA (Mw ~ 2 x 10^5), P/D = 8000; 4 isotactic PMA (Mw ~ 2 x 10^5), P/D = 4000.

spectra are drawn for both atactic PMA - AO and isotactic PMA - AO systems at high P/D values, fig. 23 shows that the AO bound to the isotactic polymer has a higher amount of monomer species than that bound to the atactic one.

Similar but more striking effects are shown by the isotactic polystyrene sulfonic acid-acridine orange system[131]. (1) The amount of free dye is practically null even at P/D = 1, as it can be argued by much sharper end point titration (fig. 24) as compared with the atactic PSS titration (fig. 10). (2) The extinction coefficient of monomer bound dye is much lower than the corresponding value for atactic PSS ($\simeq$ 45000 as compared to $\approx$ 54000 for atactic PSS) which in turn, is lower than the dye extinction coefficient in water (61000). (3) The stacking coefficient q_1 is lower ($q_1 \simeq 9$ for isotactic PSS at 20°C, as compared to q_1 = 30 for atactic PSS at the same temperature).

These effects can all be attributed to specific interactions between the benzene rings of PSS chain and dye molecules, in fact some kind of partial intercalation may be promoted, similar to that suggested in the case of DNA and acridine dyes[11,14,52]. It is interesting to note that crystalline isotactic polystyrene chains may arrange themselves in such a way as to have benzene rings regularly spaced at about 6.5 Å[136,137] (fig. 25) giving a very good possibility for intercalation to an aromatic ring. Such arrangement can be partially preserved in isotactic PSS solutions favouring the polyelectrolyte-dye interaction.

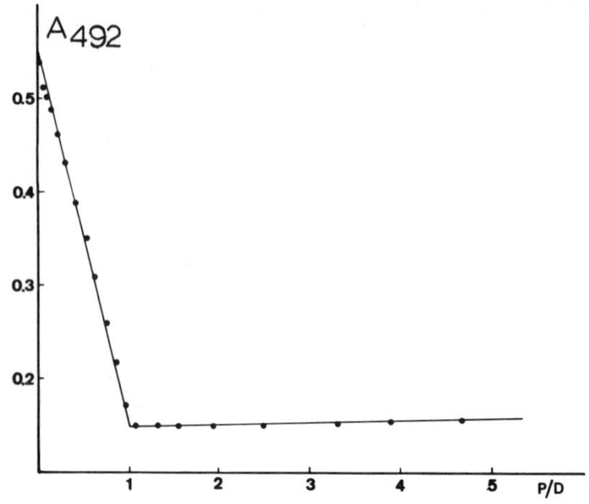

Fig. 24. Titration of Isotactic Polystyrenesulfonic Acid with 10^{-5}molar AO.

RELAXATION KINETICS EXPERIMENTS[33,42,50,71,72,76,134]

The relaxation kinetics of the dye binding process has been studied recently by Schwarz et al.[71,72,123,132,133] both from theoretical and experimental points of view. Schwarz's experiments however, were taken on polymer-dye systems exhibiting high or very high stacking effects and the kinetic process observed through T-jump experiments was the relaxation of the dye distribution equilibrium between polymer and solution, this process is dye concentration dependent.

Another possible process is the kinetics of the dye distribution along the polyelectrolyte chain following a stacking coefficient change due to the temperature jump. This second process does not involve the entire solution volume, but is specific to the polymeric phase; thus the kinetics must be expected to be independent of the dye concentration but vary with the ratio P/D.

The PSS-AO system is convenient for this process because the amount of free dye in solution is almost negligible at P/D > 80 - 100 so that any spectral change detectable under these conditions is connected with a dye distribution change inside the polymer phase. A set of T-jump experiments taken on the system PSS-AO has shown the presence of both processes[134]. At lower P/D ratio the relaxation kinetics of dye distribution between polymer and solution was detected, accordingly the reverse of mean relaxation time was found to be a linear function of the total dye concentration, at constant P/D. At higher P/D ratios the relaxation time becomes independent of total dye concentration, at constant P/D, as expected for a process involving only the inside of each

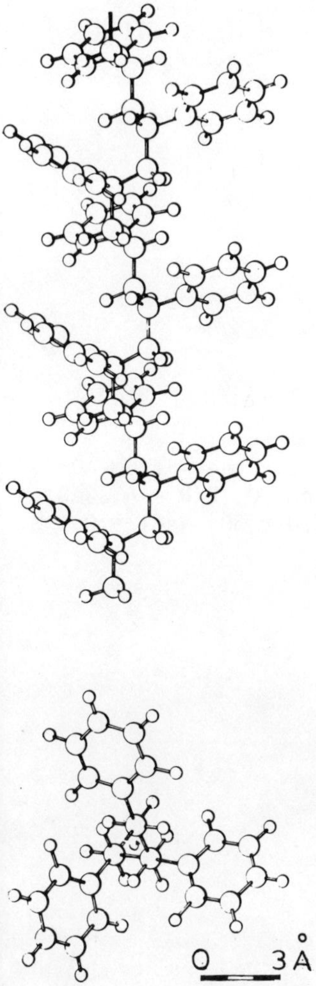

Fig. 25. Conformation of the isotactic polystyrene macromolecule in the crystalline state (side and end views) (from ref. 136).

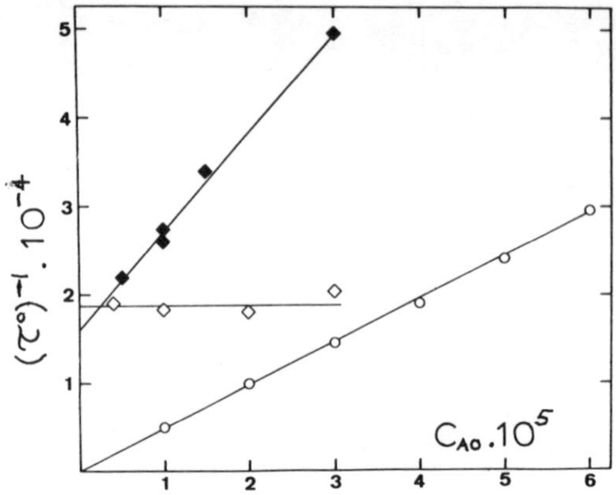

Fig. 26. Mean (reciprocal) relaxation times for the system AO-PSS at constant P/D and varying dye concentration: O, P/D = 4, NaCl = 0.04 M; ◆, P/D = 200, NaCl = 0.04 M; ◇, P/D = 50, NaCl = 0.10M. (from ref. 134).

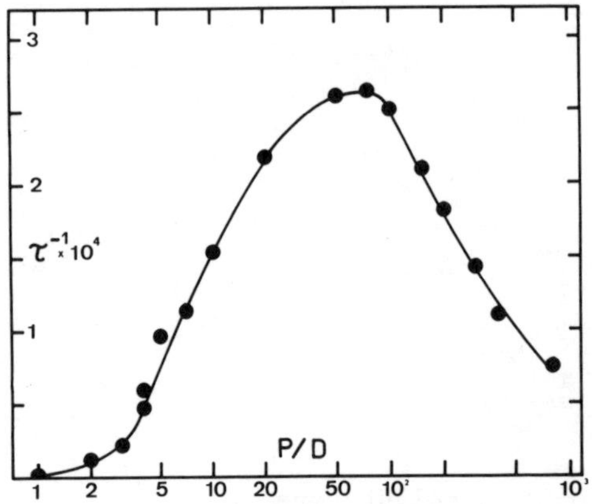

Fig. 27. T-jump experiments on the system Acridine Orange-Polystyrenesulfonic Acid. Relationship between relaxation time, τ, and P/D ratios.

polyelectrolyte molecule. At intermediate P/D ratios both kinetic processes seem to be present (see fig. 26).

As predicted by Schwarz theory the graph of the inverse relaxation time as a function of P/D at constant total dye concentration goes through a maximum (see fig. 27). A similar relaxation kinetics was recently reported by Schwarz and Klose[76] on the system proflavine-polyacrylic acid.

REFERENCES

1. The following set of references does not intend to cover all the literature on the subject.
2. L. Michaelis and S. Granick, J. Am. Chem. Soc., 67, 1212 (1945).
3. L. Michaelis, J. Phys. Chem., 54, 1 (1950).
4. G. Oster, J. Polymer Sci., 16, 235 (1955).
5. A.R. Peacocke and J.N.H. Skerrett, Trans. Farad. Soc., 52, 261 (1956).
6. W. Appel and V. Zanker, Z. Naturforschung 13B, 126 (1958).
7. R.F. Beers, D.D. Hendley, and R.F. Steiner, Nature, 182, 242 (1958).
8. D.F. Bradley and M.K. Wolf, Proc. Natl. Acad. Sci. U.S.A., 45, 944 (1959).
9. D.F. Bradley and G. Felsenfeld, Nature, 184, 1920 (1959).
10. N.S. Ranadive and K.S. Kangaonkar, Biochim. Biophys. Acta. 39, 547 (1960).
11. L.S. Lerman, J. Mol. Biol., 3, 18 (1961).
12. V. Luzzati, F. Mason and L.S. Lerman, J. Mol. Biol., 3, 634 (1961).
13. A.L. Stone and D.F. Bradley, J. Am. Chem. Soc., 83, 3627 (1961).
14. L.S. Lerman, Proc. Natl. Acad. Sci. U.S.A., 49, 94 (1963).
15. G. Weil and M. Calvin, Biopolymers, 1, 401 (1963).
16. R.E. Kay, E.R. Walwick and C.K. Gifford, J. Phys. Chem., 68, 1896, 1909 (1964).
17. D.S. Drummond, V.F.W. Simpson-Gildemeister and A.R. Peacocke, Biopolymers, 3, 135 (1965).
18. N.F. Gersch and D.O. Jordan, J. Mol. Biol., 13, 138 (1965).
19. M.J. Waring, J. Mol. Biol., 13, 269 (1965).
20. A. Blake and A.R. Peacocke, Biopolymers, 4, 1091 (1966).
21. D.S. Drummond, N.J. Pritchard, V.F.W. Simpson-Gildemeister and A.R. Peacocke, Biopolymers, 4, 971 (1966).
22. G.S. Hammes and C.D. Hubbard, J. Phys. Chem., 70, 2889 (1966).
23. D.M. Neville Jr. and D.R. Davies, J. Mol. Biol., 17, 57 (1966).
24. J.E. Scott and I.H. Willet, Nature, 209, 985 (1966).
25. B.J. Gardner and S.F. Mason, Biopolymers, 5, 79 (1967).
26. J.C. Power Jr. and W.L. Peticolas, J. Phys. Chem., 71, 3191 (1967).
27. W. Bauer and J. Vinograd, J. Mol. Biol. 33, 141 (1968).
28. W.G. Galley, Biopolymers, 6, 1279 (1968).
29. M. Gilbert and P. Claverie, J. Theor. Biol., 18, 330 (1968).
30. H. Grosjean, J. Werenne and H. Chantrenne, Biochim. Biophys. Acta., 166, 616 (1968).
31. K. Yamaoka, Biochim. Biophys. Acta, 169, 552 (1968).
32. H.J. Li and D.M. Crothers, Biopolymers, 8, 217 (1969).
33. H.J. Li and D.M. Crothers, J. Mol. Biol., 39, 461 (1969).

34. L.M. Pupa and L. Bosch, FEBS Letters, 4, 143 (1969).
35. W. Bauer and J. Vinograd, J. Mol. Biol., 47, 419 (1970).
36. J.S. Moore, G.O. Phillips, D.M. Power and J.V. Davies, J. Chem. Soc., A,1155 (1970).
37. A.L. Stone, L.G. Childers and D.F. Bradley, Trans. Farad. Soc., 66, 3081 (1970).
38. P.J. Baugh, J.B. Lawton and G.O. Phillips, J. Am. Oil. Chem. Soc., 48, 279 (1971).
39. M. Kaufman and G. Weil, Biopolymers, 10, 1983 (1971).
40. P.J. Baugh, J.B. Lawton and G.O. Phillips, J. Phys. Chem., 76, 688 (1972).
41. D.B. Dusebery and R.B. Uretz, Biophys. J., 12, 1056 (1972).
42. D.E.V. Schmechel and D.M. Crothers, Biopolymers, 10, 465 (1971).
43. E. Fredericq and C. Houssier, Biopolymers, 11, 2281 (1972).
44. T. Ito, M. Zama and J. Amagasa, Biopolymers, 11, 1583 (1972).
45. K. Yamaoka, Biopolymers, 11, 2537 (1972).
46. K. Zinner and G. Cilento, Biopolymers, 11, 1521 (1972).
47. S. Dasgupta, D.N. Misra and N.N. Dasgupta, Biochim. Biophys. Acta., 294, 38 (1973).
48. M. Domard, M. Rinaudo and R. Rinaldi, J. Chim. Phys., 1410 (1973).
49. C.H. Lee, C.T. Chang and J.G. Wetmur, Biopolymers, 12, 1099 (1973).
50. C. Steenbergen and S.C. Mohr, Biopolymers, 12, 791 (1973).
51. D.N. Goswami and N.N. Das Gupta, Biopolymers, 13, 391 (1974).
52. J.P. Schreiber and P.M. Daune, J. Mol. Biol., 83, 487 (1974).
53. S. Brenner, L. Barnett, F.H.C. Crick and A. Orgel, J. Mol. Bic 3, 121 (1961).
54. D.F. Bradley and S. Lifson,"Molecular Association in Biology," B. Pullmann, Ed. Acad. Press, New York, 261 (1968).
55. C.R. Merrill and R.W. Spencer, J. Am. Chem. Soc., 70, 3683, (1948).
56. M. Mizushima, Busseroin Kenkyu, 17, 38 (1949).
57. M.K. Pal and S. Basu, Makromol. Chem., 27, 69 (1958).
58. G. Blauer, J. Phys. Chem., 65, 1457 (1961).
59. L. Stryer and E.R. Blout, J. Am. Chem. Soc., 83, 1411 (1961).
60. G. Barone, R. Caramazza and V. Vitagliano, Ric. Sci., 32 (IIA 554 (1962).
61. P. Mukerjee and A.K. Ghosh, J. Phys. Chem., 67, 193 (1963).
62. R. Caramazza, L. Costantino and V. Vitagliano, Ric. Sci., 34 (IIA), 67 (1964).
63. R.E. Ballard, A.J. McCaffery and S.F. Mason, Biopolymers, 4, 97 (1966).
64. G. Barone, V. Crescenzi, F. Quadrifoglio and V. Vitagliano, Ric. Sci., 36, 503 (1966).
65. G.C. Hammes and C.D. Hubbard, J. Phys. Chem., 70, 1615 (1966
66. T. Soda and K. Yoshiioka, J. Chem. Soc.(Japan), 87, 324 (196
67. V. Crescenzi, F. Quadrifoglio and V. Vitagliano, J. Macromol Sci.(Chem.), A1(5), 917 (1967).

68. E.V. Anufrieva, T.M. Birshtein, T.N. Nekrasova, O.B. Ptitsin and T.V. Sheveleva, J. Polymer Sci., C16, 3519 (1968).
69. V.V. Strelko, V.A. Kanibolotskii and Z.Z. Vysotskii, Russian J. Phys. Chem., 42, 635 (1968).
70. V. Vitagliano e L. Costantino, Boll. Soc., Natur. in Napoli, 78, 169 (1969).
71. G. Schwarz and W. Balthasar, Eur. J. Biochem., 12, 461 (1970).
72. G. Schwarz, S. Kmose and W. Balthasar, Eur. J. Biochem., 12, 454 (1970).
73. V. Vitagliano and L. Costantino, J. Phys. Chem., 74, 197 (1970).
74. S. Ikeda and T. Imaea, Biopolymers, 10, 1743 (1971).
75. B.C. Myhr and J.G. Foss, Biopolymers, 10, 425 (1971).
76. G. Schwarz and S. Klose, Eur. J. Biochem., 29, 249 (1972).
77. M. Hatano, M. Yoneyma and Y. Sato, Biopolymers, 12, 895 (1973).
78. W.H.J. Stork, P.L. de Hasseth, W.B. Schippers, C.M. Kormeling, and M. Mandel, J. Phys. Chem., 77, 1772 (1973).
79. V. Vitagliano, L. Costantino and A. Zagari, J. Phys. Chem., 77, 207 (1973).
80. W.H.J. Stork, J.A.M. Van Buxsel, A.F.P.M. De Goeij, P.L. Haseth and M. Mandel, Biophys. Chem., 2, 127, 137 (1974).
81. P.H. Hillson and R.B. McKay, Nature, 210, 297 (1966).
82. G. Scheibe, A. Mareis and H. Ecker, Naturwissenschaften, 25, 474 (1937).
83. E. Rabinowitch and L.F. Epstein, J. Am. Chem. Soc., 63, 69 (1941).
84. E. Sheppard and A.L. Geddes, J. Am. Chem. Soc., 66, 1995 (1944).
85. D.R. Lemin and T. Vickerstaff, Trans. Farad. Soc., 43, 491 (1947).
86. V. Zanker, Z.physik. Chem., 199, 225, (1952).
87. Kh.L. Arvan, Doklady Akad. Nauk. SSSR, 121, 123 (1958)
88. G.R. Haugen and E.R. Hardwich, J. Phys. Chem., 67, 725 (1963).
89. K. Bergmann and C.T. O'Konski, J. Phys. Chem., 67, 2169 (1963).
90. R.B. McKay, Trans. Farad. Soc., 61, 1787 (1965).
91. W. West and S. Pearce, J. Phys. Chem., 69, 1894 (1965).
92. M.E. Lamur and D.M. Neville Jr., J. Phys. Chem., 69, 3872 (1965).
93. D.J. Blears and S.S. Danyluk, J. Am. Chem. Soc., 89, 21 (1967).
94. E. Braswell, J. Phys. Chem., 72, 2477 (1968).
95. J.F. Padday, J. Phys. Chem., 72, 1259 (1968).
96. R.E. Ballard and C.H. Park, J. Chem. Soc. A (1970), 1340.
97. P. Mukerjee and A.K. Ghosh, J. Am. Chem. Soc., 92, 6408 (1970).
98. B.H. Robinson, this publication.
99. J.F. Padday, J. Phys. Chem., 71, 3488 (1967).
100. H.A. Uedaira and Hi. Uedaira, Kolloid. Z., 194, 148 (1964).
101. M.J. Blandamer, J.A. Brivati, M.F. Fox, M.C.R. Symons and G.S.P. Verna, Trans. Farad. Soc., 63, 1850 (1967).
102. P. Mukerjee and A.K. Ghosh, J. Am. Chem. Soc., 92, 6419 (1970).
103. W. Kauzmann, Adv. Protein Chem., 14, 1 (1959).
104. G. Nemethy and H.A. Sheraga, J. Phys. Chem., 36, 3401 (1962).
105. G. Barone, L. Costantino e Vitagliano, Ric. Sci., 34 (IIA), 87 (1964).
106. W.H.J. Stork, G.J. Lippits and M. Mandel, J. Phys. Chem., 76, 1772 (1972).

107. V. Crescenzi, F. Quadrifoglio and V. Vitagliano, J. Phys. Chem., 71, 2313 (1967).
108. G. Barone, V. Crescenzi and V. Vitagliano, J. Phys. Chem., 72, 2588 (1968).
109. G.C. Kresheck and H.A. Sheraga, J. Phys. Chem., 69, 1704 (1965)
110. R.H. Stokes, J. Phys. Chem., 69, 4012 (1965).
111. G. Barone, E. Rizzo and V. Vitagliano, J. Phys. Chem., 74, 2230 (1970).
112. E.G. Finer, F. Franks and M.J. Tait, J. Am. Chem. Soc., 94, 4424 (1972).
113. L. Costantino, A.M. Liquori and V. Vitagliano, Biopolymers, 2, 1 (1964).
114. R.M. Fuoss, A. Katchalsky and S. Lifson, Proc. Natl. Acad. Sci. U.S.A., 37, 579 (1951).
115. S. Lifson and A. Katchalsky, J. Polymer Sci., 13, 43 (1954).
116. F. Oosawa, J. Polymer Sci., 23, 421 (1957).
117. G.S. Manning, J. Chem. Phys., 51, 924 (1969).
118. E. Ising, Z. Phys., 31, 253 (1925).
119. T.L. Hill, "Statistical Mechanics," McGraw Hill, New York (1956) Chapter 7.
120. B.H. Zimm and J.K. Bragg, J. Chem. Phys., 31, 526 (1959).
121. S. Lifson and A. Roig, J. Chem. Phys., 34, 1963 (1961).
122. F.W. Schneider, C.L. Cronar, and S.K. Podder, J. Phys. Chem., 72, 4563.
123. G. Schwarz, Biopolymers, 6, 873 (1968).
124. G. Schwarz, Eur. J. Biochem., 12, 442 (1970)
125. S. Lifson, J. Chem. Phys., 40, 3705 (1964)
126. M. K. Pal and M. Schubert, J. Phys. Chem., 65, 872 (1961)
127. J. C. Leyte and M. Mandel, J. Polymer Sci., A2, 1879 (1964)
128. A. M. Liquori, G. Barone, V. Crescenzi, F. Quadrifoglio, and V. Vitagliano, J. Macromol. Chem., 1, 291 (1966)
129. M. Mandel, J. C. Leyte, and M. G. Stadhouder, J. Phys. Chem., 71, 2341 (1967)
131. V. Vitagliano, L. Costantino, and R. Sartorio, to be publishe
132. G. Schwarz, Ber. Bunsenges physik. Chem., 75, 40 (1971)
133. G. Schwarz, Ber. Bunsenges physik. Chem., 76, 373 (1972)
134. V. Vitagliano, J. Phys. Chem., 77, 1922 (1973)
135. M. Yaacobi and A. Ben-Naim, J. Solution Chem., 2, 425 (1973)
136. G. Natta, P. Corradini, and I. W. Bassi, Nuovo Cimento (Suppl 15, 68
137. G. Natta, Makromol. Chem., 16, 213 (1955)

KINETIC STUDY OF COOPERATIVE BINDING REACTION OF TOLUIDINE BLUE TO POLY-α, L-GLUTAMIC ACID BY MEANS OF ELECTRIC FIELD PULSE METHOD

Tatsuya Yasunaga, Hidetoshi Takenaka, Takayuki Sano
and Yoshikuni Tsuji
University of Hiroshima, Japan

INTRODUCTION

The interactions of basic dyes with many biopolymers have been studied in view of stacking. This phenomenon of stacking has been interpreted as follows; dye molecules bound to the polymer may be sufficiently closely located on the polymer to interact with one another to form aggregates analogous to those found in free dye solution. Many static studies of stacking have been carried out[1], and a few kinetic studies have been recently reported. First kinetic study on stacking has been proceeded with by Hammes and Hubbard using T-jump method, but well-defined conclusion was not obtained.[2,3] Schwarz et al. have investigated the binding of acridine orange to poly-α-L-glutamic acid (PLGA) using a basic model in which the binding to a linear lattice of equivalent binding sites with cooperative interaction of nearest neighbours was assumed.[4-6] Their studies were performed only at pH 7.5 where PLGA is in the random coiled state, and the results were well interpreted by Schwarz's theory. In the present investigation, we attempted to elucidate the binding mechanism of Toluidine blue (TB) to PLGA in both helical and coiled states and to compare the results by means of the cooperative binding theory.

EXPERIMENTAL

The sodium salt of PLGA with a degree of polymerization (D.P) of 700 was supplied from Ajinomoto Co., Inc.. TB was purchased from Chroma Co., and was purified by washing with chloroform. The pH of the mixed solution was adjusted to the desired value by the addition of dilute NaOH or HCl solution. The pH values of the

Wyn-Jones (ed.), Chemical and Biological Applications of Relaxation Spectrometry, 467–479.

solutions were measured with a Hitachi-Horiba F-5 pH meter under the nitrogen atmosphere in order to exclude the contamination with air. The ORD measurements were carried out in the region of 200-320 nm with a Jasco ORD/UV-5. The absorption spectra were obtained with a Jasco ORD/UV 5 and Hitachi 139 spectrophotometer. The electric field pulse method apparatus used was constructed to follow the transient by an absorbance change in the sample solution. All the kinetic measurements were carried out at the electric field intensity of 50 kV/cm, and the sample cell used here has the light path of 18 mm and the electrode separation of 3mm. The temperature was controlled at 25°C with an accuracy of ± 0.5°C.

RESULTS AND DISCUSSION

The absorption spectra of PLGA-TB system under the various polymer to dye ratio (P/D) at pH 8.0 are shown in Fig. 1.

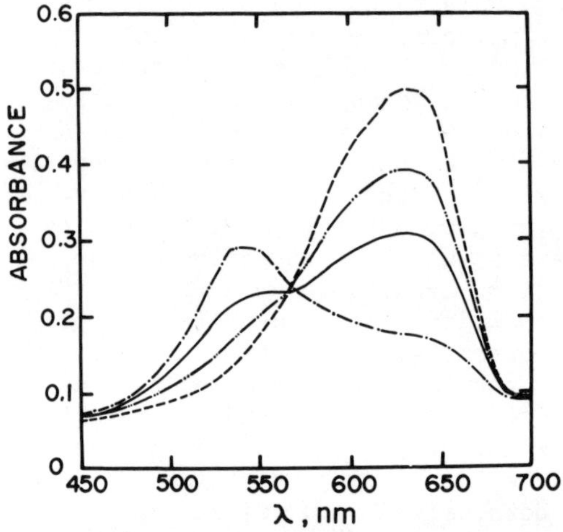

Fig. 1. The absorption spectra of TB-PLGA system at $C_A^o = 2 \times 10^{-5}$ M, pH = 8.0, and P/D = 0, ---; 0.5, —··—; 1, —; 10, —·—,

This result indicates that with increase of P/D, the metachromatic band at 540 nm becomes predominant and the band shift was not observed until P/D 10. In the region of P/D more than 10, the metachromatic band shifted to the longer wave-length i.e. negative metachromasy. This fact indicates that the dye-dye interactions on PLGA decreases with increasing P/D. The absorption spectra of PLGA-TB system with P/D = 2.5 at various pH's are shown in Fig. 2.

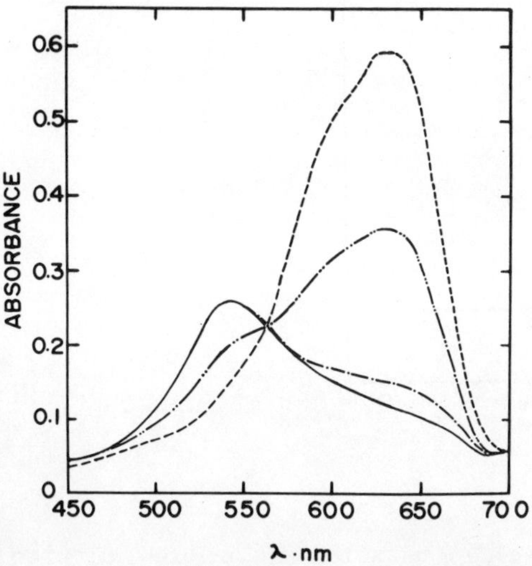

Fig. 2. The absorption spectra of TB-PLGA system at $C_A^o = 2 \times 10^{-5}$ M, P/D = 2.5, and pH = 4.5,—··— ; 6.1,—·— ; 9.2,—— . The dashed line shows the spectrum of TB solution.

With increasing pH in the region from 4 to 10, the metachromatic band increases without the band shift. Moreover, in order to examine the effect of TB to the conformation of PLGA, specific rotation measurements at 233 nm were carried out. It is found in Fig. 3 that PLGA is in helical state at pH = 5 and in random coiled state at pH = 7, and that the pH region of the helix-coil transition is not shifted by addition of TB. According to these static experiments the kinetic studies were carried out at the experimental conditions as follows; P/D = 1 to 10, and pH = 5.0 (helical state) and pH = 8.0 (random coiled state).

We first estimated the dimerization constant (K_d) of TB in the absence of PLGA. Let ε, ε_A and ε_D denote the extinction coefficient at 540 nm of the solution, the pure monomer and the pure dimer, respectively and they refer to one mole of monomer units. The following equation is generally accepted; where C_A^o is

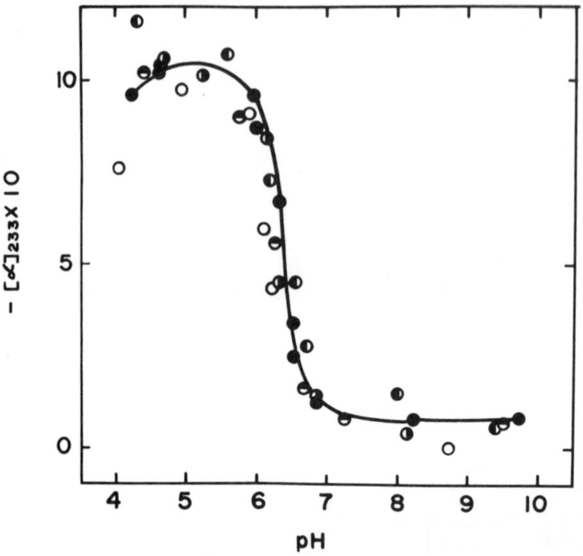

Fig. 3. The pH dependency of specific rotation in TB-PLGA solution at $C_A^o = 2 \times 10^{-5}$ M, and P/D = 3, o; 5, ◐; 10, ◑; 20, ●. The solid circle shows that of the PLGA solution at 1×10^{-4} M.

$$\sqrt{\frac{\varepsilon_A - \varepsilon}{C_A^o}} = \sqrt{\frac{2Kd}{\Delta\varepsilon}} \{\Delta\varepsilon - (\varepsilon_A - \varepsilon)\} \tag{1}$$

total weighing-in concentration of the dye and $\Delta\varepsilon = \varepsilon_A - \varepsilon_D$. Accordingly, a plot of

$$\sqrt{(\varepsilon_A - \varepsilon)/C_A^o} \text{ v.s } (\varepsilon_A - \varepsilon)$$

should yield a straight line with the intercepts $\sqrt{2K_d\Delta\varepsilon}$ on the ordinate axis and $\Delta\varepsilon$ on the abscissa axis. On this basis, we have evaluated ε_A as the best choice to obtain a straight line with the values of ε and C_A^o. Therefore, the parameters ε_D and K_d are determined from the two intercepts[5]. This plot is shown in Fig. 4. Thus, we obtain the same values $K_d = 2.5 \times 10^3$ M^{-1} at both pH = 4.6 and 7.9. The obtained K_d value is in good agreement with that of Methylene blue and Thionine which belong to Thiazine dye as TB[7].

As is shown above, addition of TB to PLGA solution caused

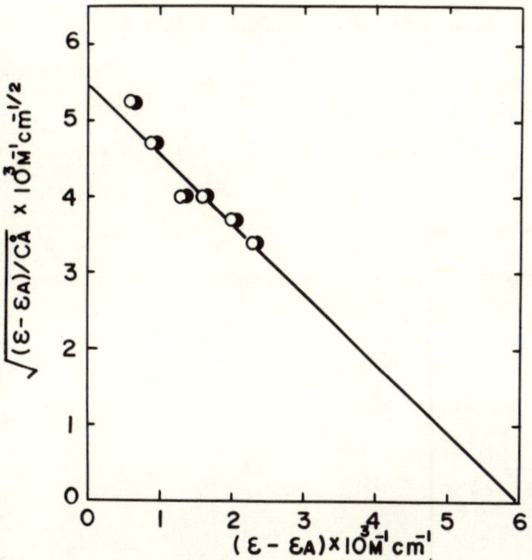

Fig. 4. Plots to determine dimerization constants (K_d) and intrinsic extinction coefficients (ε_A, ε_o) for TB solutions at pH = 4.6, ○ and 7.9, ●.

the metachromatic band at 540 nm. This phenomenon suggests that TB molecules stack on the PLGA molecules. Therefore, we should obtain the extinction coefficient ε_{st} of the stacked dye molecules. For the extinction coefficient ε of solution containing stacked dye molecules, we have at equilibrium:

$$\varepsilon = \bar{\gamma}_A \varepsilon_A + 2K_d C_A^o \bar{\gamma}^2 \varepsilon_D + \bar{\theta} g p \varepsilon_{st} \tag{2}$$

where $\bar{\gamma}_A$ is fraction of free monomeric dye, $\bar{\theta}$ is a fraction of occupied sites, g is binding sites per monomer segment of PLGA and p ≡ P/D. After further derivations, following equation is obtained[5]:

$$\varepsilon = \varepsilon_{st} + (\varepsilon_A - \varepsilon_{st})(s/K)C_A^{o-1} \tag{3}$$

where K is cooperative binding constant and s = $K\bar{C}_A$ ($\bar{C}_A$:

concentration of free monomeric dye). At strong cooperativity and medium P/D, the parameter s becomes almost constant irrespective of C_A^o. Accordingly, a plot of ε vs. $1/C_A^o$ at constant P/D should eventually result in a straight line. Then extrapolation $1/C_A^o \longrightarrow 0$ yields ε_{st} as intercept on the ordinate axis. Such experimental plots are shown in Fig. 5 for P/D = 5. Thus, we obtained $\varepsilon_{st} = 12.5 \times 10^3\ M^{-1}cm^{-1}$ in both helical and random coiled states.

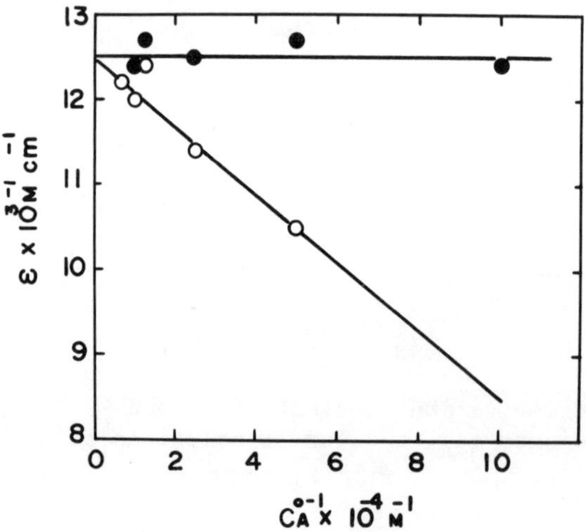

Fig. 5. Molar extinction of TB solutions as a function of the reciprocal value of the total weighing-in concentration of the dye ($1/C_A^o$) at P/D = 5; pH = 5.0, o; 8.3, ●.

The reason why ε does not depend on $1/C_A^o$ in the random coiled state is explained as follows; the K-value in Eq. (3) is just comparable with other values, as is shown below, so that the second term can be neglected as compared with the first term. The fact that ε_{st}-value turned out to be same value in both helical and random coiled states leads to the anticipation that the stacking mechanism of TB is the same scheme in two states. This is also predicted by the change in spectra as mentioned above. Another derivations yields the following relations[5]

$$\gamma_A^* + \bar{\theta}gp = 1 \qquad (4)$$

$$\gamma_A^* = \bar{\gamma}_A(1 + 2K_d C_A^o \bar{\gamma}_A) \quad (5)$$

$$\bar{\gamma}_A = (S/K)C_A^{o-1} \quad (6)$$

$$K = (\gamma_A^* + C_A^o)^{-1} + 2K_d \quad (7)$$

With the Eqs. (3), (5) and (6), γ_A^*-value was calculated, and then the plots of γ_A^*-value vs. P/D-value at constant C_A^o are represented in Figs. 6 and 7. From the intercept on the P/D-axis of the common limiting straight line for small P/D, we obtained g = 0.5 in helical state. From the γ_A^*-value of the intersection point of the auxiliary straight line with half the slope of the first one in the diagram and the experimental curve, K-value was determined by the relation (7).

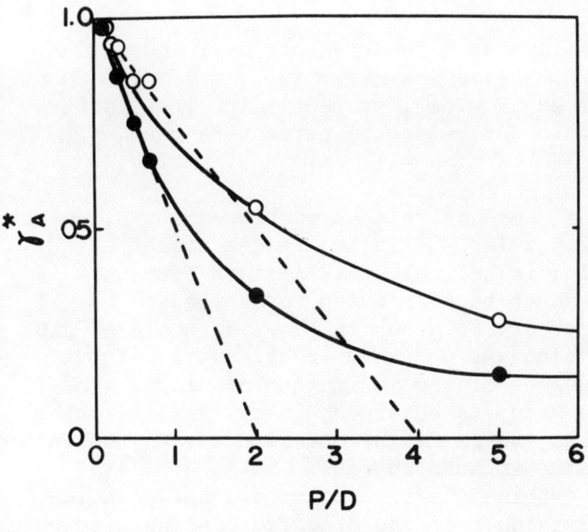

Fig. 6. Fraction of free dye as a function of small and medium polymer to dye ratio P/D. Curves refer to $C_A^o = 2 \times 10^{-5}$ M at pH = 4.5, o; 4×10^{-5} M at pH = 4.8, ●; respectively. The dashed straight lines are used to determine the parameters g and K.

As a result of this analysis $K = 1 \times 10^5\ M^{-1}$ was obtained in the

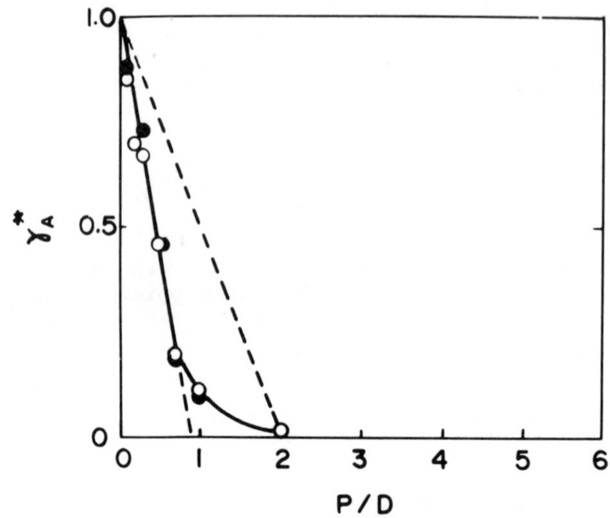

Fig. 7. Fraction of free dye as a function of small and medium polymer to dye ratio P/D. The curves refer to $C_A^o = 2 \times 10^{-5}$ M at pH = 8.8 and 4×10^{-5} M at pH = 8.3, respectively. The dashed straight lines are used to determine the parameters g and K.

helical state. After the same analysis as mentioned above, we obtained g = 1.1 and $K = 2.5 \times 10^6\ M^{-1}$ in the random coiled state. The fact that the g-value in helical state differs from that in random coiled state seems to be attributed to the reason as follows. Although the carboxylic group on the side chain of PLGA is ionized with pH, g-value was obtained in this work without taking account of the degree of the ionization of side chain. If g-value for the ionized group is obtained, it may fall into the same value in both states, which may be consistent with the fact that ε_{st} is the same value in both states.

However, g-values obtained in two states did not coincide. This fact implies the existence of the specific binding site which is also confirmed by considering the degree of ionization of PLGA.

By means of relation (4), the θ-values for various C_A^o were calculated at constant P/D and then the cooperative parameter q was obtained using the following equation[4]:

$$\frac{1 - 2\bar{\theta}}{\sqrt{\bar{\theta}(1 - \bar{\theta})}} \cdot \sqrt{\bar{C}_A} = \sqrt{q/K} - \sqrt{qK} \cdot \bar{C}_A \qquad (8)$$

Then, the q-value in helical state was turned out to be 15. In random coiled state, however it is not obtained experimentally on account of small change of $\overline{\theta}$. Therefore, approximate q value was estimated to be 100 from theoretical curve under the present experimental conditions.

In order to explain the difference of these q-values, we will take account of g-value. The fact that g-value in random coiled state is larger than that in helical state leads to the conclusion that the PLGA in random coiled state exceeds in total amount of bound dye in comparison with that in helical state. As the cooperative binding, moreover, results from the dye-dye interaction on the polymer, the strength of the interaction may depend on the density of dye. Accordingly, it is predicted that the q-value in random coiled state is larger than that in helical state.

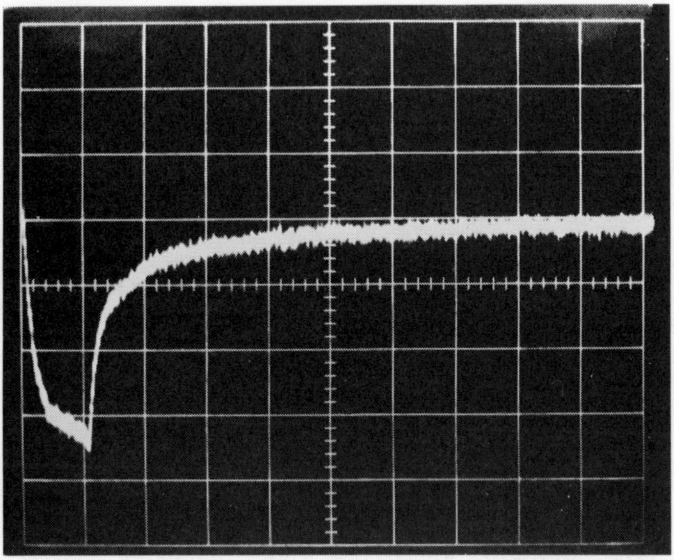

Fig. 8. Typical relaxation curve of PLGA-TB system : sweep, 20μsec; P/D = 6; pH = 4.5.

The typical relaxation curve is shown in Fig. 8. The relaxation spectrum is separated into two apparently different phase, i.e. fast change of absorbance is followed by slow change. In general, electric field applied to macromolecule induces the orientation of the macromolecule in the direction of electric field, which causes anisotropy of their optical properties. Accordingly, in this case, total change of absorbance in solution involves the dichroic effect resulting from this anisotropy in addition to the chemical relaxation effect. In order to check whether this dichroic effect is involved in present relaxation or not, the relaxation measurement was carried out at the monomer band and isobestic wavelength. When the relaxation was measured at monomer band the fast part of the relaxation curve disappeared and only slow part of it appeared, while on measuring it at isobestic wavelength only the fast part was detected. These facts show that the fast relaxation process in present curve results from anisotropy of the optical properties is caused by orientation of macromolecules and that only slow process is associated with chemical relaxation effect. In this paper, only relaxation part due to chemical reaction was analysed by use of mean relaxation time. According to the theory of cooperative binding, the binding of dye to linear polymers consists of the following two elementary steps;

$$\text{auu} + \text{A} \underset{k_D}{\overset{k_R}{\rightleftharpoons}} \text{aau} \qquad (k_R/k_D = K) \tag{9}$$

$$\text{auuu} + \text{A}_2 \underset{k'_D}{\overset{k'_R}{\rightleftharpoons}} \text{aaau} \qquad (k'_R/k'_D = K^2/K_d) \tag{10}$$

and the mean relaxation time (τ_b^*) of binding process is;

$$\frac{1}{\tau_b^*} = 2\tilde{k}_R(g/\sqrt{q})\sqrt{\bar{\theta}(1-\bar{\theta})} \cdot p \cdot C_A^o \tag{11}$$

$$\text{where} \quad \tilde{k}_R = \beta k_R + (1-\beta)k'_R \tag{12}$$

$$\text{and} \quad \beta = 1/(1 + 4K_d\bar{C}_A) \approx K/(K + 4K_d) \tag{13}$$

Since by using K and K_d values obtained above, β is given to be 0.9 and 1 for helical and random coiled state, respectively, $\tilde{k}_R$ is represented by the following approximation.

$$\tilde{k}_R \approx k_R$$

The above relaxation equation predicts that if τ_b^{*-1} is plotted against $\sqrt{\bar{\theta}(1-\bar{\theta})}.C_A^o$ at constant P/D value or against $\sqrt{\bar{\theta}(1-\bar{\theta})}.P$ at constant C_A^o value, these plots are described by straight line traversing the origin. These plots are shown in Figs. 9 and 10 in which straight lines are obtained in both cases. From the slope of the straight lines, $\tilde{k}_R$ is calculated to be 2.5 x 10^9 $M^{-1}sec^{-1}$ and 3.7 x 10^9 $M^{-1}sec^{-1}$ in helical state and coiled state, respectively. Using the approximation of $\tilde{k}_R \approx k_R$, and equilibrium constant K, k_D is calculated to be 2.5 x 10^4 sec^{-1} in helical state and

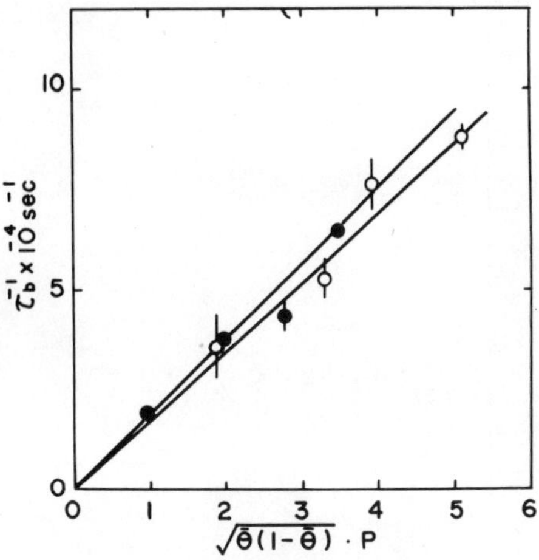

Fig. 9. Plots of the reciprocal mean relaxation time of cooperative binding τ_b^{*-1} versus $\sqrt{\bar{\theta}(1-\bar{\theta})}$ P. The straight lines refer to C_A^o = 2 x 10^{-5}M at pH = 4.7, o; 4 x 10^{-5}M at pH = 4.5, ●; respectively.

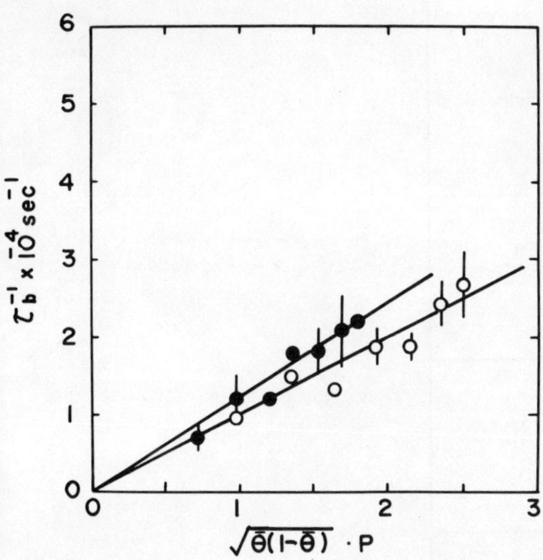

Fig. 10. Plots of the reciprocal mean relaxation time of cooperative binding τ_b^{*-1} versus $\sqrt{\bar{\theta}(1-\bar{\theta})}$ P. The straight lines refer to C_A^o = 1 x 10^{-5}M at pH = 8, o; C_A^o = 2 x 10^{-5}M at pH = 8.0, ●; respectively.

TABLE 1. THERMODYNAMIC AND KINETIC PROPERTIES OF COOPERATIVE BINDING OF TOLUIDINE BLUE TO POLY -α,L-GLUTAMIC ACID AT 25°C

	K $10^5 M^{-1}$	K_d $10^3 M^{-1}$	ε_{st} $10^3 M^{-1} cm^{-1}$	g	q	k_R $10^9 M^{-1} sec^{-1}$	k_D $10^3 sec^{-1}$
Helix	1.0	2.5	12.5	0.5	15	2.5(±0.8)	25
Random coil	25	2.5	12.5	1.1	100	3.7(±0.8)	1.5

$1.5 \times 10^3 \ sec^{-1}$ in random coiled state. The values k_R shows that the binding of TB to PLGA is diffusion controlled process. k_D-value is larger in helical state than in random coiled state. This fact suggests that TB stacked on PLGA is more stable in random coiled state than in helical state. This is also consistent with the tendency of q and K values.

CONCLUSIONS

The values of various parameters obtained in this study are shown in Table 1. A good agreement between the theory and experimental results of cooperative binding for PLGA-TB system was obtained. Particularly, it was confirmed that the theory of cooperative binding could be applied to the binding of TB to PLGA in helical state. The binding mechanism of TB to PLGA seems to be as follows; first the TB molecules bind ionically to the side chain of PLGA through a diffusion controlled process, then the bound dye is stabilized by the dye-dye interaction on the polymer. The distinct difference is shown in q, K and k_D value between the helical and random states of PLGA, i.e. TB molecule easily binds to PLGA in random coiled state than in helical state. This may be interpreted as follows; considering the quantities ε_{st} and g, it is proved that the density of bound dye molecules on the polymer is larger in random coiled state than in helical state. Accordingly the strength of dye-dye interaction on the polymer seems to be large in random coiled state.

REFERENCES

1. D.F. Bradley and M.K. Wolf, Proc. Natl. Acad. Sci. U.S., 45, 944 (1959).
2. G.G. Hammes and C.D. Hubbard, J. Phys. Chem., 70, 1615, (1966).
3. G.G. Hammes and C.D. Hubbard, J. Phys. Chem., 70, 2889, (1966).
4. G. Schwarz, Eur. J. Biochem., 12, 442 (1970).
5. G. Schwarz, S. Klose and W. Balthaser, Eur. J. Biochem., 12, 454 (1970).
6. G. Schwarz and W. Balthaser, Eur. J. Biochem., 12, 461 (1970).
7. E. Rabinowitch and L.F. Epstein, J. Amer. Chem. Soc., 63, 69 (1941).
8. L. Stryer and E.R. Blout, J. Amer. Chem. Soc., 83, 1411 (1961).

THE THERMODYNAMICS AND KINETICS OF ASSOCIATION OF ACRIDINE DYES IN AQUEOUS MEDIA : A MODEL SYSTEM FOR THE STUDY OF STACKING AND CHARGE-EFFECTS ON SELF-AGGREGATION

B.H. Robinson*, A. Seelig-Löffler and G. Schwarz**

* Chem. Labs.
Univ. of Kent,
U.K.

** Biozentrum,
Univ. of Basel,
Switzerland

INTRODUCTION

Molecular aggregation[1-4] has been the subject of numerous investigations in recent years; in biophysical chemistry, such phenomena are of particular interest because of their significance with regard to the self-organization of sub-units to large structures possessing biological activity. In this paper, we outline the methods we have employed, and the results obtained for the process of dye stacking in aqueous solution.

To enable measurements to be made conveniently by visible spectrophotometry and temperature-jump relaxation spectrometry, a member of the acridine family of dyes - acridine orange (I) - was chosen for detailed study, since aggregation beyond the dimer stage occurs readily at low concentrations ($\simeq 10^{-4}$M).

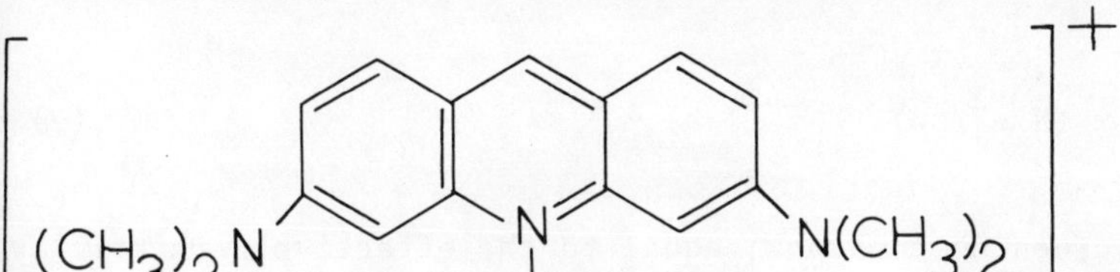

I - The Acridine Orange Cation.

The metachromacy in the visible spectrum indicates that the dye becomes stacked in a planar sandwich-type fashion as the concentration of the dye is increased. However, the dimer structure is not especially preferred, as there is no energetic reason why higher aggregates (oligomers) should not be formed. It is possible

Wyn-Jones (ed.), Chemical and Biological Applications of Relaxation Spectrometry, 481–485.

to analyse the visible spectrum of (I) as a function of concentration, and describe in some detail:-

(i) The nature of the stacking interactions in these systems.
(ii) The effects of charge (on the dye) on the tendency to self-association. (This includes an estimate of the effective permittivity (σ) or local dielectric constant in the region of the stack).

The effect on dye self-aggregation of several medium-variables have been investigated experimentally. These include: a) Temperature b) Ionic Strength (I); c) Specific salt effects and d) Water-structure perturbants (e.g. urea, dioxane).

The broad relevance of this work is that some assessment can be made of the relative importance of short-range stacking (hydrophobic and dispersive) forces and long-range (coulombic) forces as factors determining the tendency to self-association of dyes, nucleotides etc.

THERMODYNAMIC THEORY

The theoretical basis for the spectral analysis has been reported[5] in detail, and the assumptions used have been clearly stated. Briefly, the approach splits the standard free-energy change for aggregation, $\Delta G^o_{n,n+1}$, into contributions due to stacking forces ΔG^o_{st} (independent of aggregates size n) and a coulombic interaction term, $\Delta G^o_{el,n}$, which is dependent on n, as shown in equations 2 and 3.

Then, for the step-wise aggregation process,

$$c_n + c_1 \underset{}{\overset{K_{n,n+1}}{\rightleftharpoons}} c_{n+1} \qquad (n = 1 \rightarrow \infty) \tag{1}$$

$$\Delta G^o_{n,n+1} = \Delta G^o_{st} + \Delta G^o_{el,n} \tag{2}$$

where

$$\Delta G^o_{el,n} = (N_A e_o^2/\sigma a) \sum_{1}^{n} (1/n) \tag{3}$$

and σ is a screening constant, equal to the effective permittivity experienced by the stack. Allowed values are $1 \rightarrow \infty$, and a is the average separation of acridine orange units in the stack.

Three models for the aggregation process are analysed in detail. (Model 3 is intermediate between 1 and 2). They are listed in the following table.

To simplify the analysis in all three cases, a dimensionless parameter $s(= K_{12}c_1)$ is introduced, which greatly simplifies the calculations. The analytical procedure to test whether the

Model 1	Model 2	Model 3
Strong coulombic repulsion $\sigma \to 1$ $K_{n,n+1} \ll K_{12}$ $(n=2 \to \infty)$	Complete shielding of charge $\sigma \to \infty$ $K_{n,n+1} = K_{12}$	Medium repulsion $\sigma \simeq 100$ $K_{n,n+1} = K_{12}(b/n)$ where $b \sim 1.4$

experimental system most closely resembles scheme 1,2 or 3 is discussed in the paper, and is based solely on a visible spectrum analysis.

DISCUSSION OF THE RESULTS

The results show that both stacking and electrostatic forces are of importance in determining the tendency to stack. Increase of ionic strength leads to an increase in σ, and a corresponding increase in the tendency to association. This behaviour can be most easily understood in terms of the screening of positive charges in the aggregate through the preferential attraction of negative charges from the surrounding medium, resulting in a negative charge-cloud around the dye aggregate. At $I = 0.1$, $\sigma \simeq 100$, which is close to the bulk permittivity value. The approximations made in the calculations are that the positive charge is localised and the separation of the dye units is 550 pm. Molecular orbital computations[6] on the charge density distribution in the closely related dye proflavine suggest, however, that the charge is somewhat delocalized over the lower part of (I). From visible spectrum of the dye dimer, a separation of dye units of 600 pm is predicted.

The equilibrium constant, K_{12}, for dimer formation is effected in opposite ways by ionic strength and addition of co-solvents. Increase in ionic strength leads to an increase in K_{12}, while addition of small amounts (mole fraction < 0.05) of co-solvents decreases K_{12} in such a way that the order of K_{12} values is:- pure water > methanol-water > urea-water > dioxan-water. ΔH^{o}_{12} is $-38 kJ\ mol^{-1}$, suggesting that dispersive interactions are the major contributor to the stability of the dimer. This is in marked contrast to the stacking of hydrocarbon chains in water; for these systems it is thought that hydrophobic interactions, associated specifically with water structure, provide the main driving force for association (micellization).

KINETICS OF AGGREGATION

The kinetics of aggregation of acridine orange and the

amplitudes associated with this process have been studied by joule-heating temperature-jump relaxation spectrometry (with absorbance detection) as a function of both concentration and temperature in water.

The temperature-jump results obtained have been quantitatively evaluated in terms of a general theory for the kinetics of associating systems, which is discussed in a forthcoming paper[7]. No account of the theory is given in this article, since full details are given in the paper.

The relaxation times obtained from temperature-jump experiments on acridine orange self-association are complex (non-exponential), due to the presence of a multiple - equilibrium system. This results in a relaxation spectrum being observed, since the relaxation times characterizing the step-wise equilibria are similar in magnitude ($\tau_1 \simeq \tau_n$). However, an analysis based on mean relaxation times - τ^* - is feasible[8], (if certain extinction coefficient relationships are assumed), and the theory leads generally to the prediction of wavelength-dependent relaxation times for aggregation beyond the dimer stage, when absorbance detection is employed.

A kinetic and amplitude analysis was carried out for the three models discussed in section (2), and the results generally confirm the previous visible spectrum analysis that oligomers are formed in solution, and that at I = 0.1 (NaCl) in water at 25°C, Model 3 is obeyed. (i.e. $K_{n,n+1}$ slowly decreases as n is increased). It is apparent from the full paper[7] that the analysis becomes rather complicated when aggregation beyond the dimer stage is allowed, and assumptions must be made in order to achieve a solution. Similar or greater complexities might be predicted in the kinetic analysis of micellar systems.

However, a particularly simple expression relates the maximum amplitude and the total concentration of dye when a simple monomer -dimer equilibrium is assumed. In this case, K_{12} can be readily calculated without knowledge of any optical parameters (extinction coefficients) of the dye system. From a more rigorous amplitude analysis, ΔH^o_{12} and ΔS^o_{12} for dimerization can be found.

The kinetic results show that the rate constant, for step-wise aggregation of dye units at 23°C (1.0×10^9 dm^3 mol^{-1} s^{-1}) is close to but slightly less than the diffusion-controlled limiting value ($k_D = 2.4 \times 10^9$ dm^3 mol^{-1} s^{-1}).

The discrepancy can be explained in terms of solvation changes and reorientation on stacking. The enthalpy of activation, $\Delta H^{\neq}_{12}$ is close to zero, indicating that there is a negligible energy barrier to the building of stacks. (If the reaction were strictly diffusion-controlled, $\Delta H^{\neq}_{12} \simeq 13$kJ mol^{-1}). The lifetime of a dimer unit is found to be of the order of 20 μs at 23°C.

REFERENCES

1. "Molecular Association in Biology" ed. B. Pullmann, Academic, New York, 1968.

2. P. Mukerjee and A.K. Ghosh, J. Amer. Chem. Soc., 92, 6403 (1970).
3. D. Thusius, this publication.
4. J. Rassing, Adv. in Mol. Relaxation Processes, 4, 55, (1972).
5. B.H. Robinson, A. Loffler and G. Schwarz, J. Chem. Soc. Farad. Trans. 1, 56 (1973).
6. L.L. Ingraham and H. Johansen, Arch. Biochem. Biophys., 132, 205 (1969).
7. B.H. Robinson, A. Seelig- Loffler and G. Schwarz, J. Chem. Soc. Farad. Trans. 1, to be published 1975.
8. G. Schwarz, Rev. Mod. Phys. 40, 206 (1968).

ULTRASONIC ABSORPTION STUDIES IN RELATION TO CONFORMATIONAL CHANGES OF BIOPOLYMERS IN SOLUTION

R. Zana

C.R.M., C.N.R.S., 6, rue Boussingault, STRASBOURG
CEDEX 67083

INTRODUCTION

The helical conformation of polypeptides in the solid state was first reported by Pauling in 1953. It has been since recognized that polypeptides can retain this conformation in solution where changes of temperature, or pH, or solvent composition can induce a reversible conformational change from the helical conformation to a randomly coiled conformation. This change is always fairly sharp indicating a cooperative phenomenon whose thermodynamics has been extensively investigated.

On the other hand polypeptides are good model compounds for proteins and it is well established that the secondary structure of proteins includes helical segments. Also, nucleic acids adopt the well known double helix structure and other biopolymers present triple helix conformation (Collagen and some synthetic polynucleotides, for instance).

This frequent occurrence of helical structure in biopolymers, together with the fact that many biochemical reactions involve conformational changes, has led many workers to think that the kinetics of the helix-coil (H-C) transition may be of primary importance for the understanding of the kinetics of biochemical processes.

The first theoretical treatment of the kinetics of the H-C transition was reported in 1964 by Schwarz.[1] This theory concerns polypeptide solutions which offer the most simple example of such a process, and suggests the use of ultrasonic and dielectric absorption for its study. This suggestion was followed by many workers who soon reported results of ultrasonic and dielectric investigations of polypeptides, proteins and DNA solutions, which were all interpreted in terms of H-C equilibrium.

Wyn-Jones (ed.), Chemical and Biological Applications of Relaxation Spectrometry, 487–492.

The purpose of this paper is to summarize the evidence which indicates that this interpretation is, in most instances, not valid and to suggest methods which may be used to distinguish between the contributions of H-C equilibrium and of other processes to the ultrasonic absorption of biopolymer solutions.

BRIEF REVIEW OF SCHWARZ THEORY

Schwarz[1] first emphasizes that polypeptide molecules undergoing a H-C transition will respond upon a pertubation with a very large number of relaxation times, associated with the different conformations that they can adopt. An average relaxation time τ^* is defined, which is associated with the average helical content θ of the polypeptide molecules. τ^* is found to go through a maximum, τ^*_M, at mid-transition (where $\theta = 0.5$) given by $\tau^*_M = (4\sigma k_F)^{-1}$.

σ is the so-called nucleation parameter, which is related to the probability to find a residue in the helical state within a part of the chain in the coiled conformation. σ has been found to be about 10^{-3} in water and 10^{-4} in organic solvents.

k_F is the rate constant for the elementary step of growth of a helix, that is, for the formation of an intramolecular H-bond. The first aim of the kinetic study of the H-C transition is of course the evaluation of k_F. However for an estimate of the upper value of τ^*_M one can use for k_F the value of the first order rate constant obtained for H-bond formation in solution of small molecules.[2] This yields values of τ^*_M well below 1 μsec, thereby justifying the use of ultrasonic absorption for the kinetic study of the H-C transition.

The observation of a maximum of τ^* and of the excess absorption have been widely used to assign to conformational equilibria the excess absorption found with polypeptides, proteins and DNA undergoing conformational changes. However proton transfer processes[3] as well as counterion binding by polyelectrolytes[4] (most biopolymers are also polyelectrolytes) also yield absorption maxima. Methods must therefore be designed and used in order to distinguish between these various processes. We shall examine successively polypeptides, nucleic acids and proteins.

POLYPEPTIDES

1 - H.C. Transition induced by addition of salt to polypeptide solutions.

A coil to helix transition occurs upon additions of $NaClO_4$ or KSCN to a solution of poly-L-Lysine (PLL) hydrobromide at pH 6.0.[5] The plot of the absorption vs concentration of added salt shows a maximum at a salt concentration about equal to that corresponding to the transition[5] but this maximum is not due to the H-C equilibrium. This was inferred (i) from the effect of the polymer concentration on the curve absorption vs added salt concentration and (ii) from measuring the variation of absorption upon additions of KSCN, $NaClO_4$ and NaBr to PLL-hydroper-chlorate. The maximum found upon additions of KSCN or $NaClO_4$ to PLL-hydrobromide appears to be the result of two antagonistic effects[1]: an increase of absorption due to the binding of SCN^- or ClO_4^- by PLL polycations at low salt concentrations, and a decrease of absorption due to the coil to helix transition, at high salt concentration.

Thus, the excess ultrasonic absorption appears to be sensitive to the overall polymer conformation but not to the equilibrium between conformation. This sensitivity has also been observed with poly-benzyl-L-aspartate[6]. For this polypeptide the H-C transition is induced by a change of composition of the solvent (mixture of chloroform and dichloroacetic acid). The curve absorption vs solvent composition shows a rapid change but no maximum. From this result it has been inferred that $k_F < 10^9$ sec^{-1}.

2 - H-C transition induced by modification of the pH of the polypeptide solution.

Maxima or changes of τ* and of the excess absorption, induced by pH changes, have been observed for poly-L-Lysine (PLL)[7], poly-L-ornithine (PLO)[8] and poly-L-glutamic acid (PLGA)[1,9], and assigned to the H-C equilibrium. However, the side chain of these three polymers includes an amino group (PLL, PLO) or a carboxyl group (PLGA). The changes of τ* and excess absorption may therefore also be due to proton transfers.[3] In addition, counterion site binding[4] may contribute in PLGA where the H-C transition occurs when the fractional electrical charge per monomer is about 0.5.

However, the curves absorption vs pH for equimolar solutions of PLL and PDLL can by made coincident in the whole pH range.[10] Since PDLL cannot adopt the helical conformation[11], the absorption maximum found both for PLL and PDLL must be due to proton transfer. This assignment leads to values of the volume change and rate constants in excellent agreement with what may be expected for a hydrolysis equilibrium.

On the other hand, for PLO, the curves absorption vs pH determined both in H_2O and in a mixture H_2O-methanol (85/15 v/v) have been found to be very close[10] although the addition of methanol increases the helical content of PLO from 25 to 60%. Also the ultrasonic absorption titration curves are practically coincident for equimolar solutions of PLL and PLO. As for PLL, the assignment of the absorption maximum to a hydrolysis equilibrium leads to values of the volume change and rate constants in excellent agreement with what is expected for such a process.[10]

The case of PLGA is more interesting. Indeed, the maximum on the absorption vs pH curve disappears in the case of PDLGA which does not undergo the helix-coil transition.[13] Moreover using tetramethylammonium counterions, instead of sodium ions, has practically no effect on the amplitude of the absorption miximum.[13] These two results indicate that proton transfer and counterion binding cannot be responsible of the observed absorption maximum, which must therefore be assigned to the H-C equilibrium, in agreement with the results of other studies[9].

Three reasons may be found to the fact that the helix-coil equilibrium can be observed in the case of PLGA and is not detected with PLL and PLO.[10] First, the differences between the two pH's which characterize the H-C and the proton transfer equilibria is about 2.3 pH units with PLGA and only 1 and 0.5 pH units with PLL and PLO, respectively. Second, the contribution of the hydrolysis equilibrium to the ultrasonic absorption of solutions of PLL and PLO is much larger than that of the protolysis equilibrium of PLGA, because of the large difference in volume change for the two reactions (about 25 and 10 cm^3/mol., respectively). Finally the value of σ for PLGA appears to be larger than for the two others polypeptides, thereby resulting in a smaller value of τ^*_{Max} and in a larger excess absorption in the megahertz range.

PROTEINS

The excess absorption of solutions of globular proteins has been found to go through maxima, when plotted as a function of pH, both in the acid and in the alkaline range.[14-18] Several workers[14,15,17] have assigned these maxima to the conformational transitions that these proteins are known to undergo both in the acid and alkaline range. However the study of three globular proteins has revealed (i) that the amplitude and position of the absorption maximum depends only very slightly on the nature and/or stability of the protein with respect to pH[16a] and (ii) that urea has very similar effects on the ultrasonic absorption titration curves of the three proteins (β-lactoglobuline, serumalbumine and lysozyme) which greatly differ in stability with respect to urea, and of model compounds such as acetylglycine and diglycine[16b]. Therefore the observed ultrasonic absorption maxima cannot be due to a conformational equilibria. The results can be interpreted in terms of proton transfer (hydrolysis involving the side chain amino groups of the lysyl and arginyl residues in the alkaline range; protolysis involving the side chain carboxyl group of the glutamyl and aspartyl residues in the acid range). This assignment correctly predicts the value of the pH where the absorption maxima occur and the very slight dependence of their amplitude on the nature of the protein. In the case of bovine serumalbumine the absorption due to the hydrolysis equilibrium has been determined in the range 1-115MHz[19]. From this result the volume change and rate constants for this process have been evaluated, assuming that it involves only the lysyl residues of the protein, and found to be in good agreement with those determined for PLL[10]. This gives a quantitative basis to the assignment of the absorption maxima found with proteins to proton transfers.

DESOXYRIBONUCLEIC ACID (DNA)

For aqueous solutions of DNA several workers have reported an absorption maximum in the alkaline range and an increase or a maximum of absorption as the pH is decreased below 4.[17,20,21a,22] These changes of absorption, particularly that in the alkaline rang have been attributed to the double helix breakdown.[17,20] However we have shown that these changes are not affected by a thermal denaturation of the double helix, thereby excluding conformational equilibria as a cause of the observed maximum.[21a] Here again the results can be interpreted in terms of proton transfers involving protonatable groups of the base pairs. This assignment has been substantiated by recent ultrasonic absorption studies of nucleotides.[21b]

CONCLUSIONS

The studies briefly reviewed above emphasize the difficulties that may be encountered in assigning an excess ultrasonic absorption to a precise chemical process. The comparison between the results obtained with the compound investigated and with chemically related model compounds, whose excess absorption is due to a known chemical process, has proved to be extremely useful for this assignment.

In any case, ultrasonic absorption investigations in the megahertz range do not appear to be well fitted for the study of the kinetics of cooperative conformational equilibria in biopolymers. On the contrary chemical relaxation methods such as temperature jump, pressure jump and more recently electrical field jump have proved very successful in permitting the elucidation of the kinetics of many conformational equilibria involving biopolymers[23] and appear far more efficient than ultrasonic absorption in dealing with such problems.

REFERENCES

1. G. Schwarz, Ber. Bunsenges. Phys. Chem., **68**, 843 (1964) and J. Mol. Biol., **11**, 64 (1965).
2. G. Hammes et al., J. Amer. Chem. Soc., **88**, 1621 (1966); **90**, 4151 (1968); **91**, 956 (1969) and **92**, 7578 (1970).
3. K. Applegate, L. Slutsky and C. Parker, J. Amer. Chem. Soc., **90**, 6909 (1968).
4. C. Tondre and R. Zana, J. Phys. Chem, **75**, 3367 (1971).
5. R. Zana and C. Tondre, Biopolymers, **10**, 2635 (1971) and references therein.
6. R. Zana and J. Lang, Biopolymers, **12**, 79 (1973).
7. R. Parker, L. Slutsky and K. Applegate, J. Phys. Chem., **72**, 3177 (1968).
8. G. Hammes and P. Roberts, J. Amer. Chem. Soc., **91**, 1812 (1969).
9a. T. Saksena, B. Michels and R. Zana, J. Chim. Phys. Physicochim. Biol., **65**, 597 (1968); (b) K. Barksdale and J. Stuehr, J. Amer. Chem. Soc., **94**, 3334 (1972); (c) H. Inoue, J. Sci. Hiroshima Univ., Ser. A II, **34**, 37 (1970).
10. R. Zana and C. Tondre, J. Phys. Chem., **76**, 1737 (1972).
11. P. Chou and A. Scheraga, Biopolymers, **10**, 657 (1972).
12. S. Chaudhuri and J. Yang, Biochemistry, **7**, 1379 (1968).
13. R. Zana, J. Amer. Chem. Soc., **94**, 3646 (1972).
14. I. El-piner, K. Fursov and O. Zorina, Dokl. Akad. Nauk. USSR, **192**, 1160 (1970).

15. L. Kessler and F. Dunn, J. Phys. Chem., 73, 4256 (1969); ibid, 74, 2736 (1970).
16. (a) R. Zana and J. Lang, J. Phys. Chem., 74, 2734 (1970). (b) J. Lang, C. Tondre and R. Zana, ibid., 75, 374 (1971).
17. I. Elpiner, F. Braginskaya and O. Zorina, 7th Int. Cong. Acoustics, Budapest, 1971, Book of Abstracts, Vol. 2, p. 153.
18. W. D. O'Brien and F. Dunn, J. Phys. Chem., 76, 528 (1972).
19. R. Zana, J. Long, C. Tondre and J. Sturm in "Interaction of Ultrasound and Biological Tissues", J. M. Reid and M. Sikov, Ed., DHEW Publication (FDA) 73-8008 BRH-DBE 73-1,p. 21, 1972.
20. J. Lang and R. Cerf, J. Chim. Phys. Physicochim. Biol., 66, 81 (1969).
21. (a) J. Sturm, J. Lang and R. Zana, Biopolymers, 10, 2639 (1971). (b) J. Phys. Chem, 77, 2329 (1973); ibid., 78, 80 (1974).
22. W. O'Brien, C. Christman and F. Dunn, J. Acoust. Soc. Amer., 52, 1251 (1972).
23. D. Goldstack and P. Waern, Can. J. Chem., 49, 1267 (1971); D. Porschke, Eur. J. Biochem., 39, 117 (1973) and references therein; T. Yasunaga, this volume.

KINETIC STUDY OF HELIX-COIL TRANSITION IN AQUEOUS SOLUTIONS OF POLY-α, L-GLUTAMIC ACID BY MEANS OF ELECTRIC FIELD PULSE METHOD

Tatsuya Yasunaga, Yoshikuni Tsuji, Takayuki Sano and Hidetoshi Takenaka
University of Hiroshima, Japan

INTRODUCTION.

The dynamics of the helix-coil transition in polypeptides have been studied extensively by means of various methods such as temperature jump[1], ultrasonic absorption[2], and dielectric relaxation method[3]. But in these studies, there remains some problems in the assignment of relaxation to the helix-coil transition.

Previously, it has been found out by the authors that when high electric field is applied on the aqueous solution of PLGA, the conductivity of solution is relaxed[4]. From the dependences of the relaxation times on pH and on polymer concentration, it has been suggested that its relaxation phenomena may be due to the helix-coil transition of PLGA. However, the detailed consideration has not been done whether other such mechanisms can be applied. In this paper, further experiments have been carried out for assignments of relaxation mechanism. From the dependences of relaxation time on pH, polymer concentration, kinds of counterion and electric field density, helix-coil transition was clarified as the most probable mechanism and the activation parameters of the helix-coil transition was determined.

EXPERIMENTAL

The sodium salt of PLGA with degree of polymerization (D.P) of 250 and 700 was supplied from Ajinomoto Co. Inc.. The samples were dialysed against deionized water for two days and then lyophilized and dried at 50°C for four hours in vacuum just prior to the preparation of stock solutions (about 1×10^{-4} molar residue concentration) which were stored in a refrigerator at

Wyn-Jones (ed.), Chemical and Biological Applications of Relaxation Spectrometry, 493–504.

about 5°C. The sample solutions were prepared by diluting a stock solution with deionized distilled water to the desired concentration. The tetraalkylammonium-PLGA were prepared by the use of the cation-exchange resin (Dowx 50 WX 8). Completeness of ion exchange was checked by flame test.

For pH adjustment, HCl, NaOH, tetramethylammonium and tetrabutylammonium hydroxides were used. The pH measurement was carried out with Hitachi-Horiba F-5 type pH meter at room temperature. All the chemicals used were reagent grade and were used without further purification. ORD measurements were carried out in the wavelength range 225 to 310 nm with JASCO ORD/UV-5 spectropolarimeter using 5 cm quartz lidded cell. The temperature (15 - 45°C) was controlled by circulating water to a water jacketed cell holder from temperature controlled water bath (within ± 0.3°C).

In order to evaluate the helix content from ORD data, the following equation has been used:

$$\Theta = \frac{[m']_{\lambda_0} - [m']^c_{\lambda_0}}{[m']^h_{\lambda_0} - [m']^c_{\lambda_0}}$$

where $[m']_{\lambda_0}$ is the measured values of the reduced residue rotation at the minimum point of the trough in the vicinity of 233 nm. $[m']^h_{\lambda_0}$ and $[m']^c_{\lambda_0}$ are the reduced residue rotations for completely helical state and for completely coiled state, respectively. These values were calculated from the following equations by Warashina and Ikegami[5], as the function of temperature T (°C)

$$[m'(T)]^h_{233} = -18400 + 66T$$

$$[m']^c_{233} = -2000$$

The electric field pulse method apparatus used here is similar to that originally constructed by Ilgenfritz[6], and the transients are followed by a conductance change with a Wheatstone bridge. The bridge consists of a sample cell, a reference cell, and a series of carbon resistors. The distance between the two electrodes of the sample cell is 0.3 cm and that of the reference cell is variable.

The duration of the high voltage pulse applied to the bridge is 10 to 20 μsec. and the rise and decay time of the applied electric field is much faster than 0.1 μsec. The electric field intensity in the cell is usually about 30 kV/cm.

RESULTS AND DISCUSSION

The relaxation was observed only in the solution of pH range 5.5 to 8.0. The relaxation spectrum was characterized by a single relaxation and the conductivity of the solution increased with the electric field. The typical relaxation curve is shown in Fig. 1. Fig. 2 shows the pH dependence of the relaxation time of NaPLGA with D.P 250 at various polymer concentrations. Fig. 3 shows the pH dependence of the helix content of NaPLGA in order to examine the relationship between observed relaxation process and helix-coil transition. These figures show that the relaxation times have maximum value in the vicinity of the midpoint of the helix-coil transition and is independent of polymer concentration.

The same results were obtained not only in solution of NaPLGA with D.P 700, but also in the solution of tetraalkyl-ammonium-PLGA with D.P 250. These features are shown in Figs. 4, 5 and 6 with corresponding pH dependence of helix content. In Figs. 2, 4 5 and 6, the solid lines are the same shapes with exception of slightly different positions of maximum values of relaxation times. The pH dependence of relaxation time at various temperatures is shown in Fig. 7, and the corresponding transition curve obtained from ORD data is shown in Fig. 8. At every temperature, relaxation time has the same characteristics as mentioned above. Fig. 9 shows the pH dependence of relaxation time under the various applied electric field densities (E. F. D.) in the cell. When the applied E. F. D. was decreased, relaxation amplitude was decreased rapidly, but as can be seen in Fig. 9, relaxation time was independent of the applied E. F. D. Summarizing the experimental results, 1. The relaxation was observed only in solutions of pH range 5.5 to 8.0 which correspond to about transition region of the helix-coil transition by ORD measurement, and the conductivity of solution increased with the electric field. 2. The relaxation time shows maximum value near the mid-point of the helix-coil transition obtained from ORD data, and is independent of polymer concentration. These features are not affected by the D.P or the kind of counterion.

These characteristics of relaxation time suggest that the relaxation phenomena should be ascribed to the helix-coil transition. More detailed investigations were, however, carried out if some other possible mechanisms can be applied. When high electric field is applied upon polyelectrolyte solution, the electrolyte conductivity may be affected in two different ways. One of them is the change of conductivity due to change of the mean mobility of the charge carriers caused by an orientation of long axis of polyion in direction of the field (Orientation field effect)[7]. The other is an increase of the conductivity due to the dissociation of electrolyte (Dissociation field effect). In the relaxation caused by the Dissociation field effect, ion binding of counterion to polyion and proton transfer reaction of γ-carboxyl group in side chain are involved. These relaxation processes will be examined below in detail.

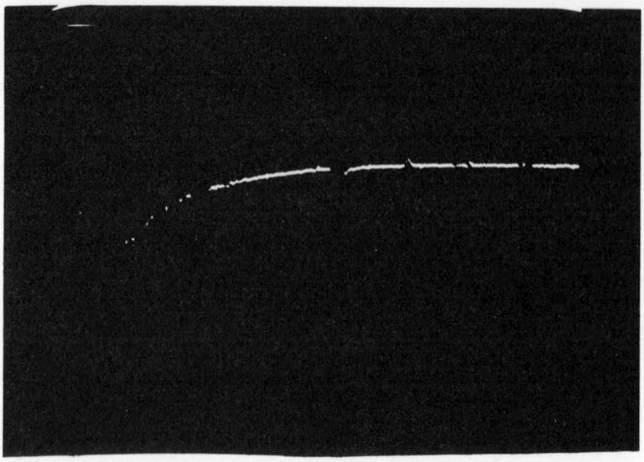

Fig. 1. The typical relaxation spectrum in 5 x 10^{-5}M NaPLGA (DP 250) solution at 25°C, pH 7.22, sweep 2 μsec/div.

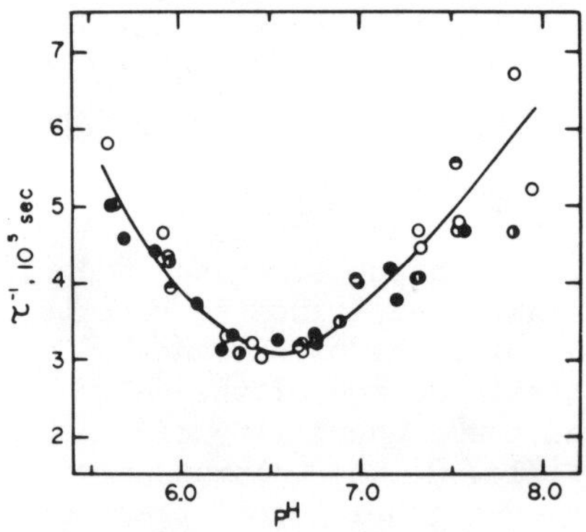

Fig. 2. 1/τ dependence on pH of various polymer concentrations at 25°C; NaPLGA (DP 250), (●) 5 x 10^{-5}M; (◑) 7 x 10^{-5}M; (o) 1 x 10^{-4}M; (◒) 2 x 10^{-4}M.

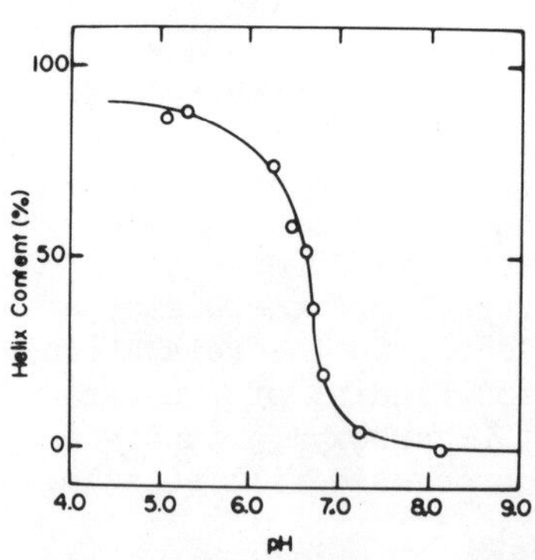

Fig. 3. pH dependence of helix content of NaPLGA (DP 250) at 25°C, concentration 1 x 10^{-4}M.

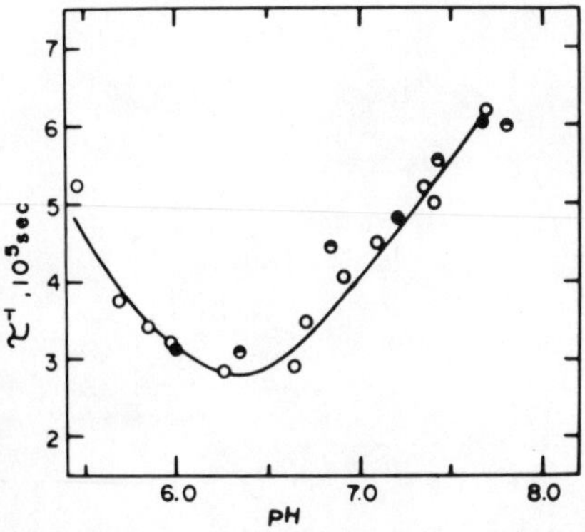

Fig. 4a. $1/\tau$ dependence on pH of various polymer concentrations at 25°C, NaPLGA (DP 700), (●) 5 x 10^{-5}M; (o) 1 x 10^{-4}M; (◓) 2 x 10^{-4}M

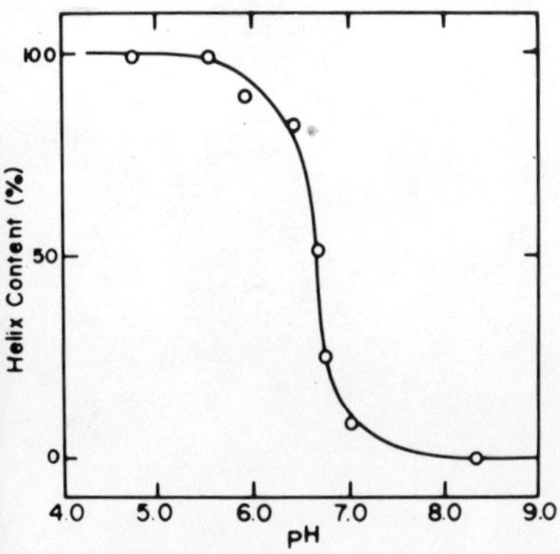

Fig. 4b. pH dependence of helix content of NaPLGA (DP 700) at 25°C, concentration 1 x 10^{-4}M

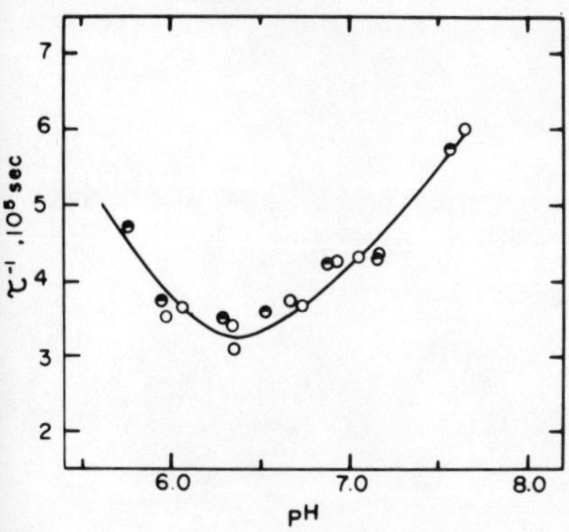

Fig. 5a. $1/\tau$ dependence on pH of various polymer concentrations; tetramethylammonium-PLGA (DP 250), (o) 1 x 10^{-4}M; (◓) 2 x 10^{-4}M

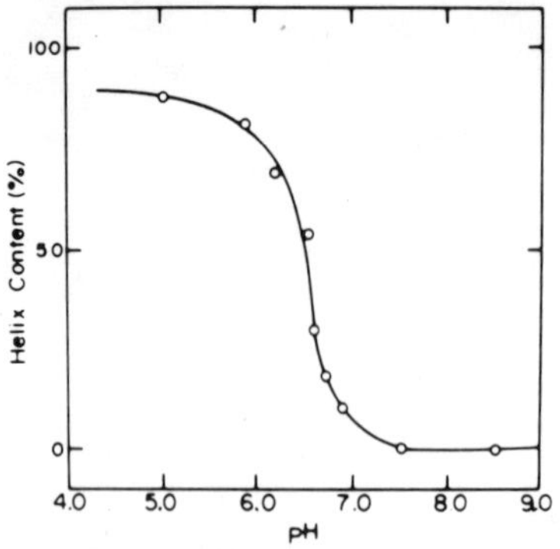

Fig. 5b. pH dependence of helix content of tetramethylammonium-PLGA (DP 250) at 25°C, concentration 1 x 10^{-4}M.

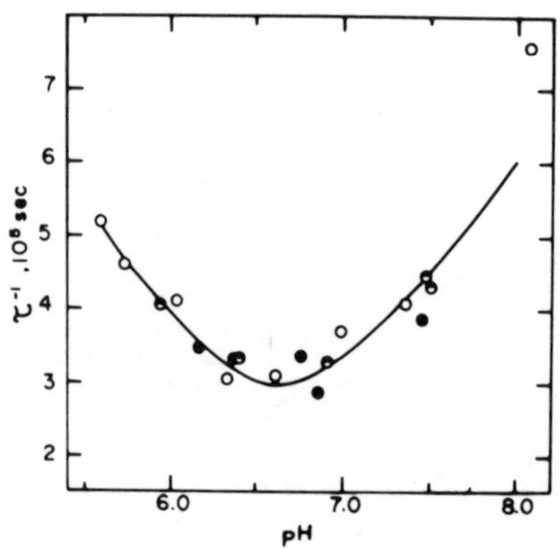

Fig. 6a. $1/\tau$ dependence on pH of various polymer concentrations, at 25°C, tetrabutylammonium-PLGA (DP 250), (●) 5 x 10^{-5}M; (o) 1 x 10^{-4}M; (◐) 2 x 10^{-4}M.

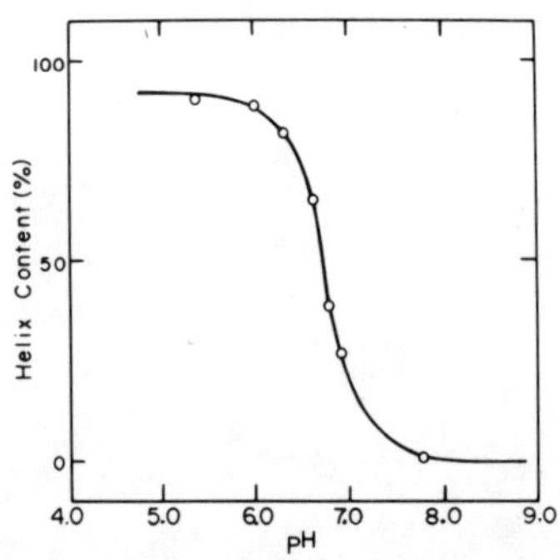

Fig. 6b. pH dependence of helix content of tetrabutylammonium-PLGA (DP 250) at 25°C, concentration 1 x 10^{-4}M.

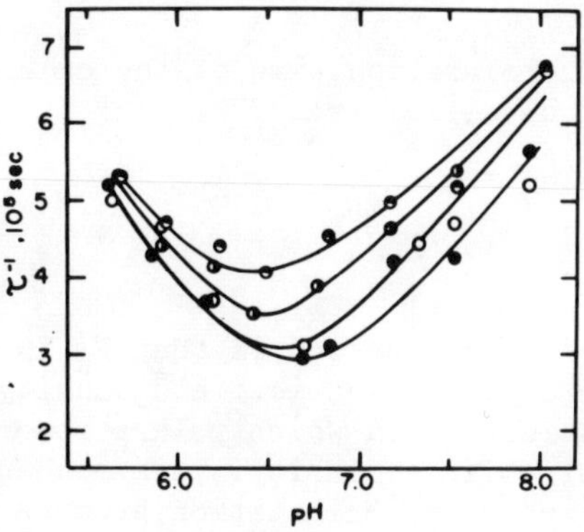

Fig. 7. pH dependence of $1/\tau$ at various temperatures; NaPLGA (DP 250), concentration 1×10^{-4}M; (●) 15°C, (o) 25°C, (◑) 35°C, (◓) 45°C.

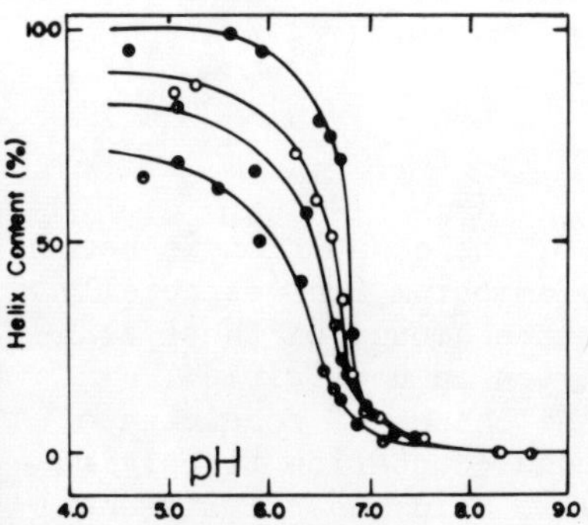

Fig. 8. pH dependence of helix content of NaPLGA (DP 250) at various temperatures; (●) 15°C, (o) 25°C, (◑) 35°C, (◓) 45°C.

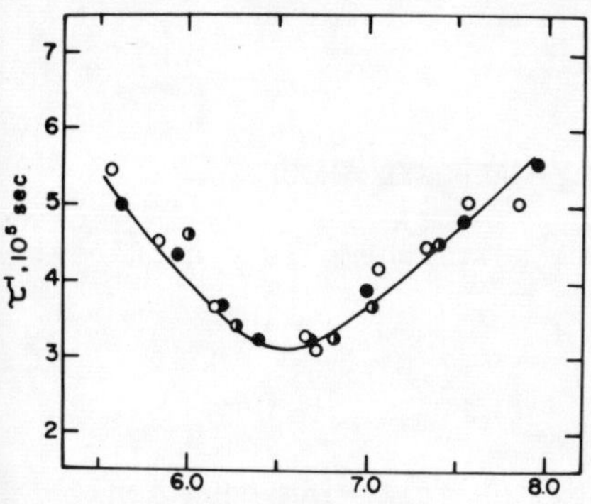

Fig. 9. pH dependence of $1/\tau$ at various applied electric field densities; NaPLGA (DP 250), concentration 1×10^{-4}M, (◑) 22 kV/cm; (●) 38 kV/cm; (o) 61 kV/cm.

1. ORIENTATION FIELD EFFECT

According to Eigen and Schwarz[7], relaxation time of the orientation process at high electric field densities is:

$$\tau_E = 1.68 \times \frac{kT}{\alpha_{\parallel} D_r} \times E^{-2}$$

The most characteristic properties of this equation is that τ_E is inversely proportional to the square of the electric field density. This fact is contrary to the present results in which relaxation time is independent of applied electric field density. This mechanism is ruled out more definitely by measuring orientation process directly by electric birefringence measurements. Preliminary experiments show that rise time of electric birefringence is two times faster than the present relaxation time and decrease considerably with applied electric field density. These facts show that the present relaxation process is not ascribed to the orientational field effect.

2. ION BINDING OF COUNTERION

In the solution of polyelectrolyte, the counterion is bound around the polyion. When the tetraalkylammonium ion (especially tetrabutylammonium ion) is counterion, the amount of these counterions bound electrostatically to polyion is negligible[8]. Wissbrum and Patterson[9] have studied the change of conductance associated with the perturbation of bound counterion to polyion under high electric field for polyacrylic acid and polystyrensulfonic acid. According to them, relaxation time for this process is dependent on both the applied electric field density and the D.P but not on the degree of neutralization of polyion. These characteristics of relaxation time contradict the present results. Consequently, this process should be ruled out for the present relaxation mechanism.

3. THE PROTON TRANSFER REACTION OF γ-CARBOXYL GROUP

For proton transfer equilibrium, next scheme is considered;

$$-RCOOH \underset{k_{-1}}{\overset{k_1}{\rightleftharpoons}} -RCOO^- + H^+$$

The relaxation time τ is expressed by the following equation

$$\frac{1}{\tau} = k_{-1}\{(RCOO^-) + (H^+)\} + k_1 = k_{-1}\{\alpha C_0 + (H^+)\} + k_1$$

where α is the degree of ionization and C_o is the total polymer concentration. Since α in pH region in which the helix-coil transition occurs is thought to be in the range of 0.2 to 0.5 under the present experimental condition[5,10], above equation is simplified to be

$$\frac{1}{\tau} \approx k_{-1}\alpha C_o + k_1$$

This equation shows that the reciprocal relaxation time increases linearly with total polymer concentration at constant pH and with pH at constant polymer concentration. This result contradicts the observed results. Thus, the possibility that the proton transfer reaction at γ-carboxyl group is responsible for the observed relaxation can be ruled out. But the above consideration does not always mean that the proton transfer reaction is not induced by the high electric field. In fact when the high electric field was applied upon the solution of PLGA containing pH indicator, it was found that considerable amounts of proton was released from polyions. These released protons may be responsible for increasing of the conductivity in present observed relaxation.

In contrast with above consideration, present relaxation phenomenon agrees satisfactorily with the helix-coil transition of PLGA. Schwarz[11] has developed Zimm-Bragg model to the kinetic theory of the helix-coil transition. According to Schwarz's model, the mean relaxation time of the helix-coil transition in the vicinity of the midpoint of the helix-coil transition is expressed by the following equation:

$$\frac{1}{\tau} = k_f\{(s' - 1)^2 + 4\sigma\}$$

where k_f and s' are the forward rate constant and the equilibrium constant for the helix growth, respectively, and σ is the nucleation parameter. This equation predicts that the mean relaxation time has the maximum value at the midpoint of the transition and is independent of polymer concentration. These facts are consistent with observed relaxation. At the midpoint of the helix-coil transition, $s' = 1$, so that above equation is simplified to

$$\tau^{-1} = 4\sigma k_f$$

where the value of σ is $(3\pm2) \times 10^{-3}$ for PLGA from titration experiments by Snipp et al.[12], and is independent of temperature[13]. Accordingly, introducing the maximum relaxation times experimentally determined to the above equation, the rate constant of helix growth, k_f, is obtained at various temperatures. The calculated values are given in Table 1. These values are comparable to those obtained by other investigators, e.g. 4.4×10^7 sec^{-1} for Poly-α-D-glutamic acid at 30°C by Inoue[2a] and $(8\pm5) \times 10^7$ sec^{-1} for PLGA in 0.03 M NaCl at 37°C by Barksdale and Stuehr[2b].

TABLE 1. THE RELAXATION TIME AND THE RATE CONSTANT OF HELIX GROWTH AT VARIOUS TEMPERATURES

T (^{o}C)	$1/\tau_{max}$ (10^{5} sec^{-1})	k_F (10^{7} sec^{-1})
15	2.9	2.4
25	3.1	2.6
35	3.5	2.9
45	4.1	3.4

According to Eyring's absolute rate process, the rate constant k_f is related to the activation enthalpy $\Delta H^{\neq}$ and entropy $\Delta S^{\neq}$ as follows:

$$k_f = \frac{kT}{h} \exp \frac{(\Delta H^{\neq} - T\Delta S^{\neq})}{RT}$$

and then $\Delta H^{\neq}$ and $\Delta S^{\neq}$ were calculated to be 1.5 Kcal/mole and -9.8 e. u., respectively, from ln $(k_f \frac{h}{kT})$ vs. 1/RT plot which was illustrated in Fig. 10. These values, which are quite different from those of usual reactions, may be reasonable for the helix-coil transition, becuase the rate of the helix-coil transition seems to be primarily controlled by steric restriction that a favourable steric position of the CO and NH group must reach through the suitable internal rotation[14]. Schwarz[15] has theoretically examined the electric field induced helix-coil transition of polypeptides with no dissociation group and has successfully applied to helix-coil transition of poly-γ-benzylglutamate in ethylene dichloride and dichloroacetic acid mixtures by means of dielectric relaxation method.[3] Recently, Kikuchi and Yoshioka[16] have observed anomalous birefringence signal in transition region of poly-L-lysine hydrobromide in water-methanol mixture by electric birefringence method, and have attributed to the transition from the charged helix to the charged coil. In addition, they have discussed the mechanism of the electric field induced helix-coil transition of polypeptides with charged group. The similar mechanism is used here, i.e. when the high electric field is applied to the solution of PLGA, the counterion atmosphere moves towards one

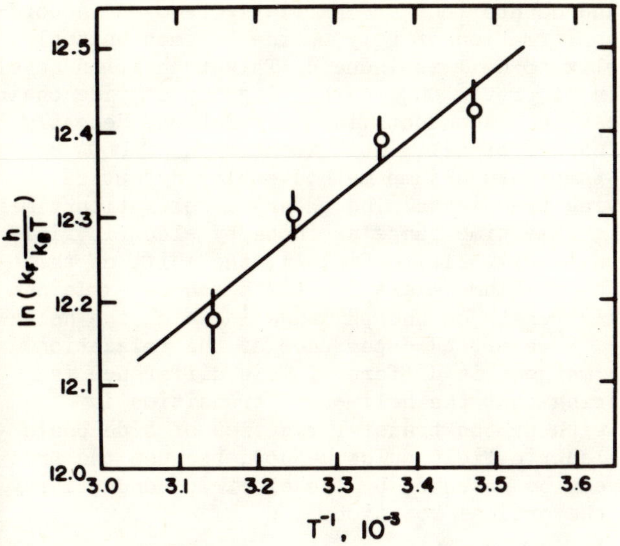

Fig. 10. $\ln(k_f \cdot \frac{h}{k\,T})$ vs. 1/T plot.

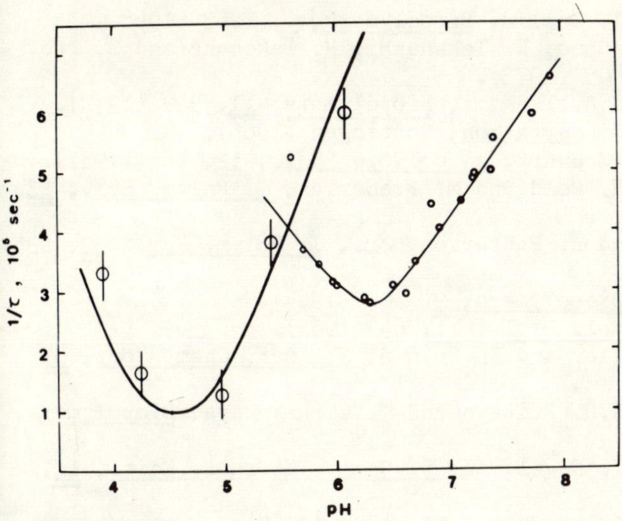

Fig. 11. 1/τ dependence on pH of NaPLGA by electric field pulse method (o) and T-jump method (ϕ).

end of polymer chain and counterions are partially removed. Accordingly, since helical conformation of polypeptide becomes unstable, the transition from helix to coil is induced. This transition accelerates the dissociation of proton on γ-carboxyl group of side chain, which causes the increasing of conductivity in solution. Recently, we have studied the kinetics of helix-coil transition of PLGA by means of the improved temperature-jump method employing optical rotation to the following transients. The observed relaxation times fall essentially in the same time range as those by electric field pulse method as is seen in Fig. 11. In Fig. 11, the shift of the pH value which shows the maximum relaxation times can be explained as the effect of ionic strength on the pH range in which the helix-coil transition occurs. However, pH dependence of the relaxation times obtained by both methods is different. This difference is anticipated in considering that the helix-coil transition is coupled complicatedly with proton transfer reaction of side chain in the experiment of electric field pulse method, because the proton transfer reaction can be detected by the electric conductivity but cannot be done by the optical rotation.

REFERENCES

1. R. Lurmy, R. Legare and W.G. Miller, Biopolymers, 2, 484 (1964).
2. (a) H. Inoue, J. Sci. Hiroshima Univ., Ser. A-11, 34, 37 (1970).
   (b) A.D. Barksdale and J.E. Stuehr, J. Amer. Chem. Soc., 94, 3334 (1972).
3. G. Schwarz and J. Seelig, Biopolymers, 6, 1263 (1968).
4. T. Yasunaga, T. Sano, K. Takahashi, H. Takenaka and S. Ito, Chem. Letters, 405 (1973).
5. A. Warashina and A. Ikegami, Biopolymers, 11, 529 (1972).
6. G. Ilgenfritz, Dissertation, Gottingen (1966).
7. M. Eigen and G. Schwarz, J. Colloid. Sci., 12, 181 (1957).
8. H.P. Gregar, D.H. Gold and M. Frederick, J. Polym. Sci., 23, 467 (1957).
9. K.F. Wissbrum and A. Patterson, Jr., J. Polym. Sci., 33, 235 (1958).
10. A.L. Jacobson, Biopolymers, 2, 207 (1964).
11. G. Schwarz, J. Mol. Biol., 11, 64 (1965).
12. R. Snipp, W. Miller and R. Nylund, J. Amer. Chem. Soc., 87, 3547 (1965).
13. V.E. Bychkova, O.B. Ptitsyn and T.V. Barkskaya, Biopolymers, 10, 2161 (1971).
14. G. Schwarz and J. Engel, Angew. Chem. Internat. Edit., 11, 568 (1972).
15. G. Schwarz, J. Phys. Chem., 71, 4021 (1967).
16. K. Kikuchi and K. Yoshioka, Biopolymers, 12, 2667 (1973).

ELECTRIC FIELD JUMP STUDY OF THE HELIX-COIL TRANSITION OF AQUEOUS POLY-α,L-GLUTAMIC ACID

A.L. Cummings and Edward M. Eyring

Department of Chemistry, University of Utah, U.S.A.

ABSTRACT. Preliminary kinetic results of a spectrophotometric E-jump study of aqueous poly-α,L-glutamic acid are reported.

An interesting conductimetric E-jump relaxation method study of the helix-coil transition of aqueous poly-α,L-glutamic acid (PLGA) by Yasunaga and coworkers[1] prompted us to explore the possibility of duplicating their results spectrophotometrically. Since they attributed the success of their conductimetric measurements to a coupling of rapid ionization of sidechain carboxyl group protons to the helix-coil transition, we tried coupling a visibly colored acid-base indicator to the protolytic reactions with the expectation of observing the same minimum in a plot of τ^{-1} vs. pH. at pH 5.6 that Yasunaga _et al_. had reported. Their results incidentally agreed well with those of Barksdale and Stuehr.[2]

We chose as our indicator bromocresol purple (5',5"-dibromo-o-cresolsulfonephthalein) (BCP) for which the pKa = 6.47 and the ion recombination rate constant $k_R = 7.7 \cdot 10^{10} M^{-1} sec^{-1}$ for the equilibrium $HIn^- \rightleftarrows H^+ + In^{2-}$ at 25° near zero ionic strength.[3] An expected advantage of a negatively charged sulfonephthalein indicator is that it should have less tendency to adhere to the negatively charged PLGA chain than acridine dyes both by virtue of Coulombic repulsion and a non-planar molecular structure less suited to intercalation.

Figure 1 depicts our experimental results. The PLGA sample (Fox Chemical Co., Los Angeles) had a M.W. of 50,000 to 100,000. The λ = 565 nm light from a pulsed[4] arc lamp was passed through the sample cell of a previously described[5] spectrophotometric E-jump apparatus. The apparent minimum in the τ^{-1} vs. pH plot at

Wyn-Jones (ed.), Chemical and Biological Applications of Relaxation Spectrometry, 505–508.

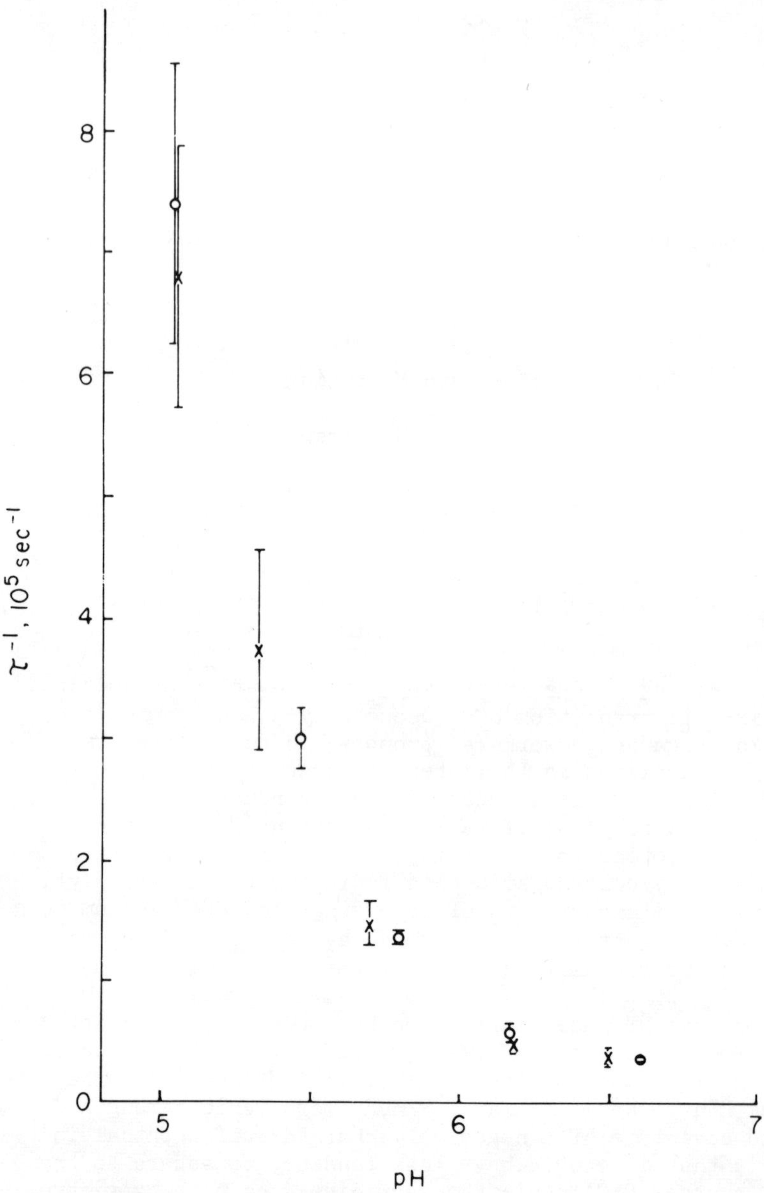

Fig. 1. pH dependence of electric field jump inverse relaxation tim in aqueous solutions of poly-α, L-glutamic acid, 6.3 X 10^{-4} residue M, with bromocresol purple indicator at 3 X 10^{-5}M in the case of circles and 1.5 X 10^{-5}M in the case of crosses. The vertical error bars indicate standard deviations. Compare with Fig. 2 of Yasunaga et al.[1]

pH ~ 6.5 disagrees with the result of Yasunaga *et al.*[2] both with respect to the minimum value of τ^{-1} and the pH at which this minimum occurs. {Elsewhere in these proceedings Yasunaga has reported a new value of the pH at the minimum τ^{-1} that agrees better with that of Fig. 1, but his minimum value of $\tau^{-1}_1 \sim 3.4 \times 10^5\ sec^{-1}$, is till markedly higher than the minimum τ^{-1} of ~ $3.5 \times 10^4\ sec^{-1}$ in Fig. 1. Yasunaga's revised pH may be attributable to a more carefully prepared PLGA sample.}

At pH 6.5 and a total BCP concentration of 1.5×10^{-5}M the relaxation time of the equilibrium $HIn^- \rightleftharpoons H^+ + In^{2-}$ alone would be ~1.5 μsec. Thus it is clearly impossible to make the approximation in this case that the indicator achieves equilibrium very much more rapidly than either the protolytic side chain equilibrium or the equilibration of the helix-coil transition. Furthermore, the relative slowness of our observed relaxation times compared to those of Yasunaga suggests that a binding of dye to polymer may be occuring that somehow slows the helix-coil transition.

The presumption that the helix-coil transition is responsible for the observed oscilloscope traces is consistent with the direction of the deflections of the experimental relaxations. An intense electric field may be expected to cause a further dissociation of an ionic equilibrium. Thus the relaxation observed here after the termination of the high field pulse should correspond to an ion association process. The transition of PLGA from random coil to helix withdraws dissociated protons from solution, and this process must induce dissociation of BCP and production of more In^{2-}. Production of In^{2-} is precisely what we observe (increasing absorbance at 565 nm) when the high field is removed from the sample cell.

Were the relaxation attributable to a PLGA-BCP binding equilibrium we would expect a dye concentration dependence of τ^{-1} that is not observed. Furthermore, binding of negatively charged dye to negatively charged PLGA would not involve charge neutralization but would deplete the number of mobile charge carriers in solution and would therefore be favored by the removal of the intense electric field. If bound In^{2-} absorbed less strongly at $\lambda = 565$ nm than does In^{2-} in solution, we would expect a decrease in absorption at 565 nm when the high voltage pulse is terminated if dye binding occured. This we also do not observe. However, we have not observed a change in the equilibrium visible absorption spectrum when PLGA is added to BCP solutions (buffered at pH 6), so this latter argument against dye binding is highly speculative.

While we intend to explore further the PLGA-BCP sample system, it is now clear that a spectrophotometric E-jump experiment on PLGA alone using optical rotatory detection would provide more convincing helix-coil transition kinetic data. Recent papers[6,7] make interesting suggestions regarding the execution of such an optical rotation experiment.

ACKNOWLEDGEMENT. This work was sponsored by Grant AFOSR 73-2444 from the Directorate of Chemical Sciences, Air Force Office of Scientific Research and by National Institute of Arthritis, Metabolism and Digestive Diseases Grant AM 06231-12.

REFERENCES

1. T. Yasunaga, T. Sano, K. Takahashi, H. Takenaka and S. Ito, "Chemistry Letters (Japan), 405 (1973).
2. A.D. Barksdale and J.E. Stuehr, J. Am. Chem. Soc. **94**, 3334 (1972).
3. D.J. Lentz, J.E.C. Hutchins and E.M. Eyring, J. Phys. Chem., **78**, 1021 (1974).
4. S.L. Olsen, L. Paull Holmes and E.M. Eyring, Rev. Sci. Instrum. **45**, 859 (1974).
5. S.L. Olsen, R.L. Silver, L.P. Holmes, J.J. Auborn, P. Warrick, Jr., and E.M. Eyring, Rev. Sci. Instrum., 1247 (1971).
6. R. Lumry and R. Legare, Anal. Chem., **41**, 551 (1969).
7. M. Dourlent, J.F. Hogrel and C. Helene, J. Am. Chem. Soc., **96**, 3398 (1974).

ULTRASONIC RELAXATION IN SOLUTION OF POLYSTYRENE AND N-POLYVINYL-PYRROLIDONE

Jørgen Rassing* and E. Wyn-Jones**

INTRODUCTION

The results presented in this contribution have been published elsewhere[1,2]. Polystyrene of molecular weights 44,700, 98,200, 173,000, and 867,000 dissolved in N,N-dimethylformamide shows an ultrasonic relaxation in the frequency range 1 - 100 MHz. The analysis of the relaxation spectrum shows that two is the minimum number of relaxation times required in order to describe the data. The data fit shows that both relaxation times are independent of the overall concentration of polymer and the molecular weight. Furthermore the relaxation amplitude increases linearly with the polymer concentration (g/l) for both relaxations.

INTERPRETATION OF THE RESULTS

The above behaviour of the relaxation parameters indicates that the origin of both relaxation processes is intramolecular, in nature, and the most probable molecular phenomenon is associated with conformational changes in the polymer chain. The fact that the relaxation parameters are not influenced by the change in molecular weights indicates that the conformational change in question takes place in segments of the polymer chain in a non-coupled way.

---

*Roskilde University Center, Denmark.

**University of Salford, G.B.

Wyn-Jones (ed.), Chemical and Biological Applications of Relaxation Spectrometry, 509–511.

A kinetic interpretation requires a reaction mechanism. It is possible to formulate a multistep mechanism based on a detailed picture of the motions and packing of all the segments in the polymer chain. However, lack of structural information makes it very difficult to derive explicitly the relaxation equation in terms of the mechanism and, so far, this derivation has not been carried out successfully. If the approximation that the intramolecular distances between the benzene rings in the polymer chain reflect the packing of the segments and that the change in these distances as a function of time reflects the motion of segments, then the conformational transition may be described in terms of a three state mechanism. The three states of different energy are defined in the following way. State III of highest energy is made up of all the triads with neighbouring benzene rings in the closest position with respect to each other. State II is made up of the remaining diads of benzene rings which are in the closest position with respect to each other. Finally, State I of lower energy is made up of all the remaining benzene rings. The relaxation is then attributed to the transition between the three states according to the following scheme.

$$\text{I} \overset{k_{11}}{\rightleftharpoons} \text{II} \overset{k_{21}}{\rightleftharpoons} \text{III} \qquad (1)$$

The transition from one state to another corresponds to an internal rotation around a C-C bond in the polymer chain.

The kinetic analysis is based on the further approximation that ΔH^o for the transition II $\rightleftharpoons$ III is larger than that of I $\rightleftharpoons$ II. Under these conditions the data are completely described with the final result that $k_{1I} = 3.8.10^7 s^{-1}$ and $k_{2I} = 4.7.10^8 s^{-1}$ where k_{1I} and k_{2I} are the forward rate constants.

DISCUSSION

The existence of the states I, II and III in a polystyrene solution has been inferred by Bovey[3] in connection with light scattering, viscosity and NMR measurements on the conformational changes in the molecules. By means of internal rotation about the single bonds in the polymer chain these states are considered to be in fast dynamic equilibrium with each other. However, as pointed out by Bauer[4], a rotation about a single bond in a polymer would force one part of the molecules to have large movements through the solvent, and this is not in accordance with the time range measured. Thus a second rotation about an adjacent bond in the opposite sense is very probable as this only affects a small part of the molecule. It is interesting to note that states I and III in the present model correspond to the two states considered by Bauer.

The values of the obtained rate constants compare well with those found for internal rotation in simple ethanes[5]. Recent experiments[5] have shown that atactic and syntactic polystyrene cause identical relaxation spectra. This is to be expected, however, because the relaxation is essentially caused by internal rotation around the C-C bond. If there happens to be a difference between the solutions made from the two polymers it must be related to different populations of the three states. This does not necessarily change the transition rate constants. If the conformational changes in the polymer chain take place by internal rotation about C-C bonds in the normal way, then one expects that the equilibrium distribution between the different states in solution is independent of the initial structure of polystyrene in the solid state.

N-polyvinyl pyrrolidone dissolved in dioxane causes one relaxation only. The behaviour of the relaxation time, the relaxation amplitude, and the sound velocity is in agreement with the above conclusions for conformational changes in the polymer chain.

CONCLUSION

The ultrasonic relaxation observed in vinyl polymers may be attributed to conformational changes in the polymer chain. The elementary reaction is an internal rotation around the C-C bond in the polymer chain. These rotations take place in a highly uncoupled way, and consequently the relaxation spectrum does not depend on the length of the polymer chain in the molecular weight range 44,700 - 867,000.

REFERENCES

1. J. Rassing, Acta Chem. Scand. 25, 1506 (1971)
2. W. Ludlow, E. Wyn-Jones and J. Rassing, Chem. Phys. Let. 13, 477 (1972).
3. F.A. Bovey, "Polymer conformation and configuration" Academic Press, New York, p.75 (1969).
4. H.J. Bauer, H. Hassler and M. Immendorfer, Discussions Faraday Soc. 49, 238 (1970).
5. W. Ludlow, personal information.

THE ORIGIN OF ULTRASONIC RELAXATION IN POLYMER SOLUTIONS

A. M. North

Department of Chemistry and Applied Chemistry,
University of Strathclyde, Glasgow, Scotland

In a consideration of the way in which a non-electrolyte polymer molecule can give rise to ultrasonic relaxation we shall treat only conformational changes which cause sound absorption either through viscothermal mechanisms or through a contribution to the relaxing specific heat. These two are, in one sense, equivalent since the frequency dependence of the viscosity of polymer solutions is described in terms of contributions from normal modes of motion, which are themselves cooperative conformational changes in the flexible polymer chain and, as such, may also contribute to the relaxing specific heat.

The problem, then, resolves to a consideration of three conformational processes. These are:-

1. The very localised conformational change which may occur in typical units or segments of the polymer chain, the behaviour of each segment being assumed to be independent of that of any other segment.
2. A similar conformational change taking place in units or segments close to chain ends and which, therefore, exhibits different energetics from that of units more characteristic of chain 'interiors'.
3. Cooperative movements in which minute partial conformational change in a number of units sums to a total alteration as described by the normal modes of motion.

The interpretation of an observed relaxation, in terms of these processes, is normally carried out by analysis of the frequency and amplitude of relaxation, in combination with consideration of the way in which the relaxation departs from ideal behaviour. In this context it must be emphasised that

Wyn-Jones (ed.), Chemical and Biological Applications of Relaxation Spectrometry, 513–514.

almost all observations to date (whether carried out on polymers monodisperse or heterodisperse in molecular weight) yield relaxation parameters broader in the frequency plane than expected for an ideal case.

In any consideration of relaxation mechanism based upon an analysis of relaxation curve shape, the point must be made that very often errors of measurement are such that the analysis is meaningful only when carried out with the minimum number of additional distribution parameters. In this context one can either discuss a 'distribution of relaxation times' characterised by a single empirical parameter (as is the case in Cole-Cole and Cole-Davidson representations of dielectric and related phenomena) or select two relaxation processes as the minimum number, which, in turn, involves an additional two parameters in the fitting of experimental results.

Studies of monodisperse polystyrene fractions[1] have shown that the relaxation amplitude and frequency varies with molecular weight, indicating without any shadow of doubt that the conformational behaviour of groups close to chain ends differs markedly from those in the 'interior' of polymer chains. In addition studies of low molecular weight paraffins[2] show that normal modes may be, in fact, acoustically active, and indeed the viscous absorption of polymeric materials will be frequency dependent in the KHz to MHz frequency range. In any study made of monodisperse polymer, it is therefore necessary to subtract off the contribution to the absorption from the chain end processes and from normal mode viscothermal processes before obtaining an absorption characteristic of 'central' unit conformational change. When this is done it is found that allowance for chain end effects reduces the high frequency absorption and that normal mode effects reduce somewhat the low frequency absorption so that the underlying molecular weight independent process (which we believe to be true conformational change) approaches much more closely ideal behaviour. Of course, when the polymer sample is formed from a distribution of molecular weights, it is extremely difficult to make properly weighted allowance for end group (high frequency) and normal mode (low frequency) effects, so that non-ideal relaxation will almost certainly be observed. Indeed, because of the uncertainties in experimental measurement, and because due allowance for these two effects may become quite appreciable, we prefer an analysis based on a single additional relaxation parameter to that based on two separate relaxation processes.

REFERENCES

1. M.A. Cochran, J.H. Dunbar, A. M. North and R.A. Pethrick, J. Chem. Soc., Faraday Trans, II, **70**, 215 (1974).
2. M.A. Cochran, P.B. Jones, A.M. North and R.A. Pethrick, J. Chem. Soc., Faraday Trans. II, **68**, 1719 (1972).

MOTION IN AMORPHOUS POLYMERS AND VISCOUS LIQUIDS

Graham Williams

Edward Davies Chemical Laboratories,
University College of Wales, Aberystwyth, SY23 1NE.

INTRODUCTION. The motion of molecules in the solid and liquid states may be studied using such techniques as dielectric relaxation[1-4], NMR relaxation[5-7], infra-red and Raman spectroscopy[8-10] inelastic neutron scattering[11-13], inelastic light scattering[14-19] and viscoelastic relaxation[4,20-21]. The different experiments probe different aspects of the molecular motion, as has been discussed by Berne[22]. The orientational motions of molecules are studied using dielectric relaxation, infra-red and Raman vibration-rotation band shapes, and in certain depolarized light scattering experiments, while the inelastic neutron scattering, polarized light scattering and viscoelastic relaxation techniques reflect, in the main, translational motions. Much interest has been shown in the very fast reorientational and translational motions of small molecules in the liquid state using the NMR, infra-red and Raman and inelastic light scattering and neutron scattering techniques. The time scale for such studies is, in most cases $< 10^{-10}$s. However, molecular motion may extend over a much longer time-scale than this. For example[23] the dielectric relaxation time of polypropylene oxide in the amorphous solid state changes from $\sim 2 \times 10^{-9}$s to $\sim 2 \times 10^{3}$s on cooling from 273 K to 200 K. Low frequency dielectric relaxation processes may be observed for certain rotator phase solids (e.g. ice[24], pentachlorotoluene[25], pentachloronitrobenzene[26]), the alcohols[3,27-29], many non-associated super-cooled liquids[21,30-42] and for amorphous and partially crystalline polymers[4].

For crystalline rotator phase solids, molecular reorientations take place cooperatively as envisaged in the simulation of Brot and Darmon[25] for 1,2,3 trichloro-trimethyl benzene. For the alcohols the breaking and reforming of hydrogen bonds are an essential part of the reorientational process of the molecules. The dielectric relaxations of supercooled (non-associated) liquids and

Wyn-Jones (ed.), Chemical and Biological Applications of Relaxation Spectrometry, 515–526.

amorphous solid polymers are qualitatively rather different from those in crystalline rotator phase solids and the alcohols, in that extremely large apparent activation energies are observed for the main process (the α process) and the loss curves have a broad and characteristically asymmetric contour. The present account gives a brief resume of the principal features of the dielectric relaxations in amorphous solid polymers and in supercooled, non associated, liquids, indicating the apparently strong common features between the two systems. The similarities have been emphasized by Johari and Smyth and by Johari and Goldstein as a result of their work on a wide range of supercooled liquids[32-36,43].

In addition to a consideration of molecular reorientation in amorphous solid polymers and supercooled liquids, we mention the possibility of the detection of chemical relaxation and of studying conformational relaxation or chemical reactions in supercooled liquids by dielectric methods.

AMORPHOUS SOLID POLYMERS

The dielectric relaxation of crystalline and amorphous solid polymers has been fully described by McCrum, Read and Williams[4]. Further aspects have been reviewed by McCall[6], Ishida[44] and by Williams and Watts[45,46]. In most amorphous polymers the magnitude of the α (glass transition) process greatly exceeds the magnitude of the β relaxation process, the α process occurring at a lower frequency than the β process at a given temperature and pressure. The total magnitude[4] of the α and β processes is related to $Ng\mu^2$, where N is the number of dipoles per unit volume, μ is the dipole moment of a repeat unit and g is a factor which takes into account the orientational correlations between chain dipoles. This magnitude is an equilibrium property of the dipolar solid. The frequency location and the shape of the dielectric dispersion and absorption curves, however, give information on the dynamics of the polymer chains in the solid state.

Whilst no general rule applies to all polymers, many exhibit an α relaxation whose frequency-temperature location is characterized by the Williams-Landel and Ferry (WLF) equation[4,47] and whose contour resembles that of the Davidson-Cole relation[28], being broader on the higher frequency side of the loss peak. Examples are polyethylene-terephthalate[4,48] where the dipole is located in the main chain, and polyvinyl acetate[4,49] where the dipole is associated with the side chain. The similarity in the contour of the dielectric absorption curves for several polymers has been discussed by Williams, Cook and Hains[50]. It has been shown that the dielectric data for polymers of widely different chemical structure may be fitted by the empirical relaxation function of Williams and Watts[51,52]

$$\phi(t) = \exp - (t/\tau_0)^{\bar{\beta}} \qquad (1)$$

where $0<\bar{\beta}\leq 1$. The relationship between the complex dielectric permittivity $\varepsilon^*(i\omega) = \varepsilon'(\omega) - i\varepsilon''(\omega)$ and the relaxation function $\phi(t)$ is the superposition relation[4,51],

$$\frac{\varepsilon^*(i\omega) - \varepsilon_\infty}{\varepsilon_o - \varepsilon_\infty} = \int_0^\infty \left[\frac{-d\phi(t)}{dt}\right] \exp - i\omega t \, dt \qquad (2)$$

The experimental data[50-52] for various polymers give $\bar{\beta}$ in the range 0.4 to 0.6. This empirical approach, as with those[4] of Davidson and Cole, Cole and Cole and of Fuoss and Kirkwood, gives no indication of the molecular behaviour of the polymers. The molecular interpretation involves the time-dependent dipole moment correlation function[53] for the chain, and will be considered below.

It is of interest to note that pressure has an important effect on the dielectric α relaxation. Studies of polymethyl acrylate[54], polypropylene oxide[23], poly-n-butyl methacrylate[55], polyethyl methacrylate and polyethyl acrylate[56,45], polynonyl methacrylate and polylauryl methacrylate[57] and several other alkyl methacrylate polymers[58] and for polyvinyl chloride, polyvinyl acetate, polychlorotrifluoroethylene and polyethylene terephthalate[59] have been made over a range of temperature and applied pressure - up to 3000 atm. The α process moves to lower frequencies as the pressure is increased at a rate $(\partial \log f_m/\partial p)_T \sim 2 \times 10^{-3}$ atm^{-1}, where f_m is the frequency of maximum dielectric loss. The shape of the α -loss curve, ε" - vs - log f, is only slightly changed with temperature and pressure.

The β relaxation in most amorphous polymers is rather small in magnitude and is characterized by broad loss curves (½ width ~ 5 decades of frequency) and an apparent activation energy which lies in the range 10 - 20 K cal mol^{-1} being much smaller than that for the α relaxation. Pressure has only a small effect on the location of the β relaxation in polyethylene terephthalate[59,60] and polyvinyl chloride[59,61], $(\partial \log f_m/\partial p)_T$ being less than 0.5×10^{-3} atm^{-1}. However, for poly-n-butyl methacrylate and polyethyl methacrylate it has been found[55,56,58] that the magnitude of the β relaxation, at temperatures close to the glass transition at 1 atm pressure, decreases remarkably with increasing pressure, the frequency location of the loss maximum being, at the same time, quite insensitive to the pressure variation.

At higher temperatures than the glass transition temperature T_g there is a tendency for the β and α processes to merge, since they have such different apparent activation energies. The experimental data for the coalescence of α and β processes to form the (αβ) process was first described by Heydemann[62] for polymethyl methacrylate, and was studied in further detail by Williams and Edwards[55] for poly-n-butyl methacrylate and by Williams[56] for polyethyl methacrylate. Later work by Sasabe and Saito[58] provided adequate confirmation that the α and β processes coalesce at high temperatures to form a single process, the (αβ) process which carries with it all of the available relaxation strength. It was found[56,58] that pressure and temperature variation was necessary in order to provide an adequate documentation of the rather broad α, β and (αβ) processes. Sasabe and Saito[58] indicated that the (αβ) process appeared to be a continuation of the α process to higher temperatures, which suggests that the processes have a common mechanism. This has been further discussed by Williams and Watts [45,46,61].

Thus there appears to be a well-defined pattern of behaviour with regard to the dielectric relaxation of amorphous solid polymers. The relaxation is associated[4] with the apparent glass transition of the polymer and is due to the large scale microbrownian motions of the chain segments. The β relaxation is a local motion of chain segments or side chains (e.g. polyalkyl methacrylates). The (αβ) process is almost certainly the α process, extrapolated from lower temperatures, which has absorbed into itself the β process. In terms of dipole relaxation, the β process partially relaxes $<\mu^2>$, the mean square dipole moment of a chain, and that which is not relaxed is eventually relaxed by the (slower) α process. At higher temperatures, i.e. beyond that of the coalescence of α and β processes, all of $<\mu^2>$ is relaxed by the (αβ) process. The phenomenological aspects of α, β and (αβ) relaxations have been discussed by Williams and Watts[45,46,61].

In molecular terms, it is necessary to relate $\varepsilon^*(i\omega)$ to the dipolar nature of the chain. This is conveniently considered in terms of the dipole moment correlation function[53,56].

$$\Gamma(t) = \frac{<\underline{\mu}_i(o).\underline{\mu}_i(t)> + \sum_j <\underline{\mu}_i(o).\underline{\mu}_j(t)>}{\mu^2 + \sum_j <\underline{\mu}_i(o).\underline{\mu}_j(o)>} \tag{3}$$

$$\frac{(\varepsilon(i\omega) - \varepsilon_\infty)}{\varepsilon_o - \varepsilon_\infty} p(i\omega) = \int_o^\infty \left(\frac{\{-d\Gamma(t)\}}{dt} \right) \exp - i\omega t \, dt \tag{4}$$

where $p(i\omega)$ is an internal field factor[53,63] $\underline{\mu}_i(t)$ is the dipole moment for a representative dipole i in a chain at time t, $<\underline{\mu}_i(o).\underline{\mu}_i(t)>$ is the autocorrelation function for the reorientation of dipole i, <> indicating a time or ensemble average, and $<\underline{\mu}_i(o).\underline{\mu}_j(t)>$ is the cross-correlation function for the relative reorientation of dipoles j and i, where j and i are in the same chain. The sum is taken over all dipoles in the chain with the restriction that $j \neq i$. Thus the dielectric α, β and (αβ) relaxations are to be associated with the behaviour of all the auto and cross-correlations in eq. (3), suggestive of a very complex situation. Fortunately for flexible chains the importance of the cross-correlations becomes decreasingly significant as the dipoles are separated down the chain[63], so it is possible, that on average, $\Gamma(t)$ is determined largely by the autocorrelation term and perhaps cross-correlation terms extending to $|j - i| \sim 4$. The observations that the relaxations in several polymers have quite similar contours for $\varepsilon''(\omega)$ - vs - log f, and the contour is unaffected by composition changes in styrene-p-chlorostyrene copolymers[64] may be taken[63] as evidence that (a) the <u>time</u>-dependence of auto and cross- correlation terms in eq. (3) are quite similar and (b) that the time dependence of the auto-correlation term is quite similar in several polymers, suggesting a common mechanism for the motion. Further[63] it is suggested that this autocorrelation function has a natural non-exponential dependence upon time due to the cooperative nature of the (α) process and may not be regarded primarily as a chain connectivity phenomenon or as arising from a distribution of relaxation times, as is usually supposed. We shall consider the mechanism for the α relaxation in more detail below. With regard to the β relaxation, the fact that it is so broad in the frequency domain means that no simple mechanism can be given for the process. However, the β process in polymers containing the dipoles rigidly attached to the main chain shows a different response to pressure variation compared with that for polymers having dipoles located in the side chain, e.g. polyalkyl methacrylates, and this may be taken as partial evidence[56,61] that the β relaxation in the latter polymers are to be associated with the local reorientation of side groups.

SUPERCOOLED AND OTHER VISCOUS LIQUIDS

Although it has been known for many years[1,30,65] that molecular liquids, e.g. certain alkyl halides, may be supercooled to form highly viscous liquids which exhibit dielectric relaxation at low frequencies, the works of Denney[30], Berberian and Cole[31], Johari and Smyth[32,33] and Johari and Goldstein[34,35] have demonstrated the apparent generality of the behaviour for supercooled non-associated dipolar liquids, i.e. they exhibit well-defined α and β relaxation processes. Johari and Goldstein[34,35] emphasized that the α and β relaxations observed in such simple systems as chlorobenzene/decalin and chlorobenzene/pyridine appeared to be quite similar to the α and β relaxations in polymers. Thus whereas in solid amorphous polymers it has been usual to think of α and β relaxations arising as a consequency of chain connectivity, such an approach is clearly inappropriate for the simpler supercooled liquids.

A detailed account of the dielectric relaxation of supercooled non associated liquids has been given recently[66]. It is found as a result of the work of Denney[30], Berberian and Cole[31], Johari, Smyth and Goldstein[32-35] and of Williams and co-workers[38-42] that the similarity between the relaxations of solid polymers and supercooled non associated liquids is demonstrated qualitatively in most cases and quantitatively in many cases: i.e. similarity with regard to the existence of α, β and (αβ) relaxations and their temperature variation, the shape of the plots ε'' - vs - log f for both the α and β relaxations.

From a molecular dipole point of view, the difference between supercooled liquids and polymers resides in the nature of the cross-correlation terms in eq. (3). For polymers the cross-correlations exist primarily within a chain, assuming inter-chain correlations of orientations to be rather small, while for simple molecular liquids the correlations (if any) will be between molecules. For dilute solutions of a dipolar solute in a non-polar solvent the dipolar cross-correlations will be negligible, e.g. for dilute chlorobenzene/decalin mixtures[35] or for phthalic anhydride in o-terphenyl[38,39], and since such systems exhibit α and β relaxations, clearly one requires a mechanism which will generate the two relaxations in the absence of cross-correlations - both for the supercooled liquids and for amorphous polymers.

MECHANISM FOR THE α, β AND (αβ) RELAXATIONS

Historically, the first successful molecular model for the reorientation of dipoles in a liquid medium is that of Debye (see refs. 1 - 4) in which a dipolar molecule is assumed to reorientate in a viscous continuum in accord with a rotational diffusion equation. The model, which does not include the inertial aspects of motion, results in a single relaxation time process of relaxation time $\tau = 3\eta V/kT$. This simple theory has had a reasonable success[1-3] in relating the observed relaxation times of simple liquids to molecular quantities, but fails quantitatively due to difficulties with the viscosity term η and due to the fact that inertial factors are of great importance for fast ($>10^{11}$ Hz) molecular reorientations[53]. For viscous liquids or for solid amorphous polymers the inertial factors should be unimportant since the relaxation frequencies are usually less than 10^6 Hz. However, the viscosity factor might appear to be an insuperable obstacle for solid polymers. This is not quite the situation, however, as a result of the generalization of the Debye approach to include the case of a viscoelastic continuum around the dipole, by Di Marzio and Bishop[67]. They obtain the following expression for $\varepsilon^*(i\omega)$

$$\frac{\varepsilon^*(i\omega) - \varepsilon_\infty}{\varepsilon_o - \varepsilon_\infty} = \left\{1 - i\omega\tau(\omega)\frac{(\varepsilon_o + 2)}{(\varepsilon_\infty + 2)}\right\}^{-1} \qquad (5)$$

where $\tau(\omega) = 3\eta(\omega)V/kT$, V is the volume of a molecule, and $\eta(\omega)$ is the dynamic viscosity, which is, in turn related to the storage modulus $G(\omega)$ according to $\eta(\omega) = G(\omega)/i\omega$. Equation (5) was used by Di Marzio for the dielectric α relaxation in certain polyalkyl methacrylates. Insertion of the known dynamic mechanical relaxation behaviour, $G(\omega)$ into eq. (5) together with assumed values for V yielded Cole-Cole diagrams which are of Davidson-Cole type, in reasonable agreement with the experimental data. Clearly this approach may also be used for the α relaxation in supercooled molecular liquids. However, it should be emphasized that this is essentially a macroscopic approach to the reorientation of dipoles. The dipole is assumed to reorientate in a viscoelastic continuum, and this may not be a reasonable approximation for a dipolar group within a polymer chain. We note at this point that the dielectric relaxation time and the steady flow viscosity have quite similar dependencies upon temperature for certain supercooled liquids (e.g. o-terphenyl solutions[41]) but the effective molecular diameters obtained using the Debye theory are far too small to be realistic. A similar result was noted earlier in a comparison of the NMR correlation time with viscosity for o-terphenyl[68].

Recently[69] Shore has considered the model of a linear array of coupled rotations where a given rotator may reorientate in the potential field of its neighbours. For given conditions for the relative rates of internal and overall motion for the chain it is found that the dipolar correlation function for a chain containing only a small number of dipoles takes the form $\phi(t) = \exp - (t/\tau)^{\frac{1}{2}}$ being very similar to the observed results for the α process in solid polymers, as discussed above, and for several supercooled liquids[33,39].

It seems likely that the mechanism for the α relaxation in both supercooled liquids and in solid polymers is associated with large scale fluctuations in the density of macroscopic regions of the material. These will occur at a rate which is very dependent upon temperature and applied pressure, as evidenced by the large apparent activation energy and activation volume for the α process[30-43, 45, 46, 54-61]. One approach to molecular relaxation in a fluctuating environment has been given by Anderson and Ullman[70] who consider that a representative dipole may reorientate via Stockastic volume fluctuations in its local environment. A variety of contours for the relaxation, dependent upon parameters of the model, were obtained. A feature of this theory is that it does predict a loss-curve of Davidson-Cole type, which has the following molecular interpretation. In the 'short-time' region of the process (high frequency side of the loss-curve) the molecules have a variety of reorientation probabilities, whereas in the 'long time' region the molecules, on average, experience many environmental changes prior to relaxation, hence tend to exhibit the same correlation time: i.e. a tendency to single relaxation time behaviour at 'long time' (low frequencies). Another approach, involving the ideas of cooperative motion is due to McDuffie and Litovitz[71,72]. Clearly these models[69-72] may be applicable to the α relaxation in both supercooled liquids and to amorphous solid polymers.

One approach, which has been quite successful for the visco-elastic relaxation in supercooled liquids, is that of defect-diffusion, due to Glarum[73] and Phillips, Barlow and Lamb[74]. A defect is assumed to diffuse through the medium, relaxing dipoles in its path. The correlation function for the reorientation of a dipole has the form[74]

$$\phi(t') = \phi_o(t') \quad \left\{(1 - 2t')\ \phi_o(t') + 2(t'/\pi)^{\frac{1}{2}}\right\} \tag{6}$$

where $t' = t/\tau_d$, τ_d is the defect-diffusion time, and $\phi_o(t') = \exp(t')\ \mathrm{erfc}(t')^{\frac{1}{2}}$. Eq. (6) refers to the relaxation of a dipole by the nearest and next nearest neighbour defects. On Fourier transformation, eq. (6) gives a dielectric relaxation curve which is fitted rather well by the Williams-Watts empirical relation (see eq. 1, 2 above) with $\bar{\beta} = 0.514$. Since a number of amorphous solid polymers and supercooled liquids resemble, quite closely, this behaviour (see e.g. ref. 50) the defect-diffusion approach may be a useful way of imagining the α process in these systems. Its application is, however, limited quantitatively[21,42].

With regard to the mechanisms of the β and (αβ) relaxations the discussion of the β process cannot be taken beyond that which has been given above, while the (αβ) relaxation has the same basic mechanism as the α process which is discussed above.

CHEMICAL RELAXATION

The dielectric relaxation behaviour discussed above is wholly concerned with the reorientation relaxation of dipole moment vectors, eq. (3). For systems in which chemical changes take place, e.g. $A \rightleftharpoons B$, one might ask if the kinetics of the conversions $A \rightleftharpoons B$ might be detected in a dielectric experiment. This has been considered by Anderson and Smyth[75], Schwarz[76,77], Williams[78] and Scheider[79]. The result is that if molecular reorientation occurs at a faster rate than chemical exchange, then the dipoles will be relaxed by the reorientation mechanism alone, and chemical relaxation will not be detected. Goulon[80] has discussed dielectric data for certain liquids, including biacetyl, in terms of chemical exchange and molecular reorientation. Internal conformational changes, however, may be detected by dielectric and mechanical relaxation techniques, and many examples, including cyclohexane derivatives attached to polymer chains, are given by Heijboer[81].

DIFFUSION CONTROLLED CHEMICAL REACTIONS

The experimental dielectric data for supercooled liquids discussed above[66] relate to the reorientations of the constituent dipolar molecules. In highly viscous liquids it is difficult to imagine that reorientational and translational motions of molecules occur on very different time-scales, thus it is unlikely that the separation of a time dependent diffusion equation into independent rotational and translational parts is reasonable. The observation[21] that the dielectric and viscoelastic relaxations in certain viscous liquids (e.g. di-n-butyl phthalate) are quite similar also suggests that the translational and rotational motions are strongly coupled. This being so, the dielectric results for supercooled and other highly viscous liquids (excluding the alcohols) may be taken as an indication of the rate at which molecules move, on average, over molecular distances. The relaxation time for o-terphenyl, for example, may be changed[68] from 10^{-11} to 10^{+3}s on cooling from 450 K to 250 K, with the greater part of the change coming in the range 290 K - 250 K. Such a remarkable variation in molecular mobility by simple change of temperature should have an important effect on the rates of chemical reactions of materials dissolved in such supercooled media. Clearly thermal reactions are inappropriate since supercooled solutions are prepared by cooling from above the melting temperature of the medium, but it would appear that photochemical reactions might be studied in such media. We note that the dielectric relaxation time or the NMR correlation time acts as a useful guide to the molecular mobility of a medium. The macroscopic shear viscosity is, however, not necessarily a correct indicator of mobility. For non-associated supercooled liquids η reflects molecular mobility, but for polymer solutions, where η may be large for only small concentrations of a high molecular weight polymer, the entanglement of chains, and the normal modes of motion of the chains are the source of high viscosity, and the remaining small molecules may be highly mobile, even for a medium or large viscosity.

REFERENCES

1. C.P. Smyth, 'Dielectric Behaviour and Structure' (McGraw-Hill, New York) 1955.
2. C.J.F. Bottcher, 'Theory of Electric Polarization', 1st Ed. (Elsevier) 1952, 2nd Ed. 1973.
3. N.E. Hill, W.E. Vaughan and A.H. Price, 'Dielectric Properties and Molecular Behaviour' (Van Nostrand, New York) 1969.
4. N.G. McCrum, B.E. Read and G. Williams, 'Anelastic and Dielectric Effects in Polymeric Solids'. (John Wiley, New York) 1967.
5. A. Abragam, 'Principles of Nuclear Magnetism', (Oxford Univ. Press) 1961.
6. D.W. McCall, in 'Molecular Dynamics and Structure of Solids' ed. R.J. Carter and J.J. Rush (Nat. Bur. Stand. Spec. Publ. 301, 1969).

7. W.P. Slichter, in 'NMR, Basic Principles and Progress', Vol.4 (Springer-Verlog, Berlin) 1971, p.209.
8. R.G. Gordon, 'Adv. Magnetic Resonance', 1968, 3, 1.
9. K.D. Moller and W.G. Rothschild, 'Far Infra-red Spectroscopy' (J. Wiley, New York) 1971.
10. P. Van Konynenberg and W.A. Steele, J.Chem.Phys., 56, 4776 (1972).
11. P.A. Egelstaff, 'An Introduction to the Liquid State', (Academis Press, New York) 1967.
12. B.K. Aldred, G.C. Stirling and J.W. White, J.C.S., Faraday Symp. 6, 135 (1972).
13. J.W. White, in 'Molecular Spectroscopy', ed. P. Hepple (Inst. of Petroleum, London) 1972.
14. R.D. Mountain, in'Critical Reviews, Solid State Science', Vol. 1, No.1, (Chemical Rubber Co., Cleveland, Ohio) p.5, (1970).
15. H.Z. Cummins, F.D. Carlson, T.J. Herbert and G. Woods, Biophys. J., 9, 518 (1969).
16. B. Chu, Ann. Rev. Phys. Chem., 21, 145 (1970).
17. W.L. Peticolas, Fortschr. Hochpolym. Fortschr. 9, 285 (1972).
18. 'Photon Correlation and Light Beating Spectroscopy' ed. H.Z. Cummins and E.R. Pike, Pienum Press, New York and London 1974.
19. P.A. Fleury and J.P. Boon, 'Adv. Chemical Phys.', 24, 1 (1973).
20. K.F. Herzfeld and T.A. Litovitz, 'Absorption and Dispersion of Ultrasonic Waves', (Academic Press, New York) 1959.
21. See e.g. M.F. Shears, G. Williams, A.J. Barlow and J. Lamb, J.C.S. Faraday II, 70, 1783 (1974).
22. B.J. Berne in'Physical Chemistry, An Advanced Treatise. The Liquid State' (Vol. VIII B) ed. H. Eyring, D. Henderson and W. Jost, (Academic Press, New York) Chapt. 9, 1971.
23. G. Williams, Trans. Faraday Soc., 61, 1964 (1965).
24. D.W. Davidson, in 'Molecular Relaxation Processes' Chem. Soc. Special Publ. No. 20. The Chem. Soc., and Academic Press, New York, 33, 1966.
25. C. Brot and I. Darmon, J. Chem. Phys., 53, 2271 (1970).
26. A. Aihara, C. Kitazawa and A. Nohara, Bull. Chem. Soc.,(Japan) 43, 3750, (1970).
27. W. Dannhauser and R.H. Cole, J. Chem. Phys., 23, 1762 (1955).
28. D.W. Davidson, Canad. J. Chem., 39, 571 (1961).
29. G.P. Johari and W. Dannhauser, J. Chem. Phys., 50, 1862 (1969).
30. D.J. Denney, J. Chem. Phys., 27, 259 (1957).'
31. J.G. Berberian and R.H. Cole, J. Amer. Chem. Soc., 90, 3100, (1968).
32. G.P. Johari and C.P. Smyth, J. Amer. Chem. Soc., 91, 5168, 1969.
33. G.P. Johari and C.P. Smyth, J. Chem. Phys., 56, 4411 (1972).
34. G.P. Johari and M. Goldstein, J. Phys. Chem., 74, 2034 (1970).
35. G.P. Johari and M. Goldstein, J. Chem. Phys., 53, 2372 (1970).

36. G.P. Johari, J. Chem. Phys., 58, 1766, (1973).
37. J.W. Winslow, R.J. Good and P.E. Berghausen, J. Chem. Phys., 27, 309 (1957).
38. G. Williams and P.J. Hains, Chem. Phys. Lett., 10, 585 (1971)
39. G. Williams and P.J. Hains, J.C.S. Faraday Symp. 6, 14 (1972)
40. M. Davies, P.J. Hains and G. Williams, J.C.S. Faraday II, 69 1785 (1973).
41. M.F. Shears and G. Williams, J.C.S. Faraday II, 69, 608 (1973).
42. M.F. Shears and G. Williams, J.C.S. Faraday II, 69, 1050 (1973).
43. M. Goldstein, J.C.S. Faraday Symp. Chem. Soc., 6, 1, (1972).
44. Y. Ishida, J. Polymer Sci., A-2, 7, 1835 (1969).
45. G. Williams, and D.C. Watts, in 'NMR, Basic Principles and Progress', Vol. 4 (Springer-Verlag, Berlin) p.271, 1971.
46. G. Williams and D.C. Watts in 'Dielectric Properties of Polymers' ed. F.E. Karasz, Plenum Press, 1972.
47. J.D. Ferry, 'Viscoelastic Properties of Polymers' (J. Wiley, New York, 1961, 2nd ed. 1970.
48. Y. Ishida, K. Yamafuji, H. Ito and Takayanagi, Koll. Z. 184, 97 (1962).
49. Y. Ishida, M. Matsuo and K. Yamafuji, Koll. Z. 180, 108(1962).
50. G. Williams, M. Cook and P.J. Hains, J.C.S. Faraday II, 68, 1045, (1972).
51. G. Williams, and D.C. Watts, Trans. Faraday Soc., 66, 80 (1970).
52. G. Williams, D.C. Watts, S.B. Dev and A.M. North, Trans. Faraday Soc., 67, 1323 (1971).
53. G. Williams, 'Chemical Reviews' (Amer. Chem. Soc.), 72, 55, (1972).
54. G. Williams, Trans. Faraday Soc., 60, 1548, 1556 (1964).
55. G. Williams and D.A. Edwards, Trans. Faraday Soc., 62, 1329 (1966).
56. G. Williams, Trans. Faraday Soc., 62, 2091 (1966).
57. G. Williams and D.C. Watts, Trans. Faraday Soc., 67, 2793 (1971)
58. H. Sasabe and S. Saito, J. Polymer Sci., A-2, 6, 1401 (1968).
59. S. Saito, H. Sasabe, T. Nakajima and K. Yada, J. Polymer Sci, A-2, 6, 1297 (1968).
60. G. Williams, Trans. Faraday Soc., 62, 1321 (1966).
61. G. Williams and D.C. Watts, Trans Faraday Soc., 67, 1971 (1971).
62. P. Heydemann, Koll. Z., 195, 122 (1964).
63. M. Cook, D.C. Watts and G. Williams, Trans. Faraday Soc., 66, 2503 (1970).
64. J. Leffingwell and F. Bueche, J. Appl. Phys., 39, 5910 (1968).
65. W.O. Baker and C.P. Smyth, J. Amer. Soc., 61, 2063 (1939).
66. G. Williams, in 'Dielectric and Related Molecular Processes' The Chemical Society Specialist Periodical Reports, ed. M. Davies, 2, 151, (1975).
67. E.A. Di Marzio and M. Bishop, J. Chem. Phys., 60, 3802 (1974).
68. D.W. McCall, D.C. Douglass and D.R. Falcone, J. Chem. Phys., 50, 3839, (1969).
69. J. Shore, Ph.D. Thesis, Univ. of Maryland, (1974).

70. J.E. Anderson and R. Ullman, J. Chem. Phys., 47, 2178 (1967).
71. G.E. McDuffie and T.A. Litovitz, J. Chem. Phys., 37, 1699, (1962).
72. T.A. Litovitz and G.E. McDuffie, J. Chem. Phys., 39, 729, (1963).
73. S.H. Glarum, J. Chem. Phys., 33, 639 (1960).
74. M.C. Phillips, A.J. Barlow and J. Lamb, Proc. Roy. Soc., A329, 193 (1972).
75. J.E. Anderson and C.P. Smyth, J. American Chem. Soc., 85, 2094, 1963.
76. G. Schwarz, J. Phys. Chem., 71, 4021 (1967).
77. G. Schwarz in 'Dielectric and Related Molecular Processes', The Chemical Society Specialist Periodical Reports, ed. M. Davies, 1, 163 (1974).
78. G. Williams, Adv. Mol. Relaxation Processes, 1, 409, (1970).
79. W. Scheider, J. Phys. Chem., 74, 4296, (1970).
80. J. Goulon, Ph.D. Thesis, Nancy (France) (1972).
81. J. Heijboer, 'Mechanical Properties of Glassy Polymers Containing Saturated Rings', TNO (Delft) Communic. No. 435, (1972).